Grzimek's ENCYCLOPEDIA OF EVOLUTION

Grzimek's
ENCYCLOPEDIA
OF EVOLUTION

Editor-in-Chief

Dr. Dr. h.c. Bernhard Grzimek

Professor, Justus Liebig University of Giessen
Director (Retired), Frankfurt Zoological Garden, Germany
Trustee, Tanzania and Uganda National Parks, East Africa

 VAN NOSTRAND REINHOLD COMPANY

New York   Cincinnati   Toronto   London   Melbourne

Van Nostrand Reinhold Company Regional Offices:
New York   Cincinnati   Atlanta   Dallas   San Francisco

Van Nostrand Reinhold Company International Offices.
London   Toronto   Melbourne

Library of Congress Catalogue Card Number: 76-9296

Typography by Santype International Ltd., Salisbury, Great Britain

Printed and bound in Italy by Campi Editore, Foligno, Italy

Published in the United States by Van Nostrand Reinhold Company
450 West 33rd Street, New York, N.Y. 10001

English edition first published in England by Van Nostrand
Reinhold Ltd.

15   14   13   12   11   10   9   8   7   6   5   4   3   2   1

EDITORS AND CONTRIBUTORS

Editor-in-Chief

Dr. Dr. H. C. BERNHARD GRZIMEK
Professor, Justus-Liebig University, Giessen
Director (Retired) Frankfurt.

DR. MICHAEL ABS
Curator, Ruhr University — BOCHUM, GERMANY

DR. SÁLIM ALI
Bombay Natural History Society — BOMBAY, INDIA

DR. RUDOLF ALTEVOGT
Professor and Section Head, Zoological Institute, University of Münster — MÜNSTER, GERMANY

DR. RENATE ANGERMANN
Curator, Institute of Zoology, Humboldt University — BERLIN, GERMANY

EDWARD A. ARMSTRONG, M.A.
Cambridge University — CAMBRIDGE, ENGLAND

DR. PETER AX
Professor, Second Zoological Institute and Museum, University of Göttingen — GÖTTINGEN, GERMANY

DR. FRANK BACHMAIER
Zoological Collection of the State of Bavaria — MUNICH, GERMANY

DR. PEDRU BANARESCU
Academy of the Rumanian Socialist Republic, Institute of Biology,
Trajan Savulescu Institute of Biology — BUCHAREST, RUMANIA

DR. A. G. BANNIKOW
Professor, Institute of Veterinary Medicine — MOSCOW, U.S.S.R.

DR. HILDE BAUMGÄRTNER
Zoological Collection of the State of Bavaria — MUNICH, GERMANY

C. W. BENSON
Department of Zoology, Cambridge University — CAMBRIDGE, ENGLAND

DR. ANDREW BERGER
Chairman, Department of Zoology, University of Hawaii — HONOLULU, HAWAII, U.S.A.

DR. J. BERLIOZ
National Museum of Natural History — PARIS, FRANCE

DR. RUDOLF BERNDT
Director, Braunschweig Institute for Population Ecology
Heligoland Ornithological Station — BRUNSWICK, GERMANY

DIETER BLUME
Instructor of Biology, Freiherr-vom-Stein School — GLADENBACH, GERMANY

DR. MAXIMILIAN BOECKER
Zoological Research Institute and A. Koenig Museum — BONN, GERMANY

DR. CARL-HEINZ BRANDES
Curator and Director, The Aquarium, Overseas Museum — BREMEN, GERMANY

DR. A. S. BRINK
Director, Museum of Man and Science

JOHANNESBURG,
SOUTH AFRICA

DONALD G. BROADLEY
Director, Reptile Section, The National Museum of Rhodesia

UMTALI, RHODESIA

DR. HEINZ BRÜLL
Director; Game, Forest and Fields Research Station

HARTENHOLM, GERMANY

DR. BERBERT BRUNS
Director, Institute for Biology, Environment and the Protection of Life

SCHLAGENBAD, GERMANY

HANS BUB
Institute for Bird Research, Ornithological Station, Heligoland

WILHELMSHAVEN,
GERMANY

A. H. CHISHOLM

SYDNEY, AUSTRALIA

HERBERT THOMAS CONDON
Curator of Birds, South Australian Museum

ADELAIDE, AUSTRALIA

DR. EBERHARD CURIO
Docent and Director Ethology Section, Ruhr University

BOCHUM, GERMANY

DR. SERGE DAAN
Animal Physiology Laboratory, University of Amsterdam

AMSTERDAM,
THE NETHERLANDS

DR. HEINRICH DATHE
Professor and Director of the Animal Park and the Zoological
Institute of the German Academy of Sciences

BERLIN, GERMANY

DR. WOLFGANG DIERL
Zoological Collection of the State of Bavaria

MUNICH, GERMANY

DR. FRITZ DIETERLEN
Zoological Research Institute and A. Koenig Museum

BONN, GERMANY

DR. ROLF DIRCKSEN
Professor, Pedagogical High School

BIELEFELD, GERMANY

JOSEF DONNER
Biology Teacher at the Gymnasium

KATZELSDORF, AUSTRIA

DR. JEAN DORST
Professor, National Museum of Natural History

PARIS, FRANCE

DR. KLAUS DOSE
Professor, Institute for Biochemistry, Johannes Gutenberg University

MAINZ, GERMANY

DR. GERTI DÜCKER
Professor and Chief Curator; Zoological Institute of the University

MÜNSTER, GERMANY

DR. MICHAEL DZWILLO
University of Hamburg, Zoological Institute and Museum

HAMBURG, GERMANY

DR. IRENÄUS EIBL-EIBESFELDT
Professor, Institute of Human Ethology, Max Planck Institute for
Behavioral Physiology

PERCHA, STARNBERG,
GERMANY

DR. MARTIN EISENTRAUT
Professor and Director, Zoological Research Station and A. Koenig Museums

BONN, GERMANY

DR. EBERHARD ERNST
Swiss Tropical Institute

BASEL, SWITZERLAND

R.-D. ETCHECOPAR
Director, National Museum of Natural History

PARIS, FRANCE

DR. R. A. FALLA
Director, Dominion Museum

WELLINGTON,
NEW ZEALAND

DR. N. N. KARTASCHEW
Docent, Department of Biology, Lomonossow State University — MOSCOW, U.S.S.R.

DR. WERNER KÄSTLE
Oberstudienrat, Gisela Gymnasium — MUNICH, GERMANY

DR. REINHARD KAUFMANN
Field Station of the Tropical Institute, Liebig University, Giessen, Germany — SANTA MARTA, COLOMBIA, S.A.

DR. MASAO KAWAI
Kyoto University — KYOTO, JAPAN

DR. ERNST F. KILIAN
Professor, Giessen University and Catadratico University, Austral, Valdivia, Chile — GIESSEN, GERMANY

DR. RAGNAR KINZELBACH
Institute for General Zoology, University of Mainz — MAINZ, GERMANY

DR. HEINRICH KIRCHNER
Landwirtschaftsrat (Retired) — BAD OLDESLOE, GERMANY

DR. ROSL KIRCHSHOFER
Zoological Garden, University of Frankfurt a.M. — FRANKFURT am MAIN, GERMANY

DR. WOLFGANG KLAUSEWITZ
Curator, Senckenberg Nature Museum and Research Institute — FRANKFURT am MAIN, GERMANY

DR. KONRAD KLEMMER
Curator, Senckenberg Nature Museum and Research Institute — FRANKFURT am MAIN, GERMANY

DR. HEINZ-GEORG KLÖS
Professor and Director, Zoological Garden — BERLIN, GERMANY

URSULA KLÖS
Zoological Garden — BERLIN, GERMANY

DR. OTTO KOEHLER
Professor Emeritus, Zoological Institute, University of Freiburg — FREIBURG, GERMANY

DR. KURT KOLAR
Institute of Comparative Ethology, Austrian Academy of Sciences — VIENNA, AUSTRIA

DR. KLAUS KÖNIG
State Ornithological Station, Baden-Wurttemburg — LUDWIGSBURG, GERMANY

DR. ADRIAAN KORTLANDT
Zoological Laboratory, University of Amsterdam — AMSTERDAM, THE NETHERLANDS

DR. HELMUT KRAFT
Professor and Scientific Advisor, Medical Animal Clinic, University of Munich — MUNICH, GERMANY

DR. HELMUT KRAMER
Zoological Research Institute and A. Koenig Museum — BONN, GERMANY

DR. FRANZ KRAPP
Zoological Institute, University of Freiburg — FREIBURG, SWITZERLAND

DR. OTTO KRAUS
Professor, University of Hamburg and Director, Zoological Institute and Museum — HAMBURG, GERMANY

DR. DR. HANS KRIEG
Professor and First Director (Retired) Scientific Collection of the State of Bavaria — MUNICH, GERMANY

DR. HEINRICH KÜHL
Federal Research Institute for Fisheries, Cuxhaven Laboratory — CUXHAVEN, GERMANY

DR. OSKAR KUHN
Professor, formerly University Halle/Saale — MUNICH, GERMANY.

DR. HANS KUMERLOEVE
First Director (Retired), State Scientific Museum, Vienna — MUNICH, GERMANY

DR. HERBERT SCHIFTER
Bird Collection, Museum of Natural History — VIENNA, AUSTRIA

DR. REINHARD SCHMIDT-EFFING
Diplom-Geologist, Palaeontological Institute and Museum, University of Münster — MÜNSTER, GERMANY

DR. MARCO SCHNITTER
Zoological Museum, Zurich University — ZURICH, SWITZERLAND

DR. KURT SCHUBERT
Federal Fisheries Research Institute — HAMBURG, GERMANY

EUGEN SCHUHMACHER
Director, Animals Films, I.U.C.N. — MUNICH, GERMANY

DR. THOMAS SCHULTZE-WESTRUM
Zoological Institute, University of Munich — MUNICH, GERMANY

DR. ERNST SCHÜZ
Professor and Director (retired), State Museum of Natural History — STUTTGART, GERMANY

DR. LESTER L. SHORT, JR.
Associate Curator, American Museum of Natural History — NEW YORK, NEW YORK, U.S.A.

DR. HELMUT SICK
National Museum — RIO DE JANEIRO, BRAZIL

DR. GEORGE G. SIMPSON
Professor, The Simroe Foundation — TUCSON, ARIZONA

DR. ALEXANDER F. SKUTCH
Professor of Ornithology, University of Costa Rica — SAN ISIDRO DEL GENERAL, COSTA RICA

DR. EVERHARD J. SLIJPER
Professor, Zoological Laboratory, University of Amsterdam — AMSTERDAM, THE NETHERLANDS

BERTRAM E. SMYTHIES
Curator (retired), Division of Forestry Management, Sarawak-Malaysia — ESTEPONA, SPAIN

DR. KENNETH E. STAGER
Chief Curator, Los Angeles County Museum of Natural History — LOS ANGELES, CALIFORNIA, U.S.A.

DR. H.C. GEORG H. W. STEIN
Professor, Curator of Mammals, Institute of Zoology and Zoological Museum, Humbolt University — BERLIN, GERMANY

DR. JOACHIM STEINBACHER
Curator, Nature Museum and Senckenberg Research Institute — FRANKFURT A.M., GERMANY

DR. BERNARD STONEHOUSE
Canterbury University — CHRISTCHURCH, NEW ZEALAND

DR. RICHARD ZUR STRASSEN
Curator, Nature Museum and Senckenberg Research Institute — FRANKFURT A.M., GERMANY

DR. ADELHEID STUDER-THIERSCH
Zoological Garden — BASEL, SWITZERLAND

DR. ERNST SUTTER
Museum of Natural History — BASEL, SWITZERLAND

DR. FRITZ TEROFAL
Director, Fish Collection, Zoological Collection of the State of Bavaria — MUNICH, GERMANY

DR. G. F. VAN TETS
Wildlife Research — CANBERRA, AUSTRALIA

ELLEN THALER-KOTTEK
Institute of Zoology, University of Innsbruck — INNSBRUCK, AUSTRIA

DR. MICHAEL L. WOLFE
Utah State University UTAH, U.S.A.

HANS EDMUND WOLTERS
Zoological Research Institute and A. Koenig Museum BONN, GERMANY

DR. ARNFRID WÜNSCHMANN
Research Associate, Zoological Garden BERLIN, GERMANY

DR. WALTER WÜST
Instructor, Wilhelms Gymnasium MUNICH, GERMANY

DR. HEINZ WUNDT
Zoological Collection of the State of Bavaria MUNICH, GERMANY

DR. CLAUS-DIETER ZANDER
Zoological Institute and Museum, University of Hamburg HAMBURG, GERMANY

DR. DR. FRITZ ZUMPT JOHANNESBURG,
Director, Entomology and Parasitology, South African Institute for Medical Research SOUTH AFRICA

DR. RICHARD L. ZUSI
Curator of Birds, United States National Museum, Smithsonian Institution WASHINGTON, D.C., U.S.A.

Grzimek's
ENCYCLOPEDIA
OF EVOLUTION

Edited by:

GERHARD HEBERER

HERBERT WENDT

ENGLISH EDITION

GENERAL EDITOR:
George M. Narita

SCIENTIFIC EDITOR:
Erich Klinghammer

TRANSLATOR:
David R. Martinez

SCIENTIFIC CONSULTANT:
Daniel L. Hartl

ASSISTANT EDITOR:
Ruth Gennrich

EDITORIAL ASSISTANTS:
Karen Boikess
Rachel Davison

PRODUCTION DIRECTOR:
James V. Leone

INDEX:
Suzanne C. Klinghammer

CONTENTS

Foreword

Current reference works on the animal kingdom describe the fauna of our world, arranged according to contemporary zoological thinking, in all its diversity. Zoological (i.e., phylogenetic) arrangement of animal species is not only useful in helping us find our way through the many species (there are over 1,000,000 that have been described in the literature) but is also the basis of zoological (and botanical) study. Ever since the theory of evolution has been developed and refined and paleontological finds have given us new insight into the evolutionary relationships between animal species, all thorough works on animal life invoke the principles of the theory of evolution for explanatory purposes as they survey the various animal groups. The modern animal world—including mankind and its many cultures—can only be explained on the basis of its evolutionary heritage.

This is the reason we have continually dealt with the evolutionary aspects of the modern animal world throughout Grzimek's Animal Life Encyclopedia, using extinct species to help explain how extant ones arose. However, evolution could not be discussed as thoroughly in the three volumes on invertebrates and ten vertebrate volumes in the depth necessary to thoroughly explain the history of living organisms on earth. Most emphasis in these volumes is placed on the life and behavior of the animals of the present.

As the encyclopedia was being written, it became clear that stressing modern animal species would leave many questions open to innumerable readers of Grzimek's Animal Life Encyclopedia. How did life arise? Just what is meant by evolution? What role did mutation and selection play and how did they interact to produce the great diversity of life as we know it now and to the flourishing and death of so many major animal groups? What methods are used today to track a modern animal group back to its ancestral roots? What is our modern picture of the geological conditions and the flora and fauna of each of the previous geological eras mentioned in each of the thirteen volumes?

These questions can only be answered when one describes the history of animals from their very beginnings and continues from that time chronologically into the present, as evolutionists and paleontologists do. Since evolutionary thinking has come to profoundly influence the modern biological outlook on the world, just as Copernicus changed our astronomical thinking in the 16th Century and Einstein our understanding of physics in the early 20th Century, any work on the animal kingdom is incomplete unless it devotes a volume to the subject of evolution. The theory of evolution is now at the center of all biological thinking and is the basis for all biological research. It is the only scientifically acceptable and recognized theory that convincingly explains the world of living organisms.

This has led us to compile an entire volume that is devoted to the theory of evolution, and in doing so, Grzimek's Animal Life Encyclopedia goes beyond what all other comparable works have done and sets new standards in encyclopedic works on the animal kingdom. The evolution volume describes at length the geological and phylogenetic (evolutionary) factors that have influenced the development of all animal species, including of course man. In this volume the reader will find the key to the great biological processes involved in evolution, processes that can only be briefly mentioned in a systematic description of the animal world.

The editors and translator of this volume on the evolution of animal life realize that while we attempt to compile a thoroughly readable work in this volume, it is impossible to refrain from using any technical terminology. There are often no popular counterparts for terms used by professional evolutionists, biochemists, and paleontologists. However, we have attempted to explain such terms as they are used in sufficient detail to enable every reader to comprehend them in the discussion they are reading at the time. We shall try to explain difficult terminology in such a way as to make this an exciting, and not a tedious, work. Terms that denigrate animals are avoided in this work, and we feel this policy is in line with the more enlightened outlook we now have toward the animal world.

Since the subject of evolution has more ramifications than any single person can master, we have relegated each major aspect to individual evolutionary biologists and paleontologists working in those fields. We thank those who have contributed to this volume and to all those responsible for the publication of this volume.

In the individual articles here the reader will find material he has not encountered in other works. However, science never comes to a standstill; it knows no final conclusion, and each new work brings forth new information compiled from the latest research in that field. Evolutionary biology is one of the most fruitful areas of biological research, and we have purposely emphasized the latest thinking on evolution rather than traditional thought. Advances are being made on numerous fronts that increase our understanding of evolution, some of these fronts including biophysics, biochemistry, and genetics. The technique of radioactive dating, with which we have been able to determine with more precision than ever before the age of extinct animal groups and geological periods and the rate of evolutionary development of these animal groups. The reader can become acquainted with current theoretical thinking about evolution in reading this volume.

We found that to effectively acquaint the reader with the development of life, in all its diversity, we could not simply describe the individual geological periods and epochs. Instead we have begun—after some introductory material on the history of the theory of evolution and on paleontology—with the major mechanisms of evolution, emphasizing chordate evolution (the group containing vertebrates, of which man is a part). So much new information has been gained in recent years on the origin of life and the meaning of the genetic code that we no longer have to work with hypotheses and presumptions. We can now make clear statements on many aspects of these topics, statements consistent with research findings on cosmology, including the origin of our solar system and our planet. These border disciplines between physics and biology are given appropriate attention in this work.

Since animal life and its phylogenetic development are largely influenced by the flora of each geological period, it has also been necessary to deal with the evolution of plants in the individual geological periods. Plants took the first steps that led to the presence of terrestrial life, and plants created the situation permitting animals to later move onto land as well. Furthermore, it was only with the aid of plants (using tall ones to glide from and jump from) that animals were able to develop the adaptations needed to locomote through the air. Floral evolution has had a pronounced influence on the flourishing and death of many an animal group.

In describing the life of previous geological periods we have been able to place more emphasis on many extinct animals that were only briefly mentioned in the previous thirteen volumes. Some enormously significant evolutionary steps (such as movement from the sea onto land, the development of warm-bloodedness, and the conquest of air) have required several articles to properly explain the periods in which they occurred. In some chapters we have information on the life forms found in some particular, bygone era. The pictures we have of these eras are no longer based on fantasy but on solid evidence resulting from paleobiological research. As examples of classic digging sites we have chosen the finds from the Swabian and Frankish Jurassic. We discuss the origin and development of our own species, what is perhaps the most fascinating evolutionary subject of all, in special chapters devoted entirely to primates and humans. We also include a chapter on domestic animals to illustrate how man has systematically influenced the evolution of animal groups for his own use.

At this point I would like to thank Professor Gerhard Heberer of

Göttingen, who was able to share the editing of this volume with me and who not only examined individual contributions but also helped find appropriate contributors for the various sections. I also thank Dr. Erich Thenius of Vienna for all his cooperation and support.

We hope we have done all we can to enlarge on what has previously been written about evolution and to give the reader a thorough introduction into the history of life on our earth and the development and passing of previous geological epochs.

Baden-Baden, Summer 1972   HERBERT WENDT

1   The History of the Theory of Evolution

By D. S. Peters and
W. F. Gutmann

As Ernst Mayr has written, the theory of evolution is the greatest unifying theory in biology. Its influence on man's outlook toward the world and his understanding of nature can only be compared with the other major scientific advances that changed our basic ideas about the natural world and continue to exert their influence: the Copernican model of the solar system and the theory of relativity propounded by Albert Einstein. The theory of evolution is usually, and rightfully, connected with Charles Darwin. However, Darwin was not the first to postulate the phylogenetic development of organisms (including man). Also, due to the lack of knowledge about genetics in his time, Darwin was not able to formulate all the aspects of his theory correctly, and his statements about inheritance have had to be modified in light of later knowledge.

Philosophers thought about evolution as early as antiquity. When we read these early theories in view of what is known today, we find that the earliest evolutionary theories often contain only a few components of what we now call evolution. The bits and pieces that these early philosophers formulated formed a mosaic that served as a philosophical basis for the later elaboration of the evolutionary theory. Evolution was not the result of centuries of scientific progress, since we find such thinking in the writings of the people we consider to be the earliest natural scientists. These Ionic philosophers and scientists, who lived in the 6th Century B.C., were bolder in one aspect than many more recent evolutionists; they did not see a border between living and non-living matter and sought to explain the world as a single, self-developing system.

**Pre-Socratic
philosophers**

Thales of Miletus (640–584 B.C.) thought that all matter arose from water. His student Anaximandros (611–546 B.C.), in contrast, held the view that the primeval material could no longer be found. Water arose from this hypothetical basic building block, and land developed from water. Living organisms supposedly arose initially in water and later developed into terrestrial organisms. Man originated from a sharklike fish according to Anaximandros.

Although these early writings show that there was some evolutionary thought among the Ionic philosophers, it appears that their main concern was not with evolution but with identifying the primeval material from which all else arose. This line of thinking was carried on by another Miletus philosopher, Anaximenes (588–524 B.C.). He felt that air was the most important "element", stating that air was the basis of all matter. Differences that man perceived in various materials, were allegedly due to differing concentrations of air. Democritus of Abdera (ca. 460–370 B.C.) carried this line of reasoning even further, postulating that atoms were the basic building blocks of all matter. Empedocles of Akragas (ca. 490–430 B.C.) also thought about ultimate matter. He believed there were four elements: earth, water, air, and fire. His views were held to be true for centuries.

According to Empedocles, living matter developed from non-living matter, with the lower organisms arising first and the more highly developed ones coming later. He formulated a rather adventurous sort of proto-reproduction: non-living matter did not give rise to complete, living organisms, but only to parts of them (limbs, heads, etc.), and these components would later join and form entire organisms. This theory quite successfully explained some of the monstrosities of Greek mythology (such as centaurs). The most important part of Empedocles's theory for us is that they contained, along with many erroneous notions, the rudiments of the concept of natural selection, which was not completely formulated until Darwin's work in the 19th Century. According to Empedocles, only the viable forms from the random joining of body parts would continue to exist, while the non-viable ones would die off.

Lucretius (96 –55 B.C.), a Roman philosopher, also emphasized the role of natural selection. He included domestic animals in his discussions and said that human care was necessary for these animals to be able to exist. Wild animals could survive because of their own special abilities (what we today would call adaptations) making them superior to certain other species, the lesser endowed ones becoming extinct.

These early evolutionary ideas had little influence on later thinking. Natural science (and even all European culture) were much more powerfully influenced by the great philosophical systems of Plato (427–347 B.C.) and Aristotle (384–322 B.C.). These systems did not accomodate evolutionary notions. Plato and Aristotle realized that there were similarities among different animals, similarities often appearing in a gradient sort of way, but the strictly ordered animal groups were understood only in formal terms and not in phylogenetic ones. It was believed that primeval reproduction of living organisms occurred in non-living material and that some species, could be transposed into others (like, for example, a cuckoo turning into a sparrow hawk). This had little do to with evolutionary thinking.

Plato and Aristotle

From here we must make a great leap into the 18th Century. This does

Medieval attitudes

not mean there were no naturalists in the Middle Ages, only that these naturalists, who in other ways laid many foundations for modern thinking, were not concerned with evolutionary problems. The few proponents of phylogenesis in this period were outside the mainstream of thought. A theory only becomes scientifically meaningful when it can successfully explain hitherto unresolvable problems or phenomena. No one needs a theory unless there is some problem requiring theoretical explanation, and the naturalists of the Middle Ages were able to explain all phenomena they knew about with the Aristotelian model and modifications made by Christianity. It would be as wrong to reproach these scientists for neglecting evolution as it would be to reproach Galileo or Newton for not recognizing relativity. Naturally, science in the medieval period often took some erroneous, even grotesque, paths, but mistaken theoretical positions have also been taken in the 20th Century.

The beginning of modern natural science

During the 18th Century, however, problems arose that created great difficulties for previous theories. New findings gave renewed significance to some of the older, nearly forgotten, evolutionary theories. Beginning with the time that the New World was discovered and explored, the scientific world became acquainted with vastly more animal and plant species, and some of the similarities between these animals and those from the Old World were unmistakable, particularly in that those differences appearing in relatively similar animals (e.g., New World cats and Old World cats) existed in a progressive manner (e.g., fur becoming progressively longer or feet larger as one proceeded to examine animals from southerly to northerly habitats). The similarities between animals were often clearly related to their geographic distribution. How could this be explained? And what did that have to do with the progression from "primitive" to "advanced" animals, a progression reflected in the newly discovered fossil record? It was the discovery of fossils that prepared the scientific world for the development of a new theory. French naturalist Benoîet de Maillet (1659–1738), who was consul in Egypt, and Danish scholar Nicolaus Steno (actually Niels Stensen; 1638–1686) believed that the various levels in the earth's crust were the result of deposits laid down over time. This was troublesome for the Christian viewpoint that the earth was only about 5000 years old, and the suspicion that the earth was far older than the literal reading of the Bible indicated, grew stronger and eventually became a certainty.

Philosophical movements, such as the thought of Immanuel Kant (1724–1804), also modified the old scholastic, Aristotelian world view. Not every philosopher took up a phylogenetic viewpoint of course. Kant, for example, did realize that a theory of evolution was possible, calling it a "daring adventure in reasoning" (*ein gewagtes Abenteuer der Vernunft*), but he did not actually move beyond his Aristotelian shadow in this regard. Many of his famous contemporaries were not much different. Johann Gottfried Herder (1744–1803) and Charles de Bonnett (1720–1793)

Fig. 1-1. Nicolaus Steno (1638–1686).

often wrote lines that are highly suggestive of an evolutionary theory. However, the final commitment to such a theory was never made. Herder believed in the immutability of species, and in Bonnett's work there is some doubt that he actually had a phylogenetic theory in mind.

This conflict is even clearer in the works of perhaps the most prominent biologist of his time, the French count Georges Louis Leclerc de Buffon (1707–1788). He developed a rather extensive description of a phylogenesis of living organisms but, perhaps in fear of the consequences thereof, did so only as a sort of semi-serious conjecture. The works of P. L. Moreau de Maupertuis (1698–1759) also preceeded his time, for he was the first to conduct investigations in inheritance. He proved that both parents can pass on their characteristics, a finding that was a death blow to the lively debated arguments by the opposing animalculists (who believed that only the father transmits characteristics) and the ovulists (who believed that only the egg bears transmittible characters). Maupertuis also suggested that species may develop gradually via changes caused in their genetic material, these changes guided by environmental conditions and a kind of natural selection. He even recognized the role of geographic isolation in determining the formation of new species. The significance of his insight was not fully realized until quite recently.

Fig. 1-2. Georges Louis Leclerc, Comte de Buffon (1707–1788).

The fact that thinking about evolution made progress so slowly, often with many dead-ends and detours, is not surprising. Even though the old explanations had deficiencies, many observations were thought to contradict what evolutionists were propounding. The question of what caused evolution continually created difficulties for proponents of evolutionary thinking. Carl von Linné (1707–1778), the father of modern systematics, also experienced this difficulty. Initially he was convinced of the immutability of species, and his chief works reflected his theoretical bias. Later, however, he did feel compelled to accept a notion of limited development within the lower systematic units he so carefully compiled. Linné felt that changes in species could occur through hybridization (crossing). However, his notion was proven invalid by the experiments of Josef Gottlieb Koelreuter (1733–1806), who crossed different species but found that no new species were created.

Fig. 1-3. Pierre Louis Moreau de Maupertuis (1698–1759).

The prominent pre-Darwinian evolutionist Jean Baptiste Antoine Pierre de Monet, Chevalier de Lamarck (1744–1829) developed an entirely different explanation for the mechanism of evolution, the impetus that makes it go. He developed the first truly comprehensive theory of evolution, one that was meant to explain the entire world and not just living things, and in this respect Lamarck shared some similarity with the Ionian philosophers of old. According to Lamarck, everything in existence is in a constant state of flux, but living organisms tend to develop toward higher, more complex arrangements. This flow of life, he said, did not move in a single direction (a straight line), since living organisms interact with the environment and are affected by their surroundings. The en-

Fig. 1-4. Carl von Linné (1707–1778).

Fig. 1-5. Jean-Baptiste de Monet, Chevalier de Lamarck (1744–1829).

Fig. 1-6. Friedrich Wilhelm Joseph von Schelling (1775–1854).

Fig. 1-7. Lorenz Oken (1779–1851).

vironment can influence the way in which a particular animal group evolves, so that the complexity of living organisms is a result of the interplay between organisms' own tendency to achieve a higher state of order and the effects of the environment. Plants respond directly to environmental pressures, while animals react according to what today we might call an "adaptation drive". Animals, so said Lamarck, somehow sense what is beneficial for them, and this sensation influences the way the animals evolve. This was what caused elks to develop prominent antlers and giraffes to develop very long necks. These changes occurred very gradually and over many generations. The basic tenet of Lamarckian evolution revolves around what we call the inheritance of acquired characteristics. An animal (say a proto-giraffe) that constantly stretches to reach food in the treetops may develop a neck slightly longer than some other animal that does not stretch its neck muscles as greatly. The original giraffe would not have inherited the long neck but acquired it through practice. In a single animal the neck would be only slightly longer than one would otherwise find, of course. However, and again this is according to the Lamarckian viewpoint, this acquired characteristic (a very slightly longer neck) would be incorporated into the genetic material of that animal and would then be passed on to its progeny. If some of these progeny also stretched into the trees, they might grow even longer necks, and they in turn would pass on the genetic trait in their progeny for somewhat longer necks. Over many generations, then, the inheritance of these acquired characteristics would eventually result in the enormously long neck we see in our modern giraffes. This was the Lamarckian interpretation of the mechanism of evolution.

Larmarck's theories caused a great deal of discussion but found few adherents. The great naturalists of the time argued with the philosophical tenets of Larmarckianism, and they argued that Lamarck's theory was not congruous with the known facts. Charles Lyell (1797–1875), one of the founders of modern geology and later a supporter of Darwin, delivered a devastating critique of Lamarckianism, and Étienne Geoffroy Saint-Hilaire (1772–1844), one of Lamarck's most prominent supporters, suffered a severe setback after losing a debate in 1830 with the world's leading zoologist, Georges Cuvier, before the Paris Academy of Science. Cuvier sharply rejected the notion of any evolutionary process.

Thus, evolutionary thinking suffered a substantial defeat at the beginning of the 19th Century and was even labeled as scientifically unsound speculation. Many natural philosophies circulating at this time (e.g., those of Friedrich Schelling and Lorenz Oken in Germany) were filled with inaccuracies that hurt other such theories. However, evolutionist thinking was not wiped out completely, but the idea of evolution was thought to be unsound in many scientific circles.

One year after the great Saint-Hilaire – Cuvier debate, the ship *Beagle* set out from England on its five-year voyage. On board was 22-year old

Charles Darwin (1809–1882), who later made the final breakthrough of a plausible theory of evolution. Although this voyage was quite a painful experience for him (since he suffered severe seasickness), Darwin collected a tremendous amount of material, which eventually served to convince him of the validity of evolution. Darwin waited a very long time before he published his theory. He said he had numerous reservations and wanted to constantly check his theoretical interpretation against new evidence. Notes piled up, and manuscripts were written, but none of these were sent off to the publisher. Darwin was also troubled by poor health, suffering from neurasthenia according to his biographers but more probably from the effects of chagas disease he contracted on his world voyage. Since he was well-to-do, Darwin led the quiet, painfully organized life of a private teacher.

Fig. 1-8. Étienne Geoffroy Sainte-Hilaire (1772–1844).

Darwin collected increasing evidence to support his theory, and he did so with great perseverance. He recognized the relationship between morphological similarity and geographical distribution within animal groups (the same observations that finally led Linné to believe that evolutionary processes do occur). The most famous examples Darwin used to illustrate evolution were the Galapagos tortoises and the Darwin finches (also of the Galapagos). When Darwin looked into domesticated animals, he noted the importance of selective breeding and also that the offspring from "pure-blooded" parents were indeed quite similar but were also visibly different. It was the fine differences that are so important for evolutionary processes, of course. The findings of geology and paleontology, both disciplines in their infancy, also supported Darwin's budding theory of evolution. Darwin received significant impetus to publish his theory when he read Thomas Malthus's book on population, in which Malthus tried to show that human beings would soon overpopulate the world if their progeny were not regulated by either disease, hunger, or birth control. Darwin agreed with Malthus's viewpoint that the same forces controlling human populations would also control animal and plant species in their struggle for existence.

Wallace and Darwin

One day Darwin received a manuscript that must have jolted him like a stroke of lightning. The author, Alfred Russell Wallace (1823–1913), was an English naturalist who had just visited southeast Asia and Indonesia. He had also made observations similar to Darwin's and had also read the Malthus book. Working entirely independent of Darwin, Wallace had developed a theory of evolution himself. It is reflective of the characters of both men that they simultaneously presented their findings before the Linné Society of London on July 1, 1858. Darwin's book, *The Origin of Species*, appeared one year later. The first edition numbered just 1250 books, which were sold out immediately. Darwin's work produced a tremendous reaction, with vigorous adherents and equally vigorous opponents of what we now recognize as an amazingly insightful work. French biologists, still under Lamarck's influence, were initially hesitant

Fig. 1-9. Alfred Russell Wallace (1823–1913).

Fig. 1-10. Karl Gegenbaur (1826–1903).

Fig. 1-11. Gregor Mendel (1822–1884).

to accept the Darwinian interpretation, while the theory found easier acceptance in England and Germany. Some of the prominent English supporters included such respected scientists as zoologist Thomas Henry Huxley (1825–1895), botanist Sir Joseph Dalton Hooker (1817–1911), and even geologist Charles Lyell, who had so severely criticized Larmarck not so many years before. The German supporters included Karl Gegenbaur (1826–1903) and Ernst Haeckel (1834–1919), about whose biogenetic work we shall later speak.

What made Darwin's thesis so convincing was the massive amount of evidence he offered, evidence he had collected over a period of decades. Nonetheless, the theory of evolution was vigorously attacked for many years, chiefly for ideological reasons. We shall not deal with post-Darwinian attacks on evolution, since the test of time has long overtaken them and they are no longer of any consequence. One flaw in Darwin's theory was the mechanism of inheritance, the same problem that had plagued all previous evolutionists. Darwin could not explain how structures and behavior were inherited and what the mechanism accounting for changes in characteristics was. In his later years he even accepted Lamarckian explanations. Once the work of Gregor Mendel (1822–1884) became known—his work on genetics was rediscovered at the turn of the century—and genetics received the vital boost it needed, it initially appeared to contradict what Darwin had said. Gradually, and in the course of many debates, the accumulation of additional knowledge produced the present picture of evolution, which in essence combines Darwin's work with the findings of modern genetics in what is called the synthetic theory of evolution. Genetics has shown us how characteristics are inherited, and that finding was the final breakthrough for evolution. The synthetic theory of evolution shows the relationships between the genetic apparatus, mutations, and natural selection, relating them to the process of evolution.

Today the theory of evolution is almost universally accepted, since it explains the world of living organisms so effectively. The process of evolution is extremely complex, and new works are continuously shedding light on certain fine aspects of the theory.

2   The Meaning of the Theory of Evolution

By D. S. Peters and
W. F. Gutmann

Science is always moving and knows no finality. The knowledge of today is the special case or even error of tomorrow. The rational-empirical philosophical basis of scientific inquiry, which creates the constant opposition of reflection (hypothesis) and experience (fact), is a theoretical building into which only those theories gain entrance that deliver the soundest explanations for observed phenomena. All theories are theoretically open to refutation, and this is also true of the theory of evolution.

We showed in our historical survey that evolutionary thinking appeared with the beginnings of natural science. However, since with increased progress in the natural sciences there were problems that theories of evolution apparently could not explain, the line of thought was initially relegated to the field of metaphysical speculation. Today the situation is quite different. For all that is presently known about biology, there is no better, more comprehensive explanation of life than the theory of evolution. There are no known natural phenomena that contradict the predictions of this theory, and there is an enormously huge body of facts supporting it. Any naturalist who does not wish to be unscientific feels compelled to accept the great theory of evolution. This does not mean, however, that the theory of evolution can be equated with unquestioned dogma. Science only produces provisional explanations and knows no dogmatic assertions. Anyone who would want to disprove the validity of the theory of evolution would have to do so with evidence. It appears at the present that no one can bring forth evidence that would disprove the theory.

Scientific theories are explanatory models used for scientific problems, but they do not solely arise from the presence of problems (i.e., unexplained phenomena). Theories often develop as a stroke of insight, and their validity is subsequently tested by comparing the predictions of the theories with observable phenomena. The theory of evolution is a prime example of the "lightning bolt" kind of theoretical formulation. Wallace formulated and wrote down the theory of evolution in one night, and that evening must have been quite an eye-opening one for Mr. Wallace! Was it not quite different with Darwin? Did he not spend decades thinking it out? No, the basic components of the Darwinian theory developed during Darwin's Galapagos visit at the very latest. Everything Darwin did after that only amounted to testing his theory, not formulating it. He was able to anticipate what many scientists after him would have to do in terms of testing his theory, and because of this Darwin could demonstrate the strength of his theory in his *Origin of Species* and could show its value as an explanatory tool. This is what we expect of theories. Evidence for theories is sometimes called proof, but the two concepts are not really identical in the strictest sense. There are no final proofs for scientific theories, for science is much more tentative than that. Proof amounts to absolute truth, and there is no absolute truth in science and no ultimate assertions.

Wallace and Darwin receive simultaneous credit for formulating the theory of evolution. Darwin's name is the better remembered because he did more to support the theory with evidence, and this is deserved, for Wallace did not go through the extensive testing that Darwin did, and it is the testing of a theory that determines whether that theory is viable.

What were the problems of species origin that older theories could not satisfactorily explain but that the theory of evolution could? The crucial phenomena were graduated similarities among living organisms, the strikingly regular way organisms had distributed themselves around the world, and the clear connection existing between these two phenomena. Furthermore, the fossil remains of plants and animals could also be arranged in an orderly way and in some cases demonstrated a progression in development that led to the structures in modern animals. All of this is convincingly explained by the theory of evolution, according to which all living organisms have a phylogenetic heritage (an historic course of development) through which all life arose. The phylogenetic process is carried out by two chief determinants: genetic variation and natural selection.

Mutation and selection

Wallace and Darwin could not understand much about the mechanisms of genetic change. Today, thanks to the tremendous growth of the science of genetics, we know a great deal about these mechanisms (although by no means everything about them!). For lack of space our discussion will be restricted to a general description and not through treatment of every aspect of genetics. Genetic material principally occurs in the chromosomes, which are in the nuclei of cells. The smallest genetic units are known as genes (more recently called cistrons, because the gene concept was originally formulated in a different way). Genes determine the appearance and other characteristics of each organism. They are arranged linearly on the chromosomes. Genetic material can be altered in many different ways. The entire chromosome contents can double or multiply even more

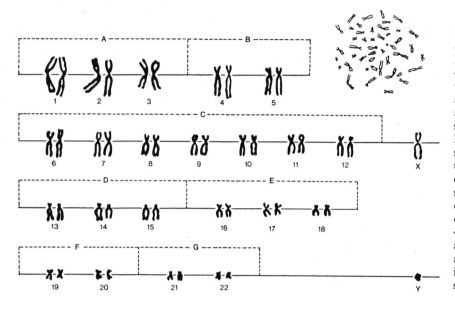

Fig. 2-1. Chromosome complement of a man. Humans normally possess 46 chromosomes, 44 of which are identical in males and females and are called autosomes or A-chromosomes. The other two, known as heterosomes or sex chromosomes, differ in that females contain two, equally large, X chromosomes, while males have one X chromosome and one Y chromosome. The 46 chromosomes are arranged and numbered in a standardized way according to shape and size, as shown here.

(this happens particularly frequently in plants), or individual chromosomes can be altered by losses or changes in certain parts thereof. The chemical structure of single genes can also change. All these changes are called mutations, although the word mutation is reserved for gene modifications alone in its strictest sense.

Mutations result in alterations of the transmission and decoding, as it were, of genetic material. When mutations occur, the appearance of an organism (its phenotype) also changes, since outward appearance is programmed and guided by the genetic apparatus (or genotype = totality of genetic elements of an organism). Changes that occur may not just be anatomical, but also behavioral, metabolic, and others. The effect of a single gene depends not only on the information carried by that gene but also by the relationship of that gene to all the other genes, and bisexual reproduction results in the mixing (or recombination) of two individuals; we can perhaps therefore understand how mutations and recombinations play such an important role in producing the diversity of life we see today, for these two phenomena result in an endless variety of genotypes and phenotypes.

Natural selection acts on all these combinations produced from bisexual reproduction. Natural selection is a stabilizing process acting in such a way that organisms better adapted to prevailing environmental conditions reproduce more successfully (i.e., in greater numbers) than those less adapted to the environment. The process can occur slowly but may take place faster than we can observe it happening. A superlative example of natural selection in action are the poison-resistant insect species. Experience has shown that insecticides typically have an initially potent effect in reducing insect populations but that the effect becomes less

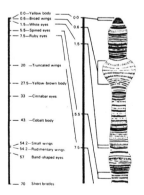

Fig. 2-2. Gene map of the X chromosome of the *Drosophila* fruit fly. The figure shows that arrangement of the genes on the chromosome strand.

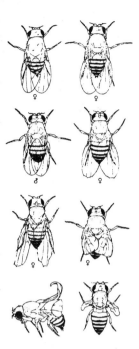

Fig. 2-3. *Drosophila melanogaster*, the fruit fly, is a superb example of the accidental appearance of mutations. The variations shown in this figure can be traced to mutations occurring in this species.

potent when the insecticide is used over and over again. The insects were merely adapting to their environment; the few that were naturally resistant to the insecticide reproduced at a higher rate than those that were not resistant, and in time the resistant strains came to predominate. Adaptation should not be confused with habituation; the former involves a change in the genetic complement of an animal species, while the latter does not.

We must assume that among the insects, some mutations had previously occurred that resulted in some being resistant to insecticides. Since at one time in history there were no insecticides, being resistant was not advantageous. The situation changed radically once insecticides like DDT were used, for virtually every insect lacking resistance to this and other similar agents died off, while the resistant animals lived and reproduced. Now resistance was a favorable attribute and was selectively advantageous for those insects having it. As long as the insecticide was used, all progeny that were resistant continued to live and reproduce also, and this caused the gradual loss of potency of the insecticide over time. A similar situation has arisen with the use of antibiotics against disease-producing organisms, since some of them exist in resistant strains. Laboratory studies must be undertaken to determine which specific resistance a strain has, and once that is found, then some other antibiotic will be used against that organism.

One of the classic examples of the force of selection is the peppered moth *Biston betularia*, whose white color provides very effective protection against predators when the moth rests on birch trees. Dark specimens have also been seen, but since such individuals stood out much more against their customary birch trees, these individuals were for a long time much less prevalent than the white variety. When industrialization dirtied the air and darkened the birch trees, the selection pressure changed and now the advantage was thrown to those moths that were darker! Soon the white *Biston* moths were the rarer of the two. The two varieties differed only by a single gene possessed by the darker one.

It was examples like the above that demonstrated the effect of mutation and selection in the course of the evolutionary process. For years scientists debated which of the two factors, mutations or selection, was the more important impetus, and usually the decision went to selection. Natural selection was even termed a "creative" quality. Actually, both phenomena complement each other. Mutations occur by chance and are indeed weeded out by the conservative process of selection, but according to the theory of evolution, selection acts merely as a sieve and does not exert a creative force of any kind. The only characteristics that can be selected are those that actually appear, and only a very, very few of all the total possible mutations appear in nature. Therefore, the mutations in a population at any specific time also play a role in determining what will happen with the evolution of that population. Non-resistant insects were wiped out when insecticides were used against them. If there were no dark *Biston* variants, the species may have disappeared entirely in industrial regions.

Evolution, therefore, is driven onward by both mutation and natural selection, the former producing genetic variety and the latter reducing genetic variety. But what actually develops? The concept of evolution deals with phylogenesis, which is the growth, through time, of an animal group. Phylogeny is distinguished from the growth of an *individual* animal, or ontogeny. A single animal does not evolve at all; what evolves is a series of generations (a species), which produces a genetic continuum and provides the kind of mixing of genes and new combinations that of necessity do not appear in a single individual.

In bisexual reproduction, which occurs in most animal life, the genetic material of two individuals is mixed. Not any two individual animals, however, can mix their genetic material together; reproduction only occurs among members of the same population. Those populations of animals in which recombination can occur and which are reproductively isolated from other populations are known as "species", or, more precisely, "biological species" (since the term species has been used in so many different ways). A biological species can be thought of as a gene pool community, the word pool referring to the sum genetic material in all the members of the population. Every gene pool protected from recombining with other gene pools can progress on its own phylogenetic path. Thus, the species level is where evolution occurs. It is the species that changes over time. This simple statement actually encompasses a wealth of concepts, of which we can only mention three:

Evolution and species

1. Biological species are true entities, but only at one moment in time; species cannot be differentiated along the time axis. This can easily be shown with the following example. We know with reasonable certainty that crocodiles and birds evolved from a common ancestor. Let us choose two species: the nightingale and the Nile crocodile. We have no difficulty distinguishing these two modern animals as members of different species, as biological species. If we now look through their phylogenetic history, predecessor by predecessor, we finally get to the common ancestor of all crocodiles and all birds, and that common ancestor was a member of one species. Each of the phylogenetic lines producing the crocodile and nightingale contains a series of reptilian-birdlike ancestors whose resemblance becomes greater as we go back in time. How, then, can we differentiate the series into species? We must realize that this is impossible, since each phylogenetic line necessarily ran from parents through their offspring, and changes that occurred were chance occurrences and not part of a master plan. It is illogical to ask how many species existed on the direct path from the crocodile-bird ancestor to the nightingale. Evolution is a continuum, not a series of species.

Biological species

Speciation

2. Speciation (the multiplication of species) happens when some gene pool is split into isolated components. That is, members of some population becomes reproductively isolated from other members of the population. In animals, geographic separation is the usual means by which reproductive isolation—and therefore speciation—comes about. In these cases one speaks of allopatric speciation (the Greek allos = other). Once geographic separation occurs, the gene flow within the population is interrupted, and each of the isolated communities henceforth continues on its own phylogenetic path. Once evolution has proceeded to such an extent that the two populations can no longer interbreed, each one has become an individual species.

Lines of descent and phylogenesis

3. It should not be difficult to understand from the above that the phylogenetic development of organisms can proceed along extremely complex pathways. Phylogenetic trees only show the actual historical process of phylogenesis. This is important to emphasize, since many people read more into these trees, with their genera, families, orders, etc., than they actually should. Systematic arrangements are practical and useful, but in terms of evolution they are always arbitrary. While the species actually exists in the sense we gave above, there is no basis for determining the "rank" and size of the so-called higher systematic units (such as, for example, mammals) on the basis of a phylogenetic tree. The "higher units" (e.g., orders and families) did not evolve as single entities, as so many people in popular discussions imply. It is common parlance, for example, to say that birds evolved from reptiles. This is misleading, since there could only have been one reptile species that was the first link in the long phylogenetic series giving rise to birds. Furthermore, that same ancestral reptilian species, to return to our example, also gave rise to the crocodile. Thus, if we reason strictly according to the descent lines on a phylogenetic tree, a crocodile is more closely related to a nightingale than to a lizard. Such an assertion is of course nonsensical. Nevertheless, even serious works will place lizards and crocodiles together with reptiles and show birds arising from them as a distinct vertebrate class. This is done because of the anatomy of these two animal classes, and anatomical plans are actually nothing more than abstractions, which under certain conditions can actually blur true relationships. Students of evolution must always be conscious of this as they attempt to examine the process of evolution and the causal factors responsible for the phylogenetic continuum.

"Chance and purpose"

At this point we would also like to discuss some of the general cultural, spiritual, and philosophical implications of the theory of evolution. Our pre-evolutionary world view, powerfully influenced by the classical philosophers, was one that attributed the diversity of life forms and their function to the presence of a grand plan operating with a purposeful goal. Once life was examined under the neutral observation of scientists, using the methodology employed to arrive at the theory of evolution, we developed an entirely different understanding. The process of evolution is not activated by some goal-oriented plan (e.g., ever better adapted animals or more and more complex animals) but is instead the result of chaotic, purely accidental changes in the genetic complement of organisms. Evolution does not proceed without direction however since natural selection influences evolutionary pathways. However, the random driving mechanism of evolution and the direction it proceeds are not two directly related phenomena causally. Thus, in spite of the fact that evolution is not directionless, there is no grand plan through which phylogenesis occurs, and there is no goal for evolution. Phylogenesis is often characterized by consistent sorts of changes because natural selection pressure often forces certain biological systems to change in a linear way. However, there are also cases in which the biotechnical possibilities are very great, and in these cases the resulting organisms display a vast diversity in their biological makeup, the result of a high number of solutions to the biological problems they face. Evolution is not characterized solely by linear development.

The preceding discussion has attempted clarifying the process of evolution in very small steps. If we are to understand how powerful phylogenetic changes over long periods of time have created the many different biological "plans" of animals, we must take a broader look at the whole process.

In the course of evolving, a biological system is subject to continuous changes in its genetic complement caused by random changes in the form of mutations. Those variants that are more efficiently adapted to the environment (i.e., those that reproduce at a higher rate) come to predominate over those that are less fit. The rate of reproduction is influenced by the efficiency of the reproductive apparatus in its environment. Anything that improves an organism's chance of reproducing and surviving will increase that organism's prevalence and will force out other, similar organisms lacking a similar advantage. Therefore, evolution is in general a process whereby some mutations out of all those created are selected out, reducing the expenditure of biological energy by eliminating the variants that are not selected. Those life habits that permit animals (and plants) to propagate viable offspring in the most efficient way will result in the selection of those animals. The term adaptation is used to denote those anatomical, behavioral, physiological, and other characteristics that contribute to an organism's viability and fertility in its encounter with the environment.

Substantial phylogenetic changes occur when, over a long period of time, mutations appear that are so adaptive they do not disappear and eventually cause some biological system to be modified in a major way as these mutations accumulate under selective pressure. In relation to the habitat, these changes provide for a reduction in energy expenditure on the part of those animals having them. Any conclusions made about the

Fig. 2-4. Homologous structures are those sharing a common phylogenetic origin. They are often recognized by anatomical similarities displayed in various species having such structures. For example, the paired fins in fishes (A. Coelacanth) and the limbs of terrestrial vertebrates (B. Lizard; C. Bird) as well as the arms and legs of humans (D) are homologous structures.

phylogenetic development of animals or plants must be made with this model in mind and must be described in terms of the adaptive significance of those changes that took place in the evolution of the groups being discussed. Use of a model is only possible with tracing the evolution of individual organs and their interrelationships, since no conclusions can be made on many of the finer details.

The process of evolution is therefore one in which the complex relationship between an organism and its environment changes over time. We can gain an appreciation of this process by examining three examples:

1. Evolution can occur in cases wherein the biological entity (the animal) stays within its habitat and does not substantially alter its life habits. Those mutations which, given the same food supply and ecological conditions, produce more economical, less energy expending, biological "machines", can eventually produce significant modifications without actually altering the functions of the organism in its environment.

2. On the other hand, organisms in which few mutations occur or only minor ones occur can conquer a new habitat (a new ecological niche) by modifying their behavior while retaining their old structures. In other words, structures can achieve new functions.

3. The third possible mode of evolution is the joint occurrence of both kinds of changes described above. Anatomical modifications and functional changes can occur simultaneously in the evolution of an animal group.

These three models are highly schematic and only describe broad classes of variation by which evolution occurs. All these types of changes are descriptions of adaptations. All components of organisms are variable, but it is important to understand that everything does not have to change simultaneously during the course of evolution. Also, change can occur without the environment changing, just as evolution can proceed slowly in organisms when the environment does become altered (as we pointed out in the three numbered examples previously). It is the relationship between the organism and the environment that determines what sort of evolutionary change will occur. If one wishes to describe phylogenetic development in a theoretically acceptable manner, the functions of the organism and the role the organism plays in the environment must be treated as strictly separate concepts but ones that are mutually influential.

For every reconstruction of the phylogenetic origin of an organism there must be some initial biological "machine" living in the environment and using biological functions for surviving in that environment. After that we consider modifications of the genotype via mutation. The original organism, through the effect of natural selection, becomes modified in a certain way corresponding to the functional demands of the environment. It evolves in such a way that later forms have greater survival chances and are assured of a higher reproduction rate than the earlier forms.

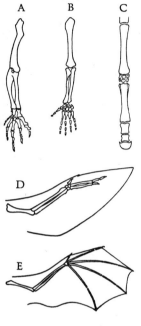

Fig. 2-5. Organs and organ systems can also be homologous, as the fore limbs of several vertebrates depicted here demonstrate.

The fore limbs of an alligator (A), man (B), and a bat (E) all have the same five-digit hand. In birds (D) and horses (C), some of these digits have become vestigial, so that birds have just three digit bones and horses have just one, the middle digit.

Similarities alone are not enough to warrant concluding that a homology exists. It is always necessary to explain these relationships in an evolutionary biological manner, meaning in terms of the adaptation process.

We have depicted and explained the mechanism of evolution in the three most important variants and have shown how conclusions on the phylogenesis of a group are drawn. There is still the question of how phylogenetic insight is gained on specific animal and plant groups. The development of each organic system and the question of the phylogenetic relationships between fairly similar animal groups and plant groups are scientific problems that must be resolved in terms of an explanation dealing with evolutionary mechanisms. Our simple models do not suffice to properly answer these questions, for they are oversimplified, and with them one could develop several theories that were compatible with evolution. A simple, certain reconstruction is impossible to develop. As in other sciences, we must make use of intuition in finding solutions to evolutionary problems. Intuitive feelings are based on the empirical evidence and collate as many facts as are relevant to the problem at hand, and these feelings are compatible with the theory of evolution and the models we projected. Any phylogenetic theory that is contradicted by the evidence has no viability.

The practical method in delving into evolutionary problems is to search for anatomical similarities between the organisms being studied and to see if the intermediate organisms between these two have been found. The theory of evolution deals with the origin of species through small mutations, so we would expect that the gradual changes occurring in evolving organism groups would show up as a series of similar organisms, with a series of intermediate forms between the ancestral species and the most highly evolved one. With modern species, the search is directed toward similarities in these intermediate forms. Organ systems in different animals that share a common phylogenetic origin are known as homologous characteristics (see Fig. 2-4). Homologies can exist even when the organisms under discussion are themselves quite different, as we show in the figure. Scientists have found it effective to search for homologies, using clearly defined distinguishing characters between organisms, and to then draw evolutionary conclusions from these homologous structure.

One of the most superb examples of a homology is found in the limbs of vertebrates. The arrangement of the bones in the paired fins of fishes and the arms and legs of terrestrial vertebrates is strikingly similar; indeed, they are homologous structures. It is also rather easy to recognize that the fore limbs of lizards, humans, the fore fins of whales, and the wings of bats correspond in their structure very closely. It is with homologies like these that biologists seek out phylogenetic data. Their presence alone is not enough to draw conclusions regarding evolution. The biologist would have to show that the paired fins in fishes were used as limbs during the transition from marine to terrestrial life and that the modifications occurring later in the fore limbs were associated with adaptations made to different habitats.

▷
This volcanic landscape probably depicts how the earth looked at an early stage before any life was present on the planet. It is presently believed that the organic building blocks from which the first living cells arose were formed from the influence of volcanic heat, electrical discharges, and solar radiation on the initially oxygen-free atmosphere of the infant earth. See Chapter 5 for additional discussion.

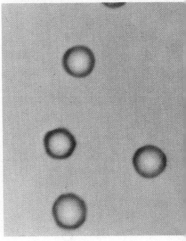

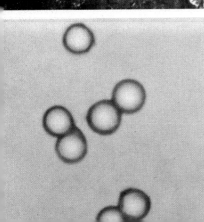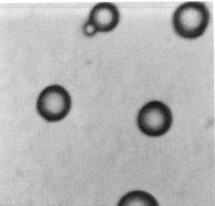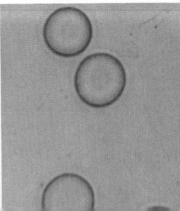

◁

These photographs were taken of a laboratory experiment demonstrating polymerization of amino acids into proteinoids (proteinlike materials) under geological conditions. From left to right and top to bottom: Hot, red-glowing lava.— A rainfall consisting of mud, ashes, and dissolved amino acids falls on the hot lava. The water boils off.—A dry crust remains behind.—Chemical reactions are initiated from the heat. They produce additional steam, and the reaction mixture becomes colored.—The mixture begins fusing.—The dehydration reaction (condensation of the amino acids) reaches its high point. All chemically active matter has been converted. —A (simulated) rainfall interrupts the reaction: the reaction products are dissolved.—Synthesis of proteinoids is ended. Most material goes into solution (i.e., dissolves) when the hot rain water strikes the lava. The surface of the lava is covered with water and a viscous residue.— When the rain water solution is examined under the microscope, one can see cell-like structures known as microspheres. Their size (here with an average diameter of about 5 microns) is dependent on how they were produced. —The microspheres associate.—Many microspheres bear tiny buds.— For comparative purposes we show human erythrocytes under the same illumination conditions and the same magnification.

A series of organ systems can only be thought of as phylogenetically related when it is possible to explain them in terms of adaptation and in a way that is compatible with the theory of evolution.

Many people have made the mistake of using morphological similarity alone to draw conclusions about phylogenesis. One problem that commonly arises when two animals share similar structures is which organism had the structure originally and in which organism was the structure a later development. The phylogenetic developmental route is another question that mere morphological similarities do not answer. Deeper investigations are needed to get at these points. Thus, evolutionary research helps to clarify some of the relationships that man has long known to exist in different species. Purely anatomical considerations are not enough evidence for thorough research, since every explanation using the theory of evolution must account for phenomena in the varied ways the theory predicts.

The need to show that phylogenetic reconstructions must be consistent with the theory of evolution is just as true with living organisms as with fossils. Anatomy is not the only source of information biologists can use with live animals, for ethologists have convincingly shown that behavior can be just as meaningful a measuring stick as anatomy in probing into evolutionary problems.

If the fossil record were complete, there would be no need to try and reconstruct phylogenetic family trees, but an evolutionary explanation would still be necessary. Actually, the fossils we have are almost always bits and pieces and not continuous sets of species or organ systems, and there are very few cases in which an entire line of descent is fully documented by fossils, in which we know the common ancestor and the evolutionary developments occurring along several lines. We do not expect the fossil situation to change much in coming years, so we can say that even with the help of fossils we have to bridge many gaps in the phylogeny of the animal and plant kingdom, using what evidence we have as best we can. One great tool we have is the ability to tell how old fossils are, a technique discussed in the following chapter.

Evolution deals not only with adult organisms but also with their embryology, covering the complete life span from the egg to the complete organism. Ernst Haeckel claimed that "ontogeny recapitulates phylogeny", meaning that as organisms develop from their embryonic beginnings, their individual development (or ontogeny) goes through a series of stages that actually reflect the phylogenetic heritage of that animal group. Haeckel was the first to suggest this on the basis of substantial evidence, but we shall only mention his notion in passing here and not discuss it in detail until Chapter 4.

Any developmental stage of an organism can be modified by a mutation, but when changes occur toward the end of embryonic development and prolong this period, the earlier embryonic stages can reflect, in part, the evolutionary development of the animal group to which that species belongs. The embryo undergoes some of the same kinds of modifications that took place as evolution occurred in that group and new species were formed from older ones. Thus, the embryo stage can unravel a number of mysteries surrounding the phylogeny of that species.

Haeckel's statement is not a law and is by no means applicable to all embryonic stages in all developing organisms. Haeckel himself, in his later years, realized that his "law" was only partly applicable, since adaptations can appear at any embryonic stage that can suppress older ones. This point deserves emphasis, since in evolution any stage of embryonic development can be modified, and there are organisms that show none of the recapitulation of phylogeny in their ontogeny, or only small bits of such a recapitulation. We must distinguish between the phylogenetically significant embryonic stages (in terms of true recapitulations) and those that are insignificant in this regard, and it is with problems like this that we apply the models described previously and scientifically acceptable explanations. In Chapter 4 we shall give a striking example of just such an inquiry. We will show that even when evolutionary conclusions are backed up by embryonic evidence, it is not enough to use morphological features alone in formulating such conclusions. There must always be a description in terms of the course of evolution and a demonstration of the adaptive value in the environment of the structures being discussed.

Every statement on the evolution of living organisms and every phylogenetic tree only makes sense when delivered with an explanation based on the theory of evolution. All phylogenetic statements, trees, and lines of descent are explanatory theories that help clarify the fossils now known and living organisms as well, and any of these theories might be discarded on the basis of future evidence. As in all problem oriented science, there is neither certainty nor a standstill in theories.

3   Introduction to Paleontology

By H. Wendt

Paleontology (from the Greek παλαιός = old, and ὄν = living organism) is the study of the life of previous geological periods. The goal of this discipline is to utilize fossils (many of which are only fragmentary) to reconstruct the development and diversification of the great animal and plant kingdoms and thus the history of life. The term "fossils" was coined by a German geologist, Georg Agricola (1494–1555), from the Latin word *fodere* (=to dig; *fossa*= trench or depression). Originally, fossils were understood to be all objects buried beneath the earth, and it was not until the 18th Century that the concept referred to the petrified remains of organisms from earlier periods in the history of the earth.

In asking about the age of various fossils, paleontology rubs shoulders with the discipline of geology, the science of the structure and changes in the earth's crust. Both fields complement each other and approach each other particularly closely in biostratigraphy, the branch of science that seeks to subdivide various earth formations and which, with the help of fossils, attempts to make relative time measurements for these formations. Since in the course of evolution numerous major animals groups have appeared, flourished, and later become extinct, each geological period has been characterized by leading members of one animal (and plant) group or another, and these key fossils have been of vital importance in dating many organisms. They were used to generate a relative time scale for organisms. More recently, the use of radioactive dating techniques has enabled us to make absolute time measurements for various organism (i.e., determine precisely how old some organism is and not just its age relative to some key fossil). Radioactive dating, which we will describe subsequently, has enabled scientists to determine the true age of rocks, deposits, and fossils and to measure the speed of development of the various classes, orders, families, and genera of flora and fauna, even individual organ systems.

Just how does a living organism becomes fossilized? And how can a fossils remain partly or wholly intact over hundreds of millions of years?

Fig. 3-1. Georg Agricola (1494–1555).

Most organisms do not fossilize, since decomposition of the tissues (through bacterial, chemical, or mechanical action) ensues shortly after death. The soft body parts are particularly susceptible to such processes. However, when a freshly killed animal rapidly gets embedded in a favorable deposit of soil and is completely closed off from the outside world, decaying processes are inhibited and the dead animal can leave an imprint of many of its details, including its softer parts. Soft organs do decompose to some extent, but the hollow spaces they leave behind are filled by deposits and are impregnated with various minerals. These inorganic deposits can impregnate the body and envelop it as well, producing a coat composed of various minerals. The coating can be removed with fine tools or the judicious use of weak acids.

Fig. 3-2. Hollow area and center of a mussel.

Whole organisms can be preserved under very unusual conditions. Emil Kuhn-Schnyder writes: "For this kind of preserving nature uses the same means employed today by the food industry: removal from air, impregnation, and drying or exposure to cold." The Siberian permafrost zone has revealed deep-frozen mammoths and woolly rhinos. Numerous toothless mastodons, elephants, camels, giant sloths, sabertooth cats, and other representatives of North America's faunal heritage have been recovered from the tar pits of La Brea in Los Angeles. The hardened resin secretions of the conifer *Pinus succinifera*, secretions which when hardened we call amber, contain a huge number of insects and other small animals of the Lower Oligocene. Many of these experienced such a sudden death that we can see them copulating, laying eggs, and catching prey.

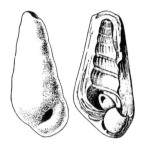

Fig. 3-3. Mummy: a snail enclosed in a calcium shell.

Emil Kuhn-Schnyder describes an interesting example of fossilization of prehistoric animals under water: these are the Eocene finds in lignite occurring in the Geisel valley near Halle, Germany. Humus acids in the lignite protected the dying organisms from decomposing via oxidation or bacterial action, and these fossils display such features as muscles, glands, internal organs (all of insects!), the shimmering hues found on insect wings, and the tissues, blood vessles, fatty cells, skin, feathers, and hairs of vertebrates. Complete dehydration can also preserve animals, and this process usually occurs in desert regions. A body buried deep in the sand was removed from any water, and these fossils have shrunken skin, ligaments, and muscles. Examples of such fossils include many dinosaurs from Upper Cretaceous North America, which have retained such details as the folds and furrows of the skin and the remains of the plants they were feeding on when they died, preserved within their intestinal tract!

Such specimens are rarities, however, and most fossils display only the hard body parts; of plants, only the cellulose typically remains. The great majority of animal fossils consists of shells and skeletons or bone fragments. Horns and chitin have not been preserved as well. The fact that an animal has hard parts does not insure that it will fossilize, however. Mollusks and vertebrate bones must also be removed from sources of

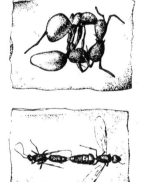

Fig. 3-4. Ants were surprised during feeding (above) and copulating (below) in their Oligocene habitat and were preserved in amber.

decay or mechanically destructive forces quickly or they, too, will be lost. The most protected deposits are in lakes, swamps, and moors, particularly shallow shelf seas. Other good fossil "habitat" is in drifting sand and dust, caves, and volcanic ash. Continental highlands are subject to too much erosion to produce good fossils, and there is little deposition material on the floor of the sea. Furthermore, the ocean floor is covered by waters that dissolve calcium shells (the water being cool and rich in carbon dioxide). Thus, most fossils known to us today stem from deposits on shallow bodies of water and from lowlands.

Requirements for fossilization

Even there, many traces of life are destroyed by decomposing from contact with the air, by being washed into rivers, and by water currents. Thus, the picture we have of prehistoric life is very fragmentary, a sort of gigantic jigsaw puzzle for which we have just a few pieces. When we encounter fossils in some layer or rocks, they are often not true life communities of many organisms that lived together at a certain time, but instead a group of organisms that drifted together after they had died and whose joint occurrence could be largely due to chance. The original hard body parts, which are usually made of calcium, are often replaced by other minerals as they fossilize. The calcium often dissolves completely, and the hollow spaces left behind can become completely filled with some other mineral, a process that obliterates the anatomy of that part of the body. However, mineral replacement can occur on a molecule-by-molecule basis, and it is these fossils that retain fine details due to the slow rate of mineral replacement.

Zoologists have described approximately 1,000,000 animal species in historic times, but the chance nature of fossilization and difficulty in locating fossil beds has resulted in the description of only 90,000–100,000 fossil species. The number of fossils found seems even smaller if we assume, with George Gaylord Simpson, that since life originated this planet has seen 500,000,000–4,000,000,000 species. There is practically no trace left of the vast number of invertebrates lacking shells or other hard parts, including ten entire phyla of animals popularly called "worms". Paleontology can tell us nothing about the evolution of all these animals, so other branches of biology (e.g., comparative anatomy and embryology) must investigate this area. Although our knowledge of prehistoric life is still in its early stages, paleontology is sufficiently advanced that we know about the evolution of many life communities and have learned about entire series of ancestral lines. Evolutionary biology has accumulated a significant amount of documentary evidence through the work of paleontologists.

Scientists and philosophers have known about fossils since antiquity. Striking mammalian fossils such as primitive elephants and cave bears were thought to be the remains of dragons, giants, and other fabled creatures, and this thought persisted into the Middle Ages and even modern times. Ancient Greek thinkers such as Xenophanes realized that some continental fossils were marine creatures, and it was from these findings that theories arose to explain how the seas changed during the earth's history and how flooding has influenced the way the earth is today. This line of thought ceased with Aristotle (384–322 B.C.). His disciples believed that fossils were primitive natural accidents, and this incorrect thinking was carried on in the Middle Ages by Avincenna (930–1037) and Albertus Magnus (1192–1280), who called fossils "games of nature" that were the creation of a *vis plastica*, a creative energy that imitated all sorts of animals and plants from primitive soil.

Fig. 3-5. Various "games of nature".

A few outsiders during the Renaissance recognized the fact that fossils were true remains of earlier organisms and not grotesque games of nature or monsters of some sort. The most important of these outsiders was Leonardo da Vinci (1452–1519), who found many fossils in canal building sites in Lombardy; he interpreted his finds as the remains of marine creatures that had once lived there. In 1517, an Italian physician and scientist named Girolamo Fracastoro (1483–1553) came to a similar conclusion. Da Vinci and Fracastoro both disputed the Biblical diluvial explanation for these fossils, a legend that was held to be fact for centuries.

Niels Stensen (1638–1686), a Danish physician and researcher, made a particularly important contribution to paleontology. He was converted to Catholicism in Italy and held important religious positions under the name Nicolaus Steno. In the Florentine collections of the Medici family, Steno saw a peculiar kind of fossil known as glossopteris (= tongue rocks) due to their tongue-shaped configuration. When a shark head was sectioned, Steno realized that these "tongue rocks" were actually fossilized shark teeth (see Fig. 3-7), and he found that the fossil teeth looked different from the teeth of modern sharks, indicating that tongue rocks were from extinct animals. Steno also compared fossil molluscan shells with modern ones and concluded that the fossil shells were the remains of an extinct fauna. Steno was also the first to find that the various layers in the earth are quite different from each other and that the fossils in the lower layers were very different from those in upper layers. Steno's work, as correct as we now know it is, was not appreciated by contemporary naturalists and its significance was not understood for another century and a half, with the work of Alexander von Humboldt.

Fig. 3-6. *Ammonites lapis*, a fossil sea urchin. It was thought to be a "game of nature" and was called pumpkin stone.

Toward the end of the 17th Century, the most wide-spread explanation for fossils was that they were petrified witnesses to the Biblical flood, a viewpoint that da Vinci had disputed in the 16th Century. The most vigorous proponents of this theory were the English physics professor John Woodward (1665–1728) and the Swiss naturalist Johann Jakob Scheuchzer (1672–1733). Even German philosopher/mathematician Gottlieb Wilhelm Leibniz (1646–1716), who vaguely suggested the possibility of transmutation of species, upheld the diluvial theory in order to maintain his teachings in agreement with the Biblical story of creation. According to this theory, the great flood completely levelled all the earth layers, and

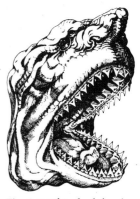

Fig. 3-7. This shark head was published in 1667 by N. Steno in his famous work.

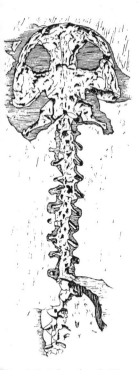

Fig. 3-8. Scheuchzer's "leg bone", to which he gave the scientific name *Homo diluvii testis* (= man, witness of the Flood).

when the waters ran back, they did so with such force that they pushed up earth and rock masses or pressed them down. The forces of the rushing water created the mountains and valleys of today, according to the diluvians. Living organisms caught in the grips of the flood were pressed into the rocks and were even transformed into rocks themselves by the tremendous pressure of the water.

In his four-volume, richly illustrated work "Physica sacra" (1731–1735), Scheuchzer attempted to unify the findings of naturalists with the explicit text of the Bible. His teachings influenced paleontological thought as well as many other disciplines, and fossils were no longer considered to be games of nature but the remains of organisms. Since the diluvians maintained that the Bible had to be literally taken for the truth, they had to maintain that there had been just one Great Flood. Pre-diluvial plants and animals could not be thought of as distinct from extant organisms. This mistaken viewpoint led, of course, to many false interpretations, which were primarily due to the fact that there were no paleontogical research techniques of any consequence at that time. Ammonites and belemnites, characteristic fossil mollusks, were joined with modern molluscan species, and the vertebrae of prehistoric fishes were thought to be the same vertebrae occurring in modern fishes. Bones of mammoths, woolly rhinoceros, and other large Pleistocene animals were allegedly the bones of unicorns, behemoths, dragons, giants, and similar mythical beings from Biblical legends. Although no one had found a living unicorn, the diluvians believed that they existed. Scheuchzer discovered the remains of a giant salamander in 1725, and he mistakenly thought that he had discovered the bones of a man drowned in the great flood; his find is shown in Fig. 3-8. Actually he was looking at an animal from the Upper Miocene.

The vocabulary of these theorists gave rise to the term Diluvium (= flood period), which is still occasionally used to denote the Pleistocene. Although the diluvians, who insisted that the Bible had to be interpreted literally, did not properly evaluate fossils, their efforts were very important for 18th Century paleontology. Wealthy patrons, including princes and high-ranking clergymen, often spent a great deal of money to obtain rare specimens for their collections. In fact, the fossil business was flourishing, and good fossils brought high prices. All these collectors were diluvians, of course. One of the first great zoologists to dig fossils in many countries, Peter Simon Pallas (1741–1811), also believed in the diluvial theory. He went to northern Asia on several extensive expeditions and found mammoth bones, the remains of a woolly rhinoceros, and other traces of giant extinct fauna. According to Pallas, northern Asia was once flooded in a catastrophic flood that embedded the remains of many large mammals in the Siberian soil.

Eventually, increasing fossil finds made it evident that all these extinct animals could not be explained by postulating a single flood. The first

Fig. 3-9. The mayor of Magdeburg, Otto von Guericke, attempted putting together a "unicorn" long before knowledge of the correlation law.

great turning point leading the way to modern paleontology and to the accurate reconstruction of prehistoric life was made in Paris. Preceding Steno by a century, Bernard Palissy (ca. 1510–1590), a glass artisan and potter, thought that the many fossilized mussel and snail shells appearing in Paris during excavations must have originated from marine organisms inhabiting some primeval ocean once covering that great city. His idea was seized by Jean Étienne Guettard in 1746. He compared the geology of northern France with southern England and discovered that the fossils in these two regions were strikingly similar. This led him to conclude that there were numerous floods in prehistoric times and not just one. Guettard also showed that the oldest specimens occurred in the deepest deposits and that the most recent ones were in the layers closest to the surface. His work indicated that there were a series of geological ages through which the earth has passed, and each one of them possessed its own, later extinct, animal world.

Georges Louis Leclerc de Buffon (1707–1788), most prominent zoologist of his time after Linné, held the same opinion. He broke with diluvial theorists by proclaiming that paleontology indicated that there were once animals on the earth that no longer exist. Buffon also rejected the idea that the earth was created in 4004 B.C. This notion was put forth in 1742 by the Irish archbishop, James Usher, using information in the Old Testament. In what has become a classic experiment Buffon melted two balls of ice and, measuring their cooling period, calculated that the earth was 74,832 years old. He thus led the way to investigating geological epochs much older than anyone had previously suspected. Toward the end of his life Buffon went a step further, and using the origin of mountains as a guideline, estimated the earth's age at several hundred thousand years or even one million years.

Later research showed that Buffon had not gone nearly far enough in his estimates! In 1862, British physicist Lord Kelvin made experiments resembling Buffon's and concluded that the earth was at least 20 and no more than 400 million years old. We shall discuss the modern techniques subsequently, stating here simply that they have enabled us to make the much more accurate estimate of the earth's age at 4,800,000,000 years.

While the groundwork for modern paleontology was being laid down in France, the English surveyor William Smith (1769–1839) developed the first means of dating earth layers. Like Guettard before him, Smith saw that there were different organisms in different geological strata. Any particular stratum could be recognized by the key fossils found within it. Thus, fossils were the illustrations in the great picture book of the earth's history and the ideal guides for those asking about the nature and age of geological formations. Smith made the first steps in this hypothetical book of ours. It was apparent that the oldest fossils were also the most primitive and that the most recent ones were also the most advanced. These were the basis for the first evolutionary theories since antiquity. Buffon had played with the

Pictures from bygone geological epochs

The age of the earth

▷
Fossil plants:
Left:
Equisetales growths:
Annularia (Carboniferous);
Leaves of calamites;
Original: Senckenberg Museum, Frankfurt, West Germany;
Right, from top to bottom:
Equisetales:
Eucalamites cruciatus;
Fossil plants showing knots and internodes (Upper Carboniferous from Saarbrücken, Germany).—
Pteridospermae or Cycadofilices: stem cross-section, leaf and seed of *Medullosa* (Upper Carboniferous, Permian); Original: Senckenberg Museum;
Ferns: *Rhacopteris elegans* (Upper Carboniferous, Czechoslovakia).

Paleontology and evolution

Fig. 3-10. Siberian mammoth as drawn by an artist in the early 18th Century.

◁
Fossil plants (Paleozoic and Mesozoic):
Left from top to bottom:
Lycopodiinae:
Surface of *Lepidodendron* (Carboniferous).—Tree fern: frond of *Psо onius* (Permian); Origiлal: Senckenberg Museum, Frankfurt, Germany.—Petrified wood: cross-section through a conifer trunk (Triassic, Arizona).
Right from top to bottom:
Lycopodiinae: *Lepidodendron aculeatum* surface showing the rhomboid figures (Upper Carboniferous Czechoslovakia); Original: Senckenberg Museum. Pteridospermae or Cycadofilices: leaf of *Glossopteris* from South African Gondwana flora (Carboniferous and Permian); Original; Senckenberg Museum.—Leaf of a palm fern: *Pterophyllum longifolium* (Upper Triassic from Lunz, Austria).

idea of evolution in believing that all animals were created according to some great unified plan. His ideas strongly influenced two of his countrymen; Étienne Geoffroy Saint-Hilaire (1772–1844) and Jean Baptiste Lamarck (1744–1829).

Both these great scientists had studied fossils and felt that species could evolve. Saint-Hilaire and Lamarck also believed that species were profoundly influenced by their habitat, climate, and other environmental factors. It was Geoffroy de Saint-Hilaire who said that the first bird must have hatched from a reptile egg. In his "Philosophie zoologique", Lamarck put paleontology a great step ahead when he remarked: "Among the fossilized remains of animals there are many organisms not known to us today. Why should they be extinct, since man could not have extirpated them? Is it not possible that these petrified individuals are the ancestors of species found today?" These first attempts to connect paleontological findings with phylogeny did not bear fruit, since the notion of evolution was rejected and sharply contested by the most prominent zoologist and vertebrate paleontologist of the time, Georges Cuvier (1769–1832).

In spite of his attitudes toward early evolution theories, Cuvier is justifiably considered to be the founder of modern paleontology. One of his most ingenious thoughts was the principle of correlation. Entire fossil skeletons are very rare finds and are not typical; more often, one encounters fragments of an entire skeleton, pieces that are difficult to put together meaningfully. According to Cuvier, all living organisms form an entity, and no part of that entity can change unless the other parts also change. Thus, one can determine what missing parts look like if not all the bones belonging to some animal are available. Convincing proof of the viability of the correlation principle was provided from a fossil dug up from Tertiary gypsum in Montmartre. Only the teeth of this animal were found. Since these teeth resembled those of modern opossums, Cuvier used his correlation principle and concluded that the fossil was a marsupial. Later the entire skeleton of the animal was recovered, and Cuvier demonstrated that these bones really did belong to a marsupial.

During this period, systematic diggings began, and fossil finds were classified just as living animals were. Like Buffon before him, Cuvier also believed that some species were extinct and that there had been a series of geological eras throughout the history of the earth during which the animal world reached increasingly greater heights of development. Instead of envisioning a slow, gradual development (i.e., as we now understand evolution), Cuvier postulated that there were recurrent catastrophes. In his most important work, "Recherches sur les ossements fossiles" (1812), he propounded his catastrophe theory in the following words: "Life on earth has frequently been destroyed by terrible events. Countless organisms have been the victims of these catastrophes. Huge floods have swallowed lowland inhabitants, and the sudden upswelling of the sea floor pushed many marine organisms on land. They are extinct forever.

They have left only traces behind, whose meaning is being studied by naturalists."

Cuvier believed that evolutionary thinking was a philosophical notion that he, as an empiricist, could simply not deal with. He wrote: "If this hypothesis is correct, one should be able to find traces of a progressive development in the animal kingdom. but until now there is no indication of that. Why haven't the insides of the earth retained any evidence of such a geneology? Because there never was such a thing, for species in earlier times were just as durable as those found today."

In one sense, Lamarck was basically not a predecessor of Darwin and evolution but a successor to a line of natural philosophers, for his viewpoint that the use or disuse of an organ would influence evolution was never proven by evidence. Although Cuvier, in his catastrophe theory, persisted in the mistaken dogma that species are immutable and therefore do not evolve, he provided the basis for modern paleontology. The greatest 19th Century paleontologists up to the publication of Darwin's work all adhered to Cuvier's pronouncements: they included Alexandre Brongniart (1770–1847), Ducrotay de Blainville (1778–1850), and Louis Agassiz (1807–1873), who became professor of zoology and geology in New Cambridge in 1846. Others influenced by Cuvier were Alcide d'Orbigny (1802–1857) and Richard Owen (1804–1892). Owen developed Cuvier's method of recognizing animals from bone fragments and reconstructing them, and the greater amount of material present by that time enabled Owen to continue Cuvier's analysis with many new fossils. Agassiz is remembered as an expert on fossil fishes, but he was also one of those scientists who recognized that the so-called diluvial period was not a flood at all but an ice age.

Investigations in many specialized aspects of paleontology not available in Cuvier's time demonstrated that there would have to be far more catastrophes than Cuvier ever postulated to explain the vast diversity of life on earth, and as more and more fossils were found the Cuvierian theory became less convincing. In his catalog of fossils compiled in 1848 and 1849, Heinrich Georg Bronn (1800–1862) listed 2050 plant species and 24,300 animal species, and this massive amount of evidence led Bronn and his contemporaries to the realization that old animal species died out gradually, not suddenly, while new species slowly developed simultaneously, not afterward.

Among geologists, the idea that the earth and its formations developed gradually was propounded shortly before Cuvier's death, but biologists ignored these pronouncements. An English private teacher, Charles Lyell (1797–1875), in his "Principles of Geology", put forth a geological theory opposing the catastrophic concept. Instead of powerful geological revolutions, there were, as Lyell interpreted the evidence, a series of very slow, gradual changes in the earth taking place over an incredibly long period of time. Lyell felt that the forces operating in the early history of the earth

Fig. 3-11. Johann Jakob Scheuchzer (1672–1733).

Fig. 3-12. Georges Cuvier (1769–1832).

Fig. 3-13. Sir Richard Owen (1804–1892).

Fig. 3-14. Louis Agassiz (1807–1873).

Fig. 3-15. Charles Lyell (1797–1875).

Fig. 3-16. Charles Darwin (1809–1882) as a young man.

were the same ones observable today. His theory provided a new basis for developing an evolutionary theory, since the concept of a gradual change in species over long periods of time would fit in beautifully with Lyell's scheme. Interestingly, Lyell himself would not adhere to an evolutionary theory! He sharply criticized Lamarckian ideas and believed that species were immutable. Darwin's magnificent theory of evolution changed his mind.

When young Charles Darwin (1809–1882) embarked on his renowned research voyage on the Beagle in 1831, he also believed in Cuvier's doctrine of species immutability. However, he was doubtful after reading Lyell's "Principles of Geology", and it was through Lyell's eyes that he looked at South American geology, with its gradual ascents and descents, mountain growths, and signs of the glacial period. He gathered a wealth of evidence supporting Lyell's contention that the forces operating on the earth's surface today are no different from the ones operating millenia previously.

When he visited the Galapagos Islands in October, 1835, Darwin first played with the idea that "species change gradually—and this question pursued me!" The flora and fauna of this island group originated on the South American continent, but these organisms were driven to the Galapagos, where they adapted to the habitat conditions there and changed to such an extent that they gave rise to new species and even genera. When Darwin investigated Tertiary fossil remains of South American ungulates, giant sloths, and other members of what was certainly an extinct world of animals, he wrote the following observation: "One is initially led to conclude that there was a catastrophe. However, after seeing that the group of extinct animals includes small and large animals from southern Patagonia, Brazil, on the Peruvian Cordilleras, in North America, and as far north as the Bering Strait, any catastrophe would have had to shake the entire world. Furthermore, an examination of the geology of La Plata and Patagonia leads to the conclusion that the present shape of the land is the result of slow, gradual changes."

Darwin found that some fossils bore a considerable resemblance to living species and were obviously related to them. Did these prehistoric animals really become extinct in the most extreme sense of the word, or should they more properly be regarded as the ancestors of our present-day organisms? From the huge amount of evidence he gathered, Darwin came to the following conclusion: "The wonderful relationship between the dead and living animals on the same continent will later shed more light on explaining the appearance of organic creatures on earth and their disappearance than any other body of facts, and I have no doubt that such a relationship truly exists."

Darwin's work, "The Origin of Species by Means of Natural Selection or The Preservation of Favoured Races in the Struggle for Life" (1859), in which evolutionary biology truly got its modern start and which had such profound implications for the natural sciences, also was of great significance for paleontology. With the publication of this work, paleontologists no longer looked merely for petrified representatives of bygone geological periods but thought in terms of finding ancestors to modern animal groups and looked for such species. It was with Darwinian thinking that the time factor began to be properly appreciated in any prehistoric scientific research. Only by thinking in terms of hitherto unimaginable time scales could one explain how individual animals and plants of today evolved from primitive forms. Paleontologists had to reformulate their attitude toward fossils from a merely descriptive one to an outlook with an evolutionary perspective.

Many prominent paleontologists were not initially prepared to do that. If Darwin's theory was correct, then generations of learned scientists were all wrong. One zoologist rejected the Darwinian theory with these words: "For centuries, capable, hard-working students of zoology have made tremendous achievements and have formulated superb systems, and their work produced a unified picture of man. Now a new man comes along who wants to destroy this so carefully constructed body of thought!" Darwin showed paleontologists how fragmentary their evidence was: "The earth crust with its many fossil remains can not be regarded as a well-filled museum but as a scanty, accidental collection often with large empty spaces." However, Darwin felt it was a prime task of paleontology to find the various "missing links", the intermediate and transition forms absent from the fossil record, "since if my theory is correct, there must have been many such intermediate species on the face of the earth in its previous history."

In 1861, the first so-called missing link was discovered. It was the primitive bird *Archaeopteryx*, an intermediate species linking reptiles with birds. Two of Darwin's most noted supporters, Thomas Henry Huxley (1825–1895) in England and Ernst Haeckel (1834–1919) in Germany, brought man into the theory of evolution. Haeckel predicted that a fossil link between anthropoid apes and humans would be found, and today a great many pre-human and early hominoid species are known. Huxley suggested that a similar phylogenetic tree existed for the horse, which progressed from five-toed protoungulates through four- and three-toed foliage-feeding primitive horses and, with continual degeneration of the fingers and toes, to the one-toed horses of today. Huxley's hypothetical model was confirmed as accurate through the work of the Russian zoologist Vladimir Onufrievitch Kovalevskii (1842–1884) and through the diggings of the American paleontologist Othniel Charles Marsh (1831–1891), although 20th Century finds have revealed that equid phylogeny is more complex than these people could have known and that besides the chief line leading to today's horses there are many blind ends.

The many new paleontological finds, whose systematic digging was inspired by Darwin, produced many more fossil species than had been

Darwin's *Origin of Species*

Fig. 3-17. Thomas Henry Huxley (1825–1895).

Fig. 3-18. Ernst Haeckel (1834–1919).

Earliest comprehensive treatises

known before. This led to a division between the study of earth formations or stratigraphy and paleontology (with its daughter disciplines paleobotany and paleozoology). The first comprehensive surveys of the animal kingdom including fossils and their phylogenetic significance became available.

In spite of the fragmentary nature of the fossil record, even with this renewed effort to locate new specimens, paleontology confirmed the theory of evolution and led to further insights into the phylogenetic pathways of living organisms. With serial cuttings of rocks on the armored jawless organisms (Agnatha) embedded there, Swedish paleontologist E. A. Stensiö found that these Silurian and Devonian creatures, the oldest vertebrates known to us, were not true placoderms but relatives of cyclostomes. The Ichthyostegalia from Upper Devonian Greenland showed that quadrupeds did not evolve from lungfishes, as was formerly thought, but from crossopterygians. Biologists had previously thought that amphibians were the immediate ancestors of mammals, but the discovery of the Theriodontia in the plains of the South African Karroo showed that mammals arose from this Permian reptile group. Finds in Germany of flying saurians and *Archaeopteryx* led scientific artists to draw pictures of them that were at first based largely on fantasy but that later could be made more accurately as paleontological and other technical methodology improved.

Dinosaurs in North America

Paleontology became well-known in North America through the pioneer efforts of Othniel Charles Marsh and Edward Drinker Cope (1840–1897), who argued with each other and who, along with Henry Fairfield Osborn (1857–1935) and many others unearthed many extensive dinosaur finds and found all sorts of extinct ungulate groups. The pampas loam of Patagonia, the area Darwin had visited, also proved to be a good source of fossils. The drawings and sculptures made by late glacial man has proven to be another source of evidence for prehistoric fauna, for these illustrate how animals like the mammoth and woolly rhinoceros looked.

Paleontology supported the theory of evolution not only in demonstrating the flourishing and disappearance of animal groups but also in showing certain regularities associated with phylogenesis. Certain generalizations were even developed; one of these, the rule of irreversibility, was formulated by a Belgian, Louis Dollo (1857–1931). According to this rule, evolution is irreversible and does not turn around, as it were. Characteristics found in some species during the course of its development and then are lost do not, at some still later time, again appear. Dollo also found that species and entire groups of animals can become extinct when they are over-specialized. When animals become so rigidly adapted to their habitat surroundings that no other environment will do for them, they die off when climatic changes occur in that habitat.

The modern field of paleobiology, founded by the Austrian paleon-

Fig. 3-19. Mammoth—ivory drawing discovered in 1865 in the cave La Madeleine (Dordogne, France).

tologist Othenio Abel (1875–1946), gives us a deep insight into the life of bygone geological periods. Paleobiologists have developed various methods that permit them to not only describe the external appearance of fossil species and their prevalence but also their habitat and the reasons for their extinction. Paleopathology has been invaluable in evaluating fossilized tracks, traces of feeding sites and excretions, and pathological changes in the hard body parts. The highly specialized field of paleo-photobiology deals only with the influence of light on fossil organisms.

Fig. 3-20. The rhinoceros *Coelodonta antiquitatis* depicted in an ice-age wall drawing on the grotto of Font-de-Gaume (Dordogne).

The Halle finds we already mentioned show that in some very rare cases we can tell something about the details of tissue structure, muscle fibers, nerve cells, and chromatophores (color-carrying bodies) in extinct animals. The field of paleohistology investigates fossils, and this field is especially important for paleobotanists, since fossilized plants often display their fine structure. The age of core samples can be established by studying the microfossils present in them, and this knowledge can be important for tapping industrially important resources like oil. A whole new field, micropaleontology, has arisen due to the significance of this.

Fig. 3-21. Giant elk in the cave Pech-Merle (France).

Paleogeography is the study of the changing geological relationships throughout the prehistoric period. Paleogeographers study land and water in different epochs, the origin and erosion of mountains, the extent of glaciation, and other phenomena. Paleoclimatology is closely related and uses fossils, soil samples, marine deposits, and above-ground formations to draw conclusions about the climate in bygone eras. We now know that there have been some very significant climatological changes, which is shown by Jurassic coal in Antarctica, the Tertiary flora remains in the arctic, and the glacier traces in South Africa, India, Australia, and a part of South America during the Permian and Triassic.

Paleontologists have to make use of other fields to determine how bygone habitats—entire ecosystems, in fact—looked, since they cannot examine these directly. They utilize morphology, zoogeography, and ecology in assigning some fossil its proper phylogenetic position. All these other disciplines, plus the refined techniques available to paleontologists, have made the field much more than a dry study of petrified bones. Its chief aim is to research the flora and fauna of prehistoric periods and to classify them in the zoological and botanical systems used for living species.

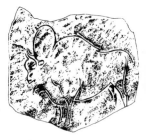

Fig. 3-22. European wisent: ice-age drawing on a calcified rock.

The development of the earth and its life has been an historical (i.e., single) event. History can be divided in two ways, however: absolute and relative. Time can be given certain units, permitting physical events to be chronologically ordered on a number scale as we, for example, measure the passage of years, and this is known as absolute measurement of time. On the other hand, we can mark the course of history by referring to major events occurring through the course of time and using these as a reference for other events, and this kind of measurement is known as relative measurement of time. An example of this would be the division of historical architectural styles such as Romantic, Gothic, Renaissance, etc.

Geological dating, by O. H. Walliser

The simplest example of time measurement on a relative basis in terms of geology is to determine which of two adjacent geological layers is the younger. In 1764, Steno determined that in most cases the higher of two layers is the more recent, but this is only true in situations where there has been little disturbance of the crust of the earth. In areas where mountains are forming, the older deposit may be found above the younger one. Steno's law is also invalid for volcanic rocks, since masses of freshly spewn lava can come to rest between other, older masses, or fuse with them. Two lava masses can only be temporally distinguished when the older masses have been altered by the heat of the newer ones. Our point here is that depth in the earth is not the sole criterion to be used for determining the relative age of deposits, and other techniques must be employed if we are to be accurate. These include various marks on the surface of the earth, such as animal impressions, marks left by objects that drifted over the site, or raindrop impressions on the ground; channels caused by running water, dry clefts (which widen toward the top), shifts in rock masses, and movements of particles from lower depths, are other important indicators. Sometimes hollow areas become partially filled with mud, and this acts as a sort of fossil hydrometer, showing us how high the mud or water level was at that site. Fossils found in life positions are also important, and such finds can be encountered readily in organisms that live in mud, like mussels. All these characteristics only tell us the age of one group of organisms relative to another. They do not explain at what time in geological history they arose.

Biochronology

If we are to organize the various layers in the earth geologically, we need as reference points some events that occurred only once and were not repeated, so that we have some directional process with which we can work. The only such reference we have is the phylogenesis of living organisms, and the use of key fossils in this regard was mentioned earlier. As we stated, William Smith was the founder of biochronology (bios = life; chronology = time measurement). One of the most important requirements in using the biochronological method is that the reference key fossils used were distributed throughout the world at the same time. On the basis of present knowledge about the speed with which a species propagates and distributes itself, it appears that this is an accurate assumption, at least when we speak in terms of long geological periods and not just a few decades. One difficulty is presented by the fact that all living organisms are dependent on their environment, and they only occur where the conditions they need also occur. Since geologists refer to individual ecosystems as facies, they speak of the facies dependency of organisms. When the facies requirements for some organism appear at point X later than at point Y, this fossil would have appeared at X later than at Y. In regions with different facies, a temporal reconstruction can only be made when the orientation of the various facies makes such a reconstruction feasible or if there is an organism present in that area that is relatively facies-independent. Some organisms were not dependent on a specific facies relationship, and the relative time at which such organisms lived is sometimes known from previous work.

The simultaneous appearance of key fossils is ensured when such fossils are found as part of a whole phylogenetic line, with their predecessors in earlier deposits.

Since we desire to make our relative time scale as fine as possible, the best key fossils would be those that are known to have lived just a very short time. Therefore, we choose fossils with the greatest evolutionary development speed, those evolving the fastest. As we show in the following chapters, there are animal groups that have hardly changed over periods of millions and even several hundred million years. They would hardly be appropriate key fossils. In contrast, other species undergo evolution very quickly, and the best key fossils were alive just 500,000 to 1,000,000 years. In absolute time scales, this span is just a portion of the margin of error, so the biochronological method of time measurement is actually much more precise than the exact radiometric dating method.

A series of now-classic key fossils was developed in the 19th Century. In the earliest geological periods they were the trilobites and graptoliths. From the Devonian to the Upper Cretaceous the key fossils were ammonites, and later mollusks and snails were the reference species. During the Paleozoic and lower Mesozoic eras, brachiopods were key fossils, and during the Cenozoic era, terrestrial vertebrates were successfully used. In response to the needs of the oil industry, which dates beds by thin core samples, paleontologists investigated the microfossils found there, especially the unicellular Foraminifera and crustaceans and found they were superb key fossils. As paleontological techniques became more refined, researchers realized that the utility of some group of organisms for biochronological work is really just a question of basic research. Even the small steps taken by slowly developing phylogenetic lines can be understood, especially with the aid of statistical methods available today. Thus, the classic key fossils of the 19th Century are still used as markers, but they have been substantially supplemented and have even been exceeded in many cases by members of almost all groups of living organisms.

The study of rock formations and petrified structures has shown that the earth has gone through an extremely long history measuring in the billions of years, and the time span that has ensued since the earliest living organisms appeared has been almost incomprehensibly long. Once scientists were aware of this, they became interested in being able to measure, in absolute terms, just how long this time span was. Estimates given at the end of the 19th Century were that the earth was between several tens of millions to hundreds of millions years old. To some geologists, these figures seemed much too small, and these geologists were right!

How was it ever possible to arrive at a specific number? In 1884, Swedish geologist G. de Geer found a method by which he could deter-

Key fossils

Dating with radioactive elements, by H. Martin

Annual rings

mine, in years, the time from the end of the last glacial period to the present, which is the most recent part of the earth's history. He counted annual rings on trees. A large ice lake had formed at the leading edge of the melting Scandinavian ice cap, and seasonal temperature changes produced a regular change in the growth rings of trees in that area with an annual cycle. In the summer, the melting ice contained a thin layer (up to several millimeters) of fine sand, and in winter, when the lake was covered with ice, trees were coated with a still thinner layer of dark mud. Thus, every double band, of new growth and the thin coating, accounted for one year of life in a tree in this region. De Geer and his colleagues found that as one moves northward from southern Sweden, the annual rings were fewer and fewer in number, meaning of course that the trees were younger, and by going far enough north one could finally find trees that had been in existence in historical times. After decades of work, de Geer and his colleagues were able to relate the prevalence of all these annual rings in different trees and to count them accurately. His work demonstrated that the ice edge was on the southern coast of Sweden 13,500 years ago. Annual rings were later used to age numerous other phenomena, such as the time of the breakthrough of the ice lake into the North Sea.

This very precise method of chronology could only be used for very short time spans in the earth's history, since layers in the earth are formed in many different ways and do not always display such a regular annual rhythm. Therefore, de Geer's methodology could not be used to ascertain the age of the world or the time span of many of its geological periods. At the turn of the century, however, the great dream of geologists to actually be able to know how long ago the great dinosaurs lived and when giant dragonflies flew over swamps was finally realized.

The first step in the fulfillment of this dream was made with the discovery of radioactivity. French physicists Henri Becquerel and Marie Curie found that certain elements emit radiation, and that the factors responsible for radiation (there are three of them) were not influenced by changes in temperature or pressure. Radiation was produced by decay within the nucleus of these radioactive substances.

Radiometric dating

Radiation could not be used to measure time until the structure of atoms was understood, for atomic structure is intimately related to the nature of radiation. The physicists studying radioactive decay gave us this understanding. Lord Ernest Rutherford made the crucial experiments and observations on radioactive decay in the early part of the 20th Century. Today we know that all chemical elements (oxygen, iron, gold, etc.) are composed of small electrical units, positively charged protons and negatively charged electrons. Another stable, building block, the neutron, lacks any electrical charge. Almost the entire mass of atoms is concentrated in the nucleus, where the protons and neutrons are located. The atomic weight is determined by the number of protons and neutrons, and atomic weight can be many times the weight of the protons alone. The nuclei are surrounded by a looser cloud of electrons, which have an extremely small mass. The chemical properties of elements are essentially determined by the structure of this electron cloud.

An element has an equal number of electrons and protons in its atoms, thus equalizing the positive and negative charge, but the number of electrons is independent of the number of neutrons, so elements can exist in more than one atomic state by having different numbers of neutrons, and these different states are called isotopes. Of course, their atomic weights vary. Isotopes are of great importance in radiometric dating, and the technique is sometimes called isotopic dating as a reflection of this importance.

The simplest and therefore lightest element of all contains atoms with one proton in the nucleus and one electron, this element being hydrogen and bearing an atomic weight of 1. Its heavier, but rarer, isotope known as deuterium (atomic weight $= 2$) has one proton and one neutron in the nucleus and a single electron. The next heavier element, helium, has two protons and two neutrons in its nucleus (and therefore an atomic weight of 4) and two electrons in its cloud. Helium is an extremely stable element, and when very heavy elements decompose they radiate helium nuclei. This kind of radiation is called alpha (α) radiation, and it was the kind that Rutherford observed. The second kind of radiation, beta (β) radiation, consists of electrons (freed from their former association with neutrons and thereby converting those neutrons into protons). The third kind of radiation, gamma (γ) radiation, consists of penetrating X-rays. The chemical nature of an element is altered by the loss of a helium nucleus or by the loss of an electron, and when such a loss occurs the element is transformed into a new, daughter, element. Daughter elements can also be radioactive and can decay into yet other elements. Radium, for example, is one of the daughter elements of the many produced by a uranium isotope. It is one of the members of a series we call a chain reaction, a series that finally ends with a stable, non-radioactive element, namely a lead isotope.

Radioactive transformations have a clocklike function when they occur at a regular rate, something like sand grains dropping through the orifice of an hour-glass. In 1902, Rutherford and Soddy found just such a regularity: of all the atoms in some sample of a radioactive element, half of these atoms decay into a specific daughter element after a definite period of time known now as the half-life of that radioactive element. Clearly, half of the original element is gone once the half-life time has been reached. For example, the uranium isotope with an atomic weight of 238 (written in physics nomenclature as U^{238}) has a half-life of 4.5 billion years, whereby the end product of its decay is a lead isotope with an atomic weight of 206 (Pb^{206}). Of 1000 atoms of U^{238}, 500 will remain after 4.5 billion years have passed, and 500 Pb^{206} atoms will have been produced. After another 4.5 billion years, just 250 atoms of U^{238} will remain from the original sample, and the number of Pb^{206} atoms will have climbed to 750. So it goes, with the regularity of a fine clock. There are half-lives of

Differential decay rates

seconds, days, millenia, and millions and billions of years for various radio-active elements and isotopes. Radioactive decay rates are particularly valu-able for geological time measurement since they are not influenced by physical or chemical changes.

Rutherford

As early as 1905, Rutherford thought it was possible to determine the age of rocks containing radioactive minerals. Any mineral containing uranium would have a certain amount of uranium and a certain amount of daughter elements, with less and less of the original uranium as the sample became older and older. If the half-life of the mother element in the sample is known, the age of the sample can be determined by com-paring how much mother element and how much daughter element(s) are present. Specifically, what can be determined is the time at which the mother element was enclosed in the growing crystal. Comparisons like this could only be made when techniques were developed for measuring and assaying the extremely small amounts of the mother and daughter isotopes that occur in natural mineral desposts.

Boltwood and Strutt

Boltwood and Strutt made the first dating determination of minerals in 1907, but their method was not very accurate. Geologist Arthur Holmes developed their technique further and applied it accurately. Imagine the surprise in the scientific world when, in the 1920s, Canadian rock samples were found that were more than 2,000,000,000 years old! Geologists had long known that these particular rocks were very old, but until then the age of the universe was the only entity thought to have an age of this magnitude. Astronomers and cosmologists have since corrected our original ideas regarding the age of the universe, and it is now (1972) esti-mated that the earth was formed approximately 4,500,000,000 to 5,000,000,000 years ago and that the oldest rocks known on earth are some 4,000,000,000 years old.

Improved dating techniques

Improved techniques helped make even greater progress with radio-active dating and permitted assaying the much weaker but far more wide-spread isotopes of potassium and rubidium. The following table shows the isotopes used for radiometric dating and their half lives:

Mother-isotope	Half-life (years)	Daughter isotope	Method
U^{238}	4.498×10^9	Pb^{206}	Uranium-lead method
U^{235}	7.13×10^8	Pb^{207}	Uranium-lead method
Th^{232}	1.39×10^{10}	Pb^{208}	Thorium-lead method
Rb^{87}	5.0×10^{10}	Sr^{87}	Rubidium-strontium method
K^{40}	1.3×10^9	Ar^{40}	Potassium-argon method
U^{234}	2.5×10^5	Th^{230}	Uranium method
Th^{230}	8.0×10^4	Ra^{226}	Ionium method
C^{14}	5.7×10^3	N^{14}	Radiocarbon method

Carbon dating

The isotopes with long half-lives are used to date very old specimens, while those with shorter half-lives are for more recent ones. Among the short-lived isotopes, the carbon isotope C^{14} has been particularly important, for the radiocarbon method permits determining the age of organic material (e.g., charcoal, bones, wood) as recent as 50,000 years of age, which is well into the last glacial period. Radiocarbon dating has made it possible to date important events like human prehistory and the beginnings of domestication and agriculture. The physicist who developed this technique, Willard F. Libby, received the Nobel Prize for his efforts.

In spite of the choice of radioactive isotopes, not all rocks can be dated. The age of a sample can only be determined when radioactive elements were incorporated into it when it was formed and when these mother elements are held in with the daughter elements. This usually occurs only in volcanic rocks formed from molten lava. Deposits containing fossils can rarely be dated since the radioactive minerals in them usually have had a long history behind them and are embedded in the sediment. One exception is the green mineral glauconite, which formed in some oceanic sands. It contains potassium and can be dated with the potassium-argon method. Thus, fossils can only be dated with radioactive techniques when they lie in sediments with lava or contain glauconite.

When these requirements are not present, the age of a sample has to be stated as some range of minimum and maximum possible age. Sometimes sediments can rest on top of some very old granite and be pushed into and through the granite when lava passes over them. The granite would mix with the sediment and the material flowing through and transporting the sediment, and the analysis of such a find would have to be aware of these kinds of effects. Sites are sought where the time difference between deposits is as small as possible. Then measurements can be taken on many rocks, but the measurements are not simple to make. For example, the following steps must be taken to date a sample of granite:

1. The sample must be very fresh, since in stormy weather various chemical processes can occur that drive off the mother or daughter isotopes.

2. A thin section of the rock has to be taken and examined under a microscope to ascertain whether the sample suffered any changes after it crystalized, since such changes could remove the daughter isotopes.

3. The sample is ground up, and the potassium and rubidium components are concentrated.

4. A binocular microscope is then used to select the cleanest rubidium and potassium bits.

5. The separation of the isotopes is accomplished in a mass spectrometer. This piece of equipment produces a high vacuum in which atoms differing by just one neutron in weight can be separated, so they can be assayed.

6. Now the dating is carried out, using some corrections.

The margin of error becomes larger as the sample being investigated is older. Laboratories in many countries throughout the world have con-

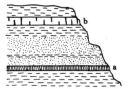

Fig. 3-23. a = lava stream; b = glauconite-sand-stone; a and b are datable.

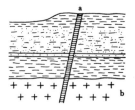

Fig. 3-24. a = lava passage; b = granite; a and b are datable.

ducted dating tests on thousands of samples from all parts of the earth and, now, even from the moon. The results of these analyses are enabling scientists to follow the course of the earth's history as they never could before and to arrange the living creatures on an absolute time scale.

4 The Phylogenetic Relationships of the Animal Kingdom

In Chapter 2, we discussed briefly that paleontology with its fossil finds cannot give a complete picture of the phylogenetic tree of the animal kingdom but that it only bridges certain gaps and gives important information on extinct animal groups with their lines of descent and intermediate forms. The fragmentary nature of the fossil record was understood as such in Darwin's and Haeckel's time. Ernst Haeckel emphasized the incompleteness of the fossil record and he stressed the importance of embryological studies in phylogenetic research. As Emil Kuhn-Schnyder puts it, Haeckel developed "a program of biological research on a greater scale than had ever been done before." In his biogenetic law, Haeckel saw not only the key to the solution of all comparative anatomy problems but, in his mind, even believed he found the most illustrative research route for learning about phylogeny.

From unicellulates to coelenterates, by H. Wendt

Biogenetic thinking had arisen in the 1820s with the work of anatomist J. Meckel (1781–1833) and zoologist Karl Ernst von Baer (1792–1876), but neither of them developed a theory of evolution. Haeckel's great biogenetic predecessor was the German zoologist Fritz Müller (1821–1897), who formulated the basic laws of biogenesis in 1864. According to Haeckel, "the sequence through which a developing individual passes in its embryological stages (a kind of development we call ontogeny), from the single cell to its fully developed state, is actually a short, compressed replay of the long series of species ancestral to that individual from the earliest geological times to the present." Simply stated, Haeckel's law of biogenesis is that ontogeny (individual development) recapitulates phylogeny (development of the species).

Biogenesis

He concluded that embryos give us the key to earlier phylogenetic stages of development of animal groups, and he enlarged on his ideas in his work *Prinzipien der generellen Morphologie der Organismen* (Principles of General Morphology of Organisms), in which he developed a monumental theoretical framework for a phylogeny of living organisms, a great plan accounting for the historical unity of both flora and fauna.

Ernst Haeckel

Today we know that embryological studies are essential for phylogenetic research. However, the various stages through which embryos pass only show us, in those cases where Haeckel's rule is accurate, what ancestral forms looked like and not how they were adaptive or what sort of habitat they lived in. The theory of evolution requires these sorts of explanations, and embryos cannot always give us the answers to questions like this, especially because some of them lack the intermediate evolutionary forms and do not conform to Haeckel's rule. In this chapter we hope to show, in a general way, how the animal kingdom evolved and what sorts of phylogenetic ties exist between various animal groups. We begin with the unicellulates, one-celled animals.

Non-nucleated unicellulates

Haeckel believed that the entire plant and animal kingdom arose from non-nucleated unicellulates, which he called moneres. Naturally, "moneres" as fossils have never been found, and no such organisms exist today. The non-nucleated unicellulates occurring today have lost their nuclei secondarily, but Haeckel's original terminology has been retained for these organisms and they are called moneres. Haeckel felt that the very first ancestors of all animals were unicellular plants, the Phytoflagellata, which gave rise to the flagellate animals (class Flagellata).

Actually, there are many purely autotrophic (photosynthesizing) flagellate species, while some have both plant and animal qualities and others feed strictly in the way of animals but whose anatomy nevertheless reflects that they evolved from plants (flagellate algae). The question immediately arises whether the animal kingdom arose from plants in a monophyletic or polyphyletic manner (i.e., from one plant group or from many of them).

From plant flagellates to animal unicellulates

Since there are many lines leading from the Phytoflagellata to animal unicellulates, it appears that the protozoa (unicellular animals) arose polyphyletically. However, all multicellular organisms seem to have arisen from the unicellulates in one evolutionary event (i.e., monophyletically): from the blastulae (packs of cells) of flagellate colonies. Some authorities, nonetheless, question whether multicellular animals could only have arisen monophyletically. All unicellular groups contain colonial species, and biologists Hadži and Steinböck have suggested that the higher organisms (with the exception of sponges) may have arisen from simple cell division and cell formation processes in the ciliates (class Ciliata), in which cell walls develop around the individual nuclei.

Most theorists, however, retain the blastaea and gastraea theories of Haeckel, which he propounded in 1874. In his scheme, the free-swimming spherical flagellate colonies, which are single-layered colonies, as we still find today in the chrysomonadins and choanoflagellates, are the ancestors of all multicellular organisms.

The early embryological stages in multicellular animals are blastulae that look much like the flagellate colony structure. We would have to assume that gelatinous supporting material appeared at very early stages in order to hold the developing multicellular animals together as blastulae. Interestingly, there are no blastulae in the lowest known extant multi-cellulates, for these animals are composed of two cell layers. These Mesozoa parasitic organisms not found in the fossil record, consist of a simple cell tube surrounding the reproductive cells. They lack an internal cavity for food uptake. It is quite possible that these parasites were once more complex than they are now and have become more simplified dur-ing their evolution. In sponges (phylum Spongia), the body wall consists of two layers, dermal and gastral, and where they meet they form an epithelium, tissue filled with water canals. Sponges have a supporting cell layer between the dermal and gastral layers, as we postulated at the begin-ning of this paragraph. This supporting layer houses horn, calcium, or silicic acid, which comprise the skeleton.

As multicellulates develop embryologically still further, an internal cavity is formed by an invagination of the blastula, and this cavity, lined by a cell layer called the entoderm, produces the gastrula stage. This is the same as what Haeckel originally called the gastraea. The cavity terminates in a mouth/anus opening, the blastopore. The doubled-layered gastrula is formed from the single-layered blastula from progressive invagination of the rear of the blastula wall, and invagination occurs on each side of the blastula opposed to the direction of swimming. These sides are known as the vegetative poles of the organism. The developing organism thus be-comes differentiated and now has a specific part of the body where food uptake and metabolizing occur. The enclosed cavity becomes a proto-intestine, lined with intestinal epithelium. The intestine also requires sup-port in the form of gelatinous matter and tissue. Haeckel's original two-layered conception is thus an oversimplification. The muscles arise during these early stages, and the muscles cannot be effective unless they have supporting tissues of their own.

One excellent model of this evolutionary stage are the coelenterates (Coelenterata or Radiata), whose present-day members are classified into the two phyla of Cnidaria and Acnidaria. Their body is also composed of two cell layers, the body-covering outer layer known as the ectoderm and the lining of the inside, the entoderm. A supporting substance is found between these two layers, and it can form a third layer (the mesoglea) by cells migrating from the outer layer or, less commonly, from the inner layer. As in Haeckel's gastraea, the coelenterate body contains a single (but often subdivided) inner cavity, and it has just one opening that serves as mouth and anus.

The phylogeny of the next important multicellulate stage, that of higher coelenterates, was beautifully researched by Haeckel in his embryo-logical studies. These animals have, between the intestine and outer wall, additional organs that form a third cell layer (see Fig. 4-5). This inner germinal layer develops from the intestine and contains the coelom, a structure for containing fluids that will be discussed subsequently.

Development of the coelom as shown in schematic cross-sections:

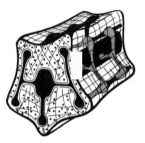

Fig. 4-1. The coelom probably originated in a gelatinous animal whose body was penetrated by intestinal canals.

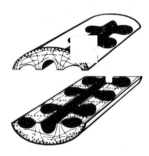

Fig. 4-2. The canals formed protrusions, which were filled with water and nutrients.

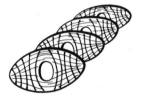

Fig. 4-3. Body support was maintained by muscles.

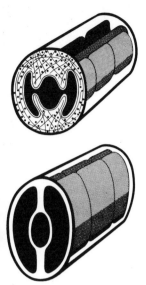

Fig. 4-4. The intestinal protrusions finally separated from the intestinal tract.

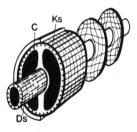

Fig. 4-5. Typical coelenterate worm shape. The coelom (C) has split the initially simple muscle lattice into an external body tube (Ks) and an internal intestinal tract (Ds). The black intestinal cavities and the coelom are fluid receptacles and give the inner muscles freedom to move. The external muscles are used to pump water in.

In most higher coelenterates, the mouth is a vestige of the primitive mouth structure, so these animals have been classified as Protostomia. A great debate is currently in progress regarding the original form of the coelom. Adolf Remane, Werner Ulrich, and other zoologists believe that it had three chambers. Thus, a slightly compartmentalized coelom was the original structure, and more complex compartmentalization (metameres) came later in evolution. The multicellular animals, using this line of thought, can be classified into three major groups:

1. Proto-coelomates, which includes a few phyla incorrectly known as worms by many laymen; 2. Bilateralia (two-sided animals), in which the mouth is still a vestigal proto-mouth but the body has bilateral symmetry; 3. Chordata (chordates) in the broadest sense of the term, a group earlier classified as Deuterostomia and in which the mouth in the proto-mouth region is a new development and in which the proto-mouth (which we recall functioned at once as mouth and anus) remains as the anus. The anus can also appear as a new structure, however.

A second theory questions this first one, which is purely schematic and really offers no explanation for why the above processes should have occurred. According to this theory, the coelom is a multi-compartmentalized structure in even the earliest organisms with a coelom, and coelenterates with few coelomic subdivisions are the result of simplification from a more complex plan. These simplifications appeared as a result of adapting to subterranean habitats. Both theoretical viewpoints acknowledge a difference between Protostomia and Deuterostomia, and each holds that the coelom originated from an invagination of the gut. We shall use the Deuterostomia as particularly convincing, telling examples of the phylogenetic relationships within the animal kingdom and of mechanisms of evolution.

Before going into that, some mention should be made of other higher coelenterates, especially those bilaterally symmetrical organisms. We shall encounter them in many different parts of this volume as we recount the evolutionary processes occurring in organisms. They achieved the height of their development in the mollusks (phylum Mollusca), which includes snails, mollusks, and cephalopods of today, and in the arthropods (phyletic group Articulata), whose members include annelids, spiders, crustaceans, centipedes, and insects. A few other phyla, particularly those with parasitic species, have undergone modifications that are so extensive their systematic position is uncertain. Certain molluscan species, such as the tentaculates and Chaetognathi, are close to the great phylogenetic branch leading on the one hand to bilaterally symmetrical animals and to the chordates on the other.

As we said previously, we shall illustrate the meaning of the theory of evolution and the phylogenetic relationships among the various animal groups by using chordates in the broad sense (i.e., as Deuterostomia). The chordate group contains the following animals: 1. Acrania, headless chordates whose most important member is *Branchiostoma*. 2. Vertebrates (Craniota or Vertebrata), a chordate group with heads and form for which the term vertebrate is a little misleading since the most primitive members lack a vertebral column. 3. Tunicates (Tunicata), which includes both sessile organisms with a large branchial cavity and free-swimming species with a similar structure. 4. Enteropneusta. 5. Pterobranchia. 6. Echinoderms. (Echinodermata), with starfishes, sea urchins, and related species. Some zoologists classify the Enteropneusta and Pterobranchia in a common phylum, Pentacoela with the Brachiata. The most recent research, however, leads us to classify Enteropneusta, Pterobranchia, and echinoderms into the group Trimera, since all have three-chambered coela. The Brachiata, if the following description is accurate, are to be regarded as a separate group of metamere protostomates.

> Chordate evolution, by W. F. Gutmann

In answer to the question of how chordates, in our broad sense evolved in the Precambrian, we can find no help from paleontologists. The so-called "lower" animals ancestral to vertebrates lacked skeletons that could fossilize and therefore, excepting those having shells, did not fossilize very much. Any explanation of chordate evolution can only be successful, as we know for theoretical reasons, if we regard the organisms as machines that changed via mutations through time, whereby those modifications affording better survival chances were retained and reappeared in later descendants.

There is a basic body plan for chordates that was a kind of phylogenetic starting point for the group. The fact that all Deuterostomata have a coelom makes our task a little easier. The internal organs are located in the coelom, which itself has a layer of epithelial cells. This kind of organ system is also found in other animal groups. The coelom is extremely important for those present-day animals that still occupy a very early stage in evolution, for the coelom is the receptacle for fluids in these wormlike animals. These creatures have a hydraulic apparatus that operates by creating pressure on the body muscles when the coelom is filled with fluid, which then deforms the body in various ways. This apparatus is responsible for the undulating locomotive pattern and for crawling found in higher animals.

> The basic chordate plan

Vertebrates, on the other hand, are not soft bags of muscles, for they are supported by a firm skeleton. It appears, then, that their ancestors lacked a firm skeletal structure and moved by means of the hydraulic system operating in coelentetrates. We could perhaps speak of a hydroskeleton in such creatures. There seems to be no other locomotion plan in all of nature's animals. Vertebrates ancestors, lacking a firm skeleton, nonetheless had mobile bodies, and the same structures they used and locomotive patterns they developed gave rise to the kinds of structures and movement patterns seen in vertebrates. These structures consist of muscles that can contract, connective tissue that holds the muscles and other organs together, and an anatomical provision for holding fluid within the body. For our movable "machine", nature apparently developed no other locomotion mode.

To more clearly explain vertebrate evolution we must go into more detail on the original anatomy of coelenterates. We shall pay particular attention to the locomotion apparatus, which as a mechanical system formed the basis of the body structure of all multi-cellular animals.

The development of the coelom

Let us envision the evolution of higher coelenterates as we portrayed it previously. Invaginations and canals formed in the gut of a very primitive, but mobile organism whose body support came chiefly from gelatinous masses. These invaginations may have been advantageous in improving the feeding efficiency of these animals. However, since all the organ systems of these primitive species were also part of the locomotion apparatus and were within the muscle sack of the animal, any enlargement of the gut must have had an influence on the animal's locomotion. The invaginations produced liquid-fluid cashions that could be deformed by muscular action. As these invaginations enlarged, a new locomotor apparatus eventually evolved over a long period of time characterized by intermediate stages of various sorts. In this new locomotive apparatus, the gut sacks performed the hydraulic function. The gut could be filled and emptied with little effort, because it did not impinge on any firm object and had little friction due to the presence of gelatinous substances in the body. This energy-saving device was advantageous and was selected for; any organisms with this sort of mutation propagated themselves in greater numbers than those without it, and this adaptation came to be a permanent feature. Those species with particularly liquid cushions were the ones that survived.

The development of a fluid-filled body led to a more economic use of body energy and thus a higher capacity physiologically. Once the gut invaginations became closed compartments, they maintained a constant volume, and the locomotory apparatus became a hydro-skeleton. It was now necessary that the muscles had to control the body at all stages of development, so instead of one great fluid chamber, a series of narrow chambers developed, and they were separated by muscle walls spanning the body in cross-section. The cavities thus took on a serial configuration, and we call them metameres (meaning similar body segments). The muscles, which previously existed as a latticelike network, became differentiated into an outer muscular tube with circular and longitudinal muscle fibers, with transverse muscles between the internal cavities, and with intestinal muscles.

The original coelenterate was a worm-shaped animal whose body has a segmental hydro-skeleton, a muscle tube filled with fluid that affected movement. The partitions between the inner cavity segments maintained the body shape and prevented the fluid pressure within the body from distributing itself randomly. The gut was isolated from the main body tube, and it, too, could move, depending on the degree the coelom was filled. The muscles used to propel the body were also separate from the gut, so gut and outer muscles did not work against each other. Once again, we see that the energy requirements for this "machine" were greatly reduced over previous body plans, and this should help us appreciate how valuable these adaptations were.

Two conditions were necessary before the gut invaginations could close and form isolated cavities. A circulatory system for blood was needed to distribute nutrients throughout the body. Liquid had always been the medium through which nutrients were transported, and what probably happened in evolution is that a number of pathways used specifically for blood developed in the intermediate cell layer between the gut and the coelom, so that nutrient-bearing fluids no longer drifted about at random but were directed along specific paths. The gut was no longer able to diffuse nutritive material throughout the body, for these creatures were more complex than earlier ones, so the presence of blood vessels acted functionally to replace the gut's function in this regard. The lining of the coelom by a continuous cell layer prevented the blood from mixing with the coelomic fluids. Without such separation, the blood could not circulate through the body wall and provide the body with food and oxygen.

The second organ system necessary to fully develop the coelenterate apparatus was an excretory system. The function of this system would be to extract waste materials from the surrounding tissues or blood and to send them out of the body. Without some canals for this purpose, the organism could (and would) die of internal poisoning. In coelenterates, the excretory system consists of metamere canals corresponding to coelomic cavities. They originate as open funnels in the coelom and terminate with an opening in the body wall to the outside. They cleanse each coelomic cavity by brushing wastes out with cilia.

Unfortunately there is no way of telling what sort of life habits our "machine", the original coelenterate, had. This kind of reconstruction of history cannot be made. The segmented hydro-skeleton would have been useful in a variety of situations and would offer advantages in numerous ecosystems. It permits locomotion via undulating snakelike movements, peristaltic-like crawling, or burrowing in the ground. Modern coelenterates display all three modes of locomotion, and they resemble primordial coelenterates to a great extent.

We cannot say anything more precisely about these primitive coelenterates, but we can investigate whether the initial body plan we outlined above could have given rise to the chordates of today. Of course, some logical explanation would also have to be made to explain the evolution of all modern coelenterates from this same original plan as well.

The six chordate groups we named earlier must be incorporated into the initial body plan organization if that model is correct. The chief question to answer is whether chordates show any signs that they evolved

▷
"Geological clock". The course of the earth's history is compared to the course of a day (inner ring) and a year (outer ring) as an explanatory device. Compared to a year, man first appeared on earth in the last five hours of the month of December (taking the previous January 1 as the day the earth was formed). Using a day as our reference, man appears on the scene one minute before midnight:
▷▷
Phylogeny of plants and animals.

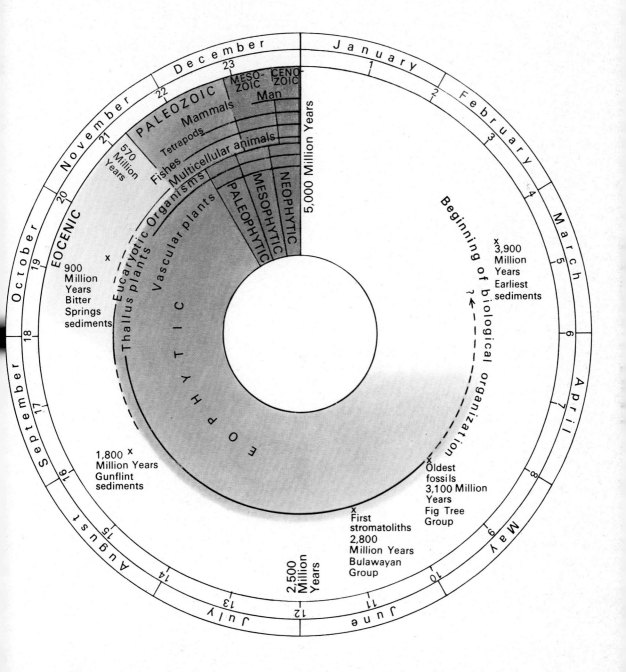

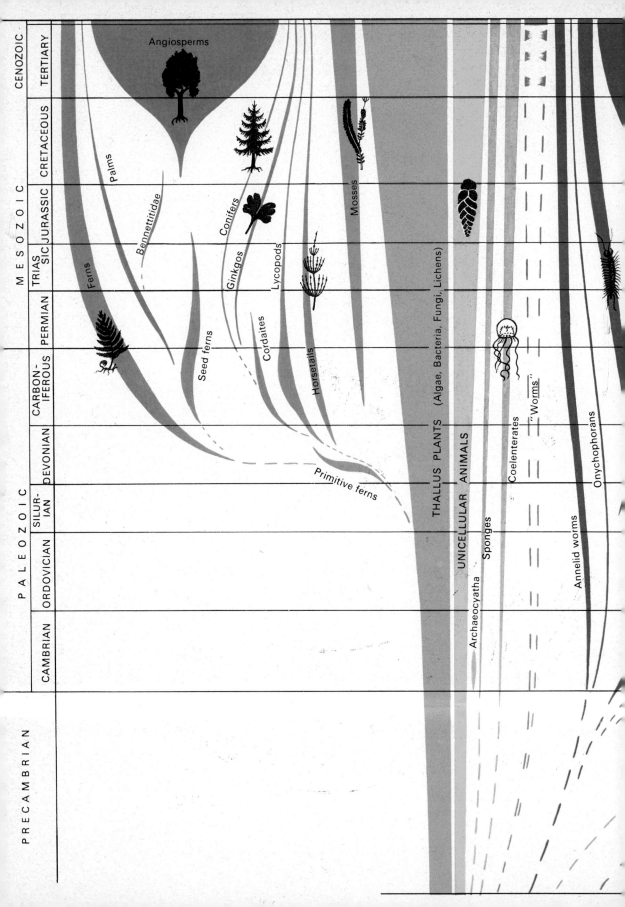

Insects

Arachnids

Crustaceans

Scaphopods

Placophores

Snails

Mollusks

Cephalopods

Brachiopods

Bryozoans

Echinoderms

Graptolithines

Tunicates

Enteropneusta

Agnatha

Acrania

Amphibians

Fishes

Reptiles

Birds

Mammals

Number of species   (Logarithmic scale)

0   10,000   50,000   100,000   200,000   300,000   400,000   500,000   1,000,000

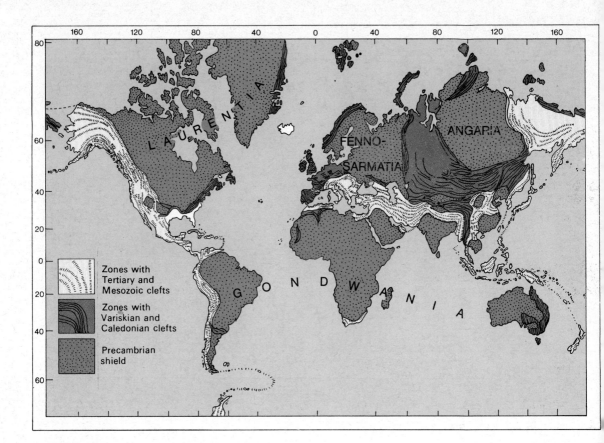

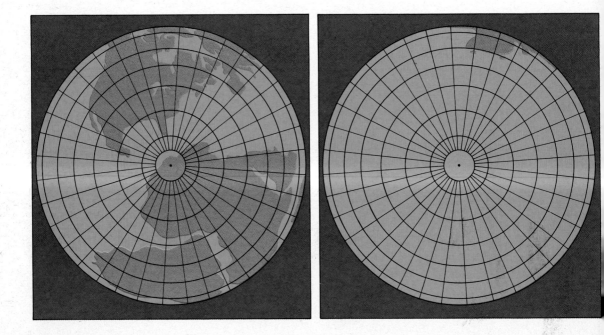

from metamere animals. They do, and the best source of evidence is found in their embryos. The most primitive existing chordate is amphioxus (*Branchiostoma*), in which the muscles and the notochord extend the length of the body. As an embryo, amphioxus has a series of metamere coelomic cavities! They form on the side of the developing gut as protrusions, and the protrusions develop into segmented muscle packets called myomeres, while the notochord is released from the roof of the gut. All other vertebrates, including humans, show clearly this metamere organization while they are developing as embryos. The middle germinal layer that eventually develops into the muscles, connective tissue, the skeleton, and the circulatory system, is segmented in metamere configurations during embryonic stages. During the course of further ontogeny, myomeres also develop from this system.

Vertebrates have yet another organ system reminiscent of the metameres of coelenterates. Each muscle packet corresponds to a renal canal, which in lower vertebrates can be fully functional. Metamere structures are found in such highly developed vertebrates as birds, mammals, and in one very special mammal, human beings. The kidneys and important portions of the reproductive organs in vertebrates are built up from metamere arranged canals, which, as they gradually grow, do not maintain their original metamere configuration.

Our model has thus received some powerful confirmation from even the highest vertebrates, for the metamere muscle packets in the coelom and the renal canals, which originally functioned to cleanse these coelomic cavities, lead us back to the model coelenterate we depicted earlier. This early coelenterate could only have been viable and mobile if it had a metamere hydro-skeleton. In other words: if it had been a worm. We can only achieve this picture of our metamere ancestor by developing the model we produced and by comparing higher vertebrate embryology with that model. We are using the kind of evolutionary reconstruction procedure we described in Chapter 2. We can only use this procedure when embryonic development is understood as a biotechnical process and one in which there is information about modern vertebrate predecessors.

Now we shall portray, using schematic models outlines in Figs. 4–11 through 4–30, the phylogeny of the vertebrates. The metamere ancestors, to summarize what we found earlier, were wormlike organisms with a hydro-skeleton. As they evolved, a process whose details escape us, these animals specialized in snakelike undulating movements for their mode of propulsion. The body was bent out laterally, and the animal could bend any part of its body, so it could move through its watery habitat by arching its body in a snakelike manner. Since longitudinal changes in the body posture brought about no advantages for swimming, the body was maintained at an unchanging longitudinal position and did not stretch. The longitudinal muscles created propulsion, and circular

◁
Above:
The date of origin and the growth of the continents.
Below:
Configuration of the continents in the Devonian period; left: southern hemisphere; right: northern hemisphere. The poles are in the middle of each map.

muscles kept the body from shortening by getting thicker. Maintaining longitudinal constancy could, therefore, only be upheld by exerting muscular effort and therefore energy.

The longitudinal muscles had a double function, propulsion and maintaining longitudinal constancy, so it seems reasonable that any modification in the body plan that relieved the muscles would prove advantageous in natural selection. The advantage that came along was the notocord, a rod of cushioning cells in the roof of the gut. The new organ was flexible and did not, therefore, interfere with the undulating style of lomotion. However, it prevented the body from losing its horizontal posture, and with continuing evolution the notochord did this more and more efficiently. The development of the notochord in the roof of the gut can be explained by the fact that the gut is in the middle of the outer muscle tube, hanging in the middle of the intermediate walls between the coelomic segments. No axis could be built in an eccentric position (i.e., not in the middle); or, it would certainly have had very few advantages. But if the notochord is at the point of balance, as it is, it maintains the position of the animal and frees the muscles from that task so they can be used only for propulsion, and all muscles functioning to maintain longitudional balance would degenerate. Those longitudinal muscles employed in developing power for locomotion could, in contrast, multiply and become even more effective.

However, these longitudinal muscles were laid down in a sequence of segments, since as myomeres they would not interfere with the transverse walls where the notochord hung. This carried the advantage that muscular work could, to some extent, be carried through the transverse walls onto the body axis. With further phylogenetic development, the notochord was freed from the gut, and a body plan arose in which the propulsion apparatus—the body axis and its associated muscles—was compressed into an area along the back of the animal. Due to the notochord axis, the body could no longer change its length and could only move by undulating. The notochord-muscle block apparatus was divided into metamere segments with transverse walls made of connective tissue, just as we see it today in Acrania and fishes. The hydro-skeleton, with its fluid receptacle, was first partially, then completely, replaced by the thickened muscles.

Thus arose the notochord-myomere apparatus of locomotion from the external muscles of the chordate ancestors. The system offered longitudinal rigidity and afforded movement via undulating motions. With still further evolutionary development, a neural tract arose, and we know it in modern vertebrates as the nervous system. It lies between the muscle blocks on the dorsal aspect (i.e., on the back), forming thereby the shortest "cable connection" between the nerves, which coordinate movement, and the muscles, which carry out the energy expenditure required for movement. A fin seam improved locomotive power still further,

Chordate development:

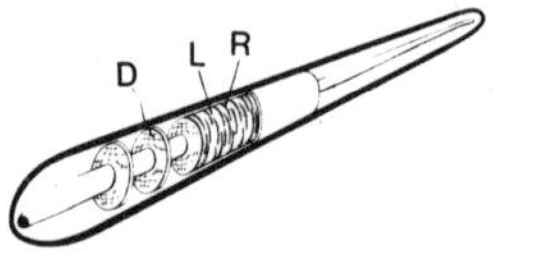

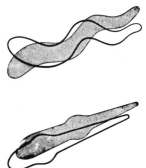

Fig. 4-6. The primitive chordate is a fluid-filled muscular tube with circular (R) and longitudinal (L) muscles held taught by the diaphragm (D).

Fig. 4-7. This metamere animal can execute undulating and peristaltic motions.

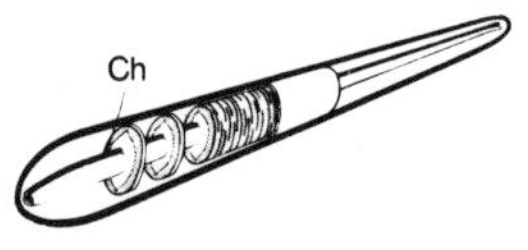

Fig. 4-8. In the transition to the true chordates, the notochord (Ch) appears in the roof of the intestinal tract as a flexible, longitudinal rod.

Notochord-myomere apparatus

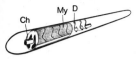

Fig. 4-9. The myomere muscle blocks (My) correspond to the earlier metameres.

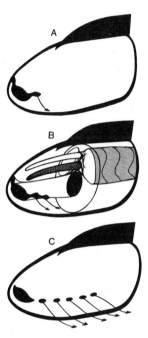

Fig. 4-10. Gill slits arise as lateral clefts in the mouth, deepen and are supported by rods.

Fig. 4-11. Typical chordate with notochord (Ch). which has separated from the roof of the gut; myomeres (My); nerve tract (Nr); and branchial region of the mouth (Kd).

enlarging the lateral surface of the body and adding to the physiological performance capability in swimming.

Once the muscles and notochord axis had been established, the coelomic transverse walls were no longer necessary, and they degenerated. As a consequence of that, the gut was no longer restricted laterally and could assume a wider variety of positions. The anus was pushed forward, and the tail appeared as a hind portion of the body lacking an intestinal tract. The intestines could now be elongated, since instead of being a straight tube it was permitted to wind about in convolutions, just as we find it today in our vertebrates.

The gill apparatus is an extremely significant structure that must be mentioned in any discussion of vertebrate evolution. The gills arose as slitted openings extending from the fore gut through the body wall and meeting the outside environment. They form a magnificent filtration system, pulling nutrients out of the water that is swept over them. It certainly is no surprise that such a structure should evolve, for the earliest vertebrates fed on plankton, the microorganisms that float about in the water. They are the tiniest source of food in the world. It would of course be a great advantage to pull these small organisms out of the water that flowed through the animal's mouth and was flushed out through the side. This could be done by having the corners of the mouth extend back further, but if they receded too far, the rigidity of the flanks of the body would be threatened. Thus, some sort of transverse membranes would have to appear, and the sides of the mouth could be modified into gill slits, and all water entering the mouth could be passed out the sides through the filtration system of the gills. This is, of course, precisely what happens in gills.

The gill apparatus was probably first used in open water. The animals with gills captured water as they swam forward, and pushed it out through the sides of the mouth where the gills were located. It is almost certain that the gill system is more recent than the myomeres, and embryological evidence supports this contention. In Acrania and lower vertebrates, it is the myomeres (muscle blocks) that appear first during ontogeny and then come the gill slits. Furthermore, in the worm-shaped ancestors of vertebrates, whose bodies were tubes subject to hydrostatic pressure, such a system would not even operate.

Since the gills were equipped with ciliated cells that could create a current without the animal having to move, it was possible for the earliest vertebrates to adopt a sessile life mode, lying motionless on the ocean floor and having a stream of water run through the gills. The selective advantage in this habit would be that energy would be conserved by such a way of life.

The tunicates specialized in lying motionless and letting their gills "do the work." These organisms, belonging to the class Ascidia, arose from chordates that held firmly to the floor. Selective pressure caused the gills to increase significantly in size, becoming the most prominent part of the body. The notochord-muscle block apparatus and the neural tract were modified, so that the body became a sacklike structure with an inlet for water and an outlet for filtered water. A protective layer made of tunicin was formed on the outer surface of the body; tunicin is a cellulose-like substance. In modern tunicates, only the very small larva contains a notochord in its tail, a fragment of a neural tract, and something resembling myomere muscle blocks. However, the gill apparatus in these larvae is already a prominent structure, and we have here an example of how certain structures have been suppressed when another happened to have much more selective value.

The Acrania have developed the sessile life mode of tunicates even further. Instead of staying at the upper surface of the ocean floor, Acrania burrow into the upper layers of the floor and use their ciliated cells to pump water through their gill systems. As we would expect, the gills in these animals achieved this by having an increased number of gill slits, and each slit was subdivided by a tongue-shaped invagination. The extensive gill apparatus did not disturb the notochord-neural tube-myomere system throughout the length of the body, but the coelom was restricted so much that the sexual organs and the segmental renal canals were forced into the peribranchial area (the hollow cavity of the gill region). The peribranchial region developed lateral folds, and in their anteriormost portion they were on top of the gill apparatus.

Our familiar amphioxus (*Branchiostoma*) is a very simple model of the vertebrate body plan; it is one of the Acrania. This has made the amphioxus of great academic importance, a creature whose acquaintance is made by every biology student. Amphioxus diverged at a very early stage from the phylogenetic pathway leading to the vertebrates. The fact that it is still alive is doubtlessly due to the fact that it spends its life burrowed in the ocean floor, although like other Acrania it can swim. Amphioxus has spent hundreds of millions of years living this way, and by living in the floor and not above it, the organism avoided competing with the many higher developed vertebrates found in the above-soil habitat and also avoided predation.

The earliest vertebrates lived in the oceans and, with the exception of floor dwellers like amphioxus, they pushed other chordates into extinction. Amphioxus and a few other outsiders are all that remain of the Acrania. However, the Acrania exhibit some traits that indicate a predisposition to pelagic life. The larval stage, known as amphioxides, lives in the water and feeds there, and the burrowing life is not assumed until reaching the adult stage. Vertebrates evolved from pelagic, not burrowing, chordates, however, the Acrania living in the soil gave rise to numerous species that were not appreciated as chordate descendants for many years. They include the Enteropneusta, Pterobranchia, and the echinoderms (Echinodermata). As we mentioned earlier, these organisms are classed

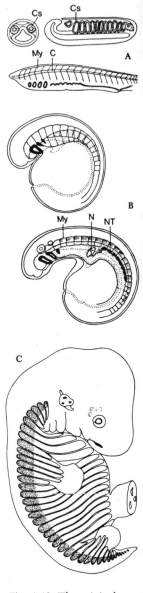

Fig. 4-12. The original metamere structure can still be seen during embryonic development of chordates.
A. *Branchiostoma* develops small coelomic cavities (Cs), which form a tight succession and from which the muscle blocks (My) arise. (C = notochord). Higher chordates (i.e., vertebrates) show a similar embryonic process with

Acrania

together as Trimera because of the triply divided coelom. This point deserves some clarification, for the reader may be confused to find that modern Enteropneusta and Pterobranchia are characterized by coela with five subdivisions and are classified with the Pentacoela as well. Morphological studies reveal, however, that the increased number of subdivisions is a secondary characteristic, i.e., one that came later in evolution, and that the original structure of these coela was with three subdivisions.

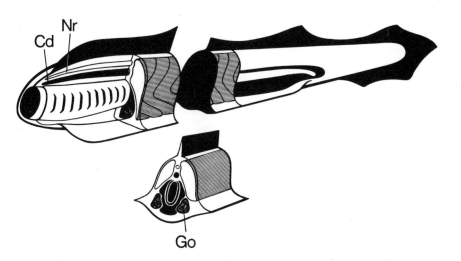

Fig. 4-13. Acrania burrow in the ground. They have an increased number of gill slits. Their gonads (Go) are in a lateral fold of the body, since the coelom has little room (Cd = notochord; Nr = nerve tract).

The Trimera

The Trimera evolved from ground-dwelling Acrania. Apparently, primitive Acrania, which initially could only burrow into the ground and do nothing else, gradually were able to move about through the floor. They arched the fore body, creating a space in front of the mouth. The muscles at the front of the body developed and finally evolved into a proboscislike digging structure. The stiff notochord was a hindrance for this kind of movement, and natural selection caused it, along with the myomeres and the neural tube, to degenerate. All that remained were a fragment of the notochord (the stomochord) behind the trunk in the newly formed neck section and a vestige of the neural tract. The gills were moved to the hind body; this was an advantage because it kept the digging trunklike structure and the gills, with their respiratory function, separated. This rear part of the body had no digging function and could not even contribute much to locomotion; it was dragged along. These early creatures had almost nothing but longitudinal muscles, the remains of the myomeres of their ancestors.

This was the way in which the three-sectioned body of these burrowing chordates developed, as we show in Fig. 4-14. The coelom was also divided in three compartments in these and in the early Enteropneusta. However, early Enteropneusta had the same feeding habits as their acranid ancestors. They took up a stream of water through the

metameres.
B. Two embryonic stages of the lamprey. The muscle blocks (My) correspond to the metamere renal canals (NT), which lead to other renal canals (N) on the coelomic cavities.
C. Higher vertebrates (mammals) during embryological development also possess muscle blocks reminiscent of the metameres of their ancestors.

mouth, and the water was passed out through the gills. The entire front portion of the body and the fore gut were covered with mucus and cilia, and these were used for food uptake.

The Pterobranchia and echinoderms also evolved from ancestors resembling Enteropneusta. Their internal anatomy, with its three-compartment structure, is the same kind found in Enteropneusta. It is

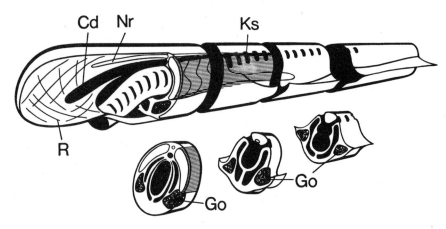

Fig. 4-14. Transition form to the Enteropneusta. The notochord (Cd) and the muscle blocks degenerate toward the posterior. A digging structure (R) forms in the front and permits the animal to burrow in the ground. The three body sections shown below depict the location of the gonads (Go) in the body wall (Nr = nerve tract; Ks = bronchial cavity).

likely that primitive Enteropneusta were able to build a tube with their mucous, and the tube led above the surface of the floor. Thus, those parts of the body used for food uptake (the mouth and neck region), increased in size and protruded outside the tube. The neck region developed into tentacles used for seizing food floating in the water. Within the cavity of

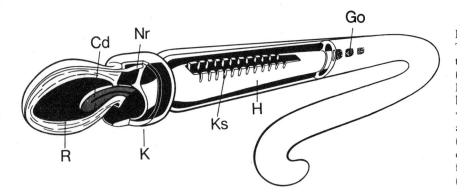

Fig. 4-15. Enteropneustid. The body has become a three-sectioned structure (R = trunk; K = collar; H = hind portion). The branchial cavity (Ks), the vestigial notochord (Cd), and the short nerve tract (Nr) are reminiscent of the corresponding structures found in their ancestors (Go = gonads).

the neck section, the tentacle had its own liquid cushion that was inflatable and could be used for food uptake. Since a stream of water no longer had to pass through the mouth, the gill slits lost their selective advantage and regressed. In modern Pterobranchia, which live in colonies composed of a great many individuals, there is just one pair of gill slits left, and this pair may even be closed.

The graptoliths, a Paleozoic group widely distributed during that era, were probably Pterobranchia. The structure of graptolith tubes resembles that of pterobranchia. Graptoliths formed large colonies of sessile animals suspended by gas-filled bubbles; they floated freely in the sea.

Pterobranchia retained the three-sectioned body form and the three-chambered coelom of the Enteropneusta. They also had a stomochord, a sign of their earlier chordate ancestry, and the stomochord was adjacent to the fore gut, just as in Enteropneusta. The presence of only longitudinal muscles in the hind body can only be explained if the Pterobranchia arose from chordates, which as we recall had only longitudinal muscles in their notochord-myomere system.

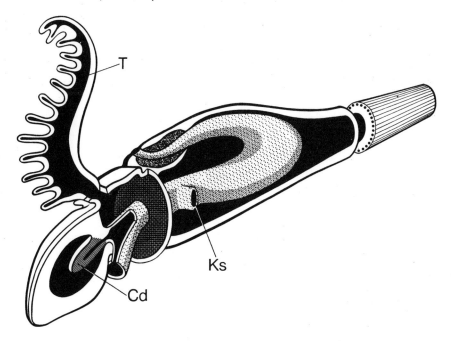

Fig. 4-16. The Protobranchia arose from organisms with the enteropneustid body plan; however, in Protobranchia the collar section has developed into a tentacle (T). Protobranchia typically have just a pair of gill slits (Ks) but retain a vestigial notochord (Cd).

Echinoderms represent a further development of the Pterobranchia body structure. Although adult echinoderms have a radiated structure, their larvae are bilaterally symmetric, an indication that their ancestors were bilaterally symmetrical coelenterates. In the larval echinoderm stage, the coelom has three chambers, just as in Enteropneusta and Pterobranchia.

Echinoderm evolution occurred in the following way: pterobranchia-like sessile species developed five tentacular rays. This sort of development would be of selective advantage since a sessile animal with a radial structure could deal with both food and enemies regardless of the direction from which they came. The hind portion of the body, as in Pterobranchia, was a very simple structure and could not grab hold of a substrate, and as a result of selective pressure a calcareous skeleton was formed to produce a very stiff body wall. Stiffening of the body caused problems only for the tentacular apparatus, since the tentacles could only be activated when the coelom was filled. That was the only way water could be pumped out of the branchial cavity through the tentacles. Any stiffening of the body wall would hamper this motion.

This biotechnical problem was solved phylogenetically by moving the tentacle-bearing coleomic cavity to the rear of the coelom, thus losing its association with the rigid body wall. Now the tentacles were liberated from the body wall and could move freely. With that development, the other parts of the body could become hardened for protective purposes.

Earliest calcium skeletons

Fig. 4-17. Transition to the echinoderms. The tentacle (T) radiates. (Go = gonads)

Primitive echinoderms were sessile, floor dwellers, but the later, more highly developed forms could move freely, using their tentacles for locomotion. We see this today in starfishes and sea urchins, who move on their tentacles (called ambulacral feet). The ambulacral system corresponds to the branchial cavity of Pterobranchia and Enteropneusta. It has a circular canal and the ambulacral feet (see Fig. 4-18). The anterior coelomic segment of the Pterobranchia and Enteropneusta became a hard canal in echinoderms, extending to the outside through the ringed canal. The hind end of the coelom surrounded the internal organs. We can find all the Pterobranchia and Enteropneusta structures in echinoderms,

Development as an
adaptive process

but without our modern evolutionary perspective we would never be able to explain this.

In Chapter 2 we described three ways in which evolution can occur, and the phylogenetic development from primitive chordates through Enteropneusta and Pterobranchia to echinoderms corresponds to the

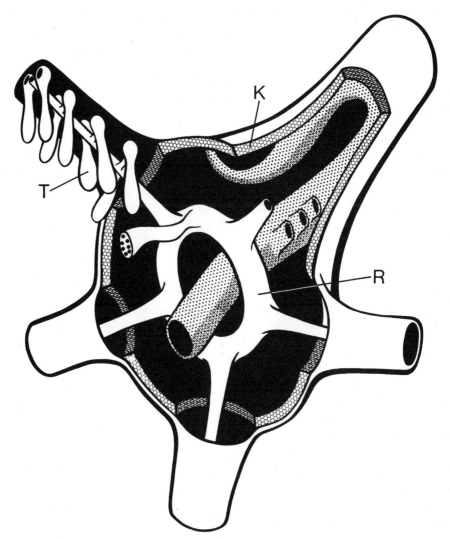

Fig. 4-18. In primitive echinoderms, a calcareous armored coat (K) forms. The tentacle apparatus (T) and ringed canal (R) are transferred into the interior of the body cavity.

third method, entering a new habitat and evolving to conform to the demands of the habitat. The biological apparatus and the way it functions underwent a change during this phylogenetic development, and the development we so briefly outlined here represents a whole series of adaptive modifications.

It would really be more appropriate to designate the vertebrates (subphylum Vertebrata) as Craniota, because the most primitive vertebrates do not have a vertebral column. However, all vertebrates have heads, and by that we mean a firm structure with large sensory organs and a brain located at the anterior end of the body. The trunk and tail sections are divided by metameric muscle blocks. We must also assume that vertebrates evolved from pelagic chordates that used undulating motions to propel themselves and that had gills for respiration and food uptake. The undulating movement of the simple Acrania was retained in the vertebrates (see Fig. 4-19), but the gills eventually became less significant and regressed, so vertebrates had fewer gill slits than Acrania. Modern Acrania, especially our friend amphioxus, are still superb models in spite of the fact that they are specially adapted to their environment.

The necessity for an undulating, snakelike locomotive movement can also be explained in mechanical terms and not just phylogenetic ones. A hydro-skeleton and the simple notochord-myomere system of the lower chordates are so ineffective at propulsion that every body segment would have to be employed to move the animal. The predisposition that permitted a head to develop was the fact that the body was modified in such a way that it could propel a stiff object like a head through the water, something that could not occur with the notochord-myomere system. The body became more efficient at locomotion by padding the last fluid-filled coelomic cavities in the myomere (i.e., the sclerocoel) with connective tissue. The muscle walls, or myospetae, were folded more prominently, and the axial skeleton became a firm supporting structure with the addition of cartilage and connective tissue. The locomotive apparatus was now more efficient and could propel an animal with a distinct head section and could thus permit such development of a head. The formation of a rigid head structure began with the development of large head sensory organs.

The anteriormost portion of the head had paired nares and a central nervous apparatus that formed in the prosencephalon, the anteriormost brain section. The animal could now smell, and after further development it acquired eyes and could see. The eyes originated as optic outgrowths from the neural tube; paired protrusions grew to the side and above the neural tube, forming dorsal eyes. Light-sensitive cells within the neural tract were moved to the optic tract, a development that makes sense since this put those cells closer to light impinging on the animal. True eyes did not exist until the optic structures were equipped with lenses to focus light. Lenses developed from parts of the body wall.

While the eyes were developing, the optical center of the brain formed in the mesencephalon, or midbrain, the brain segment just behind the prosencephalon. The nerve connections from each eye cross each other as they proceed to the brain; thus, stimulation of the right eye is transmitted to the left hemisphere of the brain, while nervous impulses from the left eye are sent to the right hemisphere. Organs of equilibrium also formed further back in the head segment. They arose from canallike

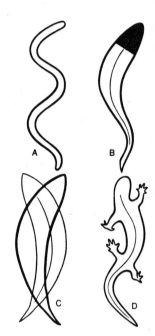

Fig. 4-19. The snakelike pattern of locomotion in terrestrial vertebrates is a primitive form that can only be understood in terms of the locomotion of vertebrate ancestors. Coelenterates (popularly called "worms") move in an undulating manner (A). After the body axis developed in fishes (B), the same mode was retained. Some fishes, especially rapid swimmers, bend their bodies so little that hardly any snakelike movement remains (C). Primitive terrestrial vertebrates retained this undulating movement from their fish ancestors; indeed, this kind of movement helped the legs move more efficiently by placing them more forward than otherwise be possible (D).

invaginations of the skin, and they were used for balancing. Later they became incorporated with hearing.

The large sensory organs in the head region could only develop if cartillage supports for the head appeared; in time, cartilaginous capsules were formed and surrounded the nose, eyes, ears, and the now enlarged

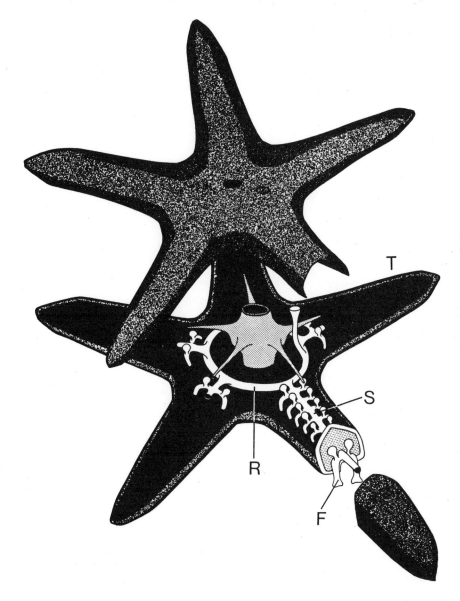

Fig. 4-20. As the anatomy of the starfish shows, the originally sessile echinoderms "fell on their faces" and became mobile, walking on their tentacles (T), which assumed the role of feet. The branchial cavity can still be seen as the circular canal (R). The true feet (F) are located on the radial canals (S).

brain. The notochord remained fairly intact in the head region and extended back to the eyes, but it was covered with cartilage and was not flexible in this part of the body.

As cartilage covered the head, the muscle blocks (myomeres) there became superfluous, and they regressed. All that remains of them now are muscles that control eye movements. Since the lateral eyes were mobile, owing to their muscular connections, the phylogenetically older dorsal eyes lost their significance and became vestigial. These dorsal eyes, which appeared long before true vision (i.e., with lenses), were used to orient the swimming animal toward light.

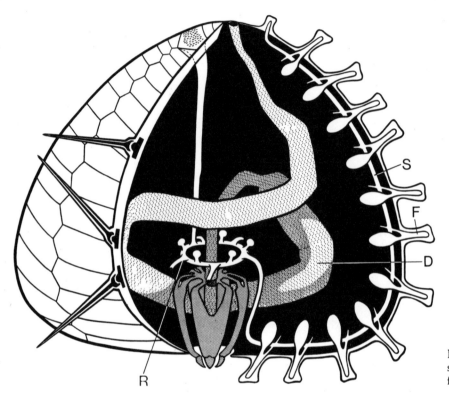

Fig. 4-21. Structure of the sea urchin. See the legend for Fig. 4-20.

The cartilage system around the brain and sensory organs formed an endocranium, which was replaced by bones later in evolution. Bones appeared after the head was fully developed, and a bony capsule surrounded the inner cartilaginous capsule, both of which helped to protect and stabilize the head. The gill apparatus, which in the Craniota was moved by powerful muscles, also received skeletal support later in evolution. The bones were preceded by a cartilaginous cushion. All these changes gave the developing vertebrate the opportunity to actually move through the water seeking its food, something it could not do before. Instead of floating or remaining sessile on some substrate, the organism could hunt for food actively, and this of course greatly increased its chances of finding food and therefore its chances of surviving and producing more offspring that continued this same life mode.

Specialized gill nerves developed to coordinate the movements of the gill system and to receive sensory input. The rearmost of these nerves, the vagus, supplied several gill arches and extended down to the heart and

Development of the Craniota (beginning with the chordate in Figure 4-11).

Fig. 4-22. In the primitive vertebrates, an external skeleton with rhomboid scales (S) arose on the muscle packets. The scales were shaped in such a way to minimize deformation of the body.

Fig. 4-23. A series of vertebrae (W) arose along the longitudinal axis of externally armored fishes.

Fig. 4-24. In teleost fishes, the internal skeleton was so effective and became so well developed that an external covering was no longer necessary, and the large, bony scales were lost.

the foregut. All the gill nerves led to the hind brain, or rhombencephalon, and this crucial area was the center of coordination for movement, feeding, respiration, and circulation of the blood. It is really no surprise that all these functions should be concentrated in one part of the brain, for they are all interrelated. The water swept in through the gills contains oxygen used for respiration, oxygen that is transported by the blood; and it contains the food that is filtered out by the gills and sent to the intestinal tract for digestion.

The most important adaptations the vertebrates possessed were their effective locomotive apparatus, method of feeding, respiration capabilities, and their nervous system with its many sensory organs for receiving sensory input, all of which were more effective than the structures doing those same jobs in the Acraniota. All this happened at a period so far back in geological history that there is no fossil evidence for these changes. Our present knowledge is based on embryological evidence. Understandably, the chordates with these adaptations had far greater survival chances than those without them, namely the Acrania. We assume that the vertebrates even pressed their less capable ancestors into extinction, excepting those with whom they did not compete, like amphioxus.

Paleozoic fossils known today include lower vertebrates that were very fishlike, all of which had a similar gill arch system. They represent an evolutionary stage of vertebrate life that preceded the acquisition of jaws, and these jawless fishes or Agnatha gave rise to the jawed fishes (superclass Gnathostomata or Pisces). Jaws came about when the front gill arches formed a movable mouth structure, enabling these fishes to seize and hold larger prey than they could obtain previously. This explains why jaws were of selective advantage and why natural selection would foster their appearance. Zoologists have not been able to determine whether it was the first or the second gill arch that developed into the jaw, but the structure of the skeleton and the nervous system shows unmistakably that the jaws did arise from gill arches of jawless fishes. Jawed fishes came to predominate over the jawless ones, a phenomenon that to some extent can be documented by the fossil record. Today there are just two jawless fish groups, neither of them very significant: lampreys and hagfishes.

Fossil documentation of vertebrate evolution begins as early fishlike creatures developed an armored covering, for the armor can fossilize fairly easily. It appears that the armor in these early fishes was purely for protection and not for body support. For example, it would seem that if armor were needed for support, the head would be one of the first parts of the body to get some, and armor should be very prominent on the head; what actually happened, as the fossil record shows, is that the entire body was covered with armor. The covering became differentiated, and diamond-shaped elements were used to form the scale covering found in the more advanced fishes, with the scales arranged in rows. It is important to remember that the notochord-axis of the body only maintains the length of the body at some fixed dimension; most of the body mass is determined by the muscle mass, which is divided into metameres penetrated only by connective tissue. Thus, the body could easily be deformed width-wise, and the scale covering maintains the integrity of the body in cross-section. Yet, and this is a magnificent example of how evolution has worked to produce an impressively efficient creation, the scales were arranged in rows, a situation that retained body integrity but permitted the flexibility needed for movement through the water. The rows of scales meet at the top and bottom of the fish's body, so they are actually rings of scales and not rows in the longitudinal sense. Anyone who looks at toys made of a series of rings can see how this configuration permits movement but gives integrity as well.

Fig. 4-25. Devonian crossopterygians led the way to terrestrial vertebrates. Their armored covering is also vestigial.

The individual scales in each row impinge on each other, of course, but they do not overlap along the front-rear direction, and this permits them to stabilize the body in cross-section. Any body deformations that do not contribute to movement are greatly inhibited by this setup. Energy is conserved, since only those body exertions that are useful must be executed, and muscular effort is not wasted on movements that do not contribute to locomotion. The evolution of scales shows that selective pressure was operating in such a way as to stabilize the propulsion apparatus. The animals with scales were thus at an advantage over those lacking them, and they drove the scaleless vertebrates almost out of existence entirely. The vast majority of fishes today have scales.

The lower jawless and primitive jawed fishes, of which we have fossils, almost always had stabilizing scales, whose significance can only be understood in evolutionary terms. As evolution continued, many fish groups later lost their rhomboid scales; this usually happened when bony supports appeared on the inside of the body along the longitudinal axis. These internal supportive elements maintained stability across the body, and the heavy external armor degenerated. The original hydro-skeleton became firmer. The formation of skeletal elements occurred in no haphazard way; skeleton appeared in such a way as to inhibit or completely prevent unnecessary movements, and it was the environment, the great *Umwelt* outside the animal, that determined which movements were adaptive and which were not.

Fig. 4-26. Primitive amphibianlike terrestrial vertebrate. Limbs developed from the paired fins. Support came exclusively from the internal skeleton.

The further development in fishes has been well documented in the fossil record. The presently predominant bony or teleost fishes (class Osteichthys), with their bony skeleton supporting the body, suppressed their older ancestors, who were equipped with a rhomboid armor scale covering. This is a fine example of survival of the fittest, and perhaps we should realize here that this phraseology does not imply that a fight of some kind was going on between the species referred to. Fishes had not been developing very long, however, until 350 million years ago when a side branch developed that gave rise to terrestrial, four-footed vertebrates.

Fig. 4-27. Reptile (a lizard). The body is still better supported by the skeleton.

Fig. 4-28. In mammals the skeleton has become a latticelike structure held by rather thin musculature.

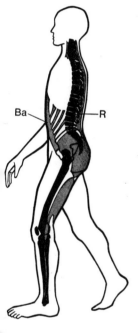

Fig. 4-29. Man walks upright, becoming a two-legged creature balancing his head on the vertebral column. The stomach muscles (Ba) and back muscles (R) maintain the body in that position.

Mutations as advances

The transition was made by crossopterygians (see Fig. 4-25), a fish group with particularly movable fins and lunglike appendages on the fore gut. Crossopterygians lived in shallow water and developed the capacity to breathe atmospheric oxygen. Mucus secretions covering the skin protected them from dehydration, and they developed the ability to survive drought periods by wallowing in mud or in very shallow water. This was the kind of animal that enabled the vertebrate group to make the great move onto land.

Land-living fishes supported themselves with their fins and moved by using their fins and by undulating the body just as they undulated it in the water. Once they were out of water, however, their bodies were subject to gravitational forces they had not experienced as intensely in their original habitat. The skeleton had to become even firmer than it was before in order to properly support the body. The amphibians with tails were the first group to develop from these land-living fishes. Their anatomy shows many signs of their fish heritage; just as in fishes, the muscles in modern amphibians are arranged in thick packets on the skeleton.

In reptiles (the next evolutionary step) and their descendants, the muscle mass begins decreasing in size. The skeleton formed a framework where the muscles could originate and which would transmit the energy originating in the muscles. The muscles were no longer forced to conform to a metamere arrangement, as in all fishes and most amphibians. They developed into larger units that spanned considerable stretches between various bones. This skeleton-muscle apparatus, which is much more distinct in birds and mammals (they are the descendants of snakelike ancestors), represents a much later stage of evolutionary development. It began, as we described earlier, with a movable outer muscle sack and led to the chordate body plan, which maintained the longitudinal integrity of the body, and finally to a locomotor apparatus of ever-firmer skeletal elements in fishes designed to reduce useless motions. It is not until the terrestrial vertebrates that we find a skeletal structure that "enforces" what the muscles perform by transmitting muscular power.

Up to a certain point, the development of the locomotive apparatus followed the path of greatest performance capacity. It is here that we can see the significance of the skeleton, which was selected over a long period of time from the many random mutations that appeared in organisms. The development of the locomotive apparatus corresponds to model 1 we described in Chapter 2. We must also remember that, only in water could the notochord-myomere apparatus have developed, for it could not sustain a terrestrial animal. Model 3 only applies once we reach terrestrial organisms. It was here that modifications in both the environment and body structures occurred. The phylogeny of the chordates offers many examples of mechanisms of evolution: modifications that can occur within a relatively uniform environment, the changing relationships between anatomy and the environment, and the use of embryological data on drawing conclusions about how some of these processes occurred.

The highest stage of development of locomotion has been reached in mammals, the group including human beings, animals that use the skeleton-muscle system. The effectiveness of this system can only be properly explained with an evolutionary perspective. From a purely theoretical point of view, the mammalian biological capacities could have been developed with an entirely different apparatus composed of muscles, bones, and connective tissue. The pathway that happened to be chosen—from the metamere muscle blocks to the development of a dorsal spinal cord—can only be appreciated from the Darwinian standpoint, looking at this process of evolution in chordates that took place starting with the metamere worm and passing through fishlike intermediate stages.

The most important stages of vertebrate phylogenetic development occurred in the Acrania, the jawless fishes, jawed fishes, amphibians, reptiles, and mammals, and we can still see all these kinds of animals today. It is only when we bring all these species into an historical perspective and evolutionary interpretation that we can really understand why the various biological systems are the way we find them to be. The evolutionary explanation is simultaneously historical and biotechnical, since the older species determined the adaptations found in the more recent species. Evolution has caused the body plans of animals to be what they are, and it is only with Darwin's great theory of evolution, together with the knowledge of genetics that has given us the contemporary synthetic theory of evolution, that we can understand biological systems.

Biological systems in historic and biotechnic perspective

If we survey the entire development of the Deuterostomia, we will recognize that all these species—including the "advanced" vertebrates and the greatly differing Enteropneusta, Pterobranchia, and echinoderms—are descended from chordates and that these chordate ancestors were derived from coelenterate ancestors with hydro-skeletons (in short: worms). All the Deuterostomia can be regarded as chordates, using the term chordate in its broadest sense. The name Deuterostomia can be dropped. The term was coined when researchers thought that the Trimera (i.e., the Enteropneusta and Pterobranchia) were the original chordate ancestors. This position is now known to be untenable, since in the sixty years since it was proposed it has found no evolutionary explanation.

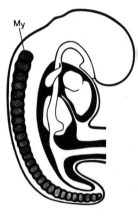

Fig. 4-30. The mammalian body, with its well-developed skeleton and muscle-covered body, arose from a progression of ancestral body plans, as did the bodies of all mammals and vertebrates. This phylogenetic heritage can still be seen in mammalian embryos, which has very distinct muscle blocks (My) that reflect mammals' ancestors.

5   The Origin of Life

By K. Dose

One of the most basic and most difficult questions is the question of where life came from. Since life without matter is inconceivable to naturalists, the question of the origin of life is intimately related to the ultimate questions of the origin of matter and the origin of the universe. Cosmological data lead to the conclusion that our universe has a definite age and, therefore, a fixed origin. This means that life has also had an origin and that the time of its origin is accessible to scientific inquiry.

That life has had an origin is generally conceded, but during the last century vigorous arguments have taken place on just how life originated. Numerous great scientists did not think the questions about the origin of life could be resolved. In an 1863 letter to Joseph Dalton Hooker, Charles Darwin wrote, "It is utter nonsense to think presently about the origin of life; one might just as well contemplate the origin of matter." For the past several decades, however, cosmologists have given this question serious consideration, and their observations and laboratory experiments have shed considerable light on formulating an answer. The origin of life is a related question to the issues of cosmology.

The biological study of living systems

In his later years, Darwin no longer had such a negative viewpoint of the utility of scientifically determining the origin of life. In a letter written in 1871, he stated that he attempted to devise a model of the evolution of life with a regular progression backward from the highest to the lowest organisms, and he finally reached the conclusion that living systems arose from non-living ones. Excited, he wrote, "It is often said that all the combinations needed to produce living things are present today just as they always were. But if (and what a gigantic if!) we could comprehend that in a small pool, with all sorts of variations in its ammonia, phosphorous salts, light, warmth, electricity, etc., the conditions were present for forming a protein that then was transformed into higher forms of living substances! Today, any such material would putrefy and would be absorbed immediately, but such putrefaction would not occur before living creatures were created."

Darwin was more than fifty years ahead of his time when he wrote that. It was not until the 1920s that Oparin (1924) and Haldane (1928) independently formulated theories to explain the origin of life. They made essentially the same statement that Darwin had made half a century before.

The scientific world in 1871 was not quite ready to research the origin of life, despite Darwin's suggestion. The predominant ideas in those days were the ones of scholars like Spinoza, Buffon, Lord Kelvin, Liebig, Arrhenius, and, especially, Louis Pasteur. They all held the viewpoint that life was as eternal as the world itself and therefore had no beginning; thus, there was no problem about the so-called origin of life, in their eyes. The idea that life on earth had originated via a process called panspermia was fairly prevalent during the last half of the 19th Century. According to this theory, espoused among others by Arrhenius, life here was transported by germ cells from other planets. According to Arrhenius, germ cells were brought to earth over tremendous distances, transported by light pressure or, as Hermann von Helmholtz thought, by meteors. For a long time, life was thought to be an entity, like energy, that radiated from matter; adherents to this thought were known as vitalists. Their concept of life contained a mystical element, and they felt that because of the mystic nature of life it was simply impossible to research living beings with scientific methods.

The extremely conservative attitude held by naturalists in the second half of the 19th Century and first quarter of the 20th Century was the result of a millenia-long struggle with the question of whether life had arisen spontaneously. The Greek philosophers who espoused spontaneous generation had a tremendous influence on later scientific thought, an influence that maintained its hold for over 2000 years! Thales of Miletus believed that living organisms could develop in mud if the mud was heated. Anaximandros, Xenophanes, Anaxagoras, Empedocles, and other great Greek philosophers taught that life arose spontaneously from what we know as lifeless matter. These archaic attitudes were understandable, for in antiquity the entire universe was thought of as a living organism. However, some philosophers (e.g., Democritus, ca. 460–362 B.C.) of that period had modern ideas. Democritus tried to explain the origin of life with atoms. He postulated that atoms of the wet, lifeless earth encountered the atoms of life and fire, and that the three combined to form a living organism. Aristotle developed this line of reasoning further and developed a comprehensive theory of the origin of life, which, over the following centuries, was modified.

According to Aristotle, living things arose from a combination of a passive element, matter, and an active element, entelechy, which represented the soul. The soul is already present, in the Aristotelian view, in the matter used to form living organisms, but only to a small extent in the earth and to a much greater extent in the water, air, and in fire. Thus,

Cosmic panspermia

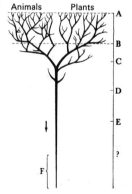

Fig. 5-1. Phylogenetic tree. A = the present; B = Cambrian/Precambrian boundary (600 million years ago); C = 1 billion years ago; D = 2 billion years ago; E = 3 billion years ago; F = first traces of life.

Life origin theories:

Fig. 5-2. Development of geese from arboreal fruits.

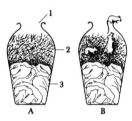

Fig. 5-3. Development of mice, according to van Helmont, from wheat (2) and dirty clothing (3) in an open vessel (1).

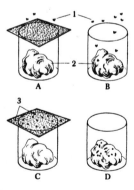

Fig. 5-4. Redi's experiment. 1 = flies; 2 = fresh meat (A with cheesecloth covering, B without this covering); 3 = fly eggs separated from meat. The rotting meat is then shown covered with cheesecloth (C) and without cheesecloth but covered with flies and maggots (D).

Aristotle believed, life arose spontaneously as worms and insects developed from morning dew, wet earth, rotting mud holes, manure, wood, hairs, sweat, and flesh. He proposed that even crustaceans and various mollusks, eels and other fishes, and even frogs and mice arose spontaneously under particular conditions from the above materials.

Medieval scientists "demonstrated" the validity of these Aristotelian theories with a series of experiments. The theory of spontaneous generation did not fall into disrepute until Aristotelian thinking lost the firm hold it had enjoyed for two millenia and accurate scientific research enjoyed greater credibility. The 1668 studies of the Italian physician Francesco Redi were among the first to make the way for a more enlightened scientific attitude. He showed that maggots developed solely from flies' eggs and not spontaneously. This gave critics of Aristotle a reliable foothold from which to attack the older theories. In 1765, the great Italian naturalist Lazzaro Spallanzani showed that microorganisms likewise could not arise spontaneously. The death blow to theories of spontaneous generation was dealt in 1862 with the experiments of Louis Pasteur. However, the organisms that Pasteur sought were modern ones. However, if we follow the Darwinian model of evolution, we are led to want to discover information about the origin of the earliest life forms, of the proto-cells that were the ancestors of the cells living today.

As Darwin indicated in 1871, the origin of life must be understood as an evolutionary process. The ultimate beginning of this process is intimately related to the origin of the universe. The development of the simplest elements (like hydrogen and other light elements) follows the evolution of stars and solar systems, and these cosmological developments are inseparable from the development of the higher elements, which arose from thermonuclear reactions occurring in stars. Some of these elements are the so-called bioelements (those used in forming living organisms), and these are of special interest to us. They are the basic material for chemical evolution (the prebiotic formation of organic substances), a process that began in the stars but was chiefly completed on earthlike planets. The micromolecules produced in chemical evolution gave rise, under the proper conditions, to macromolecules, a process that has been demonstrated in the laboratory. The macromolecules formed by heated amino acids were a kind called proteinoids, and these are of importance for their proteinlike characteristics. Proteinoids could then congregate and combine, forming still more complex structures we call protocells. The whole process of molecular evolution from the origin of the universe to these protocells can be summarized in simplified form as follows:

Origin of the universe

↓

Evolution of stars and (bio)elements

↓

Bioelements form the first organic compounds in the intergalactic clouds

↓

Differential evolution of micromolecules (e.g., amino acids) takes place on earthlike planets

↓

Under favorable geological conditions, macromolecules form. Amino acids combine to form proteinoids.

↓

Macromolecules (e.g., proteinoids) form micro-systems (protocells and organelles) immediately after they themselves have formed.

Fig. 5-5. Pasteur's experiment. 1 = the sugared yeast suspension, after boiling, remains sterile. 2 = microbe formation after breaking the neck of the flask.

Time scale of molecular and biological evolution

Today the age of the universe is estimated at between twelve and thirty billion years. Our solar system probably originated five billion years ago, and the age of the earth as a firm planet is not much less than this. There are no geological finds representating the earliest history of this planet. The oldest datable sediments found so far from the so-called fig tree system and are slightly more than three billion years old. Microfossils have been found in these ancient deposits, and these structures are believed to be the fossils of bacteria. The structures are so dissimilar in their appearance that they may actually have had an abiotic or prebiotic origin. Microfossils from the Soudan crust (2.7 billion years old), the Gunflint crust (1.9 billion years old) and the Nonesuch crust (1 billion years old) provide solid evidence that more highly developed microorganisms were on the face of the earth two to three billion years ago.

Isopronoids, alcanes, and other hydrocarbons have been isolated from these carefully prepared sediments, and these hydrocarbons could not have arisen from purely geochemical processes. Therefore it is thought that these molecular fossils were of biological origin. However, if these hydrocarbons arose nearly three billion years ago from simpler substances, the microorganisms of that period must have already had a highly developed complement of enzymes used in metabolism.

Beginning of biogenesis

If the earth's age is taken as 4.8 billion years, then approximately 1.8 billion years were required for the molecular evolution leading to the origin of life. Nearly two billion years is a time span longer than most of us can possibly comprehend, and yet to some scientists it seems too *short* a time for the necessary molecular evolution to occur! The old panspermian theories have been discarded, of course, and no one seriously maintains that living systems came to the earth three billion years ago. However, we may eventually have to conclude that the prebiotic evolution that led to these living systems must have occurred elsewhere in the universe other than on earth. It is difficult for us to believe that the environmental conditions essential for prebiotic molecular evolution were those of the Precambrian earth!

The world three billion years ago

During the time the earth assumed its present shape and configuration of geological features, the sun was also in a stable state differing little from the way it is today. The intensity of solar radiation in the uppermost layers of the atmosphere was about one-third less than it is today (see following table). But since the atmosphere of the infant earth was practically devoid of molecular oxygen, there could not have been any substantial ozone umbrella like we now have. This means that ultraviolet light with a wavelength of less than 3000 Ångströms could have reached the surface of the earth. If solar heat was the sole determinant of the earth's surface, temperature in those early geological times, the temperatures of 4.5 billion years ago would have been approximately 30°C. less than they are today.

However, warmth produced inside the earth from volcanic activity and heat radiation probably maintained the surface temperature at a very similar level to that found today. The origin of all this heat were the radioactive decay processes (or which there were three times as many as we find today) and the gravitational energy released during the aggregation of the cosmic dust cloud and the subsequent contractions of the mass we know now as earth. Some of this energy is still being released in the form of volcanic activity. Other energy transformations occur in the electric discharges we still see in our atmosphere; this electrical activity has been particularly important with regard to the origin of life, and solar radiation is the source of this energy form. The following table summarizes these various forms of energy:

Energy form	Modern earth $(cal \cdot cm^{-2} \cdot yr^{-1})\star$	Primitive earth about 4 billion (4×10^9) years ago $(cal \cdot cm^{-2} \cdot yr^{-1})\star$
Total optical radiation	262,000	170,000
UV radiation below 2000 Å	75	30
Radioactive decay to 35 km depths	15.5	47
Heat from volcanic emissions (lava)	0.15	>0.15
Electrical discharges	4	4

$\star$ calories per square centimeter per year

Starting material for life

The various components in the atmosphere of a primitive planet were vital starting material for chemical evolution. We can only make a few qualitative statements, unfortunately, about the atmospheric composition of our infant earth. Interestingly, as the following table shows, the most prevalent elements in the universe, excepting the elements helium and neon, are important bioelements:

Element	Symbol	Atomic Weight	Number of Atoms per 1000 silicon atoms in the universe
Hydrogen	H	1	25,000,000
Helium	He	4	3,800,000
Oxygen	O	16	25,000
Neon	Ne	20	14,000
Carbon	C	12	9,300

Element	Symbol	Atomic Weight	Number of Atoms per 1000 silicon atoms in the universe
Nitrogen	N	14	2,400
Silicon	Si	28	1,000
Magnesium	Mg	24	910
Sulfur	S	32	380
Argon	Ar	40	150
Iron	Fe	56	150

Bioelements as a primordial atmosphere

Within a broad temperature range around 20°C., the bioelements react chemically and form a primordial atmosphere consisting almost entirely of hydrogen (H_2), water (H_2O), ammonia (NH_3), and methane (CH_4). We must assume that during its development our "protoplant" passed through a phase characterized by higher temperatures, in which the inert gases and light compounds like water were almost entirely lost. A secondary atmosphere must have then developed from volcanically emitted gases. This new atmosphere also lacked free oxygen, but probably had, in addition to the compounds already cited, carbon monoxide (CO), carbon dioxide (CO_2), and molecular nitrogen (N_2).

Today it is believed that small molecules formed in the primordial atmosphere as a result of the energy provided by the sources shown in the previous table. Of solar radiation wavelengths, the most important here is the UV (ultraviolet) radiation shorter than 2000 Å, since this is the only wavelength range that is absorbed by the kind of atmosphere the earth had in its early history. The molecules produced from the chemical reactions triggered by UV radiation came to the earth's surface with rainfall, and when they decomposed they transferred their energy to the minerals in the rocks where they were embedded or into the sediments where they came to rest. None of these organic compounds could have been disturbed by oxidation reactions, since the atmosphere was devoid of oxygen at this time.

Figure 5-6 depicts the apparatus used by Stanley Miller in his studies on electrical discharges. Ultraviolet or ionizing radiation is sent into a re-action chamber (here Item 2). The circulating gases move the substances created from the reaction of electrical discharges and the atmosphere into a collecting chamber (Item 6), where they can be isolated from this reactive milieu.

The laboratory experiments of Miller conducted in 1953 and 1955 caused a great deal of excitement in the scientific world, since he was the first to successfully produce α-amino acids and other simple compounds using a recreated primordial atmosphere and "lightning". These amino acids, which are the building blocks of proteins and thus of all living matter, were created in much greater quantities than any other substances. They are highly adaptive compounds, in the Darwinian sense, and we can speak of their "fitness" in prebiotic evolution.

An α-amino acid has the following composition:

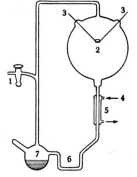

Fig. 5-6. Miller's experiment. 1 = gas supply (with methane, hydrogen, and ammonia); 2 = discharge chamber; 3 = electrodes; 4 = water; 5 = condensing tube; 6 = collecting tube for substances from the discharge chamber; 7 = boiling water, used to promote circulation of the gases.

Laboratory studies
with electrical
discharges

Amino group (NH_2)

|

Organic tail —— Carbon (C) —— Carboxyl group (COCH)

|

Hydrogen (H)

The mechanism by which these amino acids are produced as Miller produced them is still not fully understood. There may be several reaction pathways. The formation of glycine, the simplest of the aminocarbon acids, probably proceeds as follows:

Methane + water $\xrightarrow{\text{electrical discharge}}$ Formaldehyde + hydrogen
$CH_4 + H_2O$ $H_2CO + 2H_2$

Methane + ammonia $\xrightarrow{\text{elec. dis.}}$ Hydrocyanic acid + hydrogen
$CH_4 + NH_3$ $HCN + 3H_2$

Formaldehyde + hydrocyanic acid $\xrightarrow{\text{heat}}$ Aminoacetonitrile + water
+ ammonia $CH_2(NH_2)—C{\equiv}N + H_2O$
$H_2CO + HCN + NH_3$

Aminoacetonitrile + water $\xrightarrow{\text{heat}}$ Glycine + ammonia
$CH_2(NH_2)—C{\equiv}N + 2H_2O$ $CH_2(NH_2)—COOH + NH_3$

Several important electrical discharge experiments are summarized in the following table:

Author(s)	Year	Atmosphere	Products
S. L. Miller	1953 1955	CH_4, NH_3, H_2O, H_2	Amino acids, HCN, and other organic compounds
P. Abelson	1956	Various mixtures, including H_2, CO and CO_2 instead of CH_4 and $N_2 + H_2$ instead of NH_3	Amino acids, especially from a reduced atmosphere
K. Heynes, W. Walter, E. Meyer	1957	Like Miller	Amino acids and others
T. E. Pavlovskaya A. G. Passynsky	1959	Like Miller	Amino acids and others

Influence of ionizing
radiation

Rajewsky and I (Dose) conducted the first studies (1956) on the influence of ionizing radiation on various primitive atmospheres. Our results were very similar to those Miller had obtained previously. We found, however, that many more radical ions participated in the formation of individual biological building blocks, especially amino acids. The reaction mechanisms are extremely complex and have not been fully clarified, and for this reason we shall not deal more closely with them here.

The following table summarizes a number of important studies dealing with ionizing radiation:

Author(s)	Year	Radiation	Atmosphere	Products
K. Dose B. Rajewsky K. Dose	1957 1965	X-rays	CH_4, NH_3, H_2O, CO_2, N_2, CO H_2 (secondary atmosphere)	Amino acids, amines, and other organic compounds
C. Palm M. Calvin	1962	High-speed electrons	CH_4, NH_3, H_2O and H_2	Amino acids, lactic acid, HCN, urea
C. Ponnamperuma, R. M. Lemmon, R. Mariner, M. Calvin	1963	High-speed electrons	CH_4, NH_3, H_2O, some H_2	As above but also with adenine

Small organic molecules, again especially amino acids, also develop when ultraviolet radiation is sent through a primordial atmosphere. The earliest simulation studies were probably those of Groth and Suess in 1938. We summarize their and other studies in the following table: *(Ultraviolet radiation)*

Author(s)	Year	Radiation	Atmosphere	Products
W. Groth H. Suess	1938	Xenon lamp (1470 Å)	CO_2, H_2O	Formaldehyde, glyoxal
W. Groth H. v. Weysenhoff	1957	Various mercury (Hg) lines	NH_3, H_2, H_2O, simple CH com- pounds and Hg steam as a sensitizer	Amino acids, amines
T. E. Pavlovskaya A. G. Passynsky	1959	Vacuum ultraviolet	NH_3, H_2, H_2O, CH_4	Amino acids

These pioneer studies were difficult to perform experimentally, because only vacuum ultraviolet radiation (i.e., below 2000 Å; see energy form table earlier in this chapter) is taken up directly by the components of the primordial atmosphere. Very great demands are placed on the "window" through which this radiation passes as it enters the reaction chamber, and these demands are not easy to meet. These optical and chemical requirements are so complex that little research has been done on prebiotic photosynthesis with gases.

Short-term subjection of high temperatures (1000°C) on simulated primordial atmospheres has also resulted in the forming of amino acids and many other life building blocks. The first experiments of this kind were carried out by K. Harada and S. W. Fox in 1964, and we summarize work of this nature below: *(Studies with thermal energy)*

Author(s)	Year	Reaction state	Atmosphere	Products
K. Harada S. W. Fox	1964	Quartz sand, silicates; 1000°C.	CH_4, NH_3, H_2O	Numerous amino acids and their polymers
I. Oro	1965	No solid material; 1000°C.	CH_4, NH_3, H_2O	Only simple amino acids

These studies in chemical evolution, which we have only touched on in the briefest way, indicate that the primordial atmosphere of the earth contained the vital building blocks needed to produce living organisms, including amino acids, sugar, fatty acids and their derivatives, purine and pyrimidine bases, their later products (nucleosides and nucleotides), and even the porphyrines (of chlorophyll and blood pigmentation material), all in great multiplicity and produced from chemical reactions in non-living systems. They were the starting materials for more highly organized biological systems and more complex molecules.

One important question related to the abiotic formation of life building blocks has not yet been answered, and that is the question of the origin of the so-called optically active compounds.

Origin of optical activity

When a carbon atom bears four different atoms or atomic groups, an asymmetrical system develops. The compound formed can occur in two different (enantiomorphic) states, which look like mirror images of each other. All α-amino acids, with the exception of glycine, can occur in the so-called L-form and D-form. We depict below the two forms of one amino acid, alanine, one of the most important amino acids and a component of almost all protein:

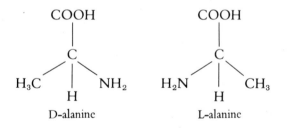

These studies in chemical evolution

Abiotic synthesis leads almost exclusively to racemic mixtures, that is, those in which the D- and L-forms of α-amino acids, sugars, etc., are in a 1:1 ratio. Biogenic compounds, those found in living organisms, usually have just one or the other form. Proteins, for example, contain only L-α-amino acids. These compounds are optically active, as we say, meaning that they rotate the plane of polarized light impinging upon them.

Assymetrical natural forces

There are numerous asymmetrical forces in nature, which could theoretically account for the fact that one specific enantiomorphic form is either produced or destroyed. It has been suggested, for example, that polarized skylight or polarized electrons produced in β-decay processes could favor the formation of one enantiomorph or the other. However, the degree of polarization of the radiations that interact with molecules is so little that significant effects with natural radiation have not yet been found.

Once a certain optically active molecular form predominated, these molecules could react with the D or L enantiomorphs of some racemic mixture and select one or the other in further reactions. One example of such a process is the characteristic disappearance of D- or L-amino acids from a supersaturated solution in which very small amounts of optically active crystals form.

The abiotic evolution of optically active compounds is thus theoretically possible, but simulation studies have not yet demonstrated with certainty that any of the above reaction pathways was of practical significance. It is therefore possible that the capability of absorbing optically active compounds and to transform them is a characteristic that did not occur in prebiotic systems but only occurred after they evolved into living cells.

For thermodynamic reasons, the probability that small building blocks would form larger molecules (e.g., peptides, proteins, polynucleotides, nucleic acids, or polysaccharides) simply by being together and being heated is very low. However, at volcanically active sites where there were lakes or an inlet of the sea, it is possible that over a period of time, very concentrated solutions could have formed and eventually developed into deposits of organic compounds. The coincidence of these chemical systems with hot volcanic rock would produce ideal chemical conditions for the polymerization of the small building blocks into huge, complex molecules. Inorganic polyphosphates, which could form from acid phosphates or ammonium phosphates at temperatures of 200–300°C, may have played a role in various reactions, for they are energy-rich compounds.

Condensation of micromolecules into oligomeres and polymeres

Workers in the laboratory of S. W. Fox in Coral Gables (U.S.A.) have produced polymers under many different reaction conditions, exposing the reaction mixture to hours of heat above the boiling point of water. The polymerization products they developed were called proteinoids due to their similarity with proteins (which arose, of course, biogenically). Proteinoids formed in many different conditions: varying with the amino acid composition of the reaction mixture, the temperature, the time of the reaction, the presence of active materials like minerals or acid. The fractions isolated by electrophoresis (a technique in which electrically charged particles are transported differentially through a fluid with a light electrical current) or chromatography (technique of separating different particles by taking advantage of their solubility characteristics in certain mediums) proved to be surprisingly uniform. This indicates that the internal organization of these large polyamino acids were greatly influenced by the reaction conditions and by the characteristics of the amino acids of which they are built.

Proteinoids

The internal organization of these polyamino acids, and of all proteins and proteinoids and nucleic acids, is most importantly determined by the

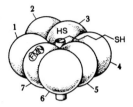

Fig. 5-7. Schematic model of the multi-enzyme complexes of fatty acid synthesis. 1, 2+4: transfer reaction; 3: condensation; 5+7: reduction; 6: dehydration.

Condensation of nucleotides

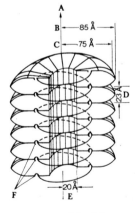

Fig. 5-8. Model of the tobacco mosaic virus. Measurements are in Ångstrom units. A = axis; B = larger radius; C = middle radius; D = height of each helical element; E = radius of the inner cavity; F = nucleic acid.

sequence of the various building blocks. These building blocks consist of amino acids or nucleotides. The importance of the order of these building blocks along the large molecule is as significant as the sequence of letters in this book. The simulation experiments used to produce the first proto-proteins have taught us that of the infinite number of amino acid sequences only a few actually appear. However, there is still a large number of combinations occurring in the first prebiotic protein-like substances (the proteinoids), and proteinoids can assume a great diversity of structural configurations.

Proteinoids can be analyzed with the same methods used for proteins. Their average molecular weight is approximately 5000. They can be decomposed into amino acids by hydrolyzing (splitting) them with acids or bases. Some proteinoids have an enzymelike (i.e., catalytic) activity. Proteinoids found thus far have had a hydrolase, decarboxylase, transaminase, catalase, and peroxidase activity.

Similar results with polycondensation of nucleotides into polynucleotides or even polymers have not yet been made. The polymers would resemble nucleic acids. Polycondensation is the formation of large molecules from simpler ones, a process that involves the release of water (hence condensation). It is possible that the geologically significant condensation conditions have not yet been duplicated in laboratories. However, it is also possible that nucleic acids (the bearers of genetic information) could not exist in the prebiotic world. Simulation studies have succeeded in producing only oligonucleotides (which have about ten building blocks). However, polysaccharides (such as polyglucose), can be produced fairly easily by heating simple sugar above 100°C, just as was done with proteinoid formation.

Some of the major studies dealing with polycondensation of amino acids, nucleotides, and sugars are summarized in the following table:

Author(s)	Year	Starting materials and conditions	Products
S. W. Fox et al.	since 1956	Mixture of all 20 amino acids, with excess of the more acid ones; over 100° for several hours	Proteinlike compounds, proteinoids
A. Schwartz S. W. Fox	1967	Cytidylic acid with adenylic acid or uridylic acid (building blocks of nucleic acids); 65°C. 1–2 hr	Compounds with usually less than five building blocks
P. T. Mora	1965	Glucose; about 150°C. for several days	Polyglucose

The principle of self-organization of biological macromolecules, by which they can form still larger macromolecular structures, is one of the most fascinating chapters in molecular biology: viruses can be broken down into protein and nucleic acid units; multienzyme complexes can reorganize themselves after being broken down, under favorable conditions. Cell organelles (organlike structures within the cytoplasm of individual cells) have the same property. The information coding this "behavior" is held within the structure of the macromolecule contributing to the process.

After molecular biologists were able to demonstrate that the thermal polymers of amino acids—proteinoids—were macromolecules bearing a certain amount of information contained within their molecular structure, it was not so surprising that most proteinoids organized themselves into more complex structures in the same way. They readily form microspheres when they are cooled and in a saturated solution; they become spherical shapes with a diameter of about 2 mm; microspheres are depicted in the Color plate, p. 42. Microspheres externally resemble coccus bacteria. Interestingly, by varying the starting material in the reaction it is possible to control whether the initial proteinoids will develop into gram-positive or gram-negative microspheres. Gram-negative, named from the color studies of the Danish bacteriologist H. Chr. Gram (1853–1938), refers to a red color, while gram-positive is a dark blue hue.

The organization of proteinoids into more complex structures

Microspheres are enclosed in membranelike envelopes. If the pH (acidity) of the surrounding water is rapidly changed, it is possible to cause the material inside the microspheres to diffuse out, leaving the membrane behind as an empty sack. Membranes are vital for all biological organisms and make it possible to have selective transport mechanisms. Under certain conditions, microspheres can form aggregations, build a temporary chemical bridge, and exchange their internal substances (which, of course, bear information). The enzymelike characteristics of proteinoids are retained after they become microspheres. Because of these characteristics, microspheres are fascinating models used to investigate metabolism and membrane transport mechanisms. Microspheres are not living systems, of course, but they are believed to greatly resemble the predecessors (protocells) of such systems.

In relation with microspheres, we must also mention coacervate systems here (see Fig. 5-9). Coacervates are materials intermediate between colloidal suspensions and precipitates. They arise, under particular pH conditions, as droplets formed from the union of materials in a colloidal suspension (e.g., gelatin and gum arabic or lecithin and gelatin), usually after being subjected to an electric charge. Coacervates are more labile structurally than microspheres, and they cannot be concentrated in a centrifuge (as can be done with microspheres). They are usually produced from the large molecules (e.g., proteins, polysaccharides, and nucleic acids) found in modern living organisms. In some respects they are useful as models, particularly in terms of investigating transport and metabolic processes. But unlike microspheres, they cannot be used as models of protocells, since they lack membranes and usually cannot be produced from non-living macromolecules.

The model for prebiotic evolution can be more or less summarized as follows:

Fig. 5-9. Coacervate drops.

Primordial atmosphere

	◁	Heat (about 1000°C.)
	◁	Electrical discharges
	◁	Ionizing radiation
↓	◁	Optical radiation

Amino acids and many other biological building blocks

| ↓ | ◁ | Additional heat (usually up to 200°C.) |

Self-organized condensation products (in the absence of a genetic code), primarily poly-amino acids (proteinoids)

| | ◁ | Self-organizing processes, especially those involving cooling of satu-rated solutions of proteinoids |
| ↓ | | |

Microspheres (and, in part, coacervates) which are models of protocells

Origin of the genetic code

This is the way prebiotic evolution occurred. However, some important questions are still open. One of these concerns the origin of the genetic code (the term used to denote genetic information held within cells).

As we previously mentioned, amino acids arrange themselves during thermal polymerization in such a way that specific structures are generated. If the amino acids were collected statically, the entire universe would not have enough matter to synthesize every possible transformation of a protein or proteinoid molecule. Experiments have shown that each polymerization act yields only a limited number of different polymers. This means that there is some sort of built-in code for specific polypeptide chains expressing itself through the interaction of the existing and added amino acids with their environment (but not with some certain genetic code). This strongly indicates that proteins were formed into the first biological systems without nucleic acids (carriers of genetic information) and that the anchoring of genetic information within nucleic acids was a later development.

Earliest life with no nucleic acids?

However, if we consider proteins and proteinoids to be primary macromolecules containing information, two more questions immediately present themselves: how do the protein(oid)s duplicate themselves during cell division? And how could the information held in the protein(oid)s be transferred to nucleic acids? Molecular biologists have as yet no answer for these basic questions. In fact, the second possibility is in direct opposition to the "dogma of molecular biology", a principle by which the information transfer from proteins to nucleic acids (i.e., the synthesis of a nucleic acid molecule with a protein molecule matrix) cannot occur in a biological system.

But what is not allowed for biological systems may be theoretically permissible in prebiotic ones. It was found that the spiral strands of polyamino acids in the α-helix contained specific mononucleotides in a certain configuration: the basic polyamino acids formed firm bonds (they included poly-L-lysine and poly-L-arginine with guanylic acid and adenylic acid; cytidylic acid formed a weak bond). Investigations with molecular models indicate that a helical (spiral, as shown in Fig. 5-10) polyamino acid like poly-L-lysine binds with adenylic acid in such as way that the adenylic acid residues formed yet another helix, with three nucleotides per amino acid residue. When the individual nucleotides formed bonds, it would be possible for a triple nucleotide group to be coded into a nucleic acid with a polypeptide (protein), just the opposite way in which the coding of amino acid structure takes place in proteins. This is still a presumption, but experimental results until now still support the notion of three nucleotides coding an amino acid.

Another hypothesis on the origin of life rests on the importance of nucleic acids. According to this hypothesis, first put forth in 1929 by the geneticist H. J. Muller, life began with the abiotic formation of one or more genes (nucleic acids). The gene hypothesis has recently been revived (after modifying it with some theoretical considerations that arose from the development of molecular biology). Muller proposed that the smallest entities of life—those that could metabolize food and reproduce themselves—were to be found in genes. In other words, nucleic acid molecules, are, according to Muller, "living" molecules since they can code protein and can multiply, and since they can also mutate they are the stuff of evolution. We lack the room to go into the details of this very intriguing hypothesis. Unfortunately, it has an important weakness: it requires, if it is valid, the first nucleic acids to be able to code the first proteins in the absence of enzymes (i.e., proteins or proteinoids). No mechanism explaining how this kind of synthesis could occur has been proposed.

A compromise has since been reached between those supporters of the protein(oid) theory and supporters of the gene theory. This compromise contains the hypothesis that protein(oid) and genetic material (nucleic acids) could have simultaneously formed the first biological systems. This compromise brings together the latest findings of molecular biology and molecular evolution and is compatible with both sets of data. However, all the knowledge we have thus far assembled does not permit us to draw a flawless picture of how the first protein(oid)s interacted with the first nucleic acids or their "ancestors". There is agreement, however, that the answers to these questions do not have to remain forever in the realm of philosophical speculation but that they can be resolved in due time as molecular biology becomes more highly developed.

Once a genetic coding system, a number of enzymes and a membrane system enveloping these developed (regardless of which of these systems came first), the whole process of cellular evolution becomes much clearer. We can assume that the abiotic development of biologically important compounds occurred while this phase of evolution was taking place. The primitive, anaerobic cells lived on as heterotrophes with organic food supplies. As this source of food gradually diminished, the earliest anaerobic organisms took up an independent (autotrophic) life. They used solar energy to power photosynthesis and to absorb carbon dioxide; thus, solar energy was being utilized to build up chemical compounds. Photosyn-

Earliest life without proteins?

Fig. 5-10. Polyamino acid helix, e.g., poly-L-lysin, and adenylic acid helices can form a double helix. The arrow show the direction the coils must be twisted to be uncoiled.

Other stages of biochemical evolution

thetic organisms developed during this phase of evolution; they were able to form compounds from molecular nitrogen (exemplified by blue-green algae).

Photosynthesizing
organisms

Photosynthesizing organisms, those that took oxygen from water for the reduction of carbon dioxide and then released oxygen into the air, led naturally to a substantial increase in the amount of atmospheric oxygen. Once a critical quantity of available oxygen was exceeded, aerobic (air-breathing) organisms could develop. Oxydative phosphorylation (an energy-releasing process involving energy-rich compounds like adenosine-triphosphate) arose as respiration of air developed. Aerobic organisms predominated over anaerobic ones, because the former could obtain more energy in the form of phosphates from the oxydative decomposition of glucose than the anaerobic organisms could from glycolysis (the break-down of glucose into lactic acid, another energy-releasing process). Bio-chemical data support the likelihood of these events transpiring. However, there is no evidence that the evolution of anaerobic and aerobic organisms had to occur in a linear, stepwise fashion as our simplified description may imply.

After this development occurred, and it probably took 2.5 billion years for it to happen and may have comprised half of the total history of the earth, there was another great leap forward in cellular evolution: cells with nuclei (i.e., eukaryotic cells) arose. Without this step, multicellular organisms would probably never have evolved. According to a widely accepted hypothesis, cells with nuclei developed from non-nucleated anaerobic cells. These prokaryotes probably lived symbiotically with the smaller, photosynthesizing (or respiring) prokaryotes. The host cells contributed the enzyme systems for fermentation (i.e., anaerobic glycolysis), while the symbionts contributed photosynthesis and respiration. Such so-called endobionts may have been the ancestors of our modern chloroplasts (the spherical pigment-bearing bodies in plant cells), or of mitochondria (the thread-shaped or spherical structures in plant and animal cells that function in respiration and metabolism). The steps by which these early cells developed nuclei are not all understood. Eukaryotic cells—modern, nucleated cells—are much more adaptable and versatile with the above-mentioned organelles than they otherwise would be. Again, we can understand how even these very early biological events were evolutionary events. Natural selection was already in full swing!

The evolutionary steps we have described thus far are all summarized in the table below. Data extend back to the Palaeozoic; there is no geological or paleontological evidence for events older than three billion years).

The lack of our complete understanding of the first half of the history of earth and all that occured in molecular and early biotic evolution during that period was one impetus for the increase in space exploration during the last decade. It was hoped that some clues to evolution could be obtained by exploring conditions on extraterrestrial bodies.

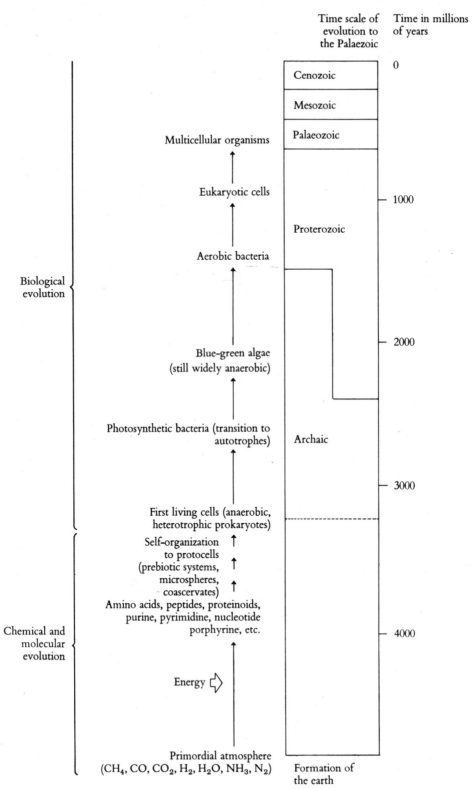

Time scale of evolution to the Palaezoic | Time in millions of years

0

Cenozoic

Mesozoic

Palaeozoic

Multicellular organisms

Eukaryotic cells

1000

Proterozoic

Aerobic bacteria

Biological evolution

2000

Blue–green algae
(still widely anaerobic)

Photosynthetic bacteria (transition to autotrophes)

Archaic

3000

First living cells (anaerobic, heterotrophic prokaryotes)

Self-organization
to protocells
(prebiotic systems,
microspheres,
coascervates)

Amino acids, peptides, proteinoids,
purine, pyrimidine, nucleotide
porphyrine, etc.

4000

Chemical and molecular evolution

Energy

Primordial atmosphere
(CH_4, CO, CO_2, H_2, H_2O, NH_3, N_2)

Formation of the earth

Likelihood of extra-terrestrial life in the solar system

So far there is no proof of extraterrestrial life. The moon flights undertaken during the Apollo program have shown that there is no life on the moon. Rocks brought back from the moon have contained only traces of organic compounds, and it is possible that those amino acids found in moon rocks got there from the effects of the rocket exhaust, fingerprints of the researchers, or other interference. Mars, Venus, and Jupiter are more appropriate subjects for investigating comparative evolution of living systems, but our technology has a few years of development to go before we can extensively study these planets.

Since the 19th Century, a great deal of excitement has been generated by numerous meteorites that have been recovered, since they contained organic compounds and organized structures. Older studies, such as the famous ones with the Orgueil Meteor, are now considered to be worthless, since the researchers contaminated their samples. Recently, however, one meteor in Australia and one found in Mexico that were practically uncontaminated were thoroughly analyzed, and the results of the analyses confirmed the concept of chemical evolution: the meteors contained, among other substances, amino acids and carbon dioxide in about the same qualitative and quantitative compositions as were produced in simulation laboratory studies in a primordial atmosphere! The composition of the meteorite extracts differed substantially from those in geological sediments, whose contents stemmed mainly from living organisms. It was also shown that the microstructures seen in meteorites were not microfossils but inorganic structures. The structure of meteorites also excludes the possibility that they came in contact with water. In these regards we have major differences between the meteorite composition and the composition of geological deposits on earth.

Extraterrestrial life outside the solar system?

If we envision how many solar systems there are in the universe and how many planets there must be, the chances that some of them bear life are really quite good. However, our chances for directly observing such phenomena are practically nil. The closest solar system to ours, Tau Ceti, is more than ten light years away. It is probably impossible for us to travel such a great distance, certainly so with present technology. Radiocommunication with beings on other planets would pose tremendous logistical difficulties; assuming there were such beings on a Tau Ceti planet, we would have to wait twenty years just to get a reply to any communication we would send.

It has been estimated that with a total population of stars numbering 10^{20} throughout the universe, there must be at least 10^9 (one billion) earthlike planets. Even if we assume that most of these are indeed inhabited, the chance that any of them have humanoid beings is vanishingly small.

6   The Early History of the Earth

The Precambrian Era, the period of the earth's early history, comprises the vast time span from the formation of the first crusts of the young planet to the beginning of the Cambrian Era, about 570 million years ago (see fold–out at the rear of this volume). Today it seems fairly certain that the earth and the other planets in the solar system formed approximately 5 billion years ago. Whether the formation of our planet was a "hot" or a "cold" process is uncertain; we are not sure whether our great globe developed from the condensation of a gaseous, heated cloud or from cold cosmic dust. However, either development pathway would lead to the increasing density of the earth toward its center and the great heat being produced there from radioactive processes. The reactions occurring at the center of the earth produced convection currents that caused materials to differentiate into the various layers we know today (see Fig. 6-1).

Today the earth is characterized by a heavy (dense) center and lighter and lighter shells as one progresses toward the outside. It is thought that the nucleus has the same composition as an iron meteorite, almost all specimens of which are composed of an alloy of 90% iron and 9% nickel. The various layers of the earth are rather clearly differentiated. Geophysical evidence has shown that earthquake tremors in the different layers occur when the speed at which the density of the layers changes very rapidly, or when some rapid change occurs within a whole aggregate. Particularly powerful tremors occur at depths between 25 and over 40 km and at 2900 km. These are presumably at the interface of different geological layers, so the earth today is divided geophysically into a core, a mantle, and a crust. The crust, which surrounds the entire earth like a thin rind, can be understood as a differentiation product of the mantle.

The early earth was subject to the same fate experienced by later crusts: it was partially uplifted and partially eroded. The series of crusts formed the layers of sediments we see today, as one crust sunk down and was covered by another. The large troughs (or geosynclines), which were long sunken regions, produced mountain ranges, which were also subject

By O. H. Walliser

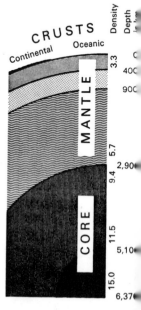

Fig. 6-1. Structure of the layers of the earth.

to erosion forces. Material pushed far into the earth was subject to great pressure and metamorphosed under these new conditions. Crystallization and even fusion can occur to rocks moved into the great depths. Because of all the forces it has been subjected to, the original crust of the earth in all probability cannot be found anymore. Since the oldest deposits known are nearly 4 billion years old, the very first crust formations probably developed much earlier.

There were several periods of mountain formation occurring during the Precambrian. No biochronological data on this early period exist, since there is no fossil record from this time period, but the chief mountain formation periods can be used to generate a time perspective for the great Precambrian events:

Chief mountain formation periods

1. About 600 million years ago: Assyntian formation in North America and Europe, corresponding to similar processes (Katangan mountain formation) in South America.

2. 850 million to 1.15 billion years ago: Grenville mountain formation in North America, Dalsland formation in Europe, Aravalli-Satpura formation in India, and Karagwe-Ankole formation in Africa.

3. 1.25–1.48 billion years ago: among other developments were the Elsonian formation in North America and Gotidian formation in Europe.

4. 1.58–1.96 billion years ago: Worldwide Hudsonian or Carelidian formation.

5. 2 to 2.2 billion years ago: Belomoridian or Marealbidian formations in America, Asia, India, and Africa.

6. 2.37 to 2.72 billion years ago: Algoman formation in North America, corresponding to the Saamidian in Europe, the Aldan in Asia, the Anshan-Sangkan in China, the Dharwar in India, and the Shamwa formation in Africa.

7. 2.87 to 3.11 billion years ago: Laurentian formation, another very widespread development. Even earlier mountain formations have been found, as for example in Africa.

Formation of shields

Mountain ranges formed in earlier folds in the earth are much more stable than the unfolded crust, and they formed the basis of our continents, with additional strengthening by newer mountain growths. Toward the end of the Precambrian, the following shields, as they are known, had developed: The Canadian shield or Laurentia in North America; the Baltic shield or Fennoscandia, extending from Scandinavia through Europe to the Urals; Angaria in Siberia; Sinia in China; the huge Gondwania in the southern hemisphere, encompassing much of what we now know as South America, Africa, India, Australia, and the Antarctic (see Color plate, p. 76).

Continental drift

We do not know whether all the larger and smaller Precambrian shields formed a single continent toward the end of the Precambrian or were separate entities. We do know that the continents changed continuously throughout geological time; this has been shown by the location of the magnetic pole, which has been used as a yardstick for determining certain geologic periods. All magnetizable minerals contained in rocks originating from the naturally heated innards of the earth orient to the magnetic pole. This orientation cannot be changed once the rocks solidify, and this original orientation can still be measured today. Thus, it is possible by taking measurements on different mineral samples to reconstruct the way in which the continents have drifted. Alfred Wegener (1880–1930) proposed that the continents are drifting away from each other, and his hypothesis was confirmed geophysically in the 20th Century.

The atmospheric constituency and the prevailing climate are major factors influencing the origin and development of life. It is doubtful that plant activity alone accounted for the total oxygen content of the atmosphere, for recent calculations indicate that ultraviolet radiation from outer space would have sufficed to produce 25% of all the oxygen in the atmosphere today by its dissociation activity of water vapor. All this oxygen would have been produced before the Middle Precambrian (since subsequent ozone formation acted as a shield against UV radiation). Since the transition from a reductive (oxygen-withdrawing) to an oxydizing (oxygen-forming) atmosphere took place during this time, we would expect that geological deposits from this period would have a higher oxygen content if the above hypothesis were correct. As it happens, these deposits indeed contain more oxygen, and therefore they deliver support to our hypothesis.

Atmosphere and climate

Rock decomposition and mechanical wear were the chief forces operating geologically during the Precambrian, predominating over chemical alterations. Mighty calcareous sediments first arose in the uppermost (i.e., most recent) Precambrian, and they were worldwide. The geological nature of these early deposits indicates that they were in a cool, moist climate. Some freezing also occurred, and the earliest evidence of this is in the Witwatersrand system of approximately 2.5 billion years ago. The traces of the glaciers formed in this period survive as glacial deposits known as tillites. They contain characteristic scratch marks caused as the glaciers slid over these rocks. Tillite deposits are often found in association with banded fine sediments that are typically associated themselves with ice edges.

The last Precambrian glacial formations were undoubtedly very significant. They occurred 650–750 million years ago and were on every continent in the world. The great climatic differences produced in this period and subsequent ones permitted a proliferation of life forms on earth, with greater variety than had ever existed before. That is to say, evolution was much more intense. The great diversification in the animal world just subsequent to the appearance of these Precambrian glacial formations can thus be understood in geological terms.

Most of the original Precambrian deposits have greatly changed since that time. The chances of finding fossils in them are very poor. Relatively

Chances of fossilization

undisturbed Precambrian deposits are restricted to rather small areas, and the further back one goes in the earth's history, the fewer are such sites that have not been altered significantly. Another logistical problem in locating Precambrian fossils is that none of the animals alive at this time had hard parts. They can only be chemically verified or, in extremely rare instances, be found preserved. Some organisms got inside some glasslike, hardening silicic acid that enveloped the organism and hardened before the animal or plant could decompose.

Because of these difficulties, it was long doubted whether any fossils at all could be located from the Precambrian. However, in the last part of the 19th Century, C. D. Walcott described Precambrian stromatolites and claimed that they were algae. During the 1920s, J. W. Gruner published his work on threadshaped microfossils he found in gravellike Precambrian rocks. These pioneer studies were greeted with doubt by other scientists but were recognized as true fossil evidence of Precambrian life by the 1950s. Since this time, there have been so many additional Precambrian finds that our knowledge of the life in this distant era is rather substantial.

The oldest fossils

The oldest known fossils are from gravelly rocks in the Swaziland super group of eastern Transvaal. In the lower part of these, the Onverwacht group, chemo-fossils have been found; their contents of such complex carbohydrates as isopronoids, alcanes, and other substances are almost certainly of biological origin. Fully preserved fossils are first known from the fig tree group, and their age is about 3.1 billion years. These are unicellular, elongated organisms comparable to bacteria, and they were given the name *Eobacterium isolatum* by E. S. Barghoorn and J. W. Schopf. Spherical unicellular organisms with hardened surfaces have also been found, and they contained organic compounds (evidence of their biological nature).

The oldest STROMATOLITES are found in the Bulawaya group of Rhodesia, deposits approximately 2.8 billion years old. These chalky, knobby fossils (see Fig. 6-2) are composed of many fine layers and may have a thickness of several centimeters to several decimeters. Their biohermic structure and composition of their carbon isotopes leaves no doubt that stromatolites were of true biological origin. They were living organisms. Their form stability permits using them for biochronological measurement, especially in the later Precambrian. Stromatolites most closely resemble modern communities of blue-green algae, which also form laminated knobby structures. Thus, photosynthesizing organisms were present as early as 2.8 billion years ago. Further evidence of this is from fossils of the Soudan iron formation in Minnesota, which are of about the same age. They are also very much like algae.

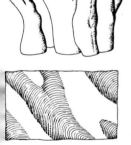

Fig. 6-2. Stromatolites. Below: in section.

Fossils in the Middle Precambrian

Fossil finds from the Middle Precambrian era are much more prevalent; we refer to a period of time 1.7 to 2.5 billion years ago. Such finds include the Witwatersrand and Transvaal super group of South Africa, the Vallen group of Greenland, the lower and middle Belchar group of Hudson Bay and the Gunflint iron formation from southern Canada. Twelve species of fossil organisms were described just from the Gunflint iron formation. They consist of unicellulates or cell fibers, divided or undivided, which in some cases are so well preserved their wall structure can be studied. Most of them are prokaryotic (i.e., non-nucleated) microorganisms belonging chiefly to various blue-green algae families (Chroococcaceae, Oscillatoriaceae, Nostocaceae). Here, in the middle Precambrian, we find oxygen-producing organisms already prevailing among the life of that period, and these were capable of photosynthesis.

The presence of various organic materials is also indicated by the greater prevalence of carbon found from the middle Precambrian. That such carbon originated from living organisms is substantiated from its carbon isotope (C^{12}/C^{13}) content, which corresponds to the proportions of those isotopes in later organisms and differs considerably from the isotopic proportions found in inorganic carbon. The enrichment of organic, biologically produced matter was great enough that this material formed whole strata that later underwent a metamorphosis and turned into graphite. We should also mention *Corycium enigmaticum*, from deposits of the Svekofennid series of central Finnland. This series is nearly 2 billion years old and has undergone considerable changes since that early time. The fossil has a diameter of up to several centimeters and has a carbon-bearing hull. A thread-shaped algae species, *Corycium* formed a ball shape, probably an adaptation to the water currents.

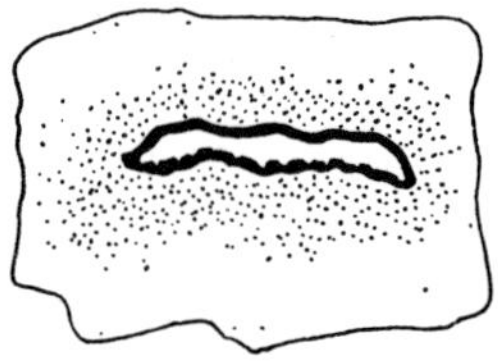

Fig. 6-3. *Corycium enigmaticum.*

Naturally, the fossil finds from the later part of the Precambrian are still more numerous. In the 1960s alone, several dozen sites bearing fossils from this time were located throughout the world. We shall only mention the finds from the Bitter Springs formation in southern Australia. Nearly fifty species of small organisms (including blue-green algae, green algae, and possibly red algae and fungi as well as fibrous bacteria and others) have been located in this 900 million year-old deposit. Most of the many blue-green algae species can even be classified with later forms (something that could not be done with earlier fossils), indicating that by this time we had "modern" algae communities. However, the possible occurrence of fungi is particularly exciting. These and other indications of the presence of cell nuclei, meiotic and mitotic cell division, demonstrate that by this period development had reached the advanced stage of eukaryotic organisms (i.e., those with nuclei). Middle Precambrian finds and other evidence suggest strongly that this development into eukaryotic organisms occurred in the period between the Gunflint and Bitter Springs formation. This is particularly important, since all multicellular animals and plants must have arisen from simple eukaryotic microorganisms.

The remains of what with complete certainty are unicellular animals have been found in the very last part of the Precambrian. These are radiolaria, which have a calcareous skeleton. The same period has yielded

Microorganisms in the Upper Precambrian

the oldest confirmed multicellular animals. Many older "fossils", upon closer analysis, have been revealed as inorganic structures, and the process of reevaluation continues and may cause us to change the above interpretations in the future. However, there is no doubt that by the end of the Precambrian there was a great diversity of life on earth; there are many tracks of worms and signs of possibly higher organisms that have been found and that date from this period.

Ediacarian fauna   The so-called EDIACARIAN FAUNA, most prevalent in the Ediacara highlands in southern Australia, are the oldest fully authenticated multicellular animals. This fauna has been found in many parts of the world, including the Kuibis quarzite of the Nama system in southwestern Africa, in the Maplewell series in Leicestershire, England, in the Vend series from the Kola peninsula, and from northern Siberia. All these deposits appear to be the same age. Tillites were apparently quite numerous in these same sediments. Even though the exact age of these fossils has not been radiometrically determined, it is clear that they are from the very last part of the Precambrian (formerly called the Upper Proterozoic, Eocambrian, or Infracambrian) and lie directly beneath the oldest Cambrian fossils.

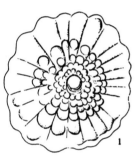

Most finds of ediacarian fauna exist only in relief, and this has been part of the reason that their systematic position is difficult to clarify. However, those few that are preserved physically are so divergent from Cambrian fauna that they must be considered as zoologically distinct species. Most of the more than two dozen ediacarian species are comparable to medusan coelenterates. They often have a center disk and an outer, broad ring (*Ediacaria*). In *Mawsonites* the inner part is divided into several lobed rings. The ridges between the individual lobes extend to the outer edge and, just as we find in modern medusas, may have been caused by contractions of the radial muscles after death.

Other, bisymmetrical, fossils like *Dickinsonia* and *Spriggina* have been classified as annelids (Annelida), but this classification has been disputed. Another species was described as early as 1930 by G. Gürich and interpreted by M. F. Glaessner as Pennatularia. H. D. Pflug has recently investigated the fully preserved fossils of the Nama system, and he concluded that they were echinoderms (Echinodermata). He also showed this for *Rangea*. The species of both fossil genera apparently lived within these deposits, and only the upper part of the body, with its characteristic feather-like marking, was uncovered.

The near future will undoubtedly lead to further discoveries and will further clarify the systematic position of these puzzling ediacarian fauna. However, a few general conclusions can already be drawn. It is certain that highly developed life forms existed by the late Precambrian, and they were widely distributed by this time. Multicellular organisms with hard internal or external skeletons were not yet present, however. The absence of calcareous skeletons is explained by postulating that in the seas at that time there was not enough calcium in solution or that there was

Fig. 6-4. Examples of Ediacara fauna:
1. *Mawsonites*; 2. *Rangea* (a in relief; b sectioned and in normal biological position).

an excess of carbon dioxide (which increases calcium solubility), which would prevent extracting calcium from the water. Evidence for this is the fact that in brachiopods, the shells contained increasing amounts of calcium carbonate during the course of the Cambrian, and the percentage of brachiopods with calcium carbonate shells increased.

However, substantial calcium deposits produced from the metabolic activities of living organisms were already present in the lower Precambrian (i.e., the stromatolite chalks) and in the Lower and Middle Cambrian were already widely distributed (archeocyathid chalks). Furthermore, the question of the development of hard parts must always be put into an evolutionary perspective. Stiff elements are used for support or protection (or perhaps for both simultaneously). They would become necessary when three-dimensional bodies encountered some unilateral force in their environment or when firm locomotive structures develop.

Protective elements must become more powerful as the weapons of enemy species become more powerful. Of course, the need for hard body parts increased with evolution. Later in evolution, strength in such body parts was increased by incorporating mineral deposits in them, replacing the purely organic substances used for protective and supporting functions. Inorganic minerals were far superior for this purpose, and so the process of natural selection caused this kind of adaptation to develop.

From all this and the fact that in the Cambrian all the important invertebrate animal groups were present, we can conclude the following: highly organized animals were in existence in the late Precambrian. The few finds we have from this distant era is not a reflection of a scarcity of species but a result of the fact that the organisms present at this time lacked the kinds of body parts that fossilize readily. Fossilizable internal and external skeletons were not yet selected in evolution. This does not exclude the possibility that the basic body plans of the various invertebrate groups developed during the Precambrian Era.

7   Paleozoic Formations

By O. H. Walliser

During the last 570 million years, the time between the Precambrian and the present, the earth has housed a vast diversity of living organisms, and for this reason we can call this whole period the Phanerozoic (the time of visible life). The division of this time span into the Paleozoic, Mesozoic, and Cenozoic is done to represent distinct phylogenetic steps in the evolution of the animal kingdom.

The Phanerozoic

The Paleozoic comprises more than half of the entire Phanerozoic, i.e., 345 of those 570 million years. Its beginning is characterized by an explosive development of living organisms, by further development and distribution of the various body plans, and by the conquest of most habitats throughout the world. However, many groups became extinct toward the end of the Paleozoic era, and they were replaced by their own descendants that were more advanced. The most important animal groups of the Paleozoic include the trilobites (subphylum Trilobita; see Chapter 9) and the graptolithines (class Graptolithina; see Chapter 9). Both occurred only in the Paleozoic, as did the Rugosa corals (see Chapter 9) and the primitive armored fishes described later in this chapter. In all phyla there were many groups that continued into later periods but flourished only during the Paleozoic. Several formations or systems are distinguished within the Paleozoic, and they are, from bottom (older) to top (younger): Cambrian, Ordovician, Silurian, Devonian, Carboniferous, and Permian. These groups are also known chronologically as periods within the Paleozoic era. Many organisms do not simply develop within one or the other period, for some develop throughout the era. We shall briefly describe the various systems in the hopes of giving a survey of the development of life during the Paleozoic era, and we shall devote attention to certain individual groups as well. We shall thus demonstrate how many groups have complex interrelationships and how their development can be related to environmental changes.

The Paleozoic's oldest system or period, the Cambrian, was named as such in 1833 by the Englishman Adam Sedgwick after the Roman name for northern Wales, *Cambria*. According to radiometric data, the Cambrian period comprises the span from 500 to 570 million years ago. If we for academic purposes compare the total history of the earth with a single year (see Color plate, p. 73), we would find that the Cambrian corresponds to the time between November 19 and November 24, placing it near the end of the total time span.

"Cambria": Roman name for northern Wales

During the CAMBRIAN PERIOD, the edges of the continents formed in the pre-Cambrian were flooded as the seas intruded on them. The high point of this transgression occurred in the Middle Cambrian. After that the seas retreated once again, in what we call a regression. The formation of the epicontinental seas on the old continents developed from the geosynclines or depressions found between the continents, on their edges, and from the major depositions.

The Caledonian geosyncline settled where the high mountains of Norway are found today. Its western edge extended to eastern Greenland, which, with North America, was much closer to Europe than it is today (see Fig. 7-1). The geosyncline settled into the trough of what later developed into the Appalachian mountains of North America. Over two kilometers of sediment was deposited at the center of this basin (in Wales), beginning with conglomerates of gravel and of quartzite in the then shallow seas. They were followed by deposits reflecting the greater depth of the trough: massive fossil-bearing silt sediments with graywacke (a fine-grained conglomerate rock). A substantial flattening occurred in the Upper Cambrian. At the edge of the great depression (in northern Scotland), quartzite deposits were supplemented by limestone formed by the ancient Archeocyatha. The deposits from epicontinental seas (which were less substantial, being only tens of meters deep) penetrated to the Baltic shield. Thus, for example, the Upper Cambrian decreased in thickness as a geological system from 800 m in Wales to 50 m in Schonen and just 15 m in Västergötland.

Fig. 7-1. Distribution of water (shaded) and land in the Middle Cambrian. Greenland and North America were still near Europe at that time.

One trough extending to the east also left Cambrian deposits in central Europe. While the Lower Cambrian is often characterized by massive Archeocyatha limestone deposits, substantial volcanic depositions developed during the Upper Cambrian. Toward the south the Caledonian geosyncline was associated with the foundation of the Tethys, a very important "Mediterranean" depression for later earth history. In the Cambrian this proto-Tethys was joined with the broad seas of Siberia and eastern Asia as well as the sinking Rocky Mountain trough in western North America.

Archaeocyatha calcium deposits

The worldwide Archeocyatha limestone deposits of the Lower and Middle Cambrian indicate that warm weather prevailed during this period, while the glacial deposits from the Upper Precambrian reflect a much colder climate during that earlier period. Thus, the Lower and Middle Cambrian were a time of a general warming trend. Many deserts must have existed in those years, since the first evaporation sediments (salt, gypsum) appear in that period in Morocco, Iran, Siberia, India, Australia, and northwestern Canada.

The first great proliferation of marine organisms occurred in the Lower Cambrian. Thus, for the animal world, the lower boundary of the Cambrian was one of the most decisive ones in the history of the world. As we stated earlier, this proliferation can largely be explained by the acquisition of hard, durable skeletal parts by organisms of that period. Hard body parts permitted the development of new, more complex anatomical structures, with new locomotory apparatus allowing these animals to radiate into hitherto uninhabited environments. The acquisition of stable supportive and protective elements gave an impetus to a still faster rate of evolutionary development: the body plans of invertebrate animals, which arose before the Cambrian, were modified in many different ways. However one must realize that the Lower Cambrian encompassed a period lasting over 30,000,000 years, about half the entire length of the Cenozoic era.

Diversification of animal life required the presence of a rich supply of plant material as food. The plant world had changed little since the late Precambrian. The greater prevalence of calcium-releasing algae and chitinous organisms indicates that the Cambrian period witnessed the flourishing of algae, organisms restricted entirely to the seas.

Animal life also developed solely in the seas. Fossils from other ecosystems—brackish and fresh water or from land—are absent in the Cambrian. The oldest, always sandy deposits of the Cambrian have revealed many fossil traces, especially those of benthonic (bottom-dwelling) animals.

Trilobites

The characteristic Cambrian animal group is that of the TRILOBITES, whose various forms have been used to subdivide the Cambrian system. If we review the complete evolution of this Paleozoic group, we find that in the Cambrian the predominant forms were primitive ones with small shields and small eyes. The only exceptions are the agnostids (order Agnostida), with a large shield and a small number of body segments. However even the primitive trilobites attained considerable sizes. With a length approaching .5 m, these trilobites were the largest Cambrian fauna. The distribution of the various trilobite genera demonstrates clearly that there were two faunal provinces no later than the Middle Cambrian. A relatively sharp border was located in the region of what later became the Appalachians. East of that lay the Atlantic or Acadobaltic province, which included northern Europe, and to the west was the Pacific province, beginning at the Rocky Mountain trough and continuing across the proto-Tethys from eastern Asia to southern Europe. Mixing of the two fauna-provinces occurred in central Europe, Siberia, South America, and Australia. Both provinces are represented in the Lower Cambrian, but the characteristic genus *Olenellus* was still cosmopolitan. The distinction later became clearer. Beginning with the Middle Cambrian, there are key fossils for each faunal province: in the Middle Cambrian we have *Paradoxides* (see Color plate, p. 128) for the Atlantic province and *Olenoides* (same Color plate) for the Pacific province, while in the Upper Cambrian the key Atlantic and Pacific province fossils are *Olenus* and *Dikelocephalus*, respectively.

The ARCHEOCYATHA (class Archaeocyatha) were another distinctive Cambrian group. They resembled corals that first appeared in the Middle Ordovician but then quickly proliferated. The resemblance was both anatomical and in terms of life habits, and archeocyatha were typical shallow water inhabitants that could proliferate under favorable conditions. In the course of time they produced reefs several hundred meters long, and the limestone deposits left on these reefs were largely produced by archeocyatha. We consider archeocyatha, like coral, to be inhabitants of warmer zones, and the prevalence of archeocyatha from California to Labrador, Scotland, central Europe, northern Africa, central Asia, and Siberia southward to Australia and the Antarctic shows that at that time the continents were characterized by climates quite different from those we find today.

In addition to the above animals, all major invertebrate groups proliferated during the Cambrian. Of the UNICELLULATES, the Upper Precambrian is characterized by radiolarians (order Radiolaria), while foraminifera (order Foraminifera) first appear in their simplest form (i.e., as spherical to conical shaped organisms with undivided chambers).

Cambrian reefs were inhabited not only by archeocyathids, the predominant species, but also by sponges such as Sphinctozoa and Inozoa and the initially rare stromatopores (order Stromatoporoidea), the oldest skeletal-producing COELENTERATES (Coelenterata). In addition to the frail medusae, coelenterates also appeared as organisms with a firm, four-sided chamber, organisms known as conularia.

ANNELIDS (phylum Annelida) and the various ARTICULATA (phyletic group Articulata) were present in large numbers, as evidenced not only by their tracks but from whole specimens recovered at the Burgess Pass in the Canadian Rocky Mountains. The Middle Cambrian slate found there was originally from very oxygen-poor stagnant water habitat, which provided superb conditions for preserving organic materials. The animals represented there as fossils include sponges, medusae, annelids, onychophorans (phylum Onychophora), and primitive articulated animals, including the oldest Xiphosura (see Chapter 9). Many of these have been recovered in such good condition that we know almost as much about them as we do about their modern descendants.

Ancient Articulata

The first true CRUSTACEANS (class Crustacea; see Chapter 9) also appeared in the Cambrian. They included the OSTRACODS (Ostracoda), often microscopically small animals protected by a double-valved housing.

Among the MOLLUSKS, snails (class Gastropoda) arose in the Lower Cambrian with bilaterally symmetric flat-shelled forms (e.g., *Scenella*), in

Mollusks

which a primitive articulation can still be seen. The first coiled species (class Bellerophontacea) arose in the later part of the Lower Cambrian. The oldest mollusks are from the Middle Cambrian. The first cephalopods (class Cephalopoda), on the other hand, appear in the Lower Cambrian (*Volborthella*). The link between these primitive molluscan species and the plectronocerates (*Plectronoceras* and others) of the Upper Cambrian has not been found, but the plectronocerates were the starting point for a sudden evolutionary development in mollusks. The prevalent hyolithids (class Hyolithida; see Chapter 9), with their small, spherical shells, were another important molluscan group.

Brachiopods

Among the BRACHIOPODS (class Brachiopoda; see Chapter 9), organisms known since the Lower Cambrian, there were as the predominant forms those with non-closing, horny or calcareous-phosphate shells. Later those with calcium shells appeared, and these brachiopods (subclass Testicardines) were still in the early stages of their evolution.

First proliferation of echinoderms

ECHINODERMS (phylum Echinodermata; see Chapter 9) also arose in the Cambrian; their presumed ancestors are known from the Upper Precambrian. All major echinoderm classes appear early in the Cambrian. In these early echinoderms, the five-rayed symmetry is either incomplete or not even present, an example of this being the bilaterally symmetric Carpoidea. Most Cambrian echinoderms were sessile forms anchored to the floor with a shaft. Only the Holothuria could move through the water.

The very "advanced" Branchiotremata were represented in the Upper Cambrian by the first graptolithines (known since the Ordovician). They were the highest form of life of the Cambrian, and no true chordates or vertebrates are known from this period. However, with the Cambrian

Conodontophorids

conodontophorids (see Chapter 9), we have organisms very close to chordates. Little is known about this group, but their toothshaped, microscopic conodonts are made of the same material as vertebrate teeth. Thus, the Cambrian saw the origin of not only all the precursors of the major vertebrate classes but also species just prior to the chordates.

Ordovician and Silurian

During the ORDOVICIAN and SILURIAN periods, 395 to 500 million years ago, the earth crust and living organisms evolved in a uniform direction. No basic differences or occurrences appeared at this time, so we shall describe both periods together. The names of the two periods arose from Charles Lapworth in 1879 and Roderick Impey Murchison in 1835, who named these systems after the Celtic Ordovician tribes in northern and eastern Wales and the Silurian tribe in Shropshire, England.

Division on the basis of graptolithines

The periods are subdivided using graptolithine evolution. The Ordovician is divided, from lower to upper layers, in the following sequence: Tremadoc, Arenig, Llanvirn, Llandeilo, Caradoc, and Ashgill. The Silurian subdivisions are: Llandovery or Valent, Wenlock, Ludlow, and Pridoli. Paleogeographic conditions during these periods were essentially identical to those of the Cambrian, especially for the further existence of the great troughs. During flood periods the seas covered the old shields once again. After the recession of the Upper Cambrian, a new transgression of the seas occurred, and it reached its high point in the Caradoc. Many parts of the Canadian shield, the Asian shields, and northern Gondwana were covered by ocean.

After a recession of the seas in the uppermost Ordovician, the Middle Silurian was the scene of the greatest advance of the seas in the entire Paleozoic. Shallow seas spread over huge areas, leaving behind deposits on what later became continental land. In the Baltic area, which was now covered by a small sea after the transgression of the Lower Cambrian, only about 350 m of sediment was deposited. This was typical of the sedimentation of the Middle Silurian, which seems small indeed in comparison to the six kilometers and more of sedimentation left in the Caledonian troughs. The deposits in the zones bordering the Baltic and southern Scandinavia consist chiefly of fossil-rich limestone, turning into slate toward the basins, which held graptolithines. The Middle Ordovician Kukruse layers have been especially significant. These likewise fossil-filled, marly sediments have so much bitumen produced by the blue-green algae that had lived there that these deposits were exploited as fuel and sources of oil.

In depressions, the deposits have a lower calcium content. The rocks found there are primarily formed from the rubble of earlier rock masses. The edges of troughs are characterized by the presence of oolithic iron ore. In contrast to the parts of the old shields covered by Ordovician-Silurian seas, geosynclines had volcanic lava that often formed substantial masses of rocks. In Caledonian geosynclines, basalt and andesite rocks had been widespread throughout the Ordovician since the Arenig and ended in the lowest parts of the Silurian. In central Europe, these rocks were limited to the Lower Ordovician and the Middle Silurian, especially in parts of Germany and Czechoslovakia.

The penetration of these lava deposits was closely related to the crust movements appearing in increasing amounts since the close of the quiet Cambrian period, and these movements finally led to a division of the geosynclines. Crust movements reached their first high point at the border of the Ordovician and Silurian. By the Upper Ordovician, a broad land zone had become visible in the eastern part of the Appalachian trough, and the northern part of this region was the site of the taconite cleft. Principal mountain-building activity in the Caledonian trough occurred at the end of the Silurian. The entire geosyncline between the Canadian shield and the Baltic shield folded and formed mountains, and this vast mountain range kept the two continents welded together for a long time. Large parts of this old mountain range can still be seen today (as, for example, in the mountains of Norway).

The distribution of fauna, prevalence of gypsum and colorful fragmented rocks, and the distribution of iron ooliths all indicate that at this time the Arctic was in the tropics, and that in fact both poles were in tropical regions. The distribution of tillites confirms this as well. During the Silurian, the climate appears to have become more equalized.

Floral development in the Ordovician and Silurian apparently had little influence on animal life. Although the first rooting plants developed in the Upper Silurian, their effect was not important until the Devonian.

GRAPTOLITHINES (see Chapter 9), which first appeared sporadically in the Upper Cambrian and had survivors into the Carboniferous, were the characteristic animals of the Ordovician and the Silurian. They have been used to classify the Ordovician and Silurian periods because they developed very quickly and permit a fine analysis of these periods. Another important fossil group were the ancient shelled cephalopods ancestral to the *Nautilus* of today and which we can call NAUTILOIDS (subclass Nautiloidea; see Chapter 9). They flourished in the Ordovician and Silurian, when they proliferated and evolved from straight-shelled species to coiled forms. The enlargement of the shell tube and the associated formation of internal chambers were important evolutionary developments in the nautiloids. Many species became extinct at the end of the Ordovician and were replaced by others, in which the shell orifice often narrowed in old age. During the Silurian, straight-shelled bactritids (Bactritida) evolved from the orthocerates (see Chapter 9), and they gave rise in the Devonian to the ammonites (see Chapters 9 and 12).

RUGOSA CORALS (subclass Rugosa) and TABULATA CORALS (subclass Tabulata) appeared for the first time in the Middle Ordovician. Together with stromatopores, they built reefs throughout the world, particularly since the Silurian period. Snails, members of the MOLUSK group, became more important than they had been in the Cambrian. However, the flat-shelled, trochospiral forms were still the predominant groups. Mussels also became more prevalent. While Ordovician fauna were taxodont, the Silurian was characterized by Anisomyoaria, the newly arisen Heterodonta, and Desmodonta.

TRILOBITES were still important Articulata, and in the Upper Ordovician they reached their high point in terms of diversity and size. Later, especially in the Silurian, trilobites became fewer in number. As they declined, OSTRACODS (subclass Ostracoda) proliferated, and they flourished in the Upper Silurian along with the orders Beyrichiacea and Leperditiacea. The Cambrian aglaspids (order Aglaspida, class Merostomata) developed in the Ordovician as a side branch of Xiphosura into the SEA SCORPIONS (order Eurypterida), a prevalent Silurian group. These well-armored, probably predatory animals, whose tracks can even be found in beach deposits, reached a length of 2 m, and during the entire Silurian only a few cephalopods equalled this length. Originally they lived in the sea, but then they penetrated brackish and even fresh water habitats. The first SCORPIONS (order Scorpiones) appeared simultaneously in the Silurian, and with the first CENTIPEDES (class Myriapoda), the first step toward conquering land probably began as well.

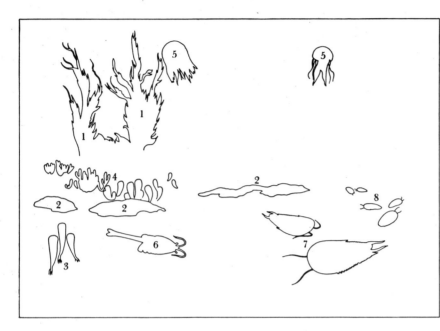

Life in a Cambrian sea: Algae (1 and 2), forming to some extent crusts and cushions; sponges (3); spongelike archeocyathids (4); medusae (5); trilobite-like Articulata (6, *Waptia*) and trilobites (7: *Paradoxides*; 8: *Ellipsocephalus*).

Among BRACHIOPODS (class Brachiopoda), the Testicardines became the predominant forms. Orthids (Orthacea) and strophomenids (Strophomenacea) were particularly prevalent. The Silurian saw the proliferation of Pentameracea, Productacea, Rhynchonellacea, Atrypacea, and Spiriferacea, animals whose newly acquired supportive elements permitted them to invade new habitats such as reefs. BRYOZOA (class Bryozoa), closely related to brachiopods, were usually colonial and exuded calcareous skeletons. They appeared in the Ordivician and proliferated quickly.

ECHINODERMS (Echinodermata) flourished. These two periods supported both primitive forms such as cystoids (Cystoidea) and groups with five symmetrical rays: blastoids, true crinoids, starfishes, and sea urchins. During the Upper Silurian, crinoids were already appearing in fossilized form!

Vertebrates made their debut during the Ordovician. Their earliest representatives were JAWLESS FISHES (superclass Agnatha; see Chapter 10), which first proliferated in the Upper Silurian, when the Acanthodii, bony fishes (see Chapter 10 for both groups), and primitive armored fishes occurred.

The geological and biological development in the Ordovician and Silurian periods show that the climatic changes, the great extent of sea division, sea transgressions and regressions, and the increasing amounts of deposition, all created new habitats, which, excepting those on land, had never been invaded before and would not be substantially altered.

This was the predisposition for the development of the various ecotypes of the world. In the Ordovician and Silurian we find that every

▷

Trilobitoidea from Middle Cambrian Burgess shale (all nearly natural size):
1. *Cheloniella*
2. *Naraoia*
3. *Yohoia*
4. *Marella*
5. *Emeraldella*
6. *Molaria*
7. *Mollisonia*
8. *Leanchoila*
9. *Burgessia*
10. *Waptia*
11. Part of *Helmetia*.

First vertebrates

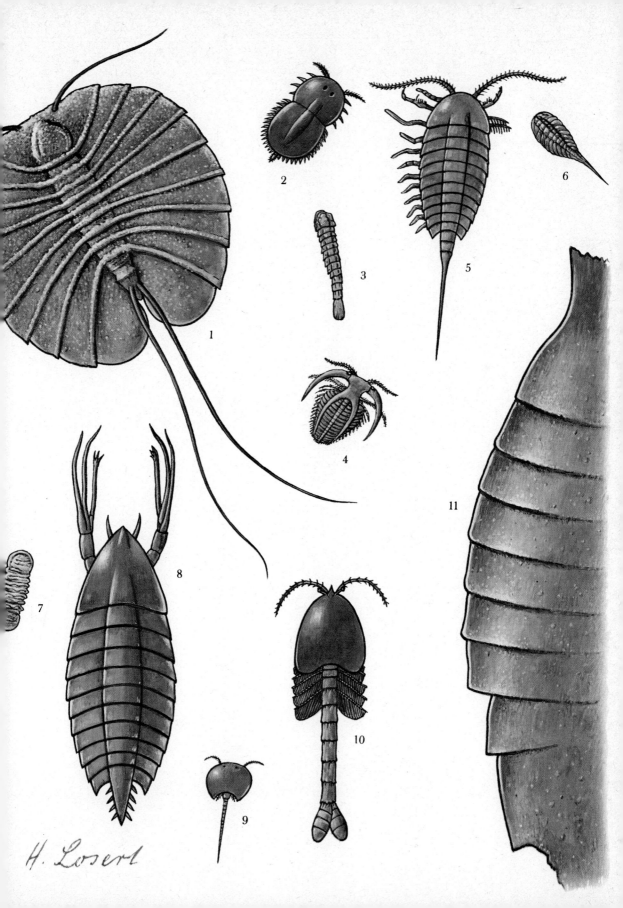

H. Losert

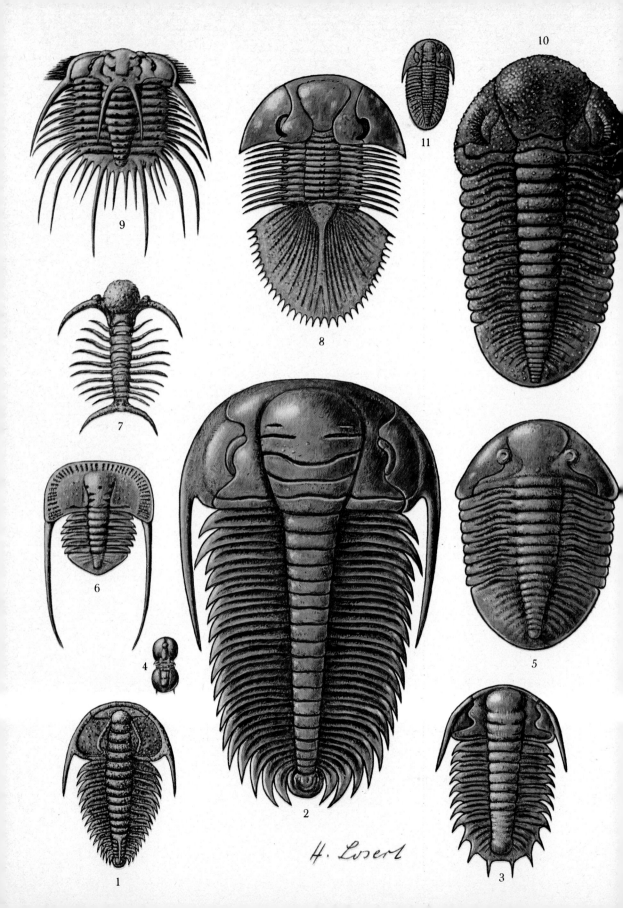

H. Losert

marine and aquatic habitat has its animal life. Fishes developed chiefly near shores. They competed there with many Articulata, especially sea scorpions. Lagoons, with their fauna, formed between the coasts and coastal reefs. The reefs themselves, which as today reached up to the surface of the water were built up by corals, stromatopores, calcareous algae, and bryozoans and were occupied by communities of Foraminifera, Articulata, snails, mussels, brachiopods, and echinoderms. In somewhat deeper, but still relatively shallow, water were the great populations of crinoids and similar organisms. They were followed by flat surfaces densely populated with brachiopods and adapted members of all the above groups. If we descend still deeper into the Ordovician/Silurian sea, we find that ground-dwelling organisms decrease rapidly. The most prevalent forms are drifting (planktonic) and swimming (nectonic) forms, such as non-sessile graptolithines and, especially, the cephalopods. It is no exaggeration to state that for all these habitats, which also characterize modern seas, there were biological counterparts to those organisms found today in the same ecosystems. Thus, this period was characterized by the attainment of the kinds of marine communities.

First marine communities

The Devonian

The DEVONIAN was named by Englishmen Roderick Impey Murchison and Adam Sedgwick in 1839 for Devonshire in southern England, where they found deposits that were more recent than the Silurian but predated the coal-bearing Carboniferous.

A large continental mass arose in the northern part of the world from the Caledonian mountain formation processes, and its southern border extended in Europe from southern Ireland across southwestern England, Belgium, Germany, central Poland, and from there further to the southwest (see Fig. 7-2). The red erosion products of Caledonian mountains assembled in the inner depression zones and at the edge of this northern continent, and in Scotland several thousand meters of "old red" sediment have collected.

The rest of Europe south of the northern continent was largely covered by the sea, as our Fig. 7-2 shows. Only in a strip extending from the central massif of France across southern Germany to Czechoslovakia may there have been scattered islands. The sea formed the widespread Variskian geosyncline. The name is derived from the Variskian tribe from the north. The Romans knew the German city of Hof as *Curia variscorum*. During the Devonian, the geosyncline underwent a subdivision caused by crust movements, and the process finally ended in the Upper Carboniferous with the formation of the Variskian mountains.

The Ardennes-Rhenish slate deposits permit us to follow the course of the Devonian in great detail. In the Lower Devonian, a major geosyncline zone was in the Taunus-Hunsrück region, where several thousand meters of clastic (fragmented) sediments have been deposited. The Hunsrück slate has been a great source of fossil fauna.

The great clastic deposits on the shelf were widely distributed. They

◁
Members of the most important trilobite groups, all shown nearly in natural size:
1. *Redlichia*
2. *Paradoxides*
3. *Olenoides*
4. *Agnostus*
5. *Harpes*
6. *Trinucleus*
7. *Deiphon*
8. *Scutellum*
9. *Odontopleura*
10. *Phacops*
11. *Phillipsia*.

now hold a characteristic fauna primarily consisting of strongly ribbed brachiopods, an indication of the presence of shallow water during the Devonian in those areas. Reefs developed along the shelf edges from the action of corals and stromatopores. Fragments from the corals and from other rocks, probably created from earthquakes, created a great deal of turbidity well into the depths of the basin. The ground-dwelling shallow-water fauna was now in foreign territory between the basin sedimentations with their fauna (which did not include benthonic species). The chief members of basin sedimentation fauna were the ammonites, calcium exuding organisms. Their deposits are thinner than those from other organisms because ammonites were found in areas that did not sink in deeply as sedimentation formed. Clefts were formed on the edge of the basin, and these were filled by volcanic material used as building blocks for other organisms. Once the lava deposits got near the surface of the water, reefs comparable to the atolls of today were built on them.

Fig. 7-2. Distribution of land and sea (1 = continental deposits; 2 = sea) in the Upper Devonian.

In connection with volcanic activity, which in the Lower Devonian produced chiefly acidic keratophyres and basic diabases, iron was produced in the Middle and Upper Devonian. Various rock types differentiated, and they were associated with habitats having their individual animal life. However, the various rock formations occurred throughout the world, since they only needed certain favorable conditions to develop. "Kell water lime" is an example of such a process. In the Harz mountains of Germany, a layer several centimeters thick of dark, chalky slate and knobby lime is found in the otherwise light cephalopod lime from the early Upper Devonian. The fauna consisted of ammonites, orthoceres, and an epiplanktonic (i.e., one attaching itself to drifting organisms) mussel (Buchiola), certain crustaceans, a deviant community with conodonts, and armored fishes. Thus, Kell water lime houses a rich diversity of fauna, and its age can be estimated to within 500,000 years, the smallest measurable time unit in the Quarternary. During this Kell water period, rock formations developed wherever there were suitable conditions for such development, and this occurred in central Germany, the Pyrenees, Morocco, and Iran.

Development of iron ore

Troughs continued to develop outside Europe. By the turn of the Silurian to the Devonian, the sea had retreated to the principal geosyncline, the Paleo-Tethys, which extended eastward to the Variskian trough, north to the Ural trough, to the circumpacific geosyncline in eastern Australia, Japan, and Siberia in the east and to North America (Rocky Mountain trough) and South America (Andes trough) in the west.

During the Lower Devonian the sea flooded from the geosynclines onto the continents. The high point of this flooding was on the Canadian shield during the late Middle Devonian. At the same time the Caledonian folded zones in the Appalachian trough raised up again, and they dispersed their rock fragments into huge deltas spreading to the west. The old Gondwana block in the south experienced more flooding along its edges than it ever had before. Devonian deposits have been found in the Amazon delta, southern Brazil, Argentina, the Falkland islands, and in the cape region of Africa. Flooding extended as far as the northwest part of Africa.

Later in the Devonian there is a distinct difference in marine animal populations, particularly in brachiopods. In the southern hemisphere, including Bolivia, Argentina, the Falklands, southern Africa, Ghana, and the Antarctic, the Malvino-Kaffric fauna predominated, and they differed from the boreal fauna of the northern continent. Climatic differences were most likely the reason for this differentiation, even though a presumed glaciation in southern Gondwana has not yet been confirmed. However, paleomagnetic measurements indicate that one of the magnetic poles was in this region, while the other was east of Japan. The Devonian equator extended from Central America to the Appalachians, England, Germany, and Arabia, into the region west of Australia. This has been confirmed by the distribution of reefs and the zone of what are called evaporites (gypsum and salts, produced from evaporation). These evaporites are nearly restricted to the Upper Devonian, so it seems that the Devonian as a whole was subject to a warming trend.

Of particular importance was the conquest of land by plants, a process begun in the Upper Silurian but which became significant in the Devonian. The plants created the setting, environmentally, for animals to become terrestrial as well. The presence of terrestrial plants influenced the mechanisms and chemistry of terrestrial climate and had geological influences on erosion and deposition.

ALGAE became more highly developed. One of the large brown algae, *Prototaxites*, attained a "trunk" diameter of up to one meter. Calcium-exuding algae, compared to those from the Silurian, declined. Of rooting plants, the predominant Lower Devonian ones were PSILOPHYTES (see Chapter 8). They disappeared during the course of the Devonian. While psilophytes originally lived in the sea and later occurred in moist depressions along continental shores, the LYCOPHYTES (see Chapter 8) penetrated further inland. They reach their initial flourishing period in the Middle Devonian, a period when treelike plants with woody trunks first appeared. The Equisetinae (see Chapter 8) also made their first appearance in the Middle Devonian. Originally as Hyenia flora, they differed little from their predecessors, but by the Upper Devonian they evolved distinct differences. FERNS also appeared in the Middle Devonian, and by the Upper Devonian they were widely distributed (Archaeopteris flora; see Chapter 8).

As a result of this movement onto land by plants, animals began radiating into the new environment. The most important of these were Articulata: scorpions, spiders, and mites. Centipedes must have lived on land by this time, too, for the wingless insects that descended from them arose in the Devonian. In the MOLLUSK group, the first terrestrial species were snails. Finally, at the turn of the Carboniferous, VERTEBRATES began radiating into terrestrial environments.

Conquest of the land by plants

The earliest terrestrial animals

The FORAMINIFERA underwent a great proliferation during the Devonian. Multicellular species appeared, and the spirally coiled endothyrids flourished from the Upper Devonian into the Lower Carboniferous. SPONGES, STROMATOPORES, RUGOSA CORALS, and TABULATA CORALS were still the most important reef builders. The Rugosa corals developed, secondarily, skeletal parts during the Devonian; Ordovician and Silurian species had lacked such structures.

TRILOBITES declined still more. A few groups, such as proetids and asteropygines, brought forth a large number of new, sometimes widely distributed and very prevalent species; but most of those trilobites present in the Silurian became extinct. Only a few phacopids and proetids survived into the Carboniferous. Of the OSTRACODS (Ostracoda), Beyrichiacea and Leperditiacea were prevalent in the Middle Devonian, especially in the border zones of continents. The Entomozoa (free-swimming forms) were good key fossils for the Upper Devonian. Well-preserved examples of Phyllopoda, Xiphosura, and Pantopoda have been found in Lower Devonian Hunsrück slate.

SNAILS and MUSSELS evolved further. However, excepting the first terrestrial snails, no new body plans evolved. The numerical ratio of more primitive and more advanced mollusks changed in favor of the latter. Tentaculites experienced just as wide a distribution as they had in the Upper Silurian. Their tapered, cylindrical shells, which were several centimeters long and bore powerful rings, probably evolved from benthonic organisms. Those with soft shells and bodies only a few centimeters long were, in contrast, either nectonic (actively swimming) or planktonic (drifting) animals.

Primitive CEPHALOPODS, the nautiloids, declined in the Devonian. Only the simple, straight orthocerates and the actinocerates were still prevalent. The bactritids of the Lower Devonian gave rise to a new, successful group, which powerfully influenced the marine fauna into the Cretaceous. These were the AMMONITES (subclass Ammonoidea). Their Paleozoic members are generally known as goniatites (see Chapter 9). Their sutures were rather simple; the shells are typically smooth and in the earliest species were not fully coiled. Ammonites proliferated after the late Lower Devonian, but this growth was ended at the outset of the Middle Devonian.

Important key ammonites are *Anetoceras*, *Mimagoniatites*, *Agoniatites*, *Anarcestes*, and related forms. The few species that reached the Upper Devonian underwent a new, greater proliferation, but this also lasted a short time, being ended by the upper boundary of the Upper Devonian. Characteristic Upper Devonian ammonites include *Manticoceras*, *Cheiloceras*, *Prolobites*, and the clymenians, in which unlike all other ammonites the shell tube is on the inner curve of the shell.

The BRACHIOPOD flourishing period continued. Advanced forms with spiral arm structures such as the spirifers were predominant. The particularly prevalent brachiopods were *Spirifer*, *Strophomena*, *Schellwienella*,

Decrease in trilobites

Increase in ammonites

▷
Some distinctive Silurian Articulata (all shown about 1/3d smaller than natural size):
1. *Ceratiocaris*, a phyllocarid from the group of higher crustaceans (Malacostraca). Sea scorpions from the Merostomata group:
2. *Pterygotus*
3. *Carcinosoma*
4. *Ctenopterus*.

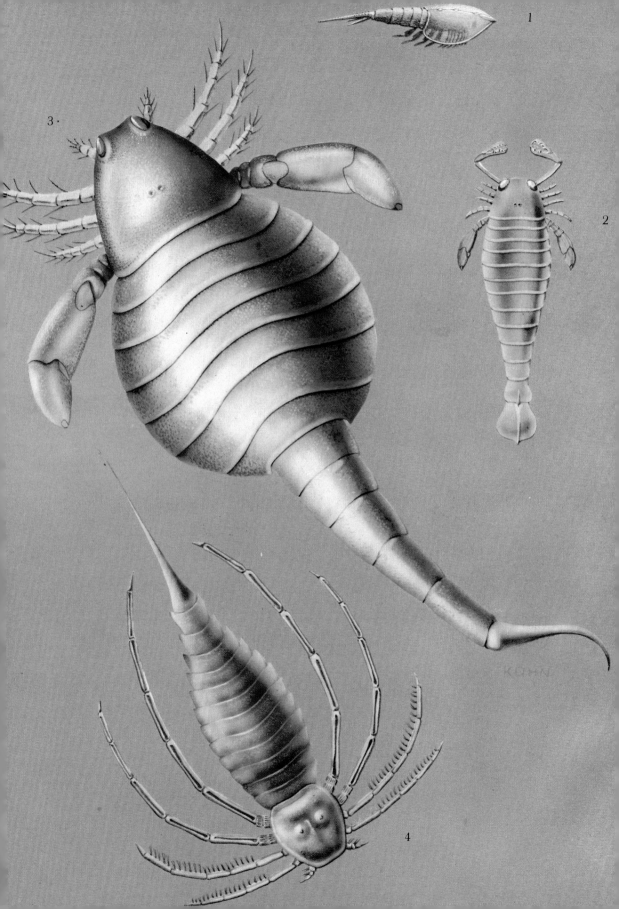

JURAS-
SIC

TRIASSIC

PERMIAN

CARBONIFEROUS

DEVONIAN

SILURIAN

ORDOVICIAN

CAMBRIAN

VALVELESS

BRACHIOPODS (INARTICULATA)

2

Orthida

4

3

1

8

7

Strophomenida

6

5

9

Rhynchonellida

11

10

15

14

Spiriferida

13

12

ARTICULATE BRACHIOPODS (ARTICULATA

Stropheodonta, Schizophoria, Atrypa, Uncinulus, and the tetrabratulacea *Stringocephalus, Rensselandia,* and *Uncites.*

Echinoderm evolution

A series of primitive ECHINODERM orders (cystoids, carpoids, edriasteroids) died out, but starfishes and sea urchins continued to develop to more modern forms, and the crinoids experienced greater propagation and proliferation than they had before. All the graptolithines, excepting the order Dendroidea, became extinct in the Lower Devonian. The conodonts developed further. A rapid anatomical transformation occurred in flat species, such as *Polygnathus* and the Upper Devonian *Palmatolepis.*

From armored fishes to "modern" fishes

VERTEBRATES developed at a relatively frantic pace. The peculiar ARMORED FISHES were particularly prominent species during Devonian vertebrate evolution, for they reached their greatest degree of diversification in the Devonian; they also went extinct during that period. Armored fishes were pushed out by "modern" species. The true BONY OR TELEOST FISHES originating in the Silurian were joined in the Lower and Middle Devonian by LUNGFISHES (Dipnoi), CROSSOPTERYGIANS OR LOBE-FINNED FISHES (Crossopterygii), and CARTILAGINOUS FISHES (Chondrichthyes). The appearance of cartilaginous fishes after teleost fishes developed shows that the former were not the ancestors of bony fishes. The first cartilaginous fishes, primitive sharks (Cladoselachii) gave rise in the Upper Devonian to true sharks. The crossopterygians gave rise to urodeles (tailed amphibians), animals that could go on land.

The Devonian is thus characterized by the development and propagation of rooting plants, the conquest of land by plants and animals, the development and first flourishing period of ammonites, and by development of fishes. It was a period of great production in both flora and fauna.

The "coal-making period"

The CARBONIFEROUS PERIOD (from the Latin *carbo* = coal) was named from the great coal deposits originating in this system that are now found in Europe and North America. The word Carboniferous means coal-making period (*ferre* = to bear). Devonian developments predisposing the Carboniferous coal-making were the development and flourishing of terrestrial plants and the Variscian mountain formation, the chief geological event of this period.

◁
Distribution and prevalence of Paleozoic brachiopods:
1. *Obolus*
2. *Lingula*
3. *Orthis*
4. *Schizophoria*
5. *Strophomena*
6. *Leptaena*
7. *Eomarginifera*
8. *Oldhamina*
9. *Pentamerus*
10. *Uncinulus*
11. *Pugnax*
12. *Atrypa*
13. *"Spirifer"*
14. *Choristites*
15. *Callispirina*
16. *Dielasma.*

During the Carboniferous period, the mountains formed from the central European geosyncline were folded more and more into the great troughs. The axis, or deepest point of the sedimenting trough, moved in the same direction as the folding. In the course of these movements, the Lower Carboniferous witnessed once again a narrow zone of volcanic activity, which in some places built up several hundred meters of basic volcanic ash. In the shallower marginal zones, reef-dwellers built up "carbon-chalk". The great basins, especially in the center, were also filled with Kulm facies. The chief portion of that was not in the form of fine muddy basin sedimentation but clastic rocks, which were transported well into the basins from turbid currents originating in the inner shelf. The inner shelf quickly retreated from the mountains as they were forming.

The high point of the folding process was reached in the Sudeten phase well into the Upper Carboniferous. The Kulm troughs, filled during the Lower Carboniferous, were raised up by this action. What was formerly the northern margin of the geosyncline sunk down as a subvariscian trough. Now called the Molasse trough, it took in the sediment loosened from the developing mountains. In doing so, the trough moved northward, and it folded from the south and thus became associated with the Variscian mountains. The Asturian cleft in the late Upper Carboniferous brought an end to this development. The sediments in the north no longer folded after the Upper Carboniferous.

Fig. 7-3. Distribution of land (2 = regions with coal formation) and sea (1) in the Upper Pennsylvanian epoch of the Carboniferous.

The Variscian cleft also influenced regions outside of middle Europe. Just south of the Variscian central zone (i.e., in the Alps), the folding process did not stop. This region was drawn into the Tethys geosyncline, which began folding at the turn of the Mesozoic and Cenozoic. The Atlas mountains arose on the south side of the Variscian sea as counterparts to similar mountain formations to the north, and from there further mountain development spread to the Antiatlas group. Mountain formation also began in the Ural trough during the Upper Carboniferous but did not reach its zenith until the Permian. The great geosynclines of Asia were also touched by the Variscian folding, thus limiting their extent. However, the Tethys trough remained intact. The Appalachian geosyncline in North America stopped developing at the turn of the Devonian to the Carboniferous. In contrast, the Cordilleran trough continued developing.

As the Molasse trough gradually flattened out, it gave impetus to the development of extensive swamps, moors, and forests in the lowlands. The vigorous plant growth was only temporarily interrupted by brief transgressions by the sea or by being covered by sedimentation. This sea-influenced development is known as paralic, in contrast to limnic or intermontane development (that occurring in basins or on the continent). Depending on the onset of mountain formative processes, the time span of coal development extended from the Upper Devonian to the Permian. Its greatest growth was realized in the Westphalian stage of the Middle Upper Carboniferous. The oldest coal deposits are from Bear Island, and they are followed by those of Spitzbergen, the Moscow basin, and the Urals. Coal formation began in marginal areas (i.e., Scotland and East Germany) during the early Upper Carboniferous, and the process rapidly spread to the other basins of central Europe. Significant (i.e., industrially significant) coal deposits during the late Upper Carboniferous developed in only a few parts of central Europe. Huge deposits formed at the same time, however, in Siberia and eastern Asia, and coal continued to form into the Permian. Still younger are the coal deposits from Brazil, South Africa, and Australia.

A moist, warm climate must have prevailed for much of the Carboniferous period along a wide belt across the face of the earth, and it appears that there was little annual variation in climate during this time. Biological

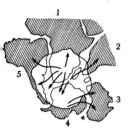

Fig. 7–4. Permocarbonian glaciation on the still closed southern continent: 1 = Africa; 2 = India; 3 = Australia; 4 = Antarctic; 5 = South America.

finds support this possibility: trees lack annual rings, and large cells produced ample tissue growth in trees. Zones of coal formation were surrounded by desert climate. This is demonstrated by salt and gypsum deposits in North America and dolomite mineral deposits on the Russian plains. In contrast, the southern continent experienced glaciation—the so-called Permocarbonic ice age—at the turn of the Carboniferous to the Permian. The earliest glacial moraines (tillites) are found in southeastern South America, South Africa, southern India, Australia, and the Antarctic. The wide distribution of glacial conditions, in contrast to the other climatic conditions, only makes sense when it is understood that the above continents were very close to each other at that time. Glacial evidence over such a large area is yet another bit of evidence for the existence of the great Gondwana southern continent. The carbon deposits in these regions arose from plants, which unlike those in the coal zone had annual rings, indicative of substantial seasonal climatic changes. To a large extent, annual rings formed after the glacial ice withdrew, a process that continued throughout the Permian, which was a warming period.

The great coal reserves of today produced in the Carboniferous period owe their existence to favorable climate and the geographic conditions needed to foster a proliferation of rooted plants. The giant (30 m high, 2 m in diameter) Lepidodendracea and Sigillariacea of this period were lycopods occurring together with Equisetales and other plants. SEED FERNS arose between the spore and seed plants. GYMNOSPERMS (Cordaitinae) also arose during this period, and at the conclusion of the Carboniferous there were even true CONIFERS with the genus *Walchia*.

The extensive swamps, forests, and plant-covered regions afforded the animal world an opportunity to proliferate, and this opportunity was particularly great for ARTICULATA. We mention the primitive spiders, whose hind body was still segmented, and after them the numerous centipedes and their relatives, of which *Arthropleura* attained a length of up to 180 cm. Others in this group included scorpions and lower insects. Insects multiplied during the Carboniferous and moved into the air, occupying yet another habitat.

Giant dragonflies with wingspans up to 70 cm developed during the Carboniferous. The predecessors of most "modern" insects arose in the late part of the Upper Carboniferous. Development of terrestrial vertebrates proceeded equally rapidly. Among amphibians, the LABYRINTHODONTS (subclass Labyrinthodontia), with some species over 5 m long, and LEPOSPONDYLI (subclass of the same name) predominated; the latter included snakelike species. BATRACHOSAURS (order Batrachosauria) arose in the early Upper Carboniferous, and they were followed shortly thereafter by the ancestral reptile group, members of the order Cotylosauria. These species developed further in the Permian and were the ancestors of all higher vertebrates.

Important modifications also occurred in water-dwelling animals. The unicellular FORAMINIFERA truly flourished. They produced, with the Fusulina, the first large foraminifera, which appeared in such great numbers they formed a substantial part of the rocks produced from their fossilized ecosystem. In COELENTERATES, the stromatopores and Tabulata coral declined severely. The Rugosa coral, on the other hand, secondarily acquired skeletal elements that proved to be of selective advantage and helped them survive longer. Of particular importance among proliferating coelenterates were the zaphrentids.

TRILOBITES were dying out rapidly, and in the Carboniferous they consisted of just a few genera (*Phillipsia, Griffithides*). EURYPTERIDS and primitive XIPHOSURA were still prevalent.

Carboniferous MOLLUSKS, particularly snails and mussels, changed little from their Devonian counterparts. Interestingly, snails and mussels invaded fresh water. Of the fresh-water mussels, the anthracosiids were widely distributed, particularly in the Upper Carboniferous. In the seas, modern species such as the pectinid, limid, and trigoniid mussels became more significant. Of cephalopods, the goniatites experienced a renewed growth period, having nearly become extinct in the late Upper Devonian. Their rapid development permits us to use them to subdivide the Carboniferous into smaller geological units. Important analytical features used in this kind of investigation include the increasing complexity of the sutures and the more prominent external features. In the Upper Carboniferous, the cephalopods with straight shells (probably bactritids) gave rise to the belemnite branch, which became widely distributed in the Mesozoic.

The most important BRACHIOPOD group were the productids (Productinae), a characteristic group of the widespread shallow seas. Other important brachiopod groups were orthids, strophomenids, and spirifers. ECHINODERMS developed into modern forms. Sea urchins, for example, now acquired firm body enclosures, in which the plates could no longer rub against each other. The number of plates diminished. In crinoids, the cups became simpler and lighter.

The last GRAPTOLITHINE descendants died during the Lower Carboniferous, along with the dendroids. CONODONTS continued to develop with a new platform type, *Gnathodus*. The primitive FISHES had become extinct, and the modern species, especially the ganoid paleoniscids, developed rapidly.

The Carboniferous, the golden age of lower rooting plants, is thus of tremendous importance for the evolution of insects, higher bony fishes, and terrestrial vertebrates.

In 1841, English geologist Murchison, working in the Ural district of Perm, named the deposits he found as belonging to the Permian period, since they were more recent than the Carboniferous but predated the Triassic period. In the mid-18th Century, J. G. Lehmann and G. C. Füchsel had undertaken the task of dividing these Permian layers, and

Flourishing of foraminiferes

The Permian

they found a Lower Permian and an Upper Permian; Murchison's terminology, of course, was later applied to these strata. The withdrawal of the seas at the end of the Variscian folding period left huge regions dry. They were either covered by the debris from new mountains or were temporarily covered by water from other geosynclines. Germany is a classic example of Permian continental, marine, and shallow sea development.

During the Lower Permian, troughs filled up from the Variscian folding processes. They ran parallel to the fold lines of the mountains, so the chief depression was in the very middle of the troughs. The movements were interrupted only briefly between the early and late Lower Permian by the weak Saal fold phase, and they again weakened at the end of the Permian. Volcanic processes were associated with this activity, especially in the Saar-Saale trough.

The north German basin

In the early Upper Permian the Arctic Sea pressed into the north German basin, and as an inland sea this basin was dependent on the incoming currents from the major sea. Relatively minor changes in the currents could strongly influence the degree of water exchange between the basin and the large sea. Four distinct cycles appeared, each of which culminated in the formation of salts and gypsum. In the first cycle, which began with gravel conglomerate layers, the basin was constricted and the depositions settling there in that oxygen-poor water contained sulfides of copper, silver, molybdenum, lead, and zinc. Bryozoans built up reefs.

The last salt cycle was covered by loam toward the end of the Permian, and these deposits gradually became part of those in the Triassic. The sea had withdrawn from the German basin, a region that repeatedly proved to be important paleogeographically.

The prominent geological event in eastern Europe was the formation of the Urals. The western geosyncline, formed during the Upper Carboniferous, was filled with over five kilometers of sediment. During the Saal phase, the Urals underwent their last folding. The preuralic trough was closed off, filled with salt deposits. For a short time there was a shallow sea in this region during the Upper Permian. The mountain formative processes, which also occurred in the Tien Shan mountains, northern Pamir, and Kwenlun, all of which produced a large continental block in Asia known as Angaraland. It joined with Fennosarmatia, the land surrounding the Caledonian and Variscian mountains.

South of this block, the Tethys continued to develop. The western Mediterranean was still continental land because of the Variscian movements. During the Permian the troughs in the region of what we now call the western Alps began enlarging. From here the Tethys extends to the east over the southern Alps, the inner parts of the Dinarian chain, Greece, Turkey, the Caucasus, Iran, and the Himalaya to Indonesia. There it ends in the old geosynclines around the Pacific.

Large inland basins on Gondwana (the southern continent) took in huge amounts of sedimentation. These included the south African Karroo and the Congo, the Brazilian Paraná basin, and the pre-Cordilleran trough in Argentina. These basins bear similarities in their rock structure and in the development of the fauna found there, especially of reptiles.

The climate can be deduced from the distribution of flora, as well as from other data. While no regional differences appeared in most of the Upper Carboniferous, these did occur in the last part of the Upper Carboniferous. On the northern continent, the *Gigantopteris* community spread from China and mixed with the European and American flora. The *Glossopteris* community arose in the southern continent, Gondwana, and during the Permian its species pressed through the Tethys to Asia, and upon mixing with the flora there created the Angara community. Thus, the equilibrium finally attained in the Triassic had been initiated in the Permian.

The appearance of the *Glossopteris* community on the southern continent in the high Upper Carboniferous, the sharp distinction between the various communities deep in the Permian, and the gradual mixing and equalizing of these different communities were all probably related to the Permocarbonian glaciation on Gondwana. Prevailing temperatures were coolest in the Middle Permian, and after that time there was a gradual warming trend. While the Lower Permian was characterized by distinct climatic zones, the differences between zones became blurred toward the end of the Permian period. The desert had become very widespread.

The Permocarbonian glaciation caused considerable differentiation among living organisms, especially on continents. This glacial period was as significant for evolution of its animals, especially reptiles, as was the glaciation that occurred in the Upper Cambrian, when life first appeared.

As in the Carboniferous, ALGAE producing calcium played a major role in building reefs during the Permian. They formed communities with BRYOZOANS and with COELENTERATES, although the latter were less important. Stromatopores and Tabulata coral disappeared almost entirely. The two-sided wall structure of the Rugosa coral was also lost during the Permian. Zaphrentids and plerophyllids led to the rayed symmetrical Madreporaria coral, which appeared in the Triassic and then predominated.

The Fusulina occurred en masse as large Foraminifera in some shallow parts of geosyncline seas, just as they did in the Upper Carboniferous. Of the BRACHIOPODS, the predominant forms in epicontinental shallow seas and elsewhere were, as before, productids, strophomenids, spirifers, and Terebratulacea. Some of the more unusual groups were the coral-shaped richthofeniids (Richthofeniidae) and the oldhaminids (Oldhaminidae), whose rear valves were covered with riblike grooves.

The TRILOBITES (subphylum Trilobita) became extinct in the Permian. CRUSTACEANS, on the other hand, proliferated. Among INSECTS, primitive groups were prevalent, but most of them became extinct before the onset of the Mesozoic. Most of the "modern" insects that undergo complete metamorphosis appeared in the Permian.

MOLLUSKS changed little compared to the Carboniferous. The primitive Bellerophontacea snails were still flourishing. In ammonites, the cleft on the edge of the chamber walls continued to develop. Ammonites and conodonts were the most important key fossils of the Permian in the geosynclines. Primitive ECHINODERMS, especially blastoids, flourished in the shallow seas. Crinoids developed many new forms, and sea urchins appeared for the first time with twenty rows of plates.

Amphibians and reptiles

The zenith of AMPHIBIAN development was already past. Their prime period ended with the beginning of the rise of reptiles. REPTILES were most prevalent from the strata of the continental basins, especially from the Karroo of South Africa. The Cotylosauria, the ancestral reptilian group, were present in several groups, of which most died out before the Triassic. Some of these attained considerable sizes. All these animals still had a closed skull roof. The opening of the roof, permitting the attachment of powerful jaw muscles, was to be a great advance for reptiles. One of the groups with this advantage were the THECODONTS (order Thecodontia), which gave rise in the Mesozoic to the large marine and terrestrial dinosaurs and to birds. Another group, the SYNAPSIDS (subclass Synapsida), were the ancestors of mammals. Numerous synapsid orders existed in the Permian. One of the prominent synapsids was the predatory lizard *Dimetrodon*, which had a dorsal sail-like structure supported by extremely long thorny processes. The therapsids (order Therapsida) acquired a palatine plate between the nasal passage and the mouth cavity, an adaptation that permitted them to breathe while they were feeding (see Chapter 11). Their teeth were the first in reptiles to be differentiated.

The Permian was therefore a period of transition. The mountain formations in the crust of the earth ended the Variscian era and at the same time introduced the new era of alpine groups. In the animal world, the primitive species disappeared, while the modern forms replacing them began proliferating. Fresh-water and terrestrial habitats were occupied during this period. The earth had arrived at this state 225 million years ago. If we compare the entire history of the earth with a single year (as we did in the Color plate, p. 73), we find that the Permian occurs on December 15, eleven days before the spread of mammals, fifteen days nineteen hours before the advent of man, and sixteen days before the present.

8   The Evolution of Plants

By W. Riegel

After the essential characteristics of plant life (e.g., photosynthesis and firm cell walls) had developed in the Precambrian, plants developed and proliferated in the Paleozoic era. The early Paleozoic saw the rise of algae and the first flourishing, followed by a decline somewhat, of phytoplankton, and in the late Paleozoic plants invaded terrestrial habitats and experienced their first crisis. Since plants form the first links in the life chain supporting animals, plant evolution has had definite effects on animal phylogeny. It has happened many times that major changes in plants have preceeded comparably significant changes in animals by half an epoch.

During the Lower Paleozoic, most plant evolution occurred at the algal stage (see Color plates, pp. 126–127) in the sea. Fossils from this period are usually calcium-exuding algae and spores with cell walls built of resistant organic matter. Only a few of these can be classified with modern algae. The total picture of algal flora in the Paleozoic is still rather fragmentary.

The transition from the Precambrian to the Cambrian apparently was not associated with any major floral evolution. Algae did not begin proliferating until the later Cambrian and the Ordovician. Calcium-producing, bottom-dwelling green algae appeared very early; they were closely related to the highly developed dasycladacean algae found today in warm seas. These algae build themselves up from a large, long base cell that supports numerous branching whorls of smaller cells. Calcium is released within the mucous-filled spaces between branches on older parts of the colony. As cells die, they become pores for the calcium structure. Within the Paleozoic, these algae developed from those with irregularly branching structures to those with a very specific organization. They first flourished in the Triassic. Spherical forms (*Mizzia*) appeared in Permian deposits. In the Paleozoic, dasycladacean spores were produced in the base cell, while in Mesozoic dasycladaceans the spores were formed at the free tips of the branches.

The CHAROPHYTES, characterized by the twirled configuration of their

Fig. 8-1. Lobes of
Prototaxites.

Dasycladacea algae

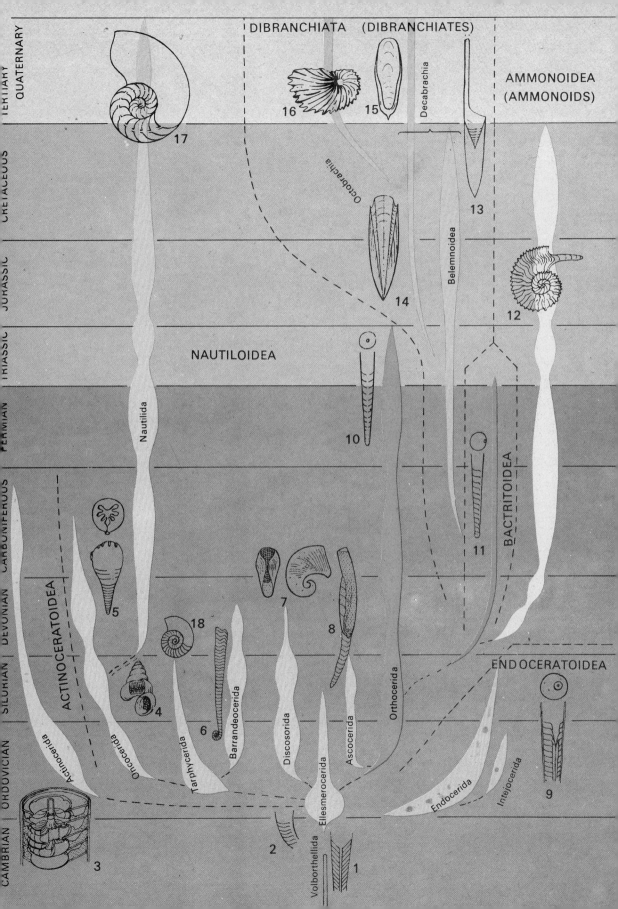

DIBRANCHIATA (DIBRANCHIATES)

AMMONOIDEA (AMMONOIDS)

NAUTILOIDEA

Decabrachia

Octobrachia

Belemnoidea

ACTINOCERATOIDEA

Nautilida

BACTRITOIDEA

ENDOCERATOIDEA

Actinocerida

Oncocerida

Tarphycerida

Barrandeocerida

Discosorida

Ellesmerocerida

Ascocerida

Orthocerida

Endocerida

Intejocerida

Volborthellida

TERTIARY
QUATERNARY
CRETACEOUS
JURASSIC
TRIASSIC
PERMIAN
CARBONIFEROUS
DEVONIAN
SILURIAN
ORDOVICIAN
CAMBRIAN

CRINOZOA ASTEROZOA ECHINOZO

PERMIAN

CARBONIFEROUS

DEVONIAN SILURIAN ORDOVICIAN

CAMBRIAN

Starfishes (Asterozoa)

Crinoids (Crinoidea)

Blastoids

Cystoids

Homalozoa

(Cystoidea)

(Blastoidea)

Eocrinoidea

Somasteroidea

Brittle stars (Ophiuroidea)

Sea urchins (Echinoidea)

...bers (Holothuroidea)

Cyclocystoidea

Ophiocystioidea

Helicoplacoidea

4

7

13

5

8

14

3

6

11

1

2

10

9

Fig. 8-2. *Baltisphaeridium* (size about 25μ).

Fig. 8-3. *Veryhachium* (size about 20μ).

rootless, leafless cell body (or thallus), left only their oogonia (egg apparatus) as fossilized remnants. For a long time, charophytes were thought to be a relatively recent group deserving a distinct phylogenetic position. However, their oogonia show that charophytes are from the Silurian, deep in the Paleozoic era. The early, primitive charophytes had egg cells that were incompletely covered by numerous straight enveloping cells. During evolution, the envelope cells spiraled in a specific direction and covered the entire egg except for a single pore, and the number of enveloping cells was reduced to five.

In RED and BROWN ALGAE, fossil finds are more the exception than the rule, so their phylogenetic history is understood very poorly, reconstructed from a few atypical species and some doubtful fragments. Whether the base tissue lobes of *Solenopora* from the Ordovician and Silurian belong to red algae is a debatable point, to cite one specific example. It is fairly certain that brown algae were represented by the laminaria order in the Silurian and Devonian with *Prototaxites* (see Fig. 8-1). These algae had stems up to several decimeters thick, and they were originally thought to be conifers. The cell fibers and tubes are woven together at the base and form fringed lobes at the top, as our figure shows. They are comparable to tree-shaped *Lessonia* brown algae of today, a variety that forms underwater forests in antarctic waters.

Our knowledge of marine phytoplankton of the Paleozoic is based chiefly on research conducted in the 1950s and 1960, but the main features of the phylogeny of these algae have been elucidated. Fossil finds have consisted of the capsule cysts primarily from unicellular planktonic algae (phytoflagellates and green algae), the walls of which were composed of extremely resistant material. There are many types of these cysts: some are spherical, some oval, some polygonal. Until their systematic positions are clarified, they are being provisionally grouped together as acritarchs. The geological history of this group began in the Precambrian with spherical forms (speromorphs). During the Cambrian, a spiny form (*Baltisphaeridium*; see Fig. 8-2) predominated. By the Ordovician period there were several new groups of these phytoplankton: polygonal (*Veryhachium*; see Fig. 8-3), spindly (*Leiofusa*), and those with complex, polygonal ornamentation (*Cymatiosphaera*) were among the new forms. A high point in their development was reached in the Silurian and, to some extent, in the Devonian as well. Thereafter the group declined, and only a few genera are known from the Carboniferous and Permian periods. Only guesses can be made about the reasons for this decline, but for some reason these algae lacked the fitness for modern ecosystems found in present-day algae. The next great algal growth spurt, which occurred in the Jurassic and Cretaceous periods, was comprised of completely different groups (diatoms, dinoflagellates, coccolithophorids), organisms that today comprise the majority of algae.

Land surfaces in the Lower Paleozoic were probably sparsely inhabited.

All evidence points to the likelihood that plants did not evolve to the stage enabling them to gain access to terrestrial habitats until the Silurian. Moving onto land was not an easy change, and many problems had to be overcome. The most challenging one was to acquire some substitute for water and to control the exchange of material with the environment. Two new materials had to be synthesized naturally: lignin, an adhering substance, and cutin, to prevent dehydration. This led to the subdivision of plants into roots and shoot structures (also known as cormophytes, from the Greek κορμός) and to the development of transport systems for water and some means of assimilation (the formation of carbohydrates from carbonic acid {H_2CO_3}). Another vital adaptation was an epidermis, a firm outside protective layer providing integrity in a terrestrial habitat. Fossils depicting for us from which algae groups the rooted plants arose are missing, as is data on comparative morphology. The ancestral groups were either highly developed green algae or brown algae, depending on the criteria used.

Certainly the oldest rooting plants were "amphibious" in that they inhabited shallow stretches of open water. A narrow, medial, woody vascular bundle permitted them to grow upright to some extent. A level shoot gave rise to several vertical shoots, which were forked and often bore sporangia (spore containers) at their tips. These plants, which lacked leaves, were PSILOPHYTES (class Psilophytinae; see Color plates, pp. 170/171 and 204). They also lacked true roots. According to O. Lignier, roots appeared later from shoots as an adaptation to living in the ground. *Rhynia* is a superb example of one of these old psilophyte ferns (see Fig. 8-4), a .5 m tall plant that formed thick grasses in, the peat of Rhynie, Scotland, its namesake, along with other psilophytes. *Rhynia* was so well preserved there that all its fine details are visible. Psilophytes are more diverse than their similar external appearance might lead us to expect, since these plants were probably ancestral to all other major groups of terrestrial plants. Some of them show possible relationships with lycopods including *Zosterophyllum* and *Taeniocrada*, whose lateral spiked sporangia had definite openings. Their flattened shoots were largely free of orifices, indicating that these parts of the plants were under water (see Fig. 8-5). *Psilophyton* is characterized by glandular-like spines and arched sporangia. Other psilophytes bear resemblances to primitive ferns. The oldest rooting plants whose age has been accurately determined (*Cooksonia*) are from Upper Silurian Czechoslovakia. Other reports of older rooting plants have not been confirmed. Using terrestrial floral spores as indices, it appears that psilophytes arose at the earliest in the Middle Silurian.

The LYCOPODS (class Lycopodiinae; see Color plates, pp. 52, 170/171, and 204) are a good example of floral evolution in the Paleozoic. Their chief characteristics are small alternating leaves, and a woody body that is star-shaped in cross-section with the sporangia located in the axes of the leaves. Modern lycopods are low, mosslike plants found in moist areas. Carboniferous species, in contrast, were stately trees existing in a large

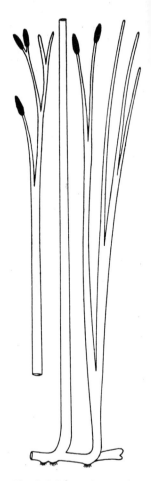

Fig. 8-4. *Rhynia* is a particularly good example of the structure of a primitive fern. Size is approximately 30 cm.

Fig. 8-5. *Zosterophyllum*, with its lateral spiked spore cases.

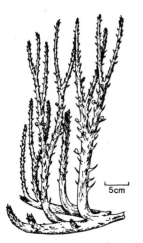

Fig. 8-6. *Drepanophycus* was widely distributed in the Lower Devonian. This plant is one of the oldest true lycopods.

numbers. They largely determined the apperance of the extensive swamp forests occurring at that time, and lycopods made a substantial contribution to the formation of coal deposits. Lycopods probably developed before the Devonian from psilophytes, as some fossil finds indicate. *Asteroxylon*, for example, is classified by some paleobotanists as a lycopod and by some as a psilophyte. The spines covering the shoots of this plant lack vascular bundles, as in psilophytes, but vascular bundled branches lead from the central woody portion, extending to the roots and probably supplying them. In this respect, *Asteroxylon* is a transition leading to *Drepanophycus* (see Fig. 8-6), a plant that is considered to be one of the oldest true lycopods. It was widely distributed during the Lower Devonian. Its simple thornlike leaves each bear one or two vascular bundles. They are distributed on plump, .5-m high shoots and sometimes bear sporangia on their upper sides. The shoots of *Protolepidodendron*, from the Lower and Middle Devonian, are similar but more delicate. Its simple, forked leaves are arranged in a spiral sequence and are attached to the shoots by cushioned structures. Sporangia, which have an opening similar to that in ferns, are usually on the upper part of the shoots on specialized leaves. The generic name indicates that these plants bore the first characteristics of leaf cushions and dense, spiraling leaf patterning found later, albeit more fully developed, in Carboniferous Lepidodentrales trees.

Duisbergia (see Color plates, pp. 170/171 and 204) is an unusual lycopod; it probably looked rather cactuslike among its Carboniferous contemporaries. It differs from all other plants of that time in forming a trunk. The "trunk", however, is club-shaped, usually undivided, and consists of a tall cylinder with several layers of intercrossing vascular bundle strands that run freely into the ground at the bottom. The upper slender part of the trunk bears a thick coat of spatulate, often lacerated leaves. The tree grew to 1 or 2 m in height, and its breadth along the ground was approximately 30 cm. The general appearance and the arrangement of the leaves in vertical rows suggests a possible relationship to the Upper Carboniferous Sigillariaceae.

Lycopods attained their treelike status by the Upper Devonian at the very latest, and these plants also developed the ability to have a secondary growth spurt in the marginal areas of the vascular bundles of the woody portion. Unlike the weedlike lycopods of the Lower Devonian, the needle-shaped leaves of lycopod trees were cast off from older shoots. This left a spiral series of scars on the trunk. *Cyclostigma* (see Color plate, p. 204) was apparently a prevalent plant in many areas. It looked like a delicate lepidodendrid tree with a height of approximately 8 m. The leaves, which bore sporangia, were modified at the ends of young shoots into cones. They display another adaptation and advancement in lycopods: heterospores. Within one cone, but different sporangia, there were male and female spores, the smaller male microspores and the larger female megaspores.

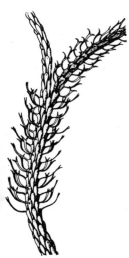

Fig. 8-7. Leaved shoot of *Protolepidodendron*.

Life in a Silurian sea:
▷▷
1. Algae
2. Bottom-dwelling coral
3. Various Rugosa corals
4. *Murchisonia*, a snail
5. *Cyclotropus*, another snail
6. A brachiopod, *Conchidium*
7. Another brachiopod, *Glasia*
8. A cephalopod, "*Orthoceras*", with a straight shell
9. *Cyrtoceras*, a cephalopod with an arched shell
10. A trilobite, *Aulacopleura*
11. A trilobite, *Cheirurus*
12. The "sea scorpion" *Eurypterus*
13. Numerous crinoids
14. A group of jawless fishes (*Phlebolepis*)

The great proliferation of lycopods in the Carboniferous reached its peak with the Lepidodendraceae (see Color plates, pp. 52 and 202/203) and Sigillariaceae (see Color plate, p. 202/203). The former have been reconstructed with the use of countless fragments to give the structure shown in Fig. 8-8. The tree bore some resemblance to a modern pine. Its powerful, unbranching trunk extended over 10 m above the ground and had a diameter of 2 m. It bore an umbrella-shaped crown whose terminal shoots and needles and cones. However, the resemblance to conifers was only superficial. Modern conifers have woody trunks with a thin bark, while the trunks of these prehistoric trees consisted almost exclusively of bark. The outer, stiffened parts of the bark acted to support the weak wooden body of these trees.

Toward the bottom, the trunk divided into four parts called stigmaria, and these can often be found intact at fossil sites. Root appendices were located spirally at the base of this trunk on forked trunk divisions; they were hollow, 1-cm thick tubes in the middle of which the vascular bundle transported nutrients on a support of connective tissue in an aqueductlike system. The crown of the tree was richly branched with many subdivisions. Needlelike leaves were on younger branches, and anatomically they resembled the leaves found on plants occurring in arid districts. This may seem rather surprising in view of the fact that Lepidodendraceae lived in swamps. This apparent contradiction can be explained by studying the roots, for they had little uptake capability; furthermore, the lack of cavities for water within the trunk and therefore the poor water supply system for the leaves required that they have a high retention capability. When the leaves were cast off from older branches,

▷
Phylogeny of fishlike species and fishes:
JAWLESS FISHES (AGNATHA):
A. Cephalaspidomorpha with cyclostomes (1: *Mayomyzon*), osteostracids (2: *Hemicyclaspis*), and anaspids. B. Pteraspidomorphi with heterostracids (3: *Anglaspis*) and thelodonts.
JAWED FISHES (GNATHOSTOMATA):
A. Placoderms with Antiarchi and Arthrodira (4: *Coccosteus*). B. Cartilaginous fishes (Chondrichthyes), including the chimaera (Holocephali) group with Iniopterygii, Chimaerida (5: *Helodus*) and Edestida, as well as elasmobranchs (Elasmobranchii) with freshwater sharks (Xenacanthida, 6: *Xenacanthus*), sharks (Selachii, 7: *Bandringa*), and primitive sharks

(Continued on p. 153.)

JAWED FISHES

AGNATHA

PLACO-DERMS

Cartilaginous fishes

ACAN-THODII

TELEOST FISHES

Chimaeras

Elasmobranchs

Spiny-rayed fishes

Lungfishes

Coelacanths

Holostei

Petromyzontida

1

2

3

4

5

6

7

8

9

10

11

12

13

Osteostraci

Anaspida

?

Heterostraci

Thelodonti

Antiarchi

Arthrodira

Iniopterygii

Chimaerida

Edestida

Xenacanthida

Selachii

Cladoselachii

Acanthodii

Chondrostei

Dipnoi

Actinistia

Onychodontida

Rhipidistia

?

Tetrapods

?

PERMIAN

CARBONIFEROUS

DEVONIAN

SILURIAN

ORDOVICIAN

CAMBRIAN

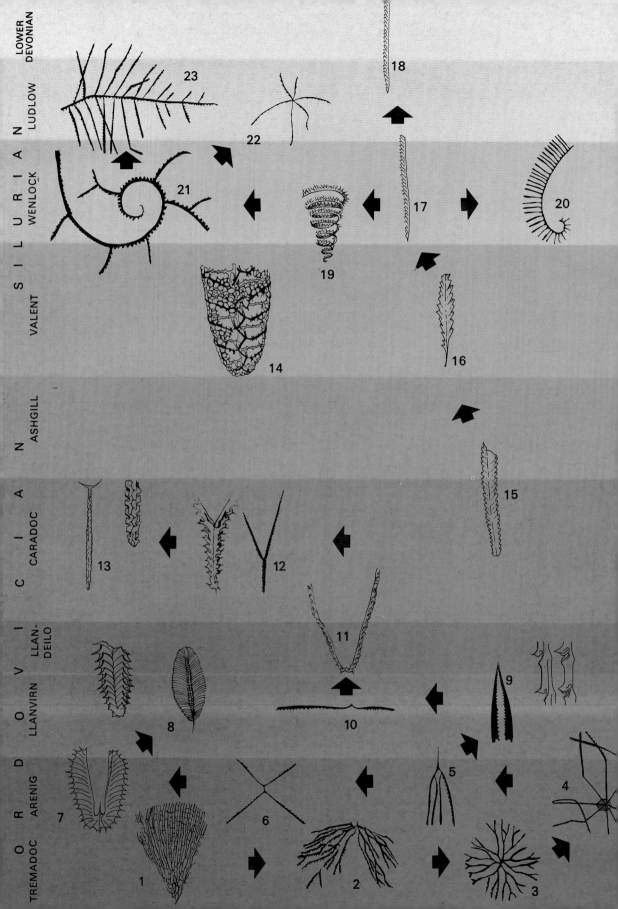

(Continued from p. 148.)

(Cladoselachii, 8: *Cladoselache*). C. Acanthodii
(9: *Acanthodes*). D. Bony
fishes (Osteichthyes),
including the Actinopterygii with Chrondostei (10:
Rhadinichthys), Holostei,
lungfishes (Dipnoi, 11:
Dipterus) and cross-
opterygians (Crossoptery-
gii) with Actinistia (=
Coelacanthida, 12:
Coelacanthus), Onychodontida, and Rhipidistia
(13: *Holoptychius*).
TETRAPODS (TETRAPODA)

◁
Distribution through time
and phylogenetic develop-
ment trends (arrows) in
some important graptoli-
thines:

diamond-shaped leaf cushions appeared on the back. In a close spiraling sequence, they covered the outside of the trunk and branches, forming a sort of armored covering. In fact, they gave the group its scientific name (from the Greek λεπιδοσ = scale).

The structure of the cushion (see Fig. 8-10) is an important feature. The leaf attachment site was located somewhat above the cushion with the central scar of the vascular bundle. Two scars within and two beneath the leaf attachment site comprise a strand system of parichnos tissue, which ventilates the inside of the bark, supplies the leaves with nutrients, and acts like the lenticells of modern trees. Lenticells refers to tissue with a great deal of intracellular space, replacing the orifice on corky plants. A ligule above the leaf attachment functions to take up water, and the absence or presence of this structure is the basis for classification of lycopods into ligulate and eligulate forms. Most treelike lycopods of the Paleozoic were ligulate.

The SIGILLARIACEANS (Sigillariaceae; see Color plate, p. 202/203) are closely related to Lepidodenraceae and resemble them in size, prevalence, and number of species. Unlike Lepidodendraceae, they had unbranched trunks or ones with one or two divisions at the top. The trunk was swollen at the base, a common phenomenon in swamp-dwelling trees. The ligamentous leaves grew to a length of 1 m and formed a thick bushy covering at the end of the trunk or its branches. The leaf attachment points were hexagonal and arranged in distinct longitudinal lines. As in Lepidodendraceae, they contained a vascular bundle and parichnos tissue. The actual leaf cushions and their scars could hardly be seen.

As far as we know, all treelike lycopods of the Carboniferous bore bisexual spores. The female megasporangia and the male microsporangia were usually held in one cone, less frequently in separate cones. In Lepidodendraceae the cones were at the tips of the youngest branches, while in sigillariaceans they were in the lower part of the leaf growth.

During the time in which they proliferated, lycopods developed female reproductive organs resembling those we find in seed plants. *Lepidocarpon* (see Fig. 8-11), from the Upper Carboniferous period, had cones in which the female megasporangia formed a nearly closed container with just a slitlike opening. Each spore case contained a single megaspore, which developed into a multicellular protonema within its enclosure. Fertilization probably took place within the capsule as well, since the sporangium and associated leaf were always one unit and were cast off together once a certain stage of maturity was attained. *Lepidocarpon* could be envisioned as a fruit that bypassed certain stages of seed development (e.g., the formation of a seed shell).

Most lycopods died out at the end of the Carboniferous. They probably could not tolerate the change from the moist Carboniferous climate to the drier climate of the Permian. Their water conducting passages were always small in proportion to their size. Only sigillariaceans survived into the Permian, and they probably led the way to the Mesozoic lycopodia, such as *Pleuromeia*, the chief lycopodian plant of sandstone. A few species did survive, even into modern times, and we know them as club mosses and selaginella (Lycopodiales and Selaginellales, respectively).

The EQUISETINAE (class Equisetinae; see Color plates, pp. 51, 170/171, and 202/203) had a similar distribution and adapted in much the same way to the environment as the lycocpods. After developing slowly in the Devonian, they flourished in the Upper Carboniferous and declined in the Permian. They were swampland inhabitants, as were lycocpods. Their major features were small leaves in a whorled configuration, shield-shaped sporangial bearers, and a radiating woody body with a substantial internal cavity. They seem to have arisen from species resembling the most primitive lycopods but developed from a branch in which the leaves gradually assumed their whorled relationship (*Protohyenia*, *Hyenia*; see Color plates, pp. 170–171). The arched spore cases of *Hyenia* and *Calamophyton* from the Middle Devonian may have been preliminary stages of what later developed into the shield-shaped sporangial bearers of the Equisetinales. In *Calamophyton* the short stem even appears to have been divided just as in equisetums, but recent investigations demonstrate that this genus belongs to earlier ferns. That means that we must begin looking for an ancestor to the equisetums all over again!

In the Upper Devonian, the time when we find the oldest confirmed equisetums, highly developed, stately treelike plants such as *Pseudobornia* (see Fig. 8-12) were already in existence. *Pseudobornia* attained a height of approximately 10 m and had a rich array of branches. In many respects it resembled the Carboniferous calamites. One to two branches spread out from each division node in the trunk (see Fig. 8-12), and the leaves on these branches were greatly ramified.

Equisetums reached their zenith in the Upper Carboniferous with the CALAMITE family (Calamitaceae; see Color plates, pp. 202/203). These articulated trees densely covered deltas. They had characteristic tubular trunks, and the centers thereof had large cavities toward which the vascular bundle tracts led. When calamite trees died, the hollow cavity was filled with depositions, and the vascular bundles were portrayed in relief. Calamites were the most widespread plants of the Carboniferous. The way in which they were fossilized has permitted us to understand the nature of their branching and to learn how the vascular bundles alternated on the trunk and branches. Various stemmy and busy calamites existed. *Stylocalamites* (see Fig. 8-13), with rigid, branchless stems, was one of the "stemmy" varieties, while *Eucalamites* (see Color plate, p. 51 and Fig. 8-13) belonged to the "bushy" group; all the nodes in the trunk bore branches. In *Calamitina* (see Color plates, pp. 202/203), nodes bearing branches were separated by several nodes without branches. Like modern equisetums, calamite stems were supported by crawling ground shoots, which were articulated like the stems and which bore true roots.

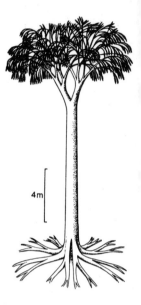

Fig. 8-8. Lepidodendraceae tree from the Upper Carboniferous.

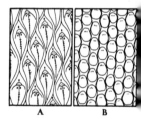

Fig. 8-9. Outside of the trunk of a Lepidodendraceae tree (A) and Sigillariacea tree (B).

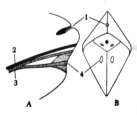

Fig. 8-10. Leaf cushion of *Lepidodendron*. A = longitudinal section. B = schematic view. 1. Ligula; 2. Leaf-trace bundle; 3. Parichnos strand; 4. Transpiration opening.

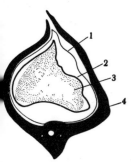

Fig. 8-11. *Lepidocarpon* in cross-section. 1. Sporangial wall; 2. Spore membrane; 3. Protonema tissue; 4. Supporting leaf.

Fig. 8-12. Twig of *Pseudo-bornia* from the Upper Devonian of Bear Island.

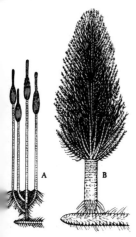

Fig. 8-13. Reconstruction of *Calamites*, an equisetum. a) *Stylocalamites*; b) *Eucalamites*.

The small, longish leaves of calamites were arranged in whorls and made a rosette pattern on the youngest branches. It was in the sporangia, however, that calamites truly revealed their diversity, for these spore cases were typically separated from the mother plant (at least they were found that way); in fact, the sporangia have only been assigned as belonging to calamites on systematic grounds! These geometrically regular but complex structures were arranged with whorls of infertile scales alternating with those of four-armed sporangial bearers, whereby both male and female sporangia appeared simultaneously (heterospores).

The immediate ancestors of calamites were in genus *Archaeocalamites* from the Lower Carboniferous period. These differed from calamites in that the vascular bundle ran through the nodules; furthermore, the leaves were very narrow and were forked.

Sphenophyllum (see Fig. 8-15) and others were members of a separate line. They occurred from the Upper Devonian to the Permian. They had whorled, cuneate leaves, which, unlike calamite leaves, had several veins. Their fossils distinctly display the three-rayed cross section of the trunk, which lacked a cavity. Fertile and infertile scales within the spore cases intermixed more than in calamites, and the spore cases seemed to rest on supporting leaves. These were probably climbing plants.

FERNS (class Filicinae; see Color plates, pp. 51/52 and 204) apparently underwent complex evolutionary processes in the Paleozoic, and there are still many open questions about their development. Descendants of psilophytes, they probably originated in the Lower Devonian. By the Middle Devonian, ferns had evolved into heterogeneous groups of fern-like growth known as old ferns. They lacked fronds in the usual sense and were a transition between the psilophytes and true ferns, and perhaps to the seed plants of the Carboniferous as well.

Walter Zimmermann has portrayed the development of the original psilophyte shoots into ferns as follows: 1. By a differential growth process, the originally equal plant branches developed into main shoots and lateral organs (e.g., leaves). 2. Ramifications developed only within one dimension (plantation). 3. Simple shoots merged together by means of connective tissue linkages, leading to the formation of leaf blades and the development of composite axes. 4. That was followed by the regression of organs (reduction).

In time two groups of old ferns arose. The older one contained the PROTOPTERIDS (see Color plate, p. 170/171), which were chiefly distributed in the Middle Devonian and soon attained the size of trees. In the small *Protopteridium*, a division into major and minor shoots was first indicated. Tree-shaped *Aneurophyton* continued the process of overgrowth and planation, and the last branches of its "frond" were in one plane. Short, compact stems were formed in *Cladoxylon* (see Fig. 8-16) by merging of several shoots, and this approached the structure occurring in younger ferns. The stem divided several times, and the younger shoots bore short, ramified leaflets and fan-shaped sporangia-bearing processes. An interesting primitive feature in this group were the uniformly structures fertile and infertile organs.

The protopteridia died out in the Lower Carboniferous, but a more recent old fern group—the COENOPTERIDS (see Color plate, p. 204)—remained until the Permian. Most of them were small, squarrose, bushy plants. Their regular branching patterns are beautifully revealed in serial sections from the dolomite and limestone deposits from Carboniferous strata. Ramifications are pairwise in many cases, and they alternate with two to the left and two to the right of the axis, oriented at a 90° angle to each other. In *Stauropteris* (see Fig. 8-17), a Carboniferous form, this pattern is repeated several times in successive branchings, forming eventually a bushy, three-dimensional frond. The final ramifications (the 5th or 6th order) are in one plane, but no laminae could form. However in the Upper Carboniferous *Etapteris*, only the first branches from the axis were in pairs and oriented at right angles; all others were in one plane and gave the frond a true fernlike appearance. Thus, the coenopterids can be considered as the fern group displaying a tendency to develop fronds through planation and merging growth, or at least as the group with the most pronounced expression of such a trend. Coenopterids contained species with homogeneous spores and those with heterogeneous spores. In the latter, megaspores were reduced to just two per sporangial case. The coenopterid group appeared to waver between evolutionary retreat and advancement.

Archaeopteris was a widely distributed tree in the Upper Devonian. The terminal branches of this plant grew together into simple fanlike structures and formed a kind of frond (see Fig. 8-18). Asexual and sexual organs appeared simultaneously on one axis, and at least some species had heterogeneous spores. Due to the frond structure, *Archaeopteris* would perhaps be considered to occupy an intermediate position in fern evolution, but recent evidence shows that there is a link between this plant and conifers. Thus, to some paleobotanists *Archaeopteris* is possibly ancestral to cordaitins and conifers from the Upper Paleozoic.

Characteristic fern foliage began proliferating in the Upper Carboniferous. Since these plants usually exist as fragments in their fossil form, they are classified on the basis of the structure of the last frond pinnule, and the vast majority of species can be divided into just a few pinnule classes (see Fig. 8-19). Among pinnate, parallel-edged leaves, a distinction is drawn between those in which the leaf is attached by a small element (see *Neuropteris* in Fig. 8-19), across its entire width (e.g., *Pecopteris*), or by having several veins grow together and form an attachment base (*Alethopteris*). Other forms were those with cuneate pinnules, such as *Sphenopteris* and *Mariopteris*. A few of these pinnules occasionally bore spore cases, while others, which have never been found with sporangia, belong to a group of seed plants we shall describe later.

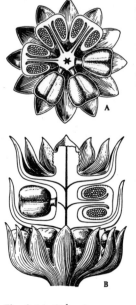

Fig. 8-14. Calamite blossoms. A: cross section; B: longitudinal section and view.

Fig. 8-15. *Sphenophyllum,* with sporangia.

Fig. 8-16. Bundling of several shoots gives *Cladoxylon* a short, compact stem.

Development of spore cases

Fig. 8-17. *Stauropteris*. Reconstruction of an infertile branch segment.

Modern ferns are classified on the basis of the structure of their spore cases. A distinction is drawn between EUSPORANGIATE FERNS (Eusporiangiatae) with spore case walls of several cell layers and LEPTOSPORANGIATE FERNS (Leptosporangiatae), with single-layered spore case walls. Both groups were present by the late Paleozoic, but their significance was the reverse of what it is today. The eusporangiate order Marattiales—occurring today in just a few tropical genera—was an important part of Carboniferous swamp forests. The group contained trees with trunks more than 10 m high and huge fronds up to 3 m long. These ferns are characterized by the fact that their spore cases have grown together into what are called synangia. Cross-sections of fossilized trunks, known as psaronia, have an unusual structure. The woody part of the plant, as in most ferns, is composed of several concentric individual bundles, whose number increases considerably from bottom to top. This part is surrounded by a thick coat of roots and leaf bases, which broaden and thus help the insufficient lower part of the stem support the plant. Unlike the stems and trunks of Lepidodendraceae and calamites, the stems of these ferns can be characterized as "leaf-root-stems". Nodules left from discarded fronds can be seen higher up on the stem. These gave rise to ramification. There are two types of eusporangiate ferns: *Megaphyton* has just two rows of fronds; in *Caulopteris* there are several rows of spiraling fronds. These ferns bore just a few crown fronds at a time.

Unlike the eusporangiate ferns, leptosporangiate spore cases bear a special opening mechanism or anulus used to disperse spores. The development of this opening and the orientation of the spore cases is of phylogenetic importance. Primitive spore cases are those not arranged in small aggregates and which have apical opening mechanisms. They appeared in the Upper Carboniferous on fronds of the *Pecopteris*-type, and those plants with such sporangia were called *Senftenbergia*. More highly developed forms also appeared in the Upper Carboniferous, including ferns like *Oligocarpia*, whose fronds were of the *Sphenopteris*-type. Their sporangia formed small aggregates and had a circular opening mechanism. The most highly developed ferns with distinct sporangia aggregates and a comblike opening are not known until the Mesozoic era, which was also the time the leptosporangiate families began proliferating.

One of the great achievements of paleobotanists has been the discovery of a widely distributed Carboniferous plant group (now extinct), which had the shape of tree ferns but reproduced like seed plants (ferns lack seeds), a group called Pteridospermae or Cycadofilices (see Color plates, pp. 51/52, 202/203, and 204). The study of this group was similar to solving a very complex puzzle, with scattered fossil organs in a large system for which we still lack all the pieces. The decisive step in the elucidation of this group was taken by two English botanists, F. W. Oliver and D. H. Scott, in the early 20th Century, when they demonstrated the relationships between various seeds, fronds, and woods (using types already known but for which many different names existed).

Two pteridosperm families (distinguished on the basis of the structure and appearance of the stem, the leaf pattern, and the nature of the seeds) existed by the Upper Paleozoic. They are grouped about their characteristic representatives, *Lyginodendron* and *Medullosa* (see Color plate, p. 51 and Fig. 8-22). The former has thin stems with a diameter of just 4 cm but also has a firm pulp surrounded by a radiating woody body. The outside of the rather thick bark was supplemented by longitudinal rows of stone cells, which bore spines. These characteristics all indicate that *Lyginodendron* was a climber. The stems bore .5 m, finely veined fronds of the *Sphenopteris*-type, which were forked once near the base. The 5-mm seeds were born on thin, spined stalks in bell-shaped casings. A "pollen chamber" was located on the free end between the seed shell and female sporangium, and male pollen was taken up in this chamber. It germinated there and fertilized the egg below, a process much like that occurring in the extant and probably related) cyadacean (palm fern) family. The male blossoms of *Lyginodendron* were in all likelihood the microsporangia also known as *Crossotheca*. They were sometimes found on the outermost ends of several *Sphenopteris*-type fronds.

The *Lyginodendron* group includes a number of other finds, including certain stems with stone cells in the outer edges of the stems but were otherwise quite uniform (*Heterangium*). Various seed organs in which two or more seeds were carried within a beaker-shaped hull have also been ascribed to this group.

The MEDULLOSA (from the Latin *medullosus* = pulp-transporting) whose tree-shaped structure and stem anatomy differs considerably from *Lyginodendron*, were a diverse group (see Color plate, p. 202/203). As in fern trees (*Psaronius*; see Color plate, p. 52), the woody body of *Medullosa* was composed of many bundles of tissue individually embedded in the pulp; their size increased toward the exterior, and in some species a true wooden cylinder was formed. On the outside, the entire woody body was enclosed by a coat of leaf bases and aerial roots. Leaf patterns in *Medullosa* were of the *Neuopteris*-and *Alethopteris*-types. The male sporangia were suspended from them, and some of these blossoms were prominent structures in bell shapes up to 4 cm in diameter, formed by the merging of hundreds of tubular spore cases into one common organ. Medullosan seeds were sometimes quite large themselves. Unlike the seeds of *Lyginodendron*, the seed shell did not merge with the female sporangium but separated when mature into an outer fleshy portion and a hard inner aspect. Most of these seeds have been found as three-sided kernels (*Trigonocarpus*), which at first look like Brazil nuts. Their prevalence and size (length up to 8 cm) are the most impressive evidence of seed formation in the Carboniferous and Permian that we have. There is virtually no doubt that pteridosperms arose from heterosporous ferns toward the end of the Devonian.

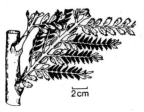

Fig. 8-18. *Archaeopteris.* Reconstruction of a fertile branch.

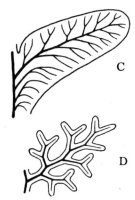

Fig. 8-19. Pinnae of Upper Carboniferous ferns: A. *Neuropteris*; B. *Pecopteris*; C. *Alethopteris*; D. *Sphenopteris*; E. *Mariopteris*.

Fig. 8-20. Seed organs of *Lyginodendron*. The seeds are about 5 mm in size.

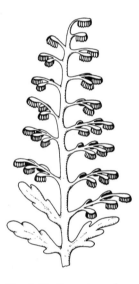

Fig. 8-21. *Crossotheca*, the male blossom of *Lyginodendron*.

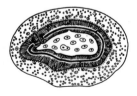

Fig. 8-22. Schematic cross section of a stem of *Medullosa*.

Seed evolution generally occurred in the following way: the number of female spores in heterosporous plants was reduced until each sporangium bore only a single female spore. In the course of further phylogenetic development, the spore itself regressed, and the egg cell remained there to be fertilized. Finally the female spore in its spore case remained on the mother plant until the germ cell of the following generation had formed. Eventually, with still further evolution, the sporangium was enclosed by a protective layer, or integument, which arose by evaginations at the base of the sporangium. The entire sequence from loose surroundings to finger-shaped outgrowths to complete envelopment of the female sporangium has been demonstrated in finds from Lower Carboniferous Scotland. Naturally, not all seed plants underwent the same evolutionary development.

As seed plants with fern features, pteridosperms at first seem to be the ideal link between true ferns and gymnosperms. However, the only modern plants known with certainty to have descended from pteridosperms are the rare palm ferns (Cycadaceae) and bennettitids (Bennettitidae). Present-day conifers (Coniferae; see Color plates, pp. 52 and 204) came from a separate old line derived either from lycocpods or primitive ferns; paleobotanists still debate this question, and the ancestors of conifers are not known for sure.

One group sharing a common origin with conifers are the CORDAITINS (Cordaitinae; see Color plates, pp. 202/203 and 204) of the Upper Carboniferous. They were the giants of the Carboniferous forests, for they grew to a height of nearly 50 m. Unbranching "fragments" of trunks more than 20 m long have been recovered. The chambered pulp of the trunk and branches and the transverse ribbing of the pulp cavities have been retained in fossils. In other respects, cordaitin trunks have a very similar structure to ancient conifers (e.g., *Araucaria*). The long, parallel-veined leaves sit on younger branches. Superficially they look like iris leaves. Differences in leaf length and relative width are used as criteria for classifying cordaitins into several species. The largest leaves have a length of approximately 1 m and a width of more than 10 cm. Many leaves often appear together, covering large surfaces on the plant. As in the medullosa, cordaitins have a two-layered seed shell, a firm inner layer and a fleshy outer layer. Due to their heart-shaped configuration, the seeds are known as *Cardiocarpus* (from the Greek καρδία = heart and καρπός = fruit).

True CONIFERS first appeared at the turn of the Carboniferous to the Permian. Their oldest representatives, *Walchia* from the Lower Permian, were relatively small and in their branching pattern and needle configuration resembled *Araucaria*. Conifers then developed rapidly into a diverse group and formed the major forest flora by the Upper Permian. The most important genera were *Ullmannia* and *Pseudovoltzia* (see Color plate, p. 204 for the former). Younger, needle-bearing branches resembling those found today, are the most frequently recovered fossils. Some branches bore long needles, while others had short ones, and these are sun and shade branches belonging to the same species. Unlike Carboniferous cordaitins, Permian conifers had distinct concentric growth zones or annual rings, and these are indices of annual climatic changes.

Investigations on female cordaitin cones and primitive conifer woods have yielded new information on the phylogeny of conifers. Fig. 8-25 illustrates the development of the female pine cone. For nearly a century, botanists argued vigorously over whether the female cones of modern conifers were individual blossoms or stands of blossoms. Paleobotanical finds now support the latter theory. In cordaitins, budlike shoots appear, two at a time, in the axes of the leaves. They cover several seed structures between spirally arranged scales. Cones developed gradually in a process of compression of the group of blossoms, degeneration and flattening of the short shoots, and modification of the scale covers, as shown in Fig. 8-25. Primitive Permian conifers display some of the possible intermediate stages and other independent developments in this process: in *Walchia* (Lower Permian), the short shoots are to some extent flattened, but they still consist of several scales and seed structures; in *Ernestiodendron* (Upper Permian), they have just one scale but several seed structures, while in *Pseudovoltzia* the short shoots consist of three scales, each with one seed. The greatest degree of regression is seen in *Ullmannia*, in which one scale covering surrounds one seed apparatus, comprising just a single shoot. Modern conifers likewise have a covering scale, fruit element, and seed on a single, vestigial shoot with leaf, and these correspond to the comparable structures in cordaitins.

The first GINKGOS (class Ginkgoinae) appeared shortly before the end of the Paleozoic, but they did not develop and proliferate as rapidly as conifers. They flourished in the Mesozoic. The forked needles of the earliest ginkgos bore only a slight resemblance to modern ginkgo leaves. However, studies on the structure of the outer leaf layer confirm the phylogenetic relationship between these trees.

Our description of terrestrial plants of the Paleozoic has largely been restricted to flora of Europe and North America. Floral evolution proceeded rather uniformly throughout the entire Paleozoic world. Clear regional differences do not appear until the Upper Carboniferous period. Carboniferous flora of the European and American Upper Carboniferous are designated as Euramerican flora. Gondwana plants were entirely different in the Upper Carboniferous. In Gondwana, the great southern continent, such major groups as Lepidodendraceae, Sigillariaceae, calamites, and large-frond ferns were absent. Characteristic Gondwana plants included woody plants with large, veined leaves, some of which (e.g., *Glossopteris*; see Color plate, p. 52) had a central vein and some of which (e.g., *Gangamopteris*) did not. Organs have been found at the base of *Glossopteris* leaves, and whether they are seeds, hermaphroditic organs, or male organs has not been clarified. These two genera were probably

Fig. 8-23. *Neuropteris* frond with male blossom.

Fig. 8-24. Cordaitin branch with several male and one female blossoms.

Fig. 8-25. Phylogenetic development of female conifer cones. A. *Cordaites* (Upper Carboniferous); B. *Ernestiodendron* (Lower Permian); C. *Pinus* (modern).

pteridosperms. Since they occurred in regions with coal deposit formations and in areas with glacial evidence, they were probably cool to temperate flora. They were temporally distributed in parts of the Carboniferous and Permian (see Color plate, p. 222/223), which is approximately comparable to the temporal distribution of northern continental flora.

The flora of what had been the Angara continent, now Siberia, were coal-forming species. These plants were chiefly represented by the cordaitin *Noeggerathiopsis*. The Cathaysia flora of southeast Asia (also primarily from the Permian) were principally characterized by the large-frond *Gigantopteris*).

Although our knowledge of Paleozoic flora may seem extensive, there are many open questions and missing links. Reproduction is one physiological aspect difficult to study in certain respects. The delicate tissues participating in sexual reproduction almost always decayed before fossilization could occur, and only sporophytes were hard enough to permit close analysis. In the most highly developed Paleozoic plants (e.g., *Lepidocarpon*), the rudimentary sexual parts on sporophytes are carried within protective seed packets, and this adaptation has permitted fairly detailed analysis of their structure. Comparable structures in more primitive terrestrial plants have never been found.

Practically nothing is known about the phylogeny of mosses and fungi, since these plants decayed so rapidly once they died. A few fortunate finds show that Hepaticae mosses (liverworts) occurred in the Upper Carboniferous. FUNGI have been found with the oldest terrestrial plants, and this indicates that their development was highly dependent on the simultaneous development of other terrestrial plants. In any event, nearly all major groups of flora had developed in the Paleozoic and thus had accomplished the most vital phylogenetic steps. We report on the distinctive flora of the Mesozoic and Cenozoic in later chapters. Flora took a step into the "modern" era with the angiosperms of the Upper Mesozoic.

9 Invertebrate Evolution in the Paleozoic Era

The extent to which we can learn about animal evolution (origin of animal groups, proliferation thereof and changes undergone in prehistoric periods) is dependent on how extensive the fossil record is for those animals. Some animal groups do not form hard body parts, and these therefore do not leave any fossil record. They will not be discussed in the subsequent chapters. These animals comprise the following phyla: Mesozoans (Mesozoa), Acnidarians (Acnidaria), flat worms (Platyhelminthes), kamptozoans (Kamptozoa), round worms (Aschelminthes or Nemathelminthes), Priapulid worms (Priapulida), sipunculid worms (Sipunculida), echiurid worms (Echiurida), water bears (Tardigrada), and Linguatulida.

By O. H. Walliser

Of the PROTOZOA (Protozoa), the only Paleozoic class for which there is any substantial record is that of RHIZOPODS (Rhizopoda). These consist of just two orders, the foraminifers (Foraminifera) and the radiolarians (Radiolaria).

Unicellulates

FORAMINIFERS (Foraminifera), which from the Jurassic have been among the most prevalent fossils appeared in the Cambrian, were still relatively rare in the Ordovician, and first flourished in the Carboniferous and Permian periods. All major foraminifer body plans evolved in the Paleozoic, undergoing the following major evolutionary changes: the earliest foraminifers were predominantly textulariins. These were so-called agglutinated forms for they mixed a foreign material (typically sand) with their own organic substance (tectin) to produce a shell. As the group evolved, calcium carbonate (or, in rare instances, silicic acid), was incorporated in the cement. Species secreting calcareous shells occurred very early, but they did not become important (with the exception of the fusulins) until the Jurassic. A second development was coiling of the shell and the formation of chambers within it. The earliest foraminifers were single-chambered organisms with a regular or irregular spherical to tubular shape. Multi-chambered shells evolved later. Coiling, either in one plane or in a spiral, was also a relatively late development.

Foraminifera

From the very beginning, foraminifers crawled along the ocean bottom

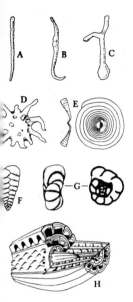

g. 9-1. Foraminifers
oranimifera):
. *Bathysiphon*;
. *Maripella*;
. *Saccorhiza*;
. *Astrorhiza*;
 Ammodiscus;
 Textularia;
. *Endothyra*;
. Schematic representation
 Triticites, showing the
ternal fusuline structure.

(they were benthonic). The number of planktonic species at any time in earth's history has never been very high. However, sessile foraminifers were not at all rare. Successive generations even built miniature reefs. Most fossil and extant foraminifers lived in shallow seas.

The oldest foraminifers were the TEXTULARIINS in the "sand-shelled" group of AMMODISCACEANS (Ammodiscacea). They had irregular, simple or branching, tubular shells and resembled the modern genus *Rhizammina*. Spherical species arose in the Ordovician; in some the openings were elongated by tubes (*Astrorhiza*), while in others the tube was connected to a small chamber (*Saccorhiza*). Coiled foraminifers appeared in the Silurian, represented chiefly by the genus *Ammodiscus*. Much later, in the Carboniferous period, complex forms with multi-chambered shells and different shell layers appeared (superfamily Lituolacea). Their chambered sand or calcareous shells display their ammodiscid heritage. *Textularia* was a significant genus; its spirally coiled initial shell section was joined to a two-layered series of chambers. The spiral juvenile shell was an evolutionary relict, and in later species it disappeared.

The FUSULINS (suborder Fusulina) were of particular importance. Numerous small pores were distributed across the entire shell except at the mouth, and these pores were the outlets for amoebic pseudopodia. The shell crystals in advanced fusulins were arranged with their optical axis perpendicular to the water surface, making the shell transparent. Thus, the group is sometimes called Vitrocalcarea (= glass calcium-shelled animals) The earliest (Ordovician) fusulins were simple spherical or tubular forms, but some had chambers. The first multi-layered ones arose in the Middle Devonian (*Semitextularia*), which were structured similarly to *Textularia*. They were followed, in the Upper Devonian, first by the family of ENDOTHYRIDS (Endothyridae), which were simple planospiral coiled, chambered species that have been used for biostratigraphic classification. They gave rise to the FUSULINACEAN superfamily (Fusulinacea), which excepting the small TETRATAXID (Tetrataxidae) group was restricted to the Carboniferous and Permian periods. Beginning in the Upper Carboniferous, they flourished to such an extent that they became reef builders in certain shallow water districts. Fusulinaceans were also planospiral coiled forms, but unlike endothyrids they had a long coil axis (much like the leaves of a cigar are wrapped about their axis). The shell was divided into chambers by partitions. Transverse and longitudinal walls sometimes developed within the chambers, and the number, orientation, and configuration of these walls differed among the various species. This great diversity in anatomy, plus differences in external anatomy and size differences from 0.5 to 70 mm, permit a fine analysis of the rapidly developing species. Within their facies, fusulinaceans are key fossils for the Upper Carboniferous and the Permian.

Of the other foraminifer groups, MILIOLINA and ROTALINA, only a few precursors arose in the Upper Carboniferous and Permian. They bore imperforate shells. The optical axis of the calcite crystals in the shell was parallel to the surface of the water, giving the shell a porcelainlike appearance rather than a glasslike one. Some systematists refer to the group as Porcellanea for this reason. The Paleozoic was a clearly distinct unit regarding foraminifer evolution, characterized by the predominance of "sand-shelled" forms, the simple structure of the Vitrocalcarea, the extinction of most fusulinacean families at the end of the Permian, and the presence of just a few porcellanean species.

RADIOLARIANS (order Radiolaria) are marine planktonic rhizopods with a calcareous internal skeleton. They existed as early as the Upper Precambrian, and in the course of geological history they have been found in deposits designated as lydites or radiolarites. Their skeletons often made a substantial contribution to the formation of these deposits, just as the case is now in the radiolarian mud of the deep seas. However, prehistoric radiolarian deposits could develop in shallow waters as well, unlike the present condition. Radiolarians were particularly prevalent during periods of volcanic activity, when the silicic acid content of the water was increased as a result of volcanic emissions.

Only two radiolarian suborders are known from the Paleozoic, Spumellaria and Nassellaria. The phylogenetic pathways of the group are not well understood in this period. Generally, however, the primitive species had rather simple skeletons. More complex forms did not appear until the Mesozoic.

Spongelike fossils have been found in the Lower and Middle Cambrian. These ARCHEOCYATHIDS (phylum Archaeocyatha) were solitary or were colonial and built reefs, and during the Cambrian they were distributed throughout the world. Archeocyathids cannot be classified with any other group. They were formerly known as Pleospongia and were thought to belong to the sponge phylum, but this was because their fossil deposits were found with sponge spicules within their hollow cavities. Actually, close investigation revealed that the spicules had fallen there from true sponges after the archeocyathids had died and were thus not part of their bodies. Since their skeletal structure deviates substantially from that of sponges, it is reasonable to grant them status as a separate phylum, which probably is between the sponges and coelenterates. The skeleton forms a spherical to conical funnel that may be over 10 cm long. In the earliest species it consisted of two walls separated by transverse septae, the animals anchored themselves to their substrate with lamellaelike outgrowths.

All archeocyathid walls had pores. This suggests that they fed by attracting water (which transported food substances) through the pores and into the central body cavity, and from there the water flowed out through the large opening at the upper end of the body. Water current was probably created by specialized cells corresponding to collar cells in sponges or by ciliated epithelium.

The archeocyathid body plan is best represented by the genera

Radiolarians

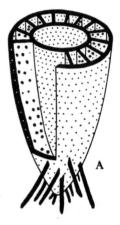

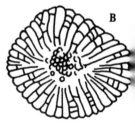

Fig. 9-2. A. Schematic diagram of a simple archeocyathid (*Ajacicyathus*); B. Cross-section through the complex structure characterizing genus *Anthomorpha*.

Archeocyathids

Archaeocyathus and *Ajacicyathus*. The latter is depicted in Fig. 9-2A. Other genera have different wall structures. Pores differ in size and arrangement, or they can be so distended that the walls can consist of just a thin network of hard parts. The same is true of the septae, which may be reinforced by additional septae running parallel to the exterior wall. In the ANTHOCYATHEANS (Anthocyathea), the central cavity is filled by a vesicular internal skeleton. These species (e.g., *Anthomorpha*; see Fig. 9-2B) can in cross-section greatly resemble coral. The MONOCYATHEANS (Monocyathea) are an exception, for they have just a single wall.

Sponges

The basic structure of fossil SPONGES or PORIFERS (phylum Spongia) does not differ from that of modern species. Since they are described at length in Volume I, we shall forego a general description here. Individual spicules have been found in various places as early as the Upper Precambrian. We know with certainty that all three sponge classes were represented by the Cambrian.

In GLASS SPONGES (class Hexactinellidae) the three-axis, silicic acid skeletal spicules are not interconnected. The skeleton usually decays after the death of the animal. Thus, whole fossil glass sponge skeletons are rare finds. The oldest fully intact species is from the Lower Cambrian to Ordovician genus *Protospongia* (see Fig. 9-3A). Since such fossils have been found in North America, Europe, and China, it is probable that the species was distributed throughout the world. The skeleton produced a ball-shaped body, which was anchored to the floor by a relatively long shaft composed of skeletal needles.

Hydnoceras (see Fig. 9-3B), an Upper Devonian to Carboniferous genus, bore a certain resemblance to the modern Venus'-flower-basket (genus *Euplectella*). The wall was composed of a thin network of regularly arranged needles. *Artraeospongia* and related species had very firm skeletons, because their needles were densely packed. Fully intact specimens of these have been found in deposits from the Lower Cambrian to the Carboniferous.

The COMMON SPONGES (class Demospongiae) includes the family BORING SPONGES (Clionidae). Boring sponges lay down a series of passages just beneath the top of calcareous shells or rocks, forming a labyrinth, and they maintain contact with the surface by means of their pores. The construction and enlarging of the labyrinthine passages is carried out by dissolving the calcium present. Clionid sponges have thus contributed substantially to the destruction of large sediment deposits. This activity has been traced as far back as the Silurian. Most common sponges, however, produce depositions. This is particularly true for LITHISTID SPONGES (order Lithistida), whose irregularly shaped, thorny skeletal needles can interlock so strongly that they do not lose their integrity after the sponge dies. Well-known Paleozoic genera include *Cnemidiastrum* (Ordovician to Jurassic), *Aulocopium* (Ordovician to Silurian; see Fig. 9-3C), and *Doryderma* (Carboniferous to Cretaceous).

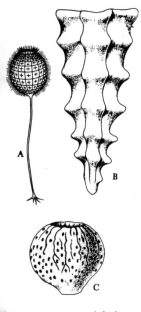

Fig. 9-3. Sponges (phylum Spongia): A. A hexactinellid (*Protospongia*); B. A siliceous sponge (*Hydnoceras*); C. A lithistid sponge, *Aulocopium*.

In Calcarea (a class), the calcite spicules are free in the soft body, and therefore the likelihood of intact fossilization is rather low. Only one genus, *Camarocladia* (a small branched form), is known from the Lower Paleozoic. Another, *Scribroporella*, existed in the Middle Devonian. Calcarean sponges were relatively uncommon in the Carboniferous and Permian as well. Only a few PHARETRONIDS (order Pharetronida) and THALAMIDS (from the exclusively fossil order Thalamida) have been found in these periods.

Of the two COELENTERATE (subsection Coelenterata) phyla, the only certain Paleozoic finds are CNIDARIANS (phylum Cnidaria). Other finds previously thought to be acnidarians (phylum Acnidaria) are now being interpreted differently. The oldest cnidarians are from Ediacarian deposits from the Upper Precambrian (see Chapter 6). Many species are seen there in relief, and they are all medusae. It is not yet possible to assign them to appropriate classes. This is even truer for the known forms from the Middle Cambrian to the Ordovician, all of which are categorized together as PROTOMEDUSAE. Many of these "medusae" may actually be spoor (or traces, tracks, droppings, etc.) of entirely different organisms!

Coelenterates

The CONULARIANS (subclass Conulata) are a Middle Cambrian to Upper Triassic group within the jellyfish or scyphozoan class. Fig. 9-4 depicts one of these. The pyramidal shell had a very thin chitin or chitin-phosphate wall. The outer surface usually had fine longitudinal and transverse ridges. A furrow formed on each of the four sides of the pyramid. The mouth in most species could be closed or at least constricted by four lateral projections of the pliable housing. Septae spread from the four walls to the middle of the body. Thus, these shells were comparable to a polyp genus (*Stephanoscyphus*) belonging to a modern deep sea medusa order (Coronata). Conularians included free-swimming forms and sessile ones, which rested on the tapered tip of the shell. Adults attained a L of 6–10 cm, but especially large ones over 20 cm have been found.

Many zoologists classify STROMATOPORES as HYDROZOANS (class Hydrozoa). One criterion supporting this classification is the (so-called trabecular) fine structure of the skeleton. It is quite unlike the skeletons of sponges and archeocyathids (although other zoologists have classified stromatopores with these groups). Stromatopores first appeared in the Cambrian, and they flourished in the Silurian and Devonian, after which time they declined rapidly. Only a few species are known from the post-Devonian period to the Cretaceous, and even these are of questionable systematic position. In the Mesozoic they played a subordinate role among the fauna of that time, but in the Paleozoic, stromatopores were among the most important reef builders. They were colonial, living in knobby or flat groups that could be one meter high; they also occurred in crust-forming or branched colonies. The calcareous skeleton had parallel lamellae on the surface, and they were supported by rods. The differences between individual species and groups do not exist in terms of external

Fig. 9-4. Schematic sketch of a conularia.

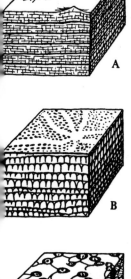

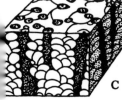

ig. 9-5. Stromatopores:
. *Stromatopora*; B. *Cla-*
rodictyon; C. *Labechia*.

Anthozoans

Rugose coral

appearance, for stromatopores could alter their external appearance significantly in adapting to one set of environmental conditions or another. Differences in the current, the substrate, or even the neighboring colony could produce external changes in stromatopores, and members of one species could look quite different in different surroundings. The crucial distinguishing feature has been the fine structure of the various species, which is analyzed by taking thin sections, as we schematically depict in Fig. 9-5. Important features include the arrangement and number of building elements and the extent to which they are developed. For example, one of the most prevalent genera, *Actinostroma*, is a good example of the basic stromatopore body plan. In *Clathrodictyon* (see Fig. 9-5B), the lamellae are very prominent, but the rods are shortened. *Stromatopora* (see Fig. 9-5A) has a spongelike, irregular skeleton, while *Labechia* (see Fig. 9-5C) has thin lamellae surrounding the vesicular cavities, and the lamellae are between powerful, thick rods.

Many species have "star roots" or astrorhiza. These are fine, vertical canals that spread out in a star shape on the lamellae and branch. These features were once used in systematics, but more recent studies have shown that these "characteristics" are in reality the spoor of wormlike animals in a commensal relationship with the stromatopore colony!

The ANTHOZOAN class (Anthozoa) includes what is now a worldwide subclass, HEXACORALS (Hexacorallia), which made its phylogenetic debut in the Mesozoic. The OCTOCORALS (subclass Octocorallia) are also largely restricted to the later eras. We know of just one Permian genera, *Trachypammia*, which may be an octocoral group. A few species from Ediacarian fauna were formerly classified with these anthozoans, but recent research has shown that, on the basis of better preserved specimens, these species belong in another phylum.

Anthozoans are represented in the Paleozoic by just two groups, which were probably confined to this era. They were the RUGOSE CORALS (subclass Rugosa) and the TABULATE CORALS (subclass Tabulata). Rugose corals differed from the Mesozoic Madreporaria coral (an order) in the arrangement of septae. The differences are shown schematically in Fig. 9-6 (A and B). Basically, rugose corals have six septae, while Madreporaria undergo a cyclical septal formation process whereby younger generations produce more septae in the cavities between the older septae. For this reason, the group is also called Cyclocorallia. In rugose coral, later septae are laid down in a pinnate pattern on the four to six original septae; thus, this group is also known as PTEROCORALLIA or TETRACORALLIA.

One general evolutionary trend among rugose coral was the degeneration of the outer wall and the building up of additional skeletal elements, either from low transverse elements (tabulae) or from arched, overlapping elements (dissepiments). The dissepiments formed a seam known as a dissepimentarium around the space (or tabularium) enclosed by the transverse elements. Transverse beams (synapticles) and posts (pali)

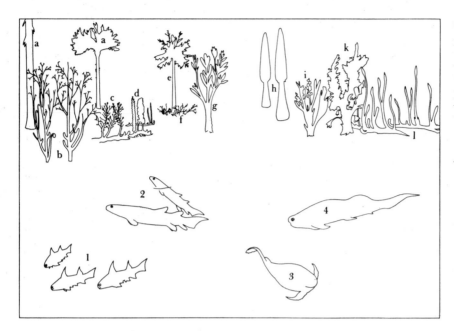

▷▷
Middle Devonian life:
Plants: Psilophytes (a, *Pseudosporochnus*), lycopods (b, *Asteroxylon*; l, *Protolepidodendron*; h, *Duisbergia*), Equisetales (c, *Protohyenia*; d, *Hyenia*), Primofilices (g, *Calamophyton*; i, *Cladoxylon*; k, *Barrandeina*; f, *Protopteridium*; e, *Aneurophyton*). Fishes: Spiny sharks (1, *Diplacanthus*), placoderms (4, the Arthrodira *Coccosteus*; 3, the antiarch *Asterolepis*) and crossopterygians (2, the rhipidistian *Osteolepis*).

▷
Development of tetrapods from rhipidistian crossopterygians
1. *Holoptychius*, one of the holoptychiids: two external nasal openings.
2. *Eusthenopteron*, an osteolepidid: one of the external nasal openings has become modified into a tear duct.
3. *Ichthyostega*, an Upper Devonian tetrapod with fish characteristics: α, vestige of the gill cover; the shoulder is separated from the head; β, fin seam with true rays; γ, scales.
4. Comparison of the fore limb skeletons of a) *Eusthenopteron* and b) an amphibian (*Eryops*): Hu = humerus; Ra = radius; Ul = ulna.
5. Tooth of an osteolepidid (*Panderichthys*), the structure of which corresponds completely to that of the tetrapod *Ichthyostega*.
▷▷▷▷
Reconstruction of the brain, nerves, and blood vessels of extinct animals:
Explanation of colors and abbreviations:
Brown: bones and ossified internal skeleton;
Orange: brain chamber or brain;

(Continued on p. 173.)

also appeared, as well as a special structural part (the columella) in the axial region. These skeletal elements displaced the septae. All skeletal parts were composed of very small individual parts with a calcium-producing center and, from that, calcite fibers radiating outward. They formed rows of plates (or trabeculae) in the septae.

Rugose corals were solitary or colonial, solitary forms prevailing in deeper waters. Branched colonies were more prevalent toward the surface, while knobby colonies were encountered in shallow agitated waters. The earliest rugose coral finds are from the Appalachian geosyncline of the Middle Ordovician. All three suborders were probably represented by this time. Ordovician rugose corals lacked dissepimentaria, which developed independently in all three suborders during the Lower Silurian. This development occurred simultaneously with the increased propagation of the rugose corals as reef builders. Another great anatomical modification occurred during the late Lower Devonian, when rugose corals underwent another stage of greater distribution of the reefs. Numerous families became extinct at the turn of the Devonian to the Carboniferous period. They were replaced by newer ones, all of which had the columella axial structure. Finally, all rugose corals, excepting the cyathaxoniids (Cyathaxoniidae) died out in the Artinsk stage of the Permian. Hexacorals apparently evolved from these; they appeared in the Mesozoic era.

The major members of the suborder STREPTELASMATINA are the genera *Petraia* (Silurian; no additional skeletal elements between the septae), *Zaphrentis* (Devonian; with dissepiments), and the colonial genera *Phillipsastrea* (Devonian) and *Hexagonaria* (Devonian to lower Carboniferous). The last one looked externally very much like the genus *Columnaria*

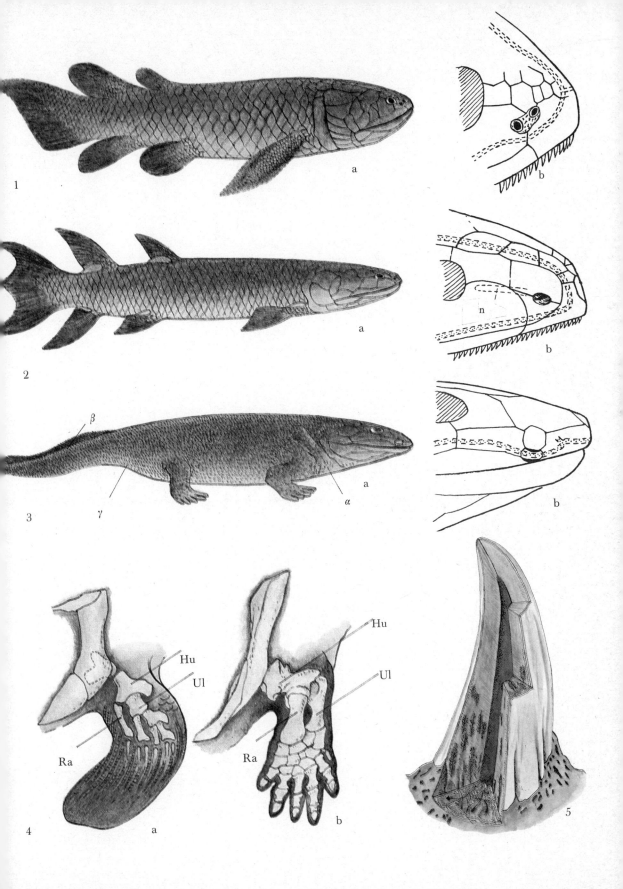

1

a

b

2

a

n

b

3

β

γ

α

a

b

4

Hu

Ul

Ra

a

Hu

Ul

Ra

b

5

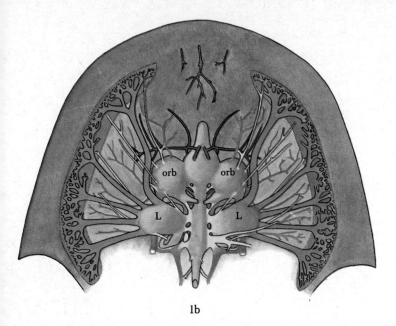

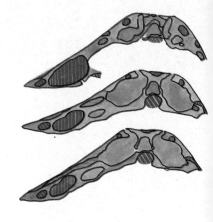

1a

1b

orb   orb

L   L

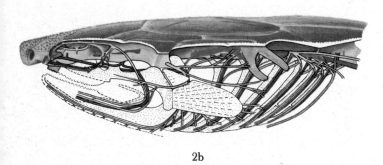

2b

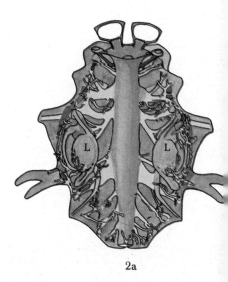

L   L

2a

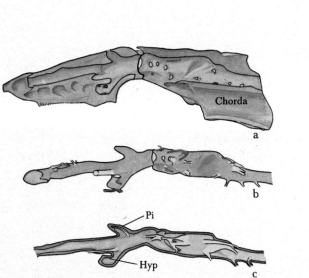

Chorda

a

b

Pi

Hyp

c

3

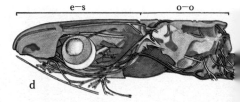

e—s   o—o

d

(Continued from p. 168.)

Yellow: nerves;
Green: nerves supplying sensory fields of the Cephalaspidomorphi;
Red: arteries;
Blue: veins;
L: equilibrium labyrinth;
Hyp: hypophysis;
orb: Optical center of the brain;
Pi: pineal organ;
e-s: ethmosphenoidal (nose-eyes) segment;
o-o: otico-occipital (ear-rear of head) segment;
A. Serial cutting method (particularly used by the Stockholm school):
1. Cephalaspidomorph jawless fish:
a) Three sections across the longitudinal axis in the hind part of the head (*Mimetaspis*).
b) Structure of the brain, nerves, and blood vessels within the head (*Kiaeraspis*).
2. Arthrodira
a) Structure of the brain, nerves, and blood vessels within the head (*Kujdanowiaspis*).
b) Hypothetical reconstruction of the gill skeletal system and its nerve and blood supply, based on the skull roof and the internal skeleton, which has the orifices through which nerves and vessels pass.
B. "Mortar" method:
3. Osteolepidid (a rhipidistian crossopterygian)
a) Section through the inner skeleton of the head (*Ectosteorhachis*).
c) Reconstruction of the brain under the assumption that the brain filled the entire cavity (*Ectosteorhachis*).
d) Inner skeleton of the head with reconstruction of the vessels between the various openings (*Eusthenopteron*).

(Devonian), a member of the second rugose coral suborder, COLUMNARIINA. The CYSTIPHILLINA included some distinctive solitary forms, which had small caps and could close themselves off (a phenomenon occurring in the Middle Devonian genus *Calceola* and the Silurian genus *Goniophyllum*). *Heterophyllia*, a genus restricted to the Visé stage of the Lower Carboniferous, was an exception. In this genus, only four septae developed, and in earlier growth stages they were branched.

The TABULATE CORAL subclass (Tabulata) consists exclusively of colonial forms. Numerous tabulae (transverse plates) transected the coral tubes. Septae were largely vestigial. The skeletal material was, as in rugose coral, built up chiefly from calcite fibers. Pores were present in the walls of some tabulate coral, giving individual organisms direct contact with each other. In others, an intermediate tissue formed between individual corals.

The oldest tabulate corals preceeded rugose corals; they occurred in the Middle Ordovician Appalachian geosyncline. *Favosites* is one of the representative genera; it occurred in branched colonies and in knobby colonies. Some of the colonies could have a diameter exceeding one meter. The individual animals were only a few millimeters thick. *Pleurodictyum* (see Fig. 9-7C) had a very uniform shape. The species *Pleurodictyum problematicum* (Lower Devonian) often lived together with a "worm", which influenced the growth of several individual coral organisms. An extensive intermediate tissue developed in *Heliolithes* (Silurian to Middle Devonian). In contrast, individual *Halysites* (Ordovician to Silurian) formed palisadelike rows, producing a complex network. *Aulopora* (Devonian) was an example of a "crawling" tabulate coral; it was only found on smooth surfaces.

With the stromatopores and rugose corals, tabulate corals were the chief reef builders during the Paleozoic. Unlike the others, they occurred in greater depths (e.g., *Favosites* and *Halysites*) and on rather muddy subtrates.

Of the ARTICULATA (phyletic group Articulata), the earliest species largely lacked fossilizable skeletal elements, and the fossil record of them is very fragmentary and is restricted to just a few groups. Almost all that is known about early articulates stems from the Middle Cambrian Burgess slate. A great deal is missing between these early species and their modern descendants, which comprises a time span of approximately 500 million years. These organisms were not confined to Burgess deposits, of course, but only the Burgess slate offered the conditions needed to preserve specimens. It appears that the articulates were a rather diverse group by the Paleozoic, and in the seas they were hardly less developed than modern forms. The few specimans retrieved from other strata therefore give us just a tiny look at the great articulate fauna of prehistoric times.

ANNELIDS (phylum Annelida) were present in the Precambrian, for there are many annelid spoors from that time. We shall subsequently discuss fossil spoor. Certain evidence of the existence of POLYCHAETES (class Polychaeta) is found in the well-known Middle Cambrian Burgess slate. A number of genera, including *Miskoia*, *Worthenella*, and *Canadia* clearly demonstrate a division of the body into individual segments. In *Wiwaxia*, on the other hand, the body was covered by elongate scales. These Cambrian genera and the few Ordovician finds available are classified in one order, the MISKOIIDS (Miskoiida). They are probably early members of the polychaete order Errantia. Characteristic of predators, the jaw of these polychaetes contained thickened chitinous chewing parts known as scolecodonts (see Fig. 9-8). Calcium was sometimes deposited in some of these elements, particularly the so-called mandibles and it was often concentrated into an enamel coat at the fore end. Many fossil polychaetes can be classified as members of modern families, based on anatomical considerations. Most scolecodonts have been found individually and not as part of entire intact jaws.

Of the SESSILE POLYCHAETES (order Sedentaria), the SERPULIDS (Serpulidae) produced calcite tubes, some of which could have an irregular configuration (as in *Serpula*) or a very characteristic form (as in *Spirorbis*), the latter being known since the Ordovician (as is the entire family). A large number of Paleozoic genera have been found, and like their modern ancestors they occurred in shallow water. The only other sessile polychaete families from the Paleozoic are the TEREBELLIDS (Terebellidae; from the Cambrian), SABELLARIIDS (Sabellariidae; from the Carboniferous), and the PECTINARIIDS (Pectinariidae; from the Permian). The pectinariids inhabited tubes composed of sand or shell fragments cemented together.

The MYZOSTOMIDS (class Myzostomida) have not been found in the Paleozoic era. However, gall-like structures on Ordovician crinoids are comparable to similar structures produced by the modern myzostomid, *Myzostoma cysticolum*, and they may have been produced by Ordovician myzostomids. The worm class Clitellata has been found in the Carboniferous. These long, slender, articulated fossils are comparable to tubifex worms (Tubificidae).

The term spoor refers to signs left by organisms on or in a sediment without leaving any skeletal elements behind. An entire discipline of biology, ichnology, deals with spoor analysis, and palichnologists study fossil spoor. Various spoor types are known, depending on the manner in which they were produced. They include signs of resting and habitation, eating (tracks left while feeding, etc.), and all kinds of tracks produced from movement. Spoor fossils are particularly important for deducing the environmental conditions under which they arose. It is exceptional to be able to assign a spoor to some specific animal, and spoor are usually assigned scientific names in their own right. We shall cite a few produced by worms, illustrating some of them in Fig. 9-9: *Chondrites* (Cambrian to Tertiary) are branched feeding passages (see Fig. 9-9A) often found in dense packs in fine, muddy sediments. *Phycodes* has been found since the

Annelids

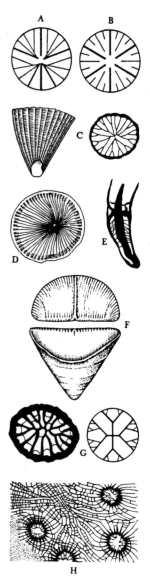

Fig. 9-6. Configuration of septae in rugose coral (A) and Madreporaria (B); C. *Petraia*; D. *Zaphrentis*; E. *Cyathaxonia*; F. *Calceola*; G. *Heterophyllia*; H. *Phillipsastrea*.

Lower Cambrian and especially in sandy depositions of the Lower Ordovician; this worm produced broomlike, outward branching eating passages. *Zoophycos* (see Fig. 9-9B), from the Devonian to the Tertiary, made spiralling tube systems. *Diplocraterion* (Lower Cambrian sandstone) likewise produced laminated structures. Tracks left by crawling have been found on the surface of strata, and some of the Cambrian spoor of this nature (e.g., *Climactichnites*; see Fig. 9-9C) are comparable to those made by modern polychaetes. Feeding spoor are particularly interesting. Many organisms had thigmotaxes (specific locomotory reactions to particular surfaces), and their feeding spoor illustrate millions of years later how they grazed off the surface on which their spoor are now fossilized.

Onychophores

ONYCHOPHORES (phylum Onychophora) are a primitive group whose body plan gave rise to that of arthropods. *Aysheaia pedunculata* (see Fig. 9-10), from the Middle Cambrian Burgess formation, is the only fossil species of this phylum. It bears a strong resemblance to a modern genus, *Peripatus*. Interestingly, this Cambrian species was a marine organism, while all modern onychophores are terrestrial.

Arthropods

ARTHROPODS (phylum Arthropoda) assuredly arose in the Precambrian, but the earliest fossil finds are Cambrian. Most present-day groups developed in the Paleozoic.

One arthropod group confined to the Paleozoic were the TRILOBITO-MORPHS (subphylum Trilobitomorpha), which includes the trilobites (Trilobita) and the trilobitoids (Trilobitoidea) or trilobitelike organisms. In both classes, the paired appendages share a common development: of the two or more pairs in front of the mouth, the first was an antenna and the second was either an antenna or a cutting tool (chelicerae). True biramous appendages were located behind the mouth.

The TRILOBITOIDS (class Trilobitoidea) include all trilobitelike organisms differing in some important way from true trilobites. With the sole exception of a Lower Devonian species (*Cheloniellon calmani*), all these finds are from the Middle Cambrian Burgess formation. *Marella* (see Color plate, p. 125) bore an armored head covering with powerful, hornlike spines extending to the back and laterally. The body had the characteristic three-part composition, but the biramous appendages were not covered with cuticular plates. This indicates that *Marella* was a swimming organism. The 12-cm *Sidneya* (see Fig. 9-11A) looked very much like the giant scorpions of a later time (discussed subsequently in this chapter). In contrast, *Burgessia* (Fig. 9-11B and Color plate, p. 125) had a shield-shaped armor covering over the head and trunk. Only the first antennae and a terminal segment (or telson) of the trunk protruded beyond the armor of this 1.5-cm trilobitoid. Its shape resembled a notostracan (Notostraca), a more recent group. *Leanchoila* (see Fig. 9-11C and Color plate, p. 125) bore a head shield and transversely segmented trunk armor and was thus very close to true trilobite appearance. However, the second antenna had been modified into a tactile and scraping organ, thus disqualifying *Leanchoila* as a trilobite and forcing its classification as a trilobitoid.

The TRILOBITES (class Trilobita) could be called the characteristic fauna of the Paleozoic, and they are certainly among the most familiar animals from this ancient period. They arose in the Cambrian and died out in the Permian, and thus never penetrated the Mesozoic era. Their shells have been found in great numbers, indicating that, like many other arthropods, trilobites molted many times in the course of their life. As in modern arthropods, the trilobite armor did not grow with the animal, so it had to be shed on occasion and replaced with a better-fitting "suit".

The name trilobite (= three-lobed animal) refers to the triple division of the body longitudinally. The trunk has a distinct axis (see Fig. 9-12) with laterally attached biramous appendages. The armor covering of the trunk is developed in accord with this structure and consists, in the middle, of an axis on which lateral plates (pleura) are attached, the plates acting to protect the appendages. Individual three-part segments of the trunk armor could be moved, for there were linkages between the armor sections. In many advanced species the trunk armor sections formed a single trunk shield, but the original subdivision into three parts can usually be seen quite clearly even in these. Anteriorly, the axis is continued as the glabella, and the original trilobite construction can often be seen here in the form of ridges in the glabella. The lateral elements of the head shield are known as cheeks, and they often terminated in the rear as spines. The edge of the cheeks was turned under (as was the edge of the trunk shield), forming a sort of pocket for the soft body.

In the head region, the delicate body could often have further protection by the presence of another plate, the hypostome. It could be pulled out of the "pocket" because the lateral cheek portion could be moved along a seam, and for this reason that portion is known as the free cheek. The suture we refer to in most cases extends from the front edge inward to the eyes, where it then runs either in front of or behind the cheek spine to the outer edge. Its course is designated as proparous or opisthoparous, respectively. The eyes usually protruded beyond the cheeks on a smooth, stalklike raised section, but these were faceted eyes and were immovable. Each one consisted of over 15,000 hexagonal or up to 400 round lenses. The appendages (see Fig. 9-12) had two linked parts, one used for respiration and the other for locomotion.

For many trilobites we know not only the adult but also the intermediate stages of growth. The growth process, as illustrated in Fig. 9-13, consisted of basically three stages designated protaspis, meraspis, and holaspis. After hatching and undergoing a few moults, the young trilobite formed an armor covering in the protaspis stage. The armor was divided but was not yet movable. In the next, meraspis, stage, a linkage appeared between the head and the rear shields. With additional moltings, more and more body segments were formed between these two parts, until the final

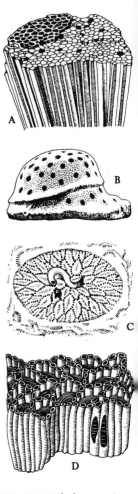

Fig. 9-7. Tabulate coral (Tabulata): A. *Favosites*; B. *Heliolithes*; C. *Pleurodictyum*; D. *Halysites*

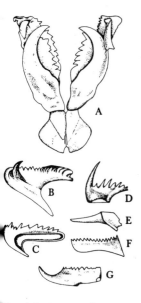

Fig. 9-8. A. Jaw of a modern annelid, *Diopatra neopolitana*; B-F. Various Paleozoic jaw fragments (or scolecodonts).

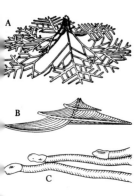

Fig. 9-9. Fossil spoor:
A. *Chondrites*;
B. *Zoophycos*;
C. *Climactichnites*.

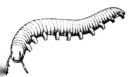

Fig. 9-10. *Aysheaia pedunculata*, an onychophoran.

number of body segments was attained in the holaspis stage. Even after that time, the trilobite could still grow, casting off its shell periodically.

We also know spoor caused by trilobites, something we said earlier is an unusual find. Movement spoor have permitted us to reconstruct the method of locomotion in trilobites. Feeding and burrowing spoor on and in deposits, respectively, have shed light on the nature of trilobite nutrition. Some species burrowed into sediment when they rested; this offered them protection and in some cases left spoor behind that were a source of joy to paleontologists. The shape of the body alone gives many clues about trilobite behavior. Most species lived on the ocean floor, searched for food at the surface, or burrowed. Others were probably good swimmers, and there were even planktonic (passively drifting) trilobites. Some species formed communities with drifting plants or graptolithines in what we call epiplanktonic communities. Many trilobites could coil up like isopods, an effective means of protecting the soft body parts from all sides and not just from the top. A number of groups had characterized coiling patterns (shown in Fig. 9-14).

During the Cambrian, trilobites evolved into a very diverse group. Size differences among adults of different species could be considerable, extending from approximately 5 mm to 75 cm (in *Uralichas ribeiroi* from the Middle Ordovician). One general trend in trilobites was increasing the size of the caudal shield at the expense of the number of trunk segments. Other changes, such as the initial increase in eye size and the subsequent degeneration of the eyes in some groups, was more a reflection of adaptation to specific new habitats than to a general trend in the group as a whole.

Trilobites are now classified into seven orders. The oldest ones are from the Lower and Middle Cambrian and are known as REDLICHIIDS (Redlichiida). They have a relatively large head shield, often with powerfully developed cheek spines, an opisthoparous facial suture, relatively small eyes whose nodule is often attached to the glabella by a ridge, a large number of trunk segments, and the absence of a true caudal shield. The key fossils include the Lower Cambrian *Redlichia* (see Color plate, p. 128), *Olenellus, Holmia* (see Fig. 9-15A); and the Atlantic genus of the Middle Cambrian, *Paradoxides* (see Color plate, p. 128), which could reach a L of up to 45 cm. These were primitive species, and after the Lower Cambrian they were followed by more modified forms, which are collectively known as AGNOSTIDS (order Agnostida; Lower Cambrian to Lower Ordovician). These were very small animals with nearly equally shaped head and caudal shields. They had just two to three body segments between the shields. Eyes, facial suture, and a hypostome were not present. Agnostids included *Eodiscus* (see Fig. 9-15B) of the Middle Cambrian and *Agnostus* of the Upper Cambrian. *Agnostus* species (see Color plate, p. 128) have been found covering entire surfaces.

The Middle Cambrian key Pacific trilobite, *Olenoides* (see Color plate,

p. 128), belonged to a third order, the CORYNEXOCHIDS (Corynexochida), which occurred only during the Cambrian. This group was characterized by a distinct, usually large, caudal shield with clear segmentation. The glabella had up to four distinct ridges; the eyes were small, and the facial suture was opisthoparous. The PTYCHOPARIID order (Ptychopariida) was a prolific one with over 800 genera occurring from the Lower Cambrian to the Middle Permian. Its oldest suborder, PTYCHOPARIINA (Lower Cambrian to Middle Ordovician), grouped about the genus *Ptychoparia* (Lower to Middle Cambrian), was characterized by a short caudal shield. The glabella did not extend to the fore edge of the semicircular head shield. A distinct ridge joined with small optical nodules with the glabella. *Olenus* (see Color plate, p. 128), which was similar but had a trapezoidal head shield, was the Atlantic key fossil of the Upper Cambrian. These were relatively small animals just a few centimeters long.

Asaphus, which had a L of up to 40 cm, was the characteristic trilobite of the Ordovician, as were all its relatives in the ASAPHIN (Asaphinae) suborder, distributed from the Upper Cambrian to the Upper Ordovician. They had large caudal shields but a reduced number of trunk segments, and the glabellar ridges were either barely distinct or absent. The ILLAENIN suborder (Illaenina) of the Ordovician to Middle Permian included, among other genera, *Illaenus* (Ordovician), whose axis can just barely be seen continuing on the caudal and head shields, and *Scutellum* (Silurian to Upper Devonian; see Color plate, p. 128), with its characteristic caudal shield whose radiating ribs are particularly prominent. These large trilobites, often exceeding 10 cm in length, were accompanied by smaller ones, some only 1 cm long like *Proetacea*. The proetacean superfamily was the only one to survive into the Devonian, and it disappeared in the Middle Permian. The major proetaceans were *Proetus* (Middle Ordovician to Middle Devonian) and *Phillipsia* (Lower Carboniferous; see Color plate, p. 128).

Some of the most striking ptychopariids were the HARPINS (Harpina; Upper Cambrian to Upper Devonian) and TRINUCLEINS (Trinucleina; Upper Ordovician to Middle Silurian). The porous head shield edge was enlarged, forming a sievelike hood. The eyes had degenerated to just two lenses in *Harpes* (see Color plate, p. 128 and Fig. 9-15C), and there was no caudal shield (pygidium). *Trinucleus* (see Color plate, p. 128) had a short trunk, a well-developed pygidium, and no eyes. It seems likely that these species sought their food within marine depositions.

The PHACOPID order (Phacopida), distributed temporally from the Lower Ordovician to the Upper Devonian, probably evolved from ptychopariids. Unlike all previous trilobites, phacopids typically had a proparous facial suture. *Cheirurus* (Lower Ordovician to Middle Devonian) had a finger-shaped labular pygidium. The delicate members of genus *Deiphon* (see Color plate, p. 128), with the inflated glabella and eyes on the leading edge of it, probably floated at the water surface, with

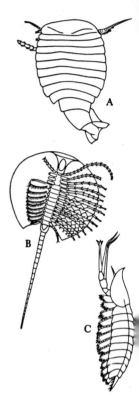

Fig. 9-11. Trilobitoids (Trilobitoidea): A. *Sidneya*; B. *Burgessia*; C. *Leanchoila*

only the glabella and the eyes protruding. Phacopids also include large, sickle-shaped eyes as in *Phacops* (Silurian to Devonian; see Color plate, p. 128). The "star tail", *Asteropyge*, is a closely related genus in which the pygidium's original segmentation is emphasized by spiny processes.

Spiny processes like these and additional ones are found in members of two other trilobite orders, the ODONTOPLEURIDS (Odontopleurida; Middle Cambrian to Upper Devonian; *Odontopleura*) and the LICHIIDS (Lichida; Middle Cambrian to Upper Devonian). The lichiids include *Ceratarges* from the Middle Devonian.

The CHELICERATE subphylum (Chelicerata) has its roots deep in the Paleozoic. One of the characteristic early chelicerates were the MEROSTOMES (class Merostomata; see Color plate, p. 133). The earliest of these, which belong to the AGLASPID order (Aglaspida), had the basic merostome structure: the fore body (known as the prosoma or cephalothorax), which arose from the merging of the head with the six adjacent thoracic segments, is covered by a uniform armor plate. It bears a pair of four-membered appendages modified into cutting devices behind which are another six pairs of running legs. The hind portion of the body (the opisthosoma or abdomen) is usually divided into movable segments and terminates in a spine-shaped tip known as the telson. In *Aglaspis* (see Fig. 9-16), an Upper Cambrian genus, the eleven to twelve abdominal segments bear paired appendages also used for locomotion. The aglaspids attained lengths of 2–21 cm. They lived in shallow seas of the early Paleozoic. The two other merostome orders evolved from them quite early.

In the GIANT SEA-SCORPIONS (order Eurypterida), the abdominal segments were likewise independently movable. However, the paired appendages attached to them were not used for locomotion but were modified into flat, gill-bearing appendages. The abdominal section was clearly divided into a fore section with seven segments (the mesosoma), an adjacent central section with five segments (the metasoma), and a spiny or plate-shaped terminal section (the telson). In front, the fore body bore a pair of three-sectioned appendages with the cutting devices, which followed the originally five running leg pairs. In many species the last pair of them was modified into a powerful paddle, as in *Pterygotus* (see Color plate, p. 133 and Fig. 9-17), distributed from the Ordovician to the Devonian. Since its running legs were very weak and the telson had become a kind of rudder, it appears quite likely that these species could swim well. In exceptional cases these animals bore elongated cutting implements. The powerful scissors bore tooth-shaped structures. *Eurypterus* (Ordovician to Devonian) had stronger running legs, a spine-shaped abdominal section, and small "scissors"-bearing appendages. In *Megalograptus* (Ordovician), the first and (especially) second running leg pairs had been modified into rather powerful organs with long double spines.

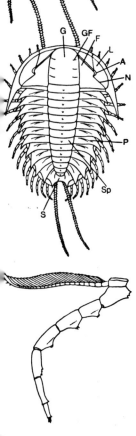

Fig. 9-12. Above: structure of a trilobite (*Olenoides*); G = glabella; GF = glabellar fold; F = rigid cheek L = free cheek; A = eye; N = facial suture; p = axis; P = pleura; Sp = tail shield. Below: Biramous appendage; the gill filaments are born on top.

Giant sea-scorpions spanned lengths of from approximately 10 cm in the smallest species to nearly 2 m in the very largest ones. The termination of the flourishing of sea-scorpions—the beginning of the Devonian—occurred at the onset of the first flourishing period of armored fishes. Thus, it appears that sea-scorpions and armored fishes competed for the same habitat. Although Ordovician sea-scorpions were exclusively marine organisms, some Silurian groups were found in brackish water, and by the Devonian nearly all of them had moved into fresh water. Their protected gills may even have permitted them to roam onto land for short periods.

The XIPHOSURID or horseshoe crab order (Xiphosura), of which a few species still survive, can be tracked back to the Silurian. Its ancestors were aglaspids. Unlike them, the number of abdominal segments steadily decreased during phylogenetic development. Furthermore, the segments tended to fuse together. The paired appendages on the hind body were leaf-shaped, as in sea-scorpions, and the hind end was swordshaped. In the early (Upper Silurian and Lower Devonian) forms, the SYNZIPHOSURINS (suborder Synziphosurina), the nine to ten abdominal segments were movable and could be coiled under the belly, acting as protection. Externally, these xiphosurids looked very much like their aglaspid predecessors. They were generally 5 cm long and usually are found with sea-scorpions. Two major genera from the bordering period between the Silurian and the Devonian were *Pseudoniscus* (see Fig. 9-18) and *Bunodes*.

In addition to these primitive forms, there were a number of species bearing a remarkable similarity to the modern xiphosurins, the limulins (suborder Limulina; Lower Devonian to the present). The fore body, with its protective shield, was relatively large. The hind end, with no more than nine, partially or fully fused segments, was shorter than the rearmost segment. Its front segments were movable (as in *Belinurus*; Upper Devonian to Lower Carboniferous), or they were all surrounded by a closed armor covering (as in *Paleolimulus*; Carboniferous to Permian). These characteristics and the movable spines on the edge of the armor were all features found in modern xiphosurids; they had therefore developed in the late Paleozoic.

Of the sixteen known ARACHNID (class Arachnida) orders, five were confined to the Paleozoic. The others are still found today. This great Paleozoic proliferation of arachnids is even more surprising in view of the fact that arachnids have very thin chitinous skin that does not preserve well. This suggests strongly that arachnids arose long before the first fossil finds.

The earliest arachnid finds, which are from the Silurian, are all SCORPIONS (order Scorpiones), animals with a relatively firm external skeleton. In these Silurian forms, such as the genera *Palaeophonus* and *Dolichophonus*, adult animals had eight mesosomal segments. The metasoma was slender and tail-shaped, consisting of five segments and the spine-shaped telson.

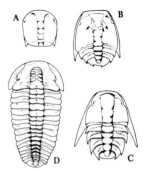

Fig. 9-13. Growth stages of the trilobite *Sao hirsuta*: A and B. Protaspis stage; C. Meraspis; D. Holaspis.

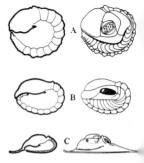

Fig. 9-14. Various means of defensively coiling the trilobite body to protect soft body parts; A. *Eocryphops*; B. *Ellipsocephalus*; C. *Harpes*.

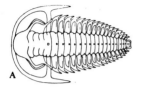

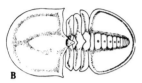

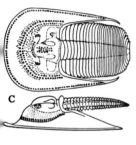

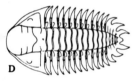

Fig. 9-15. Four different trilobites: A. *Holmia*; B. *Eodiscus*; C. *Harpes*; D. *Cheirurus*.

Pantopods

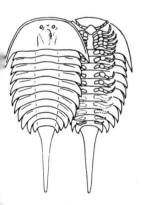

Fig. 9-16. *Aglaspis*, a merostomate, between 2 and 21 cm long.

All six pairs of fore body appendages were developed, as in modern scorpions. First came the small cutting appendages, then the pedipalpae (powerful grasping organs), and behind them four pairs of running legs. Unfortunately, the available fossil material is in such poor condition that the question on whether lung openings were present or absent cannot be answered yet. Therefore, it is an open issue whether Silurian scorpions lived on land like their Carboniferous descendants. In the latter, the first abdominal segment disappears during growth, so Carboniferous scorpions had one less segment than those from the Silurian, and therefore the Carboniferous scorpions differed substantially from their modern descendants.

TRUE SPIDERS (order Araneae) have recently been found in Lower Devonian deposits on the edge of what once was an island in the famous Hunsrück formation. The next finds are from the Carboniferous, and these spiders bear a reasonable resemblance to modern ones. Unlike them, however, Carboniferous spiders retained their segmented anatomy. Furthermore, they had more body segments (up to twelve).

The oldest MITES (order Acarina) are from Devonian Scotland. They make their next appearance in Tertiary Baltic amber. Most Paleozoic arachnids are from the Upper Carboniferous, from a period when they lived in large communities in extensive swamps with many other arthropods. Orders such as Pseudoscorpiones, Palpigradi, Solifuga, and Schizomida, as well as all modern orders, were present then. In addition there were members of five additional arthropod orders, which were confined to the late Paleozoic.

PANTOPODS (class Pantopoda or Pycnogonida) were very peculiarly shaped marine chelicerates. Only two species are known from the Lower Devonian and Upper Jurassic. The Lower Devonian species, *Palaeopantopus maucheri*, is incomplete, but it does display the characteristic features, especially on the fragmentary remains of the vestigial abdominal section. Another species, *Palaeoisopus problematicus*, from the same formation but in better condition, is very close to pantopods. However, it had a relatively long, probably segmented trunk unlike true pantopods. Furthermore, the first four leg pairs were very powerful and probably were used for swimming. A smaller pair of appendages in front of these were probably egg-bearers, as in modern pantopods. This specimen had a span of approximately 20 cm.

CRUSTACEANS (class Crustacea) were also very diverse deep into prehistoric times. They are distinguished in having two antennae and gills, and they are classified within a separate subphylum (Diantennata or Branchiata). Crustaceans have been found as early as the Cambrian. In fact, the only crustacean groups absent in the Paleozoic era were the subclasses of cephalocarids (Cephalocarida) copepods (Copepoda), mystacorarids (Mystacocarida), fish lice (Branchiura), and ascothoracids (Ascothoracida).

Phyllopods (subclass Phyllopoda) are close relatives of the LIPO-STRACANS (subclass Lipostraca) and ANOSTRACANS (subclass Anostraca), and all these are classified together as branchiopods (Branchiopoda). So far there is no evidence of branchiopods before the Tertiary. However, the Middle Devonian species *Lepidocaris rhyniensis*, for which an entire order has been developed (Lipostraca; see Fig. 9-19), was certainly very close to the branchiopods. As in branchiopods, the first antenna pair of the 3-mm animal is small, while the second antenna pair is powerfully developed. Paired appendages are on the first eleven of the nineteen abdominal segments. Only the first three appendages are lamelliform, while the others are ramified very simply.

Of the PHYLLOPODS (Phyllopoda), the NOTOSTRACANS (order Noto-straca) develop a large, more or less oval dorsal armor plate covering nearly the entire body. The trunk section of the plate was not due to fusing of trunk and head sections but to elongation of the latter. Characteristic features are the vestigial cephalic limbs (excepting the first antenna pair), the first abdominal appendages, which are modified into long, tri-branching tactile organs, the large number of adjacent, lamelliform appendages, and a greatly segmented furca (two long processes extending from the hind end and used as tactile organs). The oldest phyllopod is from the Upper Carboniferous. Like the many others found in Permian Oklahoma, this specimen belongs to the genus *Triops*, which still exists today and has hardly changed over all those millions of years. Fossil phyllopods, like their modern descendants, were fresh-water inhabitants.

The order Onychura is represented in the Paleozoic solely by the CONCHOSTRACANS (suborder Conchostraca). These crustaceans have a two-valved dorsal armor, which can cover the entire body, including its clawlike furca (see above). As in notostracans, the first pair of abdominal appendages are tactile organs. The shell was not cast off with each molt as the conchostrachan grew, for it could enlarge with the animal. Since the number of molts can vary considerably, even within a single species, the number of growth lines also varied. Some conchostracans have additional nodules or radiated ribs on the shells. The oldest find, *Rhab-dostichus*, is from marine deposits of the Upper Silurian or Devonian. Other Paleozoic finds are either from coastal regions or fresh-water deposits. In the coastal species, which lived together with marine organisms, it cannot be determined whether these conchostracans originally lived in this habitat or were transported there from fresh water. The most prevalent genera include *Isaura* (or *Cizycus*; Lower Devonian to the present) and *Leaia* (Lower Carboniferous to Upper Permian), both of which are depicted in Fig. 9-20.

The OSTRACODS (subclass Ostracoda) are an especially important crustacean group geologically. Their bivalve, typically calcified shell is hardy and has often been found in great masses in marine, brackish, and

Notostraca

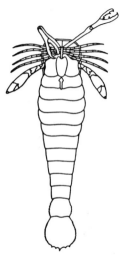

Fig. 9-17. A sea-scorpion, *Pterygotus*, seen in ventral (bellyside) aspect.

Conchostraca

(at least since the Carboniferous) in fresh-water deposits. Since they evolved rapidly, ostracods have been good key fossils and have therefore attained some notoriety among paleontologists. The body itself, which is enclosed within the shell, has rarely been fossilized, so shell characteristics are used for systematic arrangement of the ostracods. Since ostracods molt, the fossil record also contains all the growth stages. Sexual dimorphism may appear among adults. Females often have broad protrusions for the uptake of sexual products. However, the rear part of male shells in many species can also be enlarged, for these had particularly large genital organs.

Ostracods

All modern ostracod orders have been present since the Ordovician. Most of the orders existing only as fossils are confined to the Paleozoic, so there is a clear demarcation between Paleozoic ostracod formations and more recent ones. Cambrian ostracods are placed in the order ARCHAEOCOPIDA. The shell in these was less calcified and therefore more pliable. It often had spots (*Walcottella*; Middle Cambrian) or a ridged texture (*Bradoria*). The edge of the valve was long and nearly straight.

The thick, smooth shell calcified considerably in the order Leperditiida (Ordovician to Upper Devonian). Other features characterizing this group include large sphincter muscles, composed of numerous individual elements, and a long, straight valve. These ostracods reached a L of nearly 6 cm, which is considerable for such organisms. Leperditiids reached their zenith in the Middle and Upper Ordovician. *Leperditia* (Silurian to Devonian) occurred primarily in lagoons.

Fig. 9-18. A xiphosurid, *Pseudoniscus*, a horseshoe crab.

The third Paleozoic ostracod order were the BEYRICHIDS (Palaeocopida or Beyrichiida; Ordovician to Permian). They are characterized by highly sculptured shells (see Fig. 9-21C and 9-21D). The valve edge is straight and long. Sexual demorphism is well developed. *Beyrichia* was widely distributed in the Silurian and Lower Devonian and in some areas formed entire deposits. *Ctenoloculina* (Middle Devonian) is another interesting genus from the approximately 250 genera in this order. Females had a series of brood pockets on the ventral valve edge, and these opened to the outside.

Today the only extant ostracod orders are the PODOCOPIDS (Podocopida) and the MYODOCOPIDS (Myodocopida) which have been known since the Ordovician. They differ from the other ostracods in having short, arched valve margins. Their shells are often smooth, as in *Bairdia* (Silurian to present) and *Healdia* (Devonian to Permian), the two most important podocopids from the Paleozoic. The Devonian myodocopid ENTOMOZOANS were important key fossils. Their shell pattern was fingerprintlike (see Fig. 9-21F). Unlike most ostracods, which lived either in the floor of the ocean or in the uppermost sediments, entomozoans and many other myodocopids were planktonic.

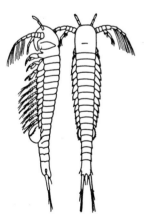

Fig. 9-19. A lipostracan crustacean, *Lepidocaris rhyniensis*.

Of the many BARNACLES (subclass Cirripedia), the only Paleozoic species are the LEPADOMORPHS (suborder Lepadomorpha). The Upper Silurian genus *Cyprilepas* had a very simple plate structure. Its calcite plates lacked growth lines. This would suggest that these primitive barnacles shed their entire shells with each molt. The second lepadomorph, *Praelepas* from the Upper Carboniferous, probably had five non-calcified plates, but they display growth lines.

In MALACOSTRACANS or HIGHER CRUSTACEANS (subclass Malacostraca), calcite armor develops and thus provided better fossilization opportunities. This subclass extends back to the Cambrian, and some of the old groups are classified with the extant LEPTOSTRACANS (order Leptostraca) within the superorder PHYLLOCARIDA. The earliest confirmed malacostracan is *Hymenocaris* from the Middle Cambrian, the only HYMENOSTRACAN (order Hymenostraca) genus. It apparently had a univalve shell. *Ceratiocaris* (see Color plate, p. 133) is a more familiar genus known to have occurred from the Lower Ordovician to the Lower Devonian but perhaps extending as far back as the Permian. It is a member of the Paleozoic order of ARCHAEOSTRACANS (Archaeostraca), one suborder of which is CERATIOCARINA. A bivalve, *Ceratiocaris* had a 16-cm shell lacking an eye nodule, but other genera did possess eye nodules, such as *Nahecaris* from the Lower Devonian.

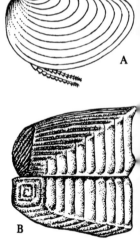

Fig. 9-20. Two concho-stracans: A. *Isaura*; B. *Leiai*.

All other higher crustaceans are grouped together as Eumalacostraca. Unlike most phyllocarids, eumalacostracans lack bivalve armor, and the abdominal section is divided into no more than six segments. The oldest known eumalacostracans are EUCARIDS (order Eucarida; Middle Devonian to Upper Carboniferous). They have extensive armor, but it is not joined with the thoracic segments. SYNCARIDS (order Syncarida) have no dorsal shield, and the first thoracic segment may be fused with the head. The oldest (Upper Carboniferous) genus is *Palaeocaris*. HOPLOCARIDS (order Hoplocarida; genera *Perimecturus* and *Archaeocaris*) occur from the Carboniferous onward; in hoplocarids, four to five trunk segments fuse with the head. Of PERACARIDS (superorder Peracarida), individuals have been found from the Upper Permian and could be classified with CUMACEANS (order Cumacea) and TANAIDACEANS (order Tanaidacea or Anisopoda). This all shows that the evolution of CRUSTACEANS (superorder Eucarida) began in the early part of the Paleozoic era.

The TRACHEATES (subphylum Tracheata) are among the most poorly fossilizable forms. However, evidence of tracheates has been found in Upper Silurian formations. The oldest finds are fragments of MYRIAPODS (class Myriapoda), and they are classified within the ARCHIPOLYPODS (Archipolypoda) subclass. The reason for this is that they differ in some important ways from the closely related DIPLOPODS (subclass Diplopoda): the head is larger than the thoracic segments and has large, close-set eyes; furthermore, the individual segments are loosely joined. *Arcantherpestes* from the Upper Carboniferous attained a L of 50 cm and was 3 cm thick. The Upper Carboniferous genus *Arthropleura* (see Color plate, p. 202/203) probably belongs in this group as well. The largest intact specimen from this genus had a L of 80 cm, and fragments that have been found indicate

Tracheates

Myriapods

that giant myriapods with a L of as much as 180 cm existed at that time. The thoracic region had overlapping plates extending laterally over the legs much as in trilobites.

Insects

INSECTS (class Insecta) descended from myriapods. The three subclasses of Apterygota (flightless insects) are distinct from the flying insects (Pterygota). The order Collembola has been found since the Middle Devonian.

FLYING INSECTS (subclass Pterygota) probably arose not long after terrestrial plants appeared. The primitive species, which could not fold their wings back, are often collectively classified as PALEOPTERANS (Palaeoptera). They include the order of PALEODICTYOPTERANS (Palaeodictyoptera; Lower Carboniferous to Permian). These were typically large insects with a moderate wingspan of 10–20 cm, but in some the wingspan was 50 cm. The mouth usually had a biting apparatus, but some species had trunklike structures probably used for sucking. Representative genera include *Stenodictya* (see Fig. 9-23) and *Lithomantis*, both Upper Carbonian insects.

MAYFLIES (division Ephemeroptera) are characterized by the unpaired tail fibers; they have been found since the Upper Carboniferous, as are DRAGONFLIES (order Odonata). Many people have seen the famous giant dragonfly *Meganeura* in museum displays of Carboniferous jungles. This huge insect, with an impressive wingspan of approximately 70 cm (see Color plate, p. 202/203), lived in the Upper Carboniferous period.

Among insects that could lay their wings back, COCKROACHES (sub-, order Blattaria) appeared quite early. The oldest finds, from the Devonian are probably cockroaches. By the Carboniferous they were widely distributed and had diverged into several groups. One genus, *Phyloblatta*, is illustrated on Color plate, p. 202/203.

STONEFLIES (order Plecoptera) also appeared in the Lower Permian, as did the ORTHOPTERANS (order Orthoptera), with a few primitive species (e.g., *Protelytron* from the Lower Permian of Kansas). They formed communities with the first BEETLES (order Coleoptera) and ALDER FLIES (order Neuroptera). Members of all three groups (i.e., SNAKE FLIES (Megaloptera and Raphidiodea) and LACE-WINGS (Planipennia)) existed at that time. SCORPIONFLIES (order Mecoptera) have also been found in the Permian. Within the scorpion fly superorder (Mecopteroidea), the only groups not found in the Paleozoic era are butterflies, flies, and fleas. By the end of the Paleozoic, every major insect group excepting wasps (order Hymenoptera) was represented. The insects are thus a good example of the development of a new body plan within a (geologically!) short period of time, when that animal group contains the anatomical predisposition or potential for entering a new habitat.

Members of all major MOLLUSK (phylum Mollusca) groups were present in the Paleozoic: placophores, monoplacophores, gastropods, scaphopods, mussels, and cephalopods. Two other groups, the hyolithids

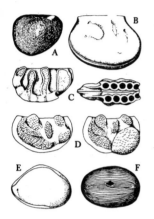

Fig. 9-21. Ostracods: A. *Bradoria*; B. *Leperditia*; C. *Ctenoloculina*, laterally and from below; D. *Beyrichia*, with the male and female on the left and right, respectively; E. *Bairdia*; F. An entomozoan.

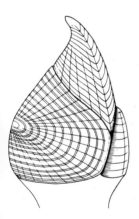

Fig. 9-22. *Praelepas*, a lepadomorph from the Upper Carboniferous.

and the tentaculatids (discussed subsequently in this chapter) were restricted to the Paleozoic era.

The PLACOPHORE class (Placophora) belongs, along with the Solenogastres and Caudofoveata, with the subphylum Aculifera and Amphineura. The only fossils from this group are placophores with subdivided calcite shells. This dorsal shell generally consists of eight plates, which overlap to such an extent they cover the entire dorsal surface of the animal. This affords adequate protection when the placophore coils up.

Fig. 9-23. *Stenodictya*, an Upper Carboniferous flying insect.

Since the only typically fossilized parts are the hard parts (in this case the individual plates), their form, texture, and structure is significant phylogenetically. The plates of most living placophores have the following structure: on the outside, as in all mollusks, there is a thin layer of organic material (the periostracum). Beneath it lies the tegmentum, an incompletely calcified layer with a labyrinth of small and large pores penetrating the chitin. The thicker, fully calcified articulamentum overlaps along with the plates. The innermost layer, found in all mollusks, is the hypostracum.

In PRIMITIVE PLACOPHORES (Palaeoloricata; Upper Cambrian to Upper Cretaceous) there is no articulamentum. Hence, the degree of overlapping of scales was not as great. In other respects, however, primitive placophores, with their eight plates, had nearly the same shape as modern placophores (Neoloricata). Among modern placophores, the Lepidopleurina have existed since the Carboniferous, while the other orders developed much later.

Fig. 9-24. A worm-shaped, elongated placophore: *Septemchiton vermiformis*.

As we already mentioned, placophores have eight plates. Deviations from this number (i.e., seven or nine plates) have often been observed, but these are found only within individuals wherein all other conspecifics have eight plates. There is only one known case of such a mutative modification proliferating throughout an entire population and becoming a distinguishing characteristic of that population, and this was in *Septemchiton vermiformis* (see Fig. 9-24) from the Upper Ordovician. As the scientific name suggests, this species was a worm-shaped elongated organism, and it bore just seven plates. It is also interesting to note that *Septemchiton* occurred in very fine sandy soil, while the other fossil finds are from regions more closely resembling the modern habitat of placophores: the shallow sea (especially in tidal zones) with a firm floor. Propagation into deeper waters is probably a more recent development. This propagation plus the great recent increase in the number of species shows that this "conservative" group, which scarcely changed during the Cambrian and since that period until now, is undergoing its flowering in the present!

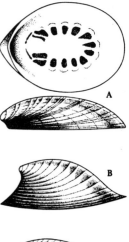

Placophores, with their bilaterally symmetrical bodies, conform to the basic, primitive body plan of mollusks. This is also true of the GASTROVERMS or MONOPLACOPHORES (class Monoplacophora). They are in the fossil record from the Lower Cambrian to the Middle Devonian, and it appeared that at that time they either died out or continued as more

Fig. 9-25. Tryblidiacea: A. *Pilina*, from the side and from the inside to show the paired muscle impressions; B. *Scenella*; C. *Cyrtonella*.

highly advanced snails. This opinion was changed in 1952 on the basis of a sensational find, when modern representatives of this group were found in deep-sea waters. At any rate, monoplacophores migrated into the deep sea during the Devonian period, and they may have been driven out of their original habitat by competitors or by enemies (predators). In the great depths of the oceans they found a home that has supported them for over 350 million years, and they have hardly evolved during that long period. The reason there is no fossil evidence of the deep-sea monoplacophore habitat is that during earth crustal movements the oceanic crust moved beneath the continental mass, creating a geosyncline and obliterating the original ocean floor where these monoplacophores had made their home.

In the cup-shaped, bilaterally symmetrical shell of the monoplacophore there are up to eight paired muscle block impressions whose arrangement seems to reflect a segmentation in the earliest species. Thus it appears that monoplacophores, like placophores, arose from a common, segmented ancestor. The oldest monoplacophore known, *Scenella*, is from the Lower Cambrian. The number of species reaches a maximum point in the Upper Cambrian and Ordovician and then declines slowly to the Middle Devonian, when the group disappears.

In the Silurian genera *Tryblidium* and *Pilina*, the apex of the shell (the point where growth begins) is at the fore edge of the shell, just as in the modern genus *Neopilina*. In other genera, however, such as *Scenella*, the apex is lifted and leans forward (see Fig. 9-25A and 9-25B). The shell arches so much that one could speak of a slight coiling tendency, something we see more strongly developed in *Cyrtonella* and related Lower Ordovician to Middle Devonian species (Fig. 9-25C). The number of paired muscle block impressions can be reduced to just two. This suggests a close relationship between this group and the bellerophontids, the first fully coiled snails (see below). While modern monoplacophores appear to be quite distinct from all other snails, the fossil record shows all intermediate forms between the two groups, and there are no fossil "links" between monoplacophores and the others. Therefore, on the basis of phylogenetic development, there is no basis for assigning monoplacophores into their own class. They could easily be classified with snails.

Further subdivisions of SNAILS or GASTROPODS (class Gastropoda) is made on the basis of the structure of the soft parts, particularly on the torsion (degree of turning) of the visceral sack and the location of the nervous system (caused by the torsion) and the placement of the gills. Since most present-day gastropods can be traced back to the Paleozoic, it is possible to classify fossil gastropods into the modern groups even though all we have of the fossils is their shells. Most Paleozoic groups still exist today.

In monoplacophores the nervous system is parallel (orthoneurous), the torsion of the visceral sack in their descendants, the gastropods, initially

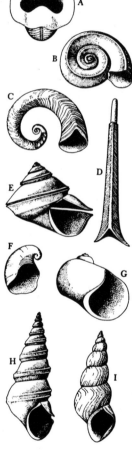

Fig. 9-26. Gastropods or snails (Gastropoda): A. *Bellerophon*; B. *Euomphalus*; C. *Lytospira*; D. *Mastigospira*; E. *Tropidostrophia*; F. *Platyceras*; G. *Naticopsis*; H. *Murchisonia*; I. *Loxonema*.

leads to a crossing-over of the two main nerve strands and hence classi-fication of the STREPTONEURAN (= cross-nerved) GASTROPODS (subclass Streptoneura). The systematic position of the primitive BELLEROPHONTIDS (suborder Bellerophontacea; Lower Cambrian to Lower Triassic) is un-certain. They possess just one pair of contracting muscles for closing the shell, and this distinguishes them from the monoplacophores. Unlike the other gastropod groups, they are coiled in a planospiral pattern and thus are orthoneurous. Many parallels in the shell structure, especially the anatomy of the cleft ligature, suggests a close relationship to the Pleuro-tomariacea, which are streptoneurans. Within the bellerophontid group there are genera in which the shell structure does not differ substantially from *Scenella* (e.g., *Helcionella*; Lower to Upper Cambrian). The shells are usually completely coiled up, and the convolutions widen to form a funnel-shaped structure (as in the namesake of the order, *Bellerophon*; Silurian to Lower Triassic) or they only slowly increase in height (as in *Cloudia*; Upper Cambrian).

The bellerophodontids gave rise to the Diotocardia (an order, formerly known as Archaeogastropoda or Aspidobranchia). *Euomphalus* (of family Euomphalidae; Lower Ordovician to Triassic) was nearly planospiral. In some closely related forms the shell could be unrolled as a secondary development (e.g., in *Lytospira* (Lower Ordovician to Silurian) or *Masti-gospira* (Middle Devonian; see Fig. 9-26D)).

The immediate predecessors of the SLIT SHELL SNAILS (family Pleuro-tomariidae) are also known from the Upper Cambrian. While in the oldest members of the family the cleft was only slightly developed, it attains a considerable length in some of the later forms (e.g., the Lower Carbonian *Tropidostrophia*; see Fig. 9-26E).

In LIMPETS (family Patellidae), unlike orthoneurous monoplacophores, the shell is not originally cup-shaped. Limpets evolved from coiled species, which is evident from the crossed longitudinal nerve fibers. However, there have only been a few Paleozoic limpet finds (Silurian to Permian), and others located in the future may cause a change in thinking about this group.

The TOP SHELL superfamily (Trochacea) includes the exclusively Paleozoic PLATYCERATES (Platyceratacea). The shell convolutions increase rapidly in girth, so that finally (e.g., in *Platyceras* of the Silurian to Permian ; see Fig. 9-26F) the shell is cap-shaped. One platycerate led a semiparasitic life on crinoids and other echinoderms.

The superfamily Neritacea includes the NERITOPSIDS, a family appearing in the Middle Devonian. The genus *Naticopsis* (Middle Devonian to Triassic; see Fig. 9-26G) was distributed throughout the world. *Mur-chisonia*, named after the British Paleontologist and stratigraphist Roderick Impey Murchison, had a spiralling, cylindrical shell (see Fig. 9-26H). The MURCHISONIACEANS (Murchisoniacea; Ordovician to Triassic), of which this genus is a member, probably diverged from pleurotomarid snails at a very early period, even though they also bear a slit in the shell.

Streptoneura

▷
Fossil marine organisms from Hünsruck slate, Germany, (Devonian). Above left: Placoderm *Lunaspis heroldi* (L about 30 cm); Above right: 32-armed starfish (*Medusaster rhenanus*). The hard parts of of the body have been con-verted into iron pyrites; Middle: Skatelike placo-derm, *Gemuendina stuertzi* (L 40 cm, width 24 cm); armor consisting of numerous small plates. The fish lived primarily on the bottom of the sea. Bottom: Jawless Pteraspidomorph, *Drepanaspis gemuendenensis* (L 35–45 cm). All originals from Heimat Museum, Bad Kreuznach, Germany.

▷▷
X-rays of paleozoic fossils. X-rays permit viewing specimens before they are prepared. Above left: An ophiuroid, *Mastigophiura grandis*; Below left: An Ordovician trilobite, *Triarthrus eatoni*, from Utica slate in Rome, New York (L ≃ 8 mm); Above right: Greatly deformed trilobite (*Phacops* sp.) caused by pressure; this is a conventional photograph; Below right: X-ray photo-graph of the same speci-men. This permits us to see the faceted eye, the fully intact intestinal tract and the appendages with their fine hair covering, probably used to burrow in the mud for food.

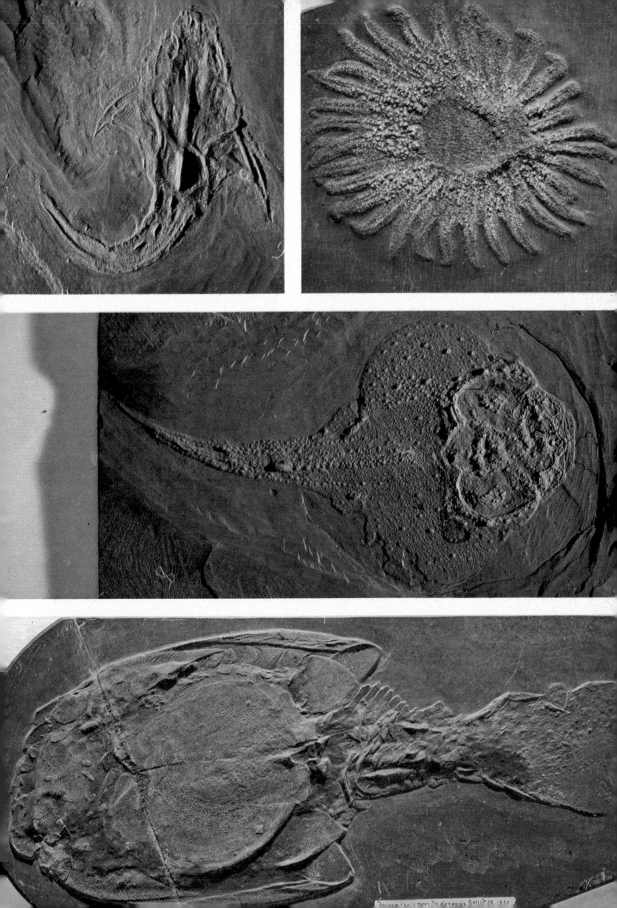

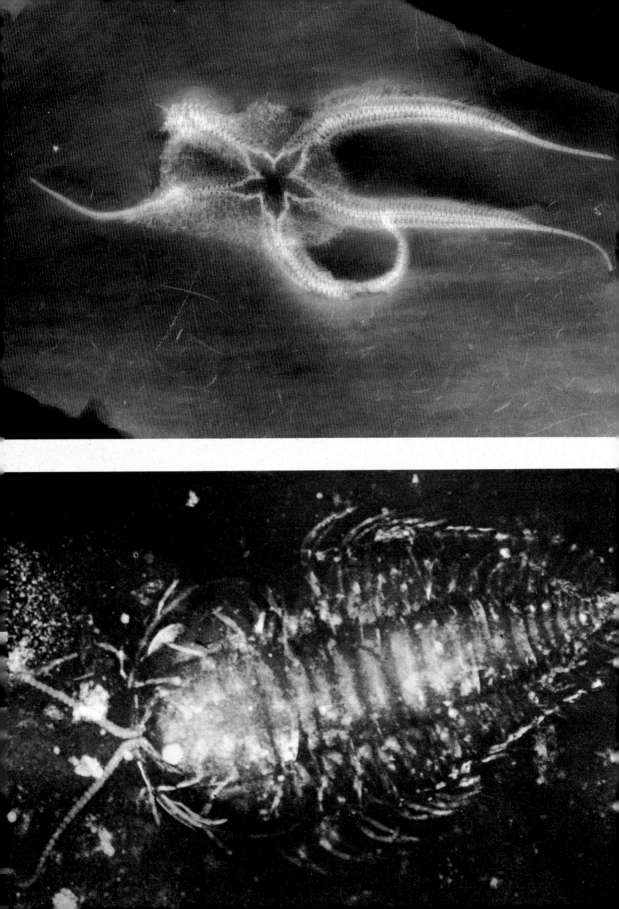

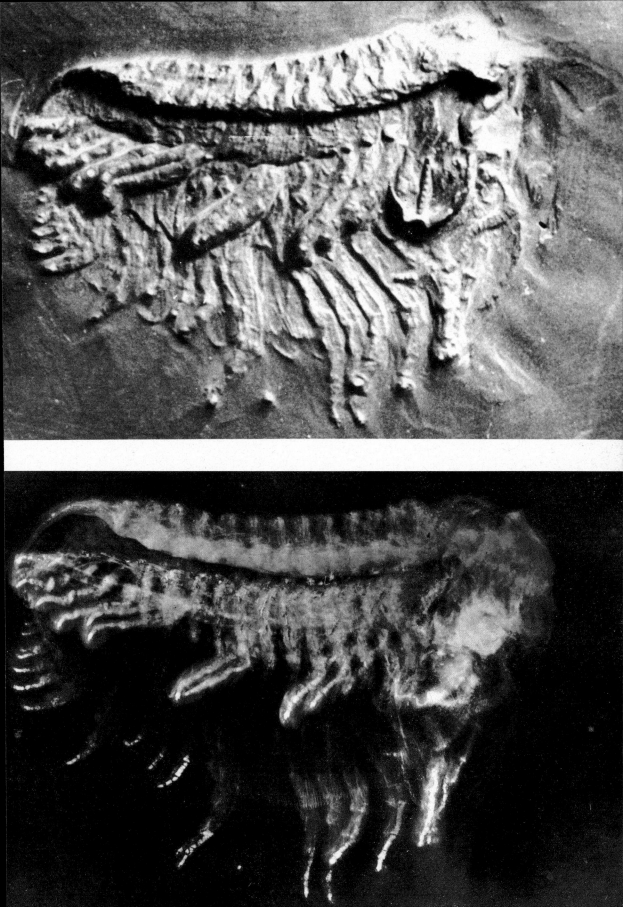

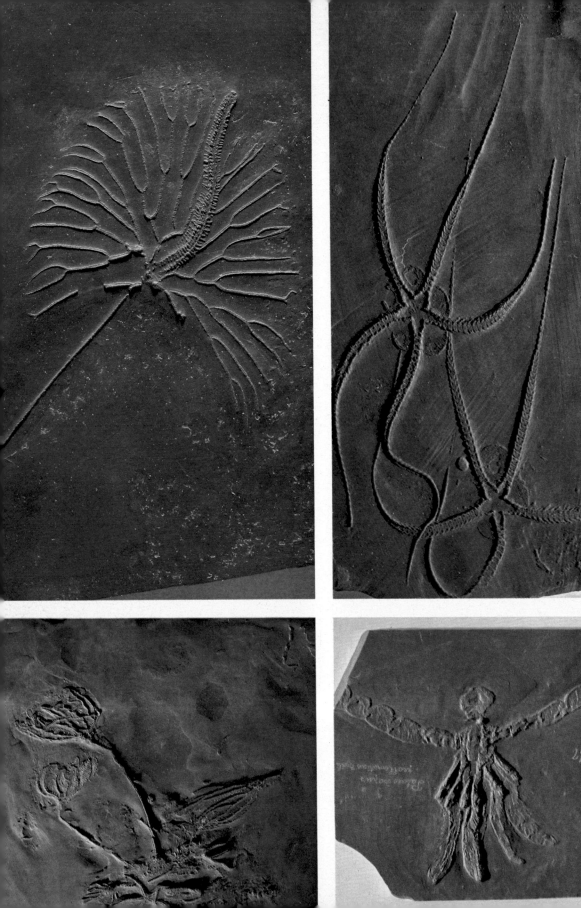

Only a few MESOGASTROPODS (suborder Taenioglossa or Mesogastropoda) appeared in the order of Caenogastropoda. The LOXONEMATIDS (Loxonematidae) occurred only from the Middle Ordovician to the Lower Carboniferous, but some closely related species reached the Jurassic. Unlike most other gastropods described until now, their tower-shaped shells lacked a slit. However, the mouth of the shell had an indentation (i.e., in *Loxonema*; see Fig. 9-26I). The TURRET snails (Turritellidae; Lower Devonian to present) are a closely related group. Only a few genera have survived into the present, just as few genera of the CYCLOPHORIDS (Cyclophoridae; Lower Carboniferous to present) lived beyond the Paleozoic.

The OPISTHOBRANCHIA (also known as subclass Euthyneura), which secondarily developed straight nerve tracts, are represented by a few genera beginning in the Carboniferous. They belong to the spindle-shaped PYRAMIDELLIDS (Pyramidellidae), while one genus (*Acteonina*) is an ACTEONID (Acteonidae).

The Paleozoic era is marked by quite a diversity in gastropods, although the upper limit of gastropod development is not so clearly delineated in the Paleozoic. A few primitive groups survived into the Triassic (e.g., the bellerophontids, euomphalids, and the murchisoniaceans). Some modern groups appeared as early as the Permian or in the Carboniferous (e.g., trochoids and streptoneurans). The Paleozoic, regarding gastropod evolution, is characterized by the rise of the monoplacophores in its early stages and then by subsequent development of the flat to tall, turret-shaped gastropods evolving from the monoplacophores. In the last part of the Paleozoic, gastropods invaded freshwater habitats and, as plants moved onto land, became terrestrial as well.

TUSK SHELLS (class Scaphopoda), with their characteristic tubular shells, adapted to a burrowing life. This specialization had developed in the Devonian, and since this time the scaphopods evolved into DENTALIIDS (Dentaliidae) and SIPHONODENTALIIDS (Siphonodentaliidae), of which the second family diverged in the Mesozoic.

The chief proliferating phase of BIVALVES (class Bivalvia) began in the early Mesozoic and coincided with the decline of the likewise bivalve brachiopods. However, all the major bivalve groups were present by the Paleozoic era. The known fossils are classified on the basis of shell features, including characters such as the hinge of the shell, muscle impressions, the course of the mantle line, and the shape, sculpture, and structure of the entire shell. Since paleontologists have been able to follow the evolution of mussels through many steps to the modern species, most fossil species can be categorized into the present systematic system, which actually is based on the development of the gills (structures that have not been retained in fossilized specimens). Of course, we do not know which of the features we shall cite were the original, primitive ones, but it seems extremely unlikely that all mollusks developed independently. The more probable phylogenetic pathway is one with many interrelated groups.

Opisthobranchia

◁
Fossil marine organisms from Hunsrück (Germany) late.
Above left: Short-shafted crinoid, *Parisangulocrinus ucumis*. The position of this specimen permits a look into the opened crown. The slightly bent anal tube can be seen in the middle;
Below left: A group of primitive echinoderms (crinoids and starfishes) that had been swept together and then fossilized;
Above right: An ophiuroid, *Furcaster palaeozoicus*. The position of the arms is evidence of water currents.
Below right: A pantopod, *Palaeoisopus problematicus*. The position of this fossil with pantopods was not confirmed until X-ray photographs were made in 1959.

The oldest confirmed bivalve species, *Lamellodonta simplex*, is from Middle Cambrian Spain (see Fig. 9-27A). Its shell has two parallel depressions, a structural type ancestral to Ordovician molluscan species whose shells are characterized by several diverging ridges (as in *Actinodonta*; see Fig. 9-27B). From here, development probably continued to the ANTHRACOSIACEANS (Anthracosiacea) and the UNIONACEANS (Unionacea), which arose in the Mesozoic era. Anthracosiaceans (see Fig. 9-27C) were the characteristic mollusks of the coal-bearing deposits of the Carboniferous and Permian periods. They evolved from marine predecessors and distributed themselves chiefly in brackish and fresh-water habitats. One group, which is probably closely related to them, are the MYOPHORIIDS (Myophoriidae; Devonian to Triassic), which are still represented today by trigoniids. Myophoriids are classified together with river and pond mussels as SCHIZODONTS (suborder Schizodonta).

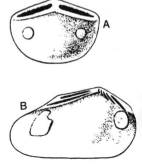

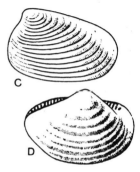

The PROTOBRANCHIA valve has numerous, uniform teeth, and this valve structure is known as a taxodont structure. It first appears in the Middle Ordovician period. In *Ctenodonta* the teeth are straight and diverge toward the middle, while in the Paleozoic NUCULIDS and NUCULANIDS (Nuculidae and Nuculanidae; Ordovician to present) the teeth are bent. The hinge in SOLEMYIDS (Solemyidae) has regressed greatly. This group arose in the Devonian. The same regressive condition is found in PRAECARDIODA (Lower Ordovician to Lower Carboniferous), which included such worldwide distributed species as *Praecardium*, *Cardiola*, and *Buchiola*. The last of the three occurred on drifting organisms and therefore led what is called an epiplanktonic life.

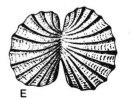

TAXODONTS (suborder Taxodonta) can also be followed as far back as the Lower Ordovician, to the *Parallelodon* group (see Fig. 9-27F). The teeth in this genus were parallel and very long. The LEPTODONTS are probably a closely related group. The suborder is known either as Leptodonta or Anisomyoaria. The relationship between taxodonts and leptodonts can be demonstrated by comparing one of the relatives of *Parallelodon* with the early PTERIOIDS (order Pterioida; Ordovician to present): in *Pterinea* (Upper Ordovician to Lower Devonian; see Fig. 9-28A), there are two parallel tooth ridges and three short teeth; in other genera from this same period lost these hinge supports, as have the modern genera.

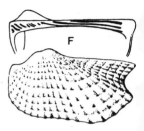

The fossil shell record confirms the theory that the taxodonts and the leptodonts belong to one related group. Other subgroups of the latter are often described in the paleontological literature: 1. MYTILIDS (Mytilidae; from the Devonian); 2. PINNIDS (Pinnidae; from the Lower Carboniferous); 3. PECTINIDS (Pectinidae; since the Ordovician), of which the semi-planktonic POSIDONIANS (Upper Devonian to Upper Cretaceous) should be mentioned, since posidonians often covered entire deposit surfaces in the Upper Devonian and Lower Carboniferous; 4. LIMIDS (Limidae; since the Lower Carboniferous). The other groups (soft-shelled clam, saddle oysters, and oysters) appeared in the Mesozoic era.

Fig. 9-27. Development of bivalves (Bivalvia) I:
A. *Lamellodonta*; B. *Actinodonta*; C. *Anthracosia*;
D. *Ctenodonta*; E. *Buchiola*;
F. *Parellelodon*.

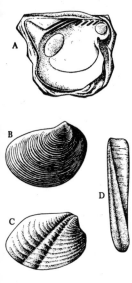

Fig. 9-28. Development of bivalves (Bivalvia) II: A. *Pterinea*; B. *Posidonia*; C. *Grammysia*; D. *Palaeosolen*.

The relationship existing between the previous mollusks and the HETERODONTS (suborder Heterodonta), another enormous group, has not yet been clarified. The oldest known heterodont, *Babinka*, stems from the Middle Ordovician. In addition to the two contracting shell muscles, *Babinka* has four additional muscles between those two major ones. This is thought to be evidence of a primitive form of segmentation, similar to what is found in monoplacophores. The next heterodont finds are from the Silurian, the period when heterodonts proliferated. From their earliest beginnings, heterodonts have not differed substantially from modern counterparts. In the Silurian period, the first LUCINACEANS (superfamily Lucinacea) appeared, while groups arising in the Devonian included the ASTARTACEANS, CARDITACEANS, and ARCTICACEANS (superfamilies Astartacea, Carditacea, and Arcticacea). All other superfamilies developed later. This is also true, with just a few questionable finds, for the entire ADAPEDONTA suborder.

TOOTHLESS BIVALVES (suborder Anomalodesmacea or Desmodonta), which were chiefly digging or boring organisms, are widely represented throughout the Paleozoic since the Ordovician by the still extant PHOLADOMYIDS (Pholadomyidae) and by the EDMONDIIDS (Edmondiidae; Devonian to Permian). The external appearance of all Paleozoic Anomalodesmaceans greatly resembles that of the modern genus *Pholadomya*, which is now found in deep seas. Extinct Anomalodesmaceans occurred in shallow seas throughout the world. The most prevalent genera were *Grammysia* (see Fig. 9-28C) in the Devonian, *Megadesmus* and *Astartila* (the latter two from the Permian). Two genera, *Orthonota* (Middle Ordovician to Middle Devonian) and *Palaeosolen* (Lower to Middle Devonian; see Fig. 9-28D), bore a striking resemblance to the modern razor shell (Solenidae) mollusks. These homeomorphic (anatomically similar) phenomena are not due to phylogenetic relationships but to living under similar environmental conditions.

The Paleozoic picture of bivalves can be summarized as follows: bivalves are known from the Middle Cambrian; their great proliferative phase, as evidenced by the appearance of all major groups, occurred in the Ordovician period. The second great proliferating phase, which was associated with the development of numerous newer families besides the older groups, took place after the Paleozoic. From that time on, however, bivalves increasingly prevailed over their ecological competitors, the brachiopods.

Cephalopods

The CEPHALOPOD class (Cephalopoda) appeared in the Paleozoic with a great number of various groups. The modern systematic divisions are unable to cope with all these primitive species, and in many cases it cannot even be determined whether the specimen at hand has two gills (as in squid) or four (as in nautilus). Ammonites, for example, have traditionally been classified with the Tetrabranchiata. However, the number of tentacles and the structure of the radula is just as indicative of their membership in the Dibranchiata. To resolve this dilemma, we place modern squid species in one subclass, Dibranchiata, and all the other fossil taxa into independent subclasses.

The central group within the phylogenetic development of the cephalopods are the NAUTILOIDS (subclass Nautiloidea). They occurred primarily in the Paleozoic, and the only remaining genus from this subclass is the chambered nautilus (*Nautilus*). The earliest nautiloids were present in the Lower Cambrian, one of the indices that divergence of the mollusks into different classes had been completed by the early Paleozoic. *Volbórthella* (see Color plate, p. 143), from the Lower and Middle Cambrian, and *Salterella* (Lower Cambrian) display the typical nautiloid features: a chambered shell, whose septae form an opening for the siphon (which extends out the rear). The shell in these two genera is straight and no more than 3 cm long. There is a great gap in the fossil record between the last *Volborthella* finds and the next younger nautiloids.

The ELLESMOCERIDS (order Ellesmocerida; Upper Cambrian to Upper Ordovician) appeared in the Upper Cambrian with two genera, the slightly arched *Plectronoceras* (see Color plate, p. 143) and the straight-shelled *Palaeoceras*. In both, the broad shell has downward arching septae. During the Lower Ordovician, a rapid evolution of this group occurred, and "suddenly" there appear nearly fifty ellesmocerid genera. These display a great variety of forms: straight or arched shells, slender to broadly conical, with a narrow or wide siphon, a smooth outer surface or one with rings, arising from an arching of the shell.

This diversity in the ellesmocerids led to the formation of closed groups during the course of the Ordovician period. One of these groups were the ONCOCERIDS (order Oncocerida; Middle Ordovician to Lower Carboniferous), which also produced diverse forms. All these cephalopods, however, are characterized by the same siphonal structure, in which supplementary ring-shaped calcium secretions can be laid down. *Oncoceras* (Middle to Upper Ordovician) is slightly arched, and the diameter of the shell in adults decreases toward the front. This constriction toward the mouth could be still further developed: in *Hemiphragmoceras* (Silurian) there is still a T-shaped opening; in *Octamerella* (Silurian; see Color plate, p. 143), the opening on the dorsal surface develops into a pinnate pattern. These animals could only extend their arms outside the shell to grab food. Another line of development led to the turret-shaped, trochospiral genera, e.g., *Foersteoceras* (Silurian; see Color plate, p. 143). *Stereotoceras* from the Devonian and related species were planospiral (coiled in a plane).

The oncocerids finally gave rise to the NAUTILIDS (order Nautilida; Lower Devonian to present), and this branch has led to the modern chambered nautilus (genus *Nautilus*; see Color plate, p. 143). Nautilids are also planospiral, but unlike oncocerids they have no subsequent depositions in the siphon. This tube is relatively thin and lies approximately in the middle of the chambers. The chamber septae can be sinusoidal on their

Nautiloids

Ellesmocerids

Fig. 9-29. Nautiloids: 1a. Suture line and 1b. shell of *Millkoninckioceras*; 2. *Vestinautilus*; 3. Suture line of *Permoceras*.

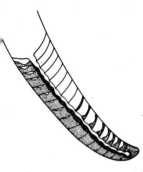

Fig. 9-30. Cross-section through the orthocerid cephalopod *Pseudocyrtoceras*. Subsequent exuded calcium in the chambers is shown by the dotted area; calcium appearing in the tube of the shell is hatched.

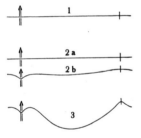

Fig. 9-31. 1. Suture lines of the orthocerid *Michelinoceras*; 2a. *Bactrites* early in life and (2b) at maturity; 3. *Lobobactrites*.

Orthocerids

edges. If the outer chamber edge, the point of contact with the shell, is rolled out on a plane surface, an undulating, often complex line results, and this is known as the suture line (see Fig. 9-29). It is used for its systematic value, along with the external shape of the shell and its sculpture.

We shall cite the following nautilids as exemplifying the order: *Millkoninckioceras* (Lower Carboniferous to Permian; see Fig. 9-29, 1a and 1b) is characterized by a smooth, round shell with a perfectly straight suture line (i.e., septae which do not have an undulating pattern). The shell only slowly increases in height. In *Permoceras* (Lower Permian), the spirals partially overlap, and the septae are so strongly curved that the suture line (see Fig. 9-29, #3) has a complexity seen usually only in ammonites. Another genus (*Pseudonautilus*), which resembled *Permoceras* in all respects, appeared in the Upper Jurassic, more than 100 million years later. One genus with a deviant shell structure, *Vestinautilus*, is from the Lower Carboniferous. The only surviving genus from all the nautilids, *Nautilus*, is all that remains of some 160 nautilid genera, 100 of which never lived beyond the Paleozoic. Nautilids flourished during the Permian, and the nautilus of today is a living relict of a fauna age millions of years—no, hundreds of millions of years—ago.

Coiled species comprised most of the TARPHYCERIDS (order Tarphycerida; Ordovician and Silurian) and BARRANDEOCERIDS (order Barrandeocerida; Middle Ordovician to Middle Devonian). Unlike nautilids, tarphycerids produced calcium secretions in the shell tube. The end of the shell is often not coiled like the front part, producing a constricted mouth opening in a genus such as *Ophioceras* (Upper Silurian). The shell of one of the well-known barrandeocerids, *Lituites* (Middle Ordovician; see Color plate, p. 143), was only coiled in the beginning portion.

DISCORIDS (order Discorida; Middle Ordovician to Middle Devonian) lacked complete coiling, and they also possessed a constricted mouth section. The shell increases in volume rapidly as one passes down the body. One of the characteristic discoid genera is *Phragmoceras* (see Color plate, p. 143) from the Middle Silurian.

The ASCOCERIDS (order Ascocerida; Middle Ordovician to Upper Silurian) were unusual in some respects. The shell consisted of two differently shaped parts (as in *Ascoceras*; see Color plate, p. 143). In the juvenile stage a "normal", only slightly arched, shell is formed. This is then joined by an attached shell in which the chambers collect on one side. After this adult shell has formed, the juvenile section is cast off. This change was probably associated with some change in life habits.

The ORTHOCERIDS (order Orthocerida; Lower Ordovician to Upper Triassic) form an important group. The genus *Michelinoceras* (see Color plate, p. 143; formerly known as *Orthoceras*), which lived during the entire time span in which the order developed, has the simplest structure of the entire group: a straight, slender shell, a conical initial chamber, hourglass-shaped septae, a thin tube in or near the middle, and no subsequent calcite secretions. Some orthocerids are slightly arched, while others have highly sculptured shells. Calcite secretions also occur in some species. They are more strongly pronounced in juvenile segments and on the ventral (belly) side than elsewhere. These secretions appear to have helped maintain stability as the orthocerid swam (assuming it maintained the shell in a horizontal position while swimming). The few preserved coloration patterns on the dorsal side also support this contention. A cross-section through the shell of *Pseudocyrtoceras* (Upper Devonian to Lower Permian) shows how the calcium deposits were arranged (see Fig. 9-30). The size of the shell varies within the orthocerid group from a few centimeters to nearly two meters (we speak here of adult size). The high point of the flourishing period (from the Ordovician to the Middle Devonian) was in the Silurian, when orthocerids lived in masses, all pointed in one direction due to the prevailing water current. They formed whole deposits. This Silurian order probably gave rise to the ammonites.

The ENDOCERATES (Endoceratoidea; Ordovician) are classified as a separate subclass. They include the largest nautiloids with shells up to 4 m long, and during the Ordovician they formed the so-called Vaginate limestone. This name is from the vaginalike insertions of the deposits in the originally very extensive siphon.

These depositions were much more complex in the inside of the siphons of ACTINOCERATES (subclass Actinoceratoidea; Ordovician to Upper Carboniferous; see Color plate, p. 143). They seem to have formed a well-developed system of small vessels. Calcium was secreted even within the chambers. This original body plan evolved rapidly after it appeared in the Ordovician and then remained nearly unchanged during the Silurian flourishing period and remained until extinction in the Upper Carboniferous.

The orthocerids gave rise to the BACTRITOIDS (Bactritoidea; Ordovician?, Silurian to Permian). The simplest bactritoid, *Bactrites* (Lower Devonian to Permian; see Color plate, p. 143), differs from the orthocerid genus *Michelinoceras* in that the initial chamber is egg-shaped and clearly distinct from the rest of the shell, and furthermore in that the siphon is located laterally. Since the rearward bend of the septal wall along the siphon borders with the outer body wall, the suture line arches backward at that point. This creates an outer lobe on the suture line sketch (see Fig. 9-31 and Chapter 12). A genus with such an outer suture line lobe appeared as early as the Lower Ordovician (*Eobactrites*). There is great doubt that this bactritoid is directly related (in a linear way) with more recent bactritoids, however. In *Lobobactrites* (Lower to Middle Devonian; see Fig. 9-31, #3), the suture line has not only an outer lobe but also a broad downward arch on the side of the front of the shell. This is known as a lateral lobe. This feature is important because it ties

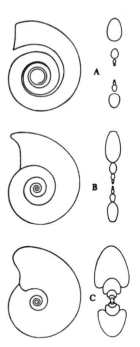

Fig. 9-32. Schematic lateral views and cross-sections illustrate increasing coiling in early ammonites: A. *Anetoceras*; B. *Gyroceratites*; C. *Agoniatites*.

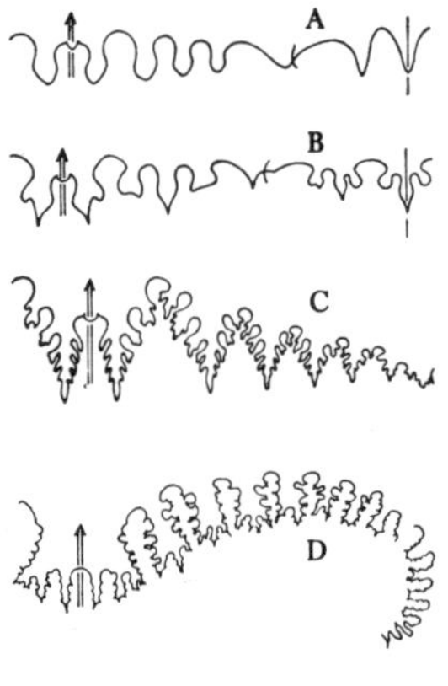

Fig. 9-33. Suture lines of four goniatites: A. *Agathiceras*; B. *Shumardites*; C. *Perrinites*; D. *Cyclolobus*.

Bactritoids

this organism in with ammonites, since primitive ammonite suture lines also display lateral lobes. A coiling tendency can be seen in *Cyrtobactrites* (Lower Devonian), with its moderately bent shell.

Another not uncommon feature should also be mentioned in connection with *Bactrites*. This genus has the simplest body plan of the subclass, just as *Michelinoceras* has in the orthocerid group. These basic types survived much longer in most cases than those more elaborate ones descended from them. In fact, each of the above two genera lived during the entire period their groups existed. Furthermore, they were more than once the starting point for yet another phylogenetic line. *Bactrites*, for example, is actually a descendant of *Michelinoceras*, and it gave rise to lines leading to the ammonites and to squid species.

AMMONITES (subclass Ammonoidea; Lower Devonian to Upper Cretaceous) owe their name to the Egyptian sun god Ammon. Heavily ribbed ammonites were thought by ancient Egyptians to be the horns of the ram, the animal symbolizing Ammon and his sacred animal. Hindus have venerated these discus-shaped ammonites as the parcheesi throwing shells of Vishnu, one of the chief Hindu gods. The earliest ammonites are more or less fully coiled bactritids. As evolution proceeded, the degree of coiling increased. The fossil record has given us the progressive nature of this development, as we see for example in the series of genera *Anetoceras-Gyroceratites-Agoniatites* portrayed in Fig. 9-32. Initially the convolutions did not touch each other, later they did, and still later (as in *Mimagoniatites* of the Lower Devonian, not shown here) they enveloped each other, a process carried still further in *Agoniatites*, in which there is no longer any space inside the shell close to the initial chamber (see right side, Fig. 9-32). This phylogenetic development is reflected in the ontogenetic development of each organism. The egg-shaped to conical initial chamber is followed by a specially developed septum. The next septum has the simplest structure of any within the entire shell, and its suture line has the outer lobe, a lateral lobe, and an inner lobe found in Paleozoic species. It is therefore three-lobed and is technically called a trilobate primary suture. The earliest ammonites are characterized by such a structural configuration. In later forms the folding of the suture line becomes more complex during the growth of the individual animal, so that phylogenetic development can be viewed ontogenetically.

Because of these rapidly changing characteristics, ammonites have been important for the study of evolution as well as key fossils. The basic structure of all systems from the Devonian to the Cretaceous can be identified with the help of ammonites. Paleozoic ammonites differ from their Mesozoic relatives in the simple undulating suture line. Furthermore, Mesozoic ammonites have four or five lobes in the first suture line vs. just three in Paleozoic species. This is the basis of identifying the latter as old ammonites or goniatites.

The first goniatites appeared in the late Lower Devonian. Some of

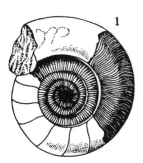

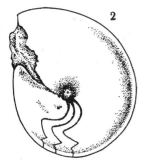

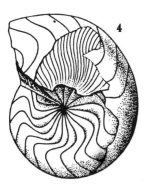

Fig. 9-34. Early ammonites: 1. *Anarcestes*; 2. *Maenioceras*; 3. *Beloceras* suture line; 4. *Tornoceras*.

▷▷
An Upper Carboniferous landscape:
Plants: Equisetum (Calamites: a. *Calamitina*; b. *Stylocalamites*; c. *Eucalamites*; i. another Equisetum, *Sphenophyllum*), lycopods (d. Lepidodendraceae trees; e. Sigillariaceae trees; n. Moor fern), tree ferns (g. *Parsonius*), pteridosperms (h. *Lyginopteris*; f. *Medullosa*) and cordaitins (1. *Eu-Cordaites*; k. *Poa-Cordaites*; m. *Dory-Cordaites*);
Animals: myriapods (3. *Acanthopleura*), scorpions (7. *Cyclophthalmus*), insects (5. *Meganeura*; 6. *Phyloblatta*, near foreground), amphibians (1. *Dolichosoma*; 2. *Urocordulys*) and early reptiles (4. *Hylonomus*).
(continued on p. 205)

these ANARCESTINES (suborder Anarcestina; Devonian) have already been mentioned. The namesake of the group is the Lower and Middle Devonian genus *Anarcestes*. The key fossil of the highest Middle Devonian, *Maenioceras*, is characterized by an additional suture line lobe. In the Upper Devonian, *Manticoceras* has a small saddle formed from the outer lobe (see Color plate, p. 224), and in *Beloceras* the suture line (see Fig. 9-34, #3) has become extremely complex and has many more lobes than in any predecessors.

The CLYMENIANS (suborder Clymeniina), an Upper Devonian group, are an exception. Their siphon is not outside but inside, and therefore there is no lobe on the outside. The Color plate, p. 224 depicts two clymenians, *Platyclymenia* on the left side and *Wocklumeria* on the right.

▷
Phylogeny of tetrapods in the Paleozoic:
1. *Osteolepis* (an osteolepidid crossopterygian, the group that gave rise to tetrapods).
AMPHIBIANS (AMPHIBIA): A. Lepospondyli, with aistopods (Aistopoda; 3. *Dolichosoma*), nectrideans (Nectridea; 4. *Urocordylus*), microsaurs (Microsauria; 5. *Microbrachis*) and Lysorophia ("protosalmander"). —B. Labryinthodonts (Labryrinthodontia), with Ichthyostegalia (2. *Ichthyostega*), Temnospondyli-like rhachitomes (Rhachitomi, ancestors of Lissamphibia; 6. *Eryops*) and Stereospondyli (7. *Gerrothorax*), Plagiosauria and Batrachosauria, such as anthracosaurs (Anthracosauria; 8. *Pholidogaster*), Seymouriamorpha (9. *Seymouria*) and Gephyrostegidea. —C. Lissamphibia (all three present orders are descendants of the rhachitomes).
REPTILES (REPTILLA): A. Anaspida, with the primitive cotylosaurs (Cotylosauria; 10. *Hylonomus* (see also Color plate, p. 202/203)). B. Araeoscelidians, with the Araeoscelidia (11. *Araeoscelis*). C. Proganosauria, with the mesosaurs (Mesosauria; 12. *Mesosaurus*). D. Lepidosaurs (Lepidosauria), with the millerosaurs (Millerosauria; 13. *Millerosaurus*) and Eosuchia. E. Archosaurs (Archosauria), with the thecodonts (Thecodontia; 14. *Ticinosuchus*). F. Synapsida, with the pelycosaurs (Pelycosauria; 15. *Dimetrodon*) and therapsids (Therapsida; forerunners of mammals) such as Deinocephalia (16. *Moschops*), Anomodontia, and theridonts (Theriodontia; 17. *Lycaenops*).

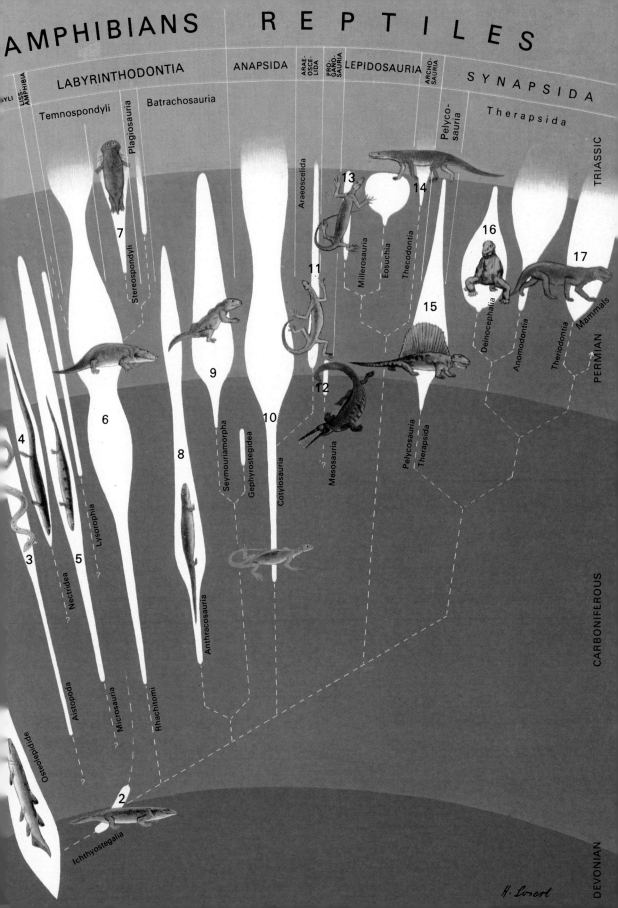

(Continued from p. 200)

LYCOPODS: 8. *Lycopodites* (weedlike lycopod; Carboniferous); 9. *Lepidodendron* (Lepidodendraceae; Carboniferous; height about 20 m); 10. *Sigillaria* (Sigillariaceae; Upper Carboniferous; height about 15 m); 11. *Selaginellites* (a selaginella; Upper Carboniferous); 12. *Cyclostigma* (Upper Devonian/Lower Carboniferous; height about 6 m); 13. *Protolepidodendron* (Devonian; height about 0.5 m); 14. *Duisbergia* (Middle Devonian; height up to 2 m).

EQUISETUM: 15. *Hyenia* (Middle Devonian; height about 0.5 m); 16. *Pseudobornia* (Upper Devonian; height about 10 m); 17. *Calamitina* (Upper Carboniferous; height about 10 m); 18. *Sphenophyllum* (Carboniferous; height about 1 m); 19. *Schizoneura* (weedy Equisetum; Permian; height several meters).

◁

Plant phylogeny during the Paleozoic.

PSILOPHYTES: 1. *Rhynia* (Lower Devonian, height about 0.5 m).

FERNS: Primofilices: 2. *Cladoxylon* (Middle and Upper Devonian, height about 0.5 m); 3. *Stauropteris* (Carboniferous; reconstruction of a frond).

Eusporangiate ferns (Maratiales): 4. *Psaronius* (Lower Permian, height about 10 m).

SEED PLANTS: Pteridosperms (Pteridospermae): 5. *Medullosa* (Upper Carboniferus, height about 5 m).—
Conifers: 6. *Ullmannia* (Upper Permian).—
Cordaitins: 7. *Eu-Cordaites* (Upper Carboniferous, height about 30 m).

The anarcestines gave rise to the GONIATITES (suborder Goniatitina; Middle Devonian to Upper Permian). *Tornoceras* (see Fig. 9-34, #4) is a simple form from the Middle and Upper Devonian. In *Cheiloceras* and the related genera *Sporadoceras* and *Imitoceras* (see Color plate, p. 224) from the Upper Devonian, *Gattendorfia* from the Lower Carboniferous, and *Hunanites* from the Permian, the suture line is more complex. Another lobe develops on both sides of the lateral lobe. A yet greater degree of suture line complexity is found in the AGATHICERATES, members of which include *Agathiceras* (Upper Carboniferous to Middle Permian), *Shumardites* (Upper Carboniferous), *Perrinites* (Middle Permian), and *Cyclolobus* (Upper Permian). A closely related genus, *Waagenoceras*, is shown on the Color plate, p. 224. In another line, which developed from *Goniatites* (Lower Carboniferous; see Color plate, p. 224) via *Gastrioceras* (Upper Carboniferous; see Color plate, p. 224) to *Adrianites* (Middle Permian; see Color plate, p. 224), the variations in suture line patterns are not nearly as pronounced.

The anarcestines (from forms such as the PROLOBITES (Upper Devonian)) also gave rise to the PROLECANITES (suborder Prolecanitina; Upper Devonian to Upper Triassic). One prolecanite, *Prolecanites* (Lower Carboniferous; see Color plate, p. 224) started a line leading to the XENODISCIDS (*Xenodiscus*, Upper Permian; see Color plate, p. 224). This small group gave rise then to the Triassic ammonites (see Color plate, p. 224), which are called ceratites. A second line, beginning with the Permian genus *Medlicottia* (see Color plate, p. 224) and extending into the Mesozoic era, ended without further descendants in the Triassic.

It is an interesting aspect of ammonite evolution that the number of genera and species experiences a decrease over a long period of time followed by a proliferative burst and yet another gradual decrease. These changes happened between the Middle and Upper Devonian, the Upper Devonian and the Lower Carboniferous, and at the end of the Permian period. Each time, a simple, relatively primitive body plan gave rise to a new line, and it was those who descended from that simple plan and were more complex that died out.

During the Carboniferous period, the bactritids gave rise to the Dibranchiata (or SQUIDS), the second of the two cephalopod subclasses still found today. The oldest dibranchiate finds are from the Upper Carboniferous, and they belong to the BELEMNITES (subclass Belemnoidea; Upper Carboniferous to Upper Cretaceous). They are characterized by a new modification; a firm calcareous body called the rostrum surrounds the straight, chambered housing inherited from the orthocerids and bactritids. Thus, unlike their predecessors, belemnites surround these hard parts with the soft body parts, just as modern squid surround their pen. After a long initial developing period, belemnites proliferated in the Mesozoic era, and they will be discussed in more detail in the chapter handling that era.

The Paleozoic era is distinguished from the following epochs by its cephalopods. During the Lower Paleozoic, primitive cephalopods were predominant, and of them all only the orthoceratids and nautilids lived past the great step into the Mesozoic. Ammonites appeared in the Devonian, beginning with the primitive goniatites. With a single exception, all goniatites are restricted to the Paleozoic.

Paleozoic deposits often contain various groups of small, often conical shells. In most cases these fossils can be classified as mollusks. However, the class within this phylum to which they belong has not yet been determined.

HYOLITHIDS (order Hyolithida) were widely distributed from the Lower Cambrian to the Middle Permian. They are characterized by bilaterally symmetrical, thin, conical shells of calcite, which can be closed by a cover and can reach a L of several centimeters. *Camerotheca* (from the order Camerothecida; Cambrian) has a similar shape but has intermediate floors in the shell, as had *Hyolithellus* (order Hyolithelminthes; Cambrian).

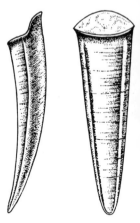

Fig. 9-35. Hyolithids (Hyolithida).

TENTACULITES occurred from the Lower Ordovician to the Upper Devonian, and today these organisms are generally classified into two orders (Tentaculida and Dacryoconarida). They are characterized by sharply tapered, conical shells, which are calcareous and are sculptured. The typical tentaculite genus, *Tentaculites* (Silurian and Devonian), is usually several centimeters in length. Prominent ringed thickened sections of the shell in this genus often form transverse floors, indicating that the genus inhabited the ground. In contrast, *Nowakia* (Upper Silurian to Upper Devonian), a characteristic representative of the DACRYOCONARIDS, often reaches a L of just several millimeters. The thin shell also has rings, and longitudinal ribs as well. The shell of *Styliolina* (Devonian) is completely smooth. These small species did not live on the floor but passively or actively swam in the open sea. Tentaculites have been known for over 150 years, but they have only recently received renewed attention, for they have proven of importance as key fossils.

The TENTACULATA phylum is of phylogenetic interest because it has several characteristics intermediate between the two phyla surrounding it, Protostomia and Deuterostomia. The placement of Tentaculata is not based on the fossil record, for the decisive criteria are the soft body parts, and these are not usually retained in fossils. The only tentaculates represented in the fossil record are those with hard body parts. One exception to this may be the phoronid worms (class Phoronidea), since recent studies by E. Voigt indicate that these "worms" may have bored certain passages found in fossils. Some Paleozoic tentaculates may have evolved from phoronids.

BRYOZOANS (class Bryozoa) can be dated with certainty from the Ordovician period. They are found in Ordovician deposits in all shallow water strata, and during the Carboniferous and Permian periods, bryozoans were important contributors to reef building. The order Cheilo-

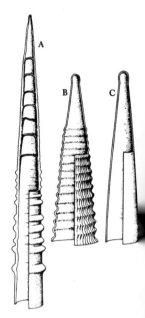

Fig. 9-36. Schematic drawing of three Tentaculata: A. *Tentaculites*; B. *Nowakia*. C. *Styliolina*.

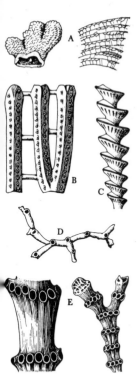

Fig. 9-37. Corallike bryozoans (Bryozoa): A. Monticulipora (full view and in
cross-section); B. Section
of Fenestella; C. Archi-
medes; D. Stomatopora;
E. Mitoclema.

Brachiopods

stomata does not appear until the Cretaceous, but the Stenostomata and Ctenostomata are represented from the beginning onward. There are two other orders from the Ordovician to the Permian: Trepostomata and Cryptostomata. Since we must assume that bryozoans and brachiopods have a common origin, we can conclude that bryozoans must have been present in the Cambrian, although at that time they did not form skeletons.

Since many massive trepostomates bear a strong resemblance to certain corals, the systematic position of some species is uncertain. Individual animals within colonies were within tubes, inside of which the animals built transverse walls as they built forward. One typical genus is *Monticulipora* (see Fig. 9-37A) from the Ordovician. The cryptostomates have a similar body plan, but individual animals possessed much shorter tubes. A spongelike tissue was often located between the individual rows. One of the most prevalent genera from the Ordovician to the Permian was *Fenestella* (see Fig. 9-37B), which built funnel-shaped structures with cross braces. The animals within the colony sat at the top of the system. *Archimedes* (see Fig. 9-37C), occurring from the Upper Carboniferous to the Permian, built a screw-shaped structure wound around a massive axis.

Only a few ctenostomate genera occurred in the Paleozoic. These were either forms whose shells and other hard parts were covered with a crust (e.g., *Marcusodictyon*; Ordovician) or those which left scratches in these calcified parts (e.g., *Ropalonaria*; Ordovician to Permian). Stenostomate colonies consisted of simple, long tubes lacking transverse voors. The group includes *Stomatopora* (see Fig. 9-37D), a genus distributed geologically from the Ordovician to the present. Other colonial structures were built by the Ordovician and Silurian *Mitoclema* (see Fig. 9-37E).

BRACHIOPODS (phylum Brachiopoda) developed into a huge, diverse group during the Paleozoic. They have a different shell structure from mollusks. Most hingeless brachiopods have a horny-phosphate shell. In all others the shells consists of two calcite layers, which is covered by a thin layer of organic material. The outer, thin calcite layer of the shell is composed of fine fibers running parallel to the surface. The thicker inner shell layer is comprised of transverse calcite crystals. In some groups the shell is interrupted by vertical canals extending outward in a trumpetlike pattern. Others, in which the inner layer of the shell has rod-shaped structures, only appear to have this patterning. These different valve characteristics are useful systematically. Other criteria used as systematic characters are the development of the arm, the shaft hole, the hinge, and muscle attachment sites, and external appearance.

The simplest structural type is found in HINGELESS BRACHIOPODS (class Ecardines or Inarticulata; Cambrian to present). Both valves lack hinges and can only be kept together by the muscles. The lophophores (specialized tactile organs) lack supporting skeleton. In LINGULIDS (order Lingulida; Cambrian to present), the stalk is in a groove between the valves. The group includes the LINGULA (genus *Lingula*; see Color plate, p. 134), an animal that has not changed noticeably in the more than 400,000,000 years since it first appeared in the Silurian period! *Obulus* (see Color plate, p. 134) is a characteristic Cambrian and Ordovician genus. In both these genera the shell is horny-phosphatic, while in *Trimerella* (see Fig. 9-38, #1), a Silurian genus, the shell has powerful calcite plates. Lamellae developed in the shell, and they were supported by a septum, providing a better muscle attachment point.

In ACROTRETIDS (order Acrotretida; Cambrian to present) the stalk penetrates in part through a hole in what is called the stalk hinge (which is ventral, in contrast to the dorsal location of the arm hinge). *Acrotreta* (Ordovician period, with related species surviving into the Upper Devonian), a small, horn-shelled genus, had a cylindrical stalk hinge resting on the arm hinge like a small cover. The CRANIDS (Craniidae; Cambrian?, Ordovician to present), which still exist today, are characterized by a calcite shell with canals, giving it the dotted appearance we described earlier.

HINGED BRACHIOPODS or ARTICULATA (class Testicardines or Articulata), which also have a calcite shell, first appeared in the Lower Cambrian with the ORTHIDS (order Orthida; Cambrian to Permian). They had either dotted or non-dotted shells. The stalk could be restricted or completely closed off by subsequent structures formed in the shell. Two small, vertical plates served as supports for the arms. The infrequent Cambrian orthids were followed by the flourishing period, the Ordovician, characterized by *Orthis* (see Color plate, p. 134), a coarse-ribbed, non-dotted genus with a long, straight shell. Other characteristic genera included the thicker *Platystrophia* (Ordovician to Silurian), the fine-ribbed *Schizophoria* (Silurian to Permian; see Color plate, p. 134), and *Dalmanella* (Ordovician to Silurian), which resembled *Orthis* but was dotted and had a shorter, slightly angular valve margin.

The STROPHOMENIDS (order Strophomenida; Lower Ordovician to Lower Jurassic), with nearly 400 genera, included those with "false-dotted" shells and was the largest brachiopod order. In *Strophomena* (Middle to Upper Ordovician; see Color plate, p. 134), the concave stalk valve was arched in the same direction as the convex arm valve. The opposite relationship is found in *Rafinesquina*, *Leptaena* (Ordovician to Devonian; see Color plate, p. 134) was a prevalent, worldwide genus in which both hinges bent in the direction of the arm valve. In *Strophodonta* (Upper Ordovician to Upper Devonian), the edge of the shell has small teeth. The genus *Chonetes* (Devonian) and related taxa are characterized by spines along the shell edge. In PRODUCTIDS (Productidae), the predominant Carboniferous group, the entire valve could be equipped with spines. See for example *Eomarginifera* on the Color plate, p. 134. *Gigantoproductus giganteus*, with a L exceeding 20 cm, was one of the largest brachiopods. *Oldhamina* (see Color plate, p. 134), a Permian genus,

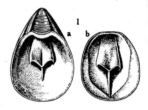

Fig. 9-38. Inarticulata: 1. *Trimerella*; 2. *Acrotreta*; in each, a and b depict the stalk and arm valves, respectively.

Articulata

Fig. 9-39. Hinge in *Strophodonta*, with a double attachment site and teeth.

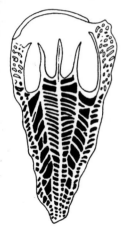

Fig. 9-40. Strophomenids: schematic section through *Richthofenia.*

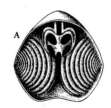

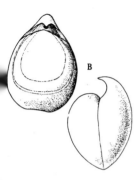

Fig. 9-41. Spirifers: Valve of *Athyris*, showing spiralling structures. A. Schematic view; B. Exterior and lateral view.

had a very unusual shape. The valve had a medial ridge on the inside, from which smaller ridges extended laterally, reaching the ends of the valves. *Richthofenia* (see Fig. 9-40) is another Permian genus. Its coral-like appearance was an adaptation to the reef habitat in which the genus occurred. Transverse surfaces were built across the shell during growth.

The PENTAMERIDS (order Pentamerida; Middle Cambrian to Upper Devonian) typically had greatly arched, non-dotted valves with a short hinge; powerful, supported muscle plates; and medial septae. The Silurian genus *Pentamerus* (see Color plate, p. 134) is representative of the group.

In RHYNCHONELLIDS (order Rhynchonellida; Middle Cambrian to present), shell rigidity is maintained in spite of the short hinge by a deep fold in the front edge of the valve. Rhynchonellids have rounded, non-dotted shells, usually with an indentation (or sinus) on the valve. The base of the gill arm is calcified. *Pugnax* (Middle Devonian to Carboniferous; see Color plate, p. 134) has a very pronounced sinus. In *Uncinulus* (Devonian) the ribs extend into long spines at the ends, and thus a netlike webbing is formed at the opening of the shell. The webbing prevents large particles from entering the shell interior.

SPIRIFERS (order Spiriferida; Middle Ordovician to Lower Jurassic) played a particularly important role. They differ from all other brachiopods in having calcified, spiral arm supports. *Atrypa* (see Color plate, p. 134), a genus with relatively fine ribs, is one of the earliest, least differentiated forms. It occurred from the Silurian to the Devonian period. *Athyris* (Devonian to Triassic) and related forms look externally like terebratulids (see below), but the former are usually somewhat broader. SPIRIFERS (Spiriferidae), the group around *Spirifer*, are particular important key fossils (see Color plate, p. 134). Other important genera include *Cyrtina* (Silurian to Permian), *Uncites* (Middle Devonian), and *Retzia* and *Meristella* from the Lower Devonian.

In TEREBRATULIDS (order Terebratulida; Lower Devonian to present) the supporting skeleton for the gill arms is composed of rather simple loop. The configuration of this loop can be used as the basis of distinguishing three major terebratulid groups. The first includes key fossils from the Lower Devonian, e.g., *Rensselaeria*, which occurred on sandy floors; and *Stringocephalus* from Middle Devonian reefs. The second group is represented by the modern terebratulids, whose external appearance is largely the same as *Dielasma* (Upper Carboniferous to Permian). The third group, to which the modern genus *Terebratella* belongs, was only represented in the Paleozoic era by a few forms (e.g., *Cryptonella*; Lower Devonian to Permian).

The Paleozoic era, in contrast with later epochs, was characterized by a much greater variety in brachiopods than was later the case. Of the 200 families known today, at least 100 of them were restricted to the Paleozoic era. All orders other than rhynchonellids and terebratulids flourished before the Mesozoic era. Today, brachiopods are highly confined in terms of their distribution and habitats in which they are found, but in the Paleozoic they were widely distributed in both respects. Outside of the reefs, brachiopods appear to have been the most prevalent shallow sea fauna, for their fossils in shallow sea depositions are the predominant ones.

ECHINODERMS (phylum Echinodermata) today include five classes. The most modern, up-to-date paleontological work (*Treatise on Invertebrate Paleontology*) adds thirteen more Paleozoic classes. If members of one of these classes could be found today, they would have to be placed in their own class, so great would the differences be between any of them and extant echinoderms. If the differences were viewed as part of the phylogenetic development of echinoderms and if the fact were appreciated that major differences can arise even in closely related groups when new body plans are being evolved, then the differences between the extinct and extant echinoderms would be somewhat exaggerated by the construction of a new class for any "extinct" echinoderms found today. At any rate, the great diversity in echinoderms is already evident in the Paleozoic era. We can only mention a few of these extinct groups.

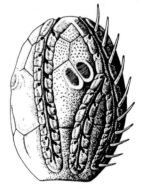

Fig. 9–42. *Callocystites*, a cystoid.

Echinoderms may have existed in the Precambrian (see Chapter 7). However, the fossils available from that distant time give us no clues about the origin and initial growth of this phylum. Members of all echinoderm classes are found in the early Paleozoic era. While all modern echinoderms have a pentaradial structure, this characteristic was barely present in the early species.

The HOMALOZOANS (Cambrian to Devonian), also known as CARPOIDS (subphylum Homalozoa), were a very peculiar group. They had a flattened, asymmetrical shell composed of irregular plates. In the genus *Dendrocystoides* from the Ordovician (class HOMOIOSTELEA; Lower Cambrian to Lower Devonian; see Color plate, p. 144), one end of the body had a stalk and the other had an arm, and both were covered with bilaterally symmetrical plates. *Ceratocystites* (Middle Cambrian, class STYLOPHORA; Middle Cambrian to Middle Devonian) had a similar structure but lacked the stalk. The opposite is true of *Trochocystites* (class HOMOSTELEA; Middle Cambrian), which had no arm but did have a short stalk. It is likely that most of these animals dwelled on the ocean floor.

CRINOZOANS (subphylum Crinozoa) includes all groups closely related to crinoids and feather stars. All of these echinoderms have radial symmetry. The spherical or cuplike body is typically covered with regularly arranged, often polygonal plates. A stalk is present throughout life or just in the juvenile stage. Arms are also developed, and these are used for feeding.

The oldest crinozoans are EOCRINOIDS (class Eocrinoidea; Lower Cambrian to Silurian), and in certain respects they are intermediate between cystoids and crinoids. The ramification of the ambulacral groove and the lateral location of the anus are cystoidlike features, while the cup plates are, as in crinoids, arranged in three rings. One of the representative eocrinoids was *Macrocystella* from the Lower Ordovician (see Color plate, p. 144), with a 23-cm cup bearing five paired arms. The long stalk was used to anchor the body by wrapping the stalk around a suitable object.

CYSTOIDS (class Cystoidea; Lower Ordovician to Upper Devonian) were a diverse group of stalked echinoderms. The cup in cystoids consisted of numerous, irregular or regular plates, most of which were porous. Brachioles (catching fingers) were on the ambulacral fields, and true arms were absent. These ambulacral fields could extend from the head nearly to the origin of the stalk (*Callocystites*; Silurian) or be restricted to the head area (as in *Cheirocrinus*; Ordovician). While the groups to which the above two genera belong have five rays, there were others with less (as few as two) rays. *Echinosphaerites* (see Color plate, p. 144), a genus occurring in masses during the Lower and Middle Ordovician period, possessed a spherical shell composed of numerous polygonal plates with two to three brachioles near the mouth. The very short stalk was used to anchor the animal.

BLASTOIDS (class Blastoidea; Silurian to Permian) differ from cystoids in having exclusive pentamerous symmetry; the fewer, uniform plates; and an absence of pores in the plates. The brachioles were in close-set rows on the five ambulacral fields, and short, folded inward invaginations (known as hydrospires) were at the edges. Hydrospires probably had a respiratory function. They terminated in two longitudinal canals per ambulacral field, and the canals extended outward to separate or common openings in the apex of the animal. Blastoids were particularly prevalent during the Lower Carboniferous period in North America and in the Permian of Timor. Representative genera include *Pentremites* (see Fig. 9-43) and *Orophocrinus* (see Color plate, p. 144).

CRINOIDS (class Crinoidea) appeared in the Lower Ordovician, and they developed into the most diverse echinoderm group of all. Most fossils had very long stalks. They often formed the major part of deposits in shallow seas, although most of these deposits are not composed of the entire animal but just their appendages (trochites). The few stalked forms among modern crinoids occur in deep seas, and most modern species lack stalks and are free-swimming organisms. Thus, modern crinoids and feather stars are placed in subclass ARTICULATA (since the Triassic) as opposed to all other Paleozoic subclasses.

In subclass CAMERATA (Ordovician to Permian), the cup plates were joined by a stiff connection. Early species had a large number of plates, but with evolution the plate number steadily decreased. The chiefly double-rowed arms (single-rowed at their tips) usually bore inward oriented brachioles or fibers. The stalks often had root-shaped processes. *Acanthocrinus* (Lower to Middle Devonian; see Color plate, p. 144) pos-

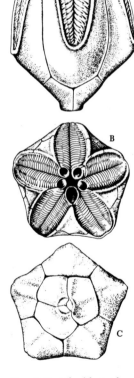

Fig. 9-43. The blastoid *Pentremites*: A. Side view; B. From above (the mouth); C. From below (with attachment point).

sessed cup plates with spines. *Scyphocrinites* (Upper Silurian to Lower Devonian) employed loboliths for swimming; these were spherical swellings at the roots, and they floated in the water. In subclass FLEXIBILIA (Ordovician to Permian), the cup was flexible, since elastic bands joined the individual plates. The subclass INADUNATA (Ordovician to Triassic), like the Flexibilia, was characterized by a stiff cup, but the arms were free. Their lower parts were not incorporated into the cup. This group includes the prevalent Middle Devonian genus *Cupressocrinites*. The inadunates were the only crinoids surviving into the Mesozoic era, and they were the ancestors of the modern members of the subclass, which arose during the Triassic.

Fig. 9-44. Schematic sectioned lobolith of the camerate *Scyphocrinites*.

Star-shaped, free-moving echinoderms are placed in the subphylum of ASTEROZOANS (Asterozoa; Ordovician to present). The oldest of the three asterozoan classes (and the one ancestral to the other two) are the SOMASTEROIDEANS (Somasteroidea), which are intermediate between crinozoans and later asterozoans. Somasteroideans are only known from the Lower Ordovician to the Upper Devonian, but the PLATASTERIDS (family Platasteriidae) of today, which are generally classified with starfishes, appear to belong to this group. The radiating arms (or ambulacral fields) have a double row of plates, which even surround the mouth area. Numerous rod-shaped elements extend outward from the plates, and they are joined by an organic membrane. Two characteristic somastoidean genera include *Chinianaster* and *Villebrunaster* (see Color plate, p. 144), both from the Lower Ordovician.

True STARFISHES (class Asteroidea; Lower Ordovician to present) of the Paleozoic differ from modern forms in only a few respects. Some groups had a larger plate number. The arms are very broad in the earliest starfishes (e.g., *Platanaster*; Middle Ordovician), and some species even had more than five arms (e.g., *Lepidasterella*; Upper Devonian). Other prominent genera include *Urasterella* (Lower Devonian; Hunsrück formation) and *Neopalaeaster* (Lower Carboniferous), shown in the Color plate, p. 144.

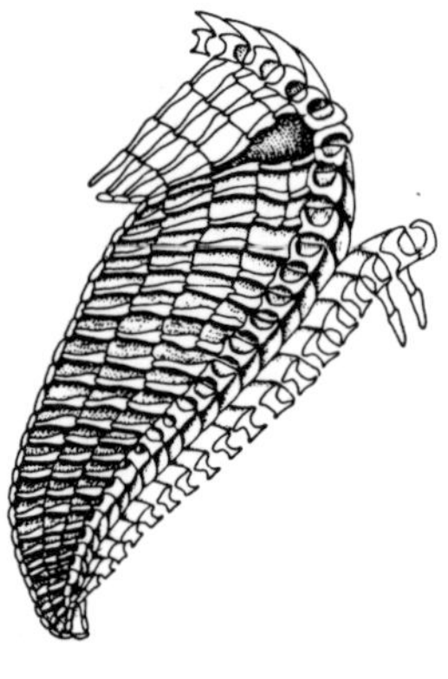

Fig. 9-45. An arm of the somasteroid *Chinianaster*.

BRITTLE STARS (class Ophiuroidea), in which the arms are clearly distinct from the central body disc, arose in the Lower Ordovician, and all brittle star orders were present by the Devonian period. Typical genera include *Bundenbachia*, *Medusaster* (with fifteen arms), and ten-armed *Kentrospondylus* (see Color plate, p. 144).

The second large group of free-living echinoderms are the ECHINOZOANS (subphylum Echinozoa; Lower Cambrian to present). In addition to sea urchins and holothuroids, there were numerous primitive Paleozoic groups deviating considerably from those found today.

Echinozoans

The HELICOPLACOIDS (class Helicoplacoidea) from the Lower Cambrian are among these very different echinozoans. These were free-living, cylindrical animals with a flexible shell composed of spiral plates. The arms were also spiral (as in *Helicoplacus*; see Color plate, p. 144).

OPHIOCYSTIOIDS (class Ophiocystioidea; Lower Ordovician to Middle Devonian) were another small group. The body was covered by stiff, large plates. A number of armlike feet covered with scales were on the lower (mouth) side of the body. The mouth opening itself was protected by five movable plates, while the anus, on top of and on the side of the body, was protected by a pyramidal structure of plates. *Volchovia* (see Color plate, p. 144), from the Lower Ordovician, is representative of the ophiocystioids.

CYCLOCYSTOIDEANS (class Cyclocystoidea; Middle Ordovician to Middle Devonian) were small, disc-shaped echinoderms with a shell composed of numerous plates. Various large plates on the body were arranged in several rings. The upper side of the body was relatively stiff, while the underside (mouth side) was relatively mobile, since the small plates found there contained little calcite (see *Cyclocystoides*, Color plate, p. 144). In EDRIOASTEROIDEANS (class Edrioasteroidea; Lower Cambrian to Lower Carboniferous), the disc- to sack-shaped, flexible shell was also composed of numerous plates. Five arms radiated from the centrally located mouth, and the arms were usually bent. The anus was on the side of the body. The similarity between these and ophiocystioids is only superficial, and edrioasteroideans were more closely related to cystoids (and then only distantly). *Edrioaster* (Middle Ordovician; see Color plate, p. 144) is representative.

Sea urchins

The development of SEA URCHINS (class Echinoidea; Ordovician to present) led during the Paleozoic from flexible shells to stiff ones. Of the five orders in subclass Perischoechinoidea, only Cidaroida lived past the Paleozoic. This subclass unites various sea urchins with different numbers of plate rows. The arms might have one to fourteen rows, while the fields between them could have two to twenty rows of plates. *Bothriocidaris* from the Middle Ordovician period, the oldest sea urchin, has the least number of plate rows. In *Melonechinus* (Lower Carboniferous; see Color plate, p. 144), on the other hand, there are six to twelve rows on the arms and in the field between them there are three to eleven rows. In cidaroids, which appeared in the Devonian, the calcite plates were overlapping like roof tiles in the earliest species, and the shell of the animal was flexible. Toward the end of the Paleozoic the shell became stiffer.

Since SEA CUCUMBERS (class Holothuroidea; Lower Devonian to present) lack a closed skeleton, they have not usually fossilized. However, their skeletal elements have been found dispersed among deposits. The only intact specimens known are from the Lower Devonian Bundenbach (Germany) formation (*Palaeocucumaria*; see Color plate, p. 144), the North American Carboniferous, and one find in a Mesozoic deposit (the Soln-hofener formation, also in Germany). These skeletal elements, called sclerites, have many different shapes, including anchor- and wheel-shaped. Holothuroids are important geologically because they consume fresh deposits and thus help prepare sedimentation.

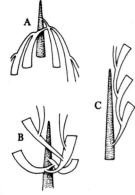

Fig. 9-46. Sprout forms in graptoliths (see text for explanation).

The Paleozoic is thus important for echinoderm evolution. All echinoderm classes of the Mesozoic and Cenozoic were present in the Paleozoic era. Within and in addition to these classes, there were primitive groups found only within the Paleozoic.

The phylum PENTACOELA is of greay phylogenetic interest, because its members share certain characteristics with echinoderms and with chordates (see W. F. Gutmann's discussion on chordate evolution in Chapter 4). Since modern representatives of this phylum have no body parts that would normally fossilize, there is hardly any possibility that the fossil record of Pentacoela can answer all the questions paleontologists have about this group. We are really in need of detailed knowledge about extant groups to clarify the position of some of the fossil pentacoelans. This kind of clarification has indeed been made with GRAPTOLITHS (Graptolithida; Middle Cambrian to Lower Carboniferous), which is classified as an independent class beside the Pterobranchia, both in the subphylum Branchiotremata. Graptoliths are so called because the surfaces they cover often look as if they were covered with writing.

Graptoliths were colonial, and it appears that their tubular shells were made of the same material used by Pterobranchia. The tubes consisted of two layers, of which the outer one was not put down until adulthood was reached. The inner layer formed semicircular segments that met on the ventral and dorsal surfaces in a zigzag line. This characteristic is also found in Pterobranchia (see Fig. 9-47). The colony (called a rhabdosome) consisted of individual cuplike structures (= thecae), which develop by sprouting from a mother cell (sicula) and later from stalk cells (stolothecae). Many highly divergent graptolith body plans have been found dating from the earliest known graptolith period. We shall only discuss two of these, because they were widely distributed and were important as key fossils (the most important key fossils, in fact, of the Ordovician and Silurian).

The DENDROID GRAPTOLITHS (order Dendroidea; Middle Cambrian to Lower Carboniferous) are distinguished by the fact that the stolothecae give rise to two different kinds of thecae, which may indicate a difference in sex. The colony in this order is always thecate (i.e., branching, with many ramifications), and neighboring branches might be joined by transverse bridges. This was true in *Dictyonema*, for example, the longest lived (Upper Cambrian to Lower Carboniferous) dendroid graptolithine genus. The genus appeared in masses in the Lowest Ordovician (see Color plate, p. 152). *Bryograptus* (see Color plate, p. 152) lacked transverse elements, and *Staurograptus* (see Color plate, p. 152) had only a few, regularly arranged branches. Thecae schemes are illustrated in Fig. 9-48.

Species like *Staurograptus* gave rise to the important graptolithine order Graptoloidea (Lower Ordovician to Lower Devonian). This order differed from the previous in that there was just one type of theca. Within the order, the form of branching and thecal sequence could vary con-

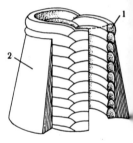

Fig. 9-47. Structure of graptolithine housing, with an inner (1) and outer (2) layer.

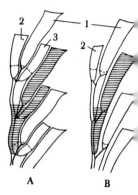

Fig. 9-48. Structure of cup in dendroids (A) and graptoliths (B). The cross-hatched sections in A and B are comparable areas. Autothecal (1) and bithecal (3) cups are those inhabited by the individual member of the colony.
2 = stolotheces.

siderably. Three types are depicted in Fig. 9-46. In A, the first theca developing from the sicula is oriented downward, as are subsequent thecae (the sicula is dotted). In B the first theca is downward, but all the others are oriented upward, and in C they are all oriented upward. Type C (i.e., Monograptidae) appeared in the Silurian. Types A and B prevailed in the Ordovician, and only a few species with these thecal types survived into the Lower Silurian. Within all groups, there were both branched and unbranched genera. The Color plate, p. 152 depicts just a few of these, and we shall not describe them individually.

Paleozoic deposits often contain layers of 5–3 mm fossils known as CONODONTS (class Conodontophora). The name (= conical-toothed) is a reference to the earliest, single-tipped species (e.g., *Distacodus*; Lower Ordovician to Middle Silurian). In others, such as *Belodus* (Ordovician), additional teeth appear, and there is a general process of increasing tooth numbers throughout conodont evolution. Other important genera include: *Plectodina* (Middle to Upper Ordovician), *Keislognathus* (Upper Ordovician), *Hindeodella* (Silurian to Triassic), *Trichonodella* (Ordovician to Devonian), and *Spathognathodus* (Silurian to Triassic). The plate-shaped forms such as *Amorphognathus* (Upper Ordovician), *Polygnathus* (Devonian to Lower Carboniferous), *Palmatolepis* (Upper Devonian), *Gnathodus* (Carboniferous), and *Gondolella* (Permian and Triassic) are all important key fossils. Some of the conodont tooth structures are shown in Fig. 9-49.

Because of their important role, there is a great deal of literature on conodonts. Their fossil remains are composed of fluorine phosphates (fluorapatite), the same mineral in vertebrate teeth. Since some acids (e.g., acetic acid and formic acid) dissolve calcium but not fluorapatite, conodonts can easily be removed from calcium deposits. One kilogram of a calcium deposit can contain hundreds or even thousands of conodont specimens. The animal that bore these various teeth has not yet been identified! However, there are some finds with conodonts in an orderly arrangement, and this fact together with statistical evaluations leads us to believe that conodont-bearers (Conodontophora) were a phylogenetic entity with different tooth types. The individual conodonts, which unlike our teeth were free, helped support the connective tissue surrounding them. Thus, the function of the conodont apparatus was probably filtration, and conodont animals were probably free swimming organisms.

None of these individual characteristics gives us a clear picture of the systematic position of the conodont group. However, if all the available information is pulled together, the picture that results is one of a free-swimming, filter-feeding animal with hard parts composed of fluorapatite. This is the kind of animal that would be close to the ancestral stock of all chordates. If conodont animals are not the chordate ancestors, but only a side branch from the true chordate ancestors, confirmation of the above picture would be of great interest to biologists. Since conodonts

9-49. A. Reconstruction of a conodont apparatus; B to N=isolated, individual conodonts; *Distacodus*; C. *Belodus*; *Plectodina*; E. *Keislognathus*; F. *Hindeodella*; *Spathognathodus*; *Trichonodella*; I. *Amorphognathus*; K. *Polygnathus*; *Gnathodus*; M. *Palmatolepis*; N. *Gondolella*.

first appear in the Cambrian, the origin of chordates is nearly as ancient as that of the other major animal groups. The finds we have of jawless fishes (Agnatha), which are present since the Ordovician, make this supposition of the antiquity of chordates even more likely.

10   Paleozoic Vertebrates

By H. P. Schultze

The oldest known vertebrates are from the Middle Ordovician (not because of poor fossilizing conditions in older strata, for the Cambrian deposits are ideal for fossilization), and they are from Harding sandstone extending from Colorado to Montana. These deposits contain only fragments, scales, and bones of vertebrates. These fossils are of fully developed jawless fishes (superclass Agnatha); specifically, they are members of the genera *Astraspis* and *Eriptychius*, which are both in the heterostracan subclass (Heterostraci) of the pteraspidomorph class (Pteraspidomorphi). Of course, these finds tell us nothing about the origin of vertebrates, their relationship to other chordates, or even to the invertebrates.

The absence of fossil vertebrates from earlier geological periods is presumably due to the fact that in those times vertebrates did not produce apatite, the characteristic mineral building block that readily fossilizes. Unless we are fortunate enough to stumble across some unusually favorable deposit with older vertebrate fossils, we must rely on reconstructing the vertebrate fossils, we must rely on reconstructing the vertebrate ancestors of the past using anatomical features of extinct species and primitive but extant species.

The fossil jawless fishes are divided into two classes with a total of five subclasses:

A. CEPHALASPIDOMORPHS (Cephalaspidomorphi): 1. Subclass OSTEO-STRACANS (Osteostraci); small, only exceptionally up to 60 cm long; head shield with characteristic sensory reception fields; the mouth and the numerous gill openings are on the underside of the head armor; the eyes are set close together, with the pineal opening between them; the internal head skeleton is ossified; the body is armed with high-set scales; the belly is flattened; the caudal fin is heterocercal (i.e., the upper lobe is larger than the lower); osteostracans lived from the Upper Silurian to the Upper Devonian; examples include *Cephalaspis* and *Hemicyclaspis*. 2. Subclass ANASPIDS (Anaspida); very small (L up to 15 cm) fishes lacking a head shield; the gill openings are on the side, and the mouth is at the end of the

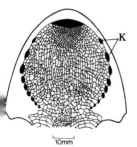

Fig. 10-1. *Hemicyclaspis*, an osteostracan cephalaspidormorph with numerous gill openings (K) on the lower side of the head shield.

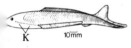

Fig. 10-2. *Jamoytius*, an anaspid cephalaspidomorph, greatly resembling a cyclostome (K = Gill openings).

body; scaled and naked forms occur, each of which in a typical fish shape; the caudal fin is hypocercal (i.e., the lower lobe is larger than the upper); Upper Silurian to Upper Devonian, with the genera *Pharyngolepis*, *Pterygolepis*, and *Jamoytius* (see Fig. 10-2). 3. Subclass CYCLOSTOMES (Cyclostomata); naked, with an eel-shaped body and no ossification; the fins form a continuous seam; the only known fossil form is *Mayomyzon* (Illinois Upper Carboniferous; see Color plate, p. 149); cyclostomes have survived into the present.

B. PTERASPIDOMORPHS (Pteraspidomorphi): 1. Subclass HETERO-STRACANS (Heterostraci); L up to 30 cm (in some cases as much as 150 cm); head armor is composed of several plates with a lateral line system; the eyes are typically lateral; the mouth is on the underside and is nearly terminal (at the very end of the body); lateral gill openings are on the rear of the head armor; the body bears scales and has a typical fish shape; the caudal fin may be multilobular or hypocercal; Middle Ordovician to Upper Devonian; examples of heterostracans are *Astraspis* (see Fig. 10-3 and Color plate, p. 241), *Eriptychius*, *Anglaspis*, *Poraspis*, *Pteraspis* and *Drepanaspis* (see Color plate, p. 189). 2. Subclass THELODONTS (Thelodonti); L 10–20 cm; no head armor; the mouth is terminal; the eyes are on the side; the head is partially flattened; the head and body are covered with numerous small scales (dermal teeth); the caudal fin is hypocercal; thelodonts probably existed in the Ordovician and are known with certainty from the Silurian to the Middle Devonian; examples are *Phlebolepis* and *Thelodus*. Some researchers assign thelodonts to an individual class on the same level as cephalaspidomorphs and pteraspidomorphs.

Unlike existing jawless fishes (i.e., cyclostomes), these "shell-skinned" animals were enclosed in a firm external skeleton composed of head armor and a scale covering over the body. In the oldest genera, *Astraspis* (see Color plate, p. 241) and *Eriptychius*, the head armor was composed of individual, small, polygonal plates; pteraspidomorphs had larger plates, and in cephalaspidomorphs the plates had fused to form a unified head shield. A few Devonian pteraspidomorphs returned to the more primitive head plate structure (e.g., *Drepanaspis* from the Devonian). Other groups apparently never had head armor. Like cyclostomes, fossil agnathans had unpaired fins or fin seams. None of them had paired fins (the condition found in jawed fishes); only occasionally were there structures behind the head bearing some resemblance to paired fins.

An important distinction between the pteraspidomorphs and cephalaspidomorphs lies in the development of the nose. Pteraspidomorphs had paired nasal sacks and paired nares, while the cephalaspidomorph nose was unpaired. This difference alone is an indication that jawed fishes with paired nares are more closely related to pteraspidomorphs, while modern agnathans (the cyclostomes) could only be closely related to cephalaspidomorphs. Erik Stensiö's contention that hagfishes evolved from the pteraspidomorphs is unlikely in view of this evidence. As in cyclo-

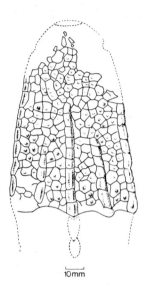

10mm

Fig. 10-3. *Atraspis* from the Middle Ordovician, a pteraspidomorph with head armor of small polygonal plates.

Fig. 10-4. *Poraspis*, a pteraspidomorph with head armor of large plates (K=gill opening).

stomes, cephalaspidomorphs possess external gill openings, which lie either on the underside of the head shield or on the side of the body. The single fossil cyclostome (*Mayomyzon*) bears a considerable resemblance to modern lampreys.

The ossified inner skeleton of the skull permits us to reconstruct the osteostracan brain, including the nerves and blood vessels (see Color plate, p. 172). In 1927, the great Swedish paleontologist Erik Stensiö published his finding that demonstrated for the first time how serial cuttings could be used to reconstruct a fossil brain cavity. In serial cutting, a piece of fossil skull is sliced, piece by piece in very thin sections, all sections being the same thickness. Although only the hard body parts remain in fossils, they contain cavities and canals in which the brain, nerves, and blood vessels had been housed. Cavities can only be used to study parts when those parts had nearly filled the respective cavity. This requirement was neatly met in early agnathans and in most Paleozoic fishes, since ossification occurred throughout most of the skull. Thus, we can study phylogenetic relationships on the basis of organ structures even when those organs have not been retained directly in the fossil record.

The results of this kind of study have shown that the cephalaspidomorph brain resembles that of modern cyclostomes. The labyrinth (or inner ear), for example, has two halfmoon-shaped arches just as we find in cyclostomes, while all other fishes have three of these arches. Only for osteostracans do we have insight into the structure of the brain, nerves, and blood vessels, for the skull was not ossified in other fossil agnathans. We know nothing at all about the soft parts of pteraspidomorphs, which would have perhaps given even more information about jawless fish phylogenetic relationships.

Fossil jawless fishes occupied numerous ecological niches. Of the cephalaspidomorphs, the osteostracans were bottom dwellers (flattened body, mouth on the underside), while anaspids, with their spindle-shaped bodies, were free swimmers. In heterostracans, development also proceeded from early bottom-dwelling forms to free-swimming animals. In some of them, the mouth was on the upper side of the body, indicating that they must have fed near the water surface. Widespread thelodonts were at least to some extent free swimming, and their scales have been recovered from many deposits. Most Paleozoic agnathans lived in coastal sea water, with the sole exception of *Mayomyzon* which probably occurred in fresh water as well, since it has been found in delta depositions of the Upper Carboniferous.

JAWED FISHES (Gnathostomata) developed from the ancestors of agnathans. In Gnathostomata, the front gill arches were modified into biting jaws, and this is one of the major distinctions between Agnatha and Gnathostomata. Another external feature by which the two groups can be readily distinguished are the paired fins (the pectoral and pelvic fins),

Reconstruction of the brain, nerves, and blood vessels

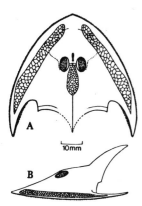

Fig. 10-5. *Cephalaspis*, a cephalaspidomorph with a unitary head shield. A=seen from above; B=lateral view.

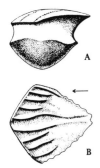

Fig. 10-6. Acanthodian scales (magnification 40x): A. Seen from the side (as they stick in the skin); B. Seen from above (as it appears on the surface of the skin).

▷▷
A Permian steppe landscape: Plants: Pteridosperms (a and b *Glossopteris*) and cordaitins (c. *Noeggerathiopsis*). Animals: mammallike reptiles (Synapsida): 1. *Galepus*; 2. *Moschops*).

which only occur in jawed fishes. The earliest jawed fishes, which were SPINY SHARKS, have been found in the Upper Silurian, the same time the cephalaspidomorphs and higher bony fishes appeared.

The SPINY SHARKS (Acanthodii) were rather unusual looking creatures. They were small (L not usually more than 30 cm), and the head and spindle-shaped body were covered with very small rhomboid scales. The head lacked bony plates; the notocord was unsegmented; the eyes were large; the mouth was terminal; the caudal fin was heterocercal. Spiny sharks occurred from the Upper Silurian to the Lower Permian. Examples of the group include the genera *Climatius, Euthacanthus, Triazeugacanthus* (see Fig. 10-7), and *Acanthodes* (see Color plate, p. 149).

All the fins in spiny sharks lacked true fin rays, but in front the fins were strengthened by a spine. The early spiny sharks had additional spine pairs between the paired pectoral fins and the paired pelvic fins.

▷
Phylogeny of Synapsids and Early Mammals.
MAMMALLIKE REPTILES or SYNAPSIDS (SYNAPSIDA): A. Pelycosaurs (Pelycosauria), with the caseoids (1. *Cotylorhyncus*) Edaphosauria, Ophiacondontia (2. *Ophiacodon*) and Sphenacodontia (3. *Dimetrodon*). —B. Therapsids (Therapsida), mammalian precursors, with the Deinocephalia (4. *Ulemosaurus*) and Anomodontia with the groups Dromasauria (5. *Galepus*), Venyukovoidea and Dicynodontia (6. *Kannemeyeria*); further, the Theriodontia, with the groups Eotheriodontia (7. *Phthinosuchus*), Gorgonopsia (8. *Scymnognathus*), Therocephalia (9. *Lycosuchus*), Bauriomorpha (10. *Bauria*), Cynodontia (11. *Cynognathus*) and Tritylodontia (12. *Oligokyphus*).
MAMMALS: A. Mesozoic mammals, including the Multituberculata (13. *Ctenacodon*), of which the haramiyids can only be provisionally included, the Triconodonta (15. *Trinacodon*), to which the Sinocodontida (15. *Sinocodon*) and Eozostrodontida probably belong, the Docodonta, the Symmetrodonta (16. *Spalacotherium*), which probably include the kuehneotheriids and the Pantotheria (17. A member of the family Dryolestidae). — B. True mammals, with the marsupials and the higher mammals.

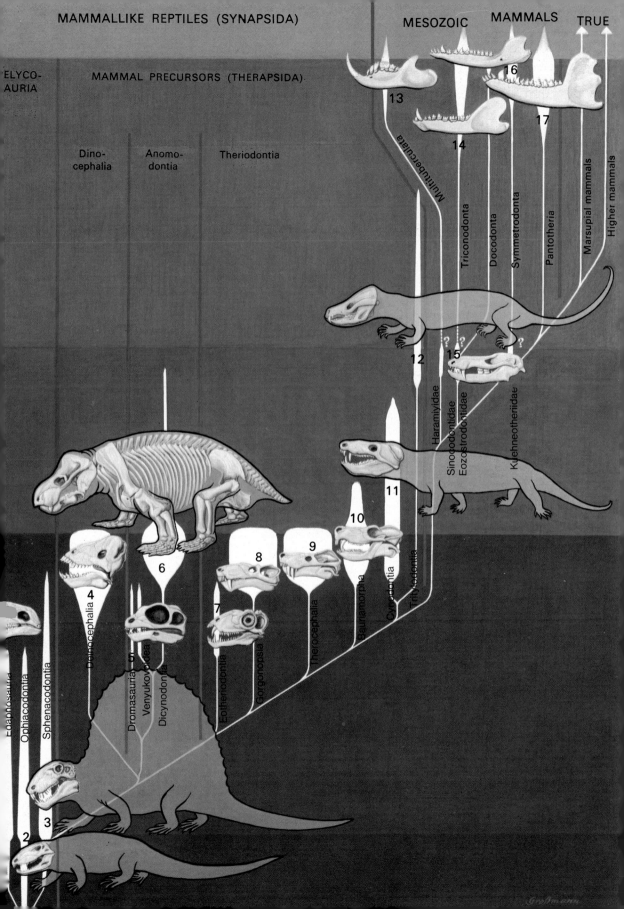

MAMMALLIKE REPTILES (SYNAPSIDA)

MESOZOIC MAMMALS TRUE

PELYCO-
SAURIA

MAMMAL PRECURSORS (THERAPSIDA)

Dino-
cephalia

Anomo-
dontia

Theriodontia

Triconodonta

Docodonta

Symmetrodonta

Pantotheria

Marsupial mammals

Higher mammals

Multituberculata

Haramiyidae

Sinoconodontidae
Eozostrodontidae

Kuehneotheriidae

Dromasauria

Venyukovioidea

Dicynodonta

Eotitanosuchia

Gorgonopsia

Theriocephalia

Bauriamorpha

Cynodontia

Tritylodontia

Dinocephalia

Eotitanosuchia

Ophiacodontia

Sphenacodontia

2 3 4 5 6 7 8 9 10 11 12 13 14 15 16 17

This may indicate that paired fins are the remains of a medial fin seam. Another distinctive feature in early spiny sharks are the teeth, which are firmly fused with the jaw bones. Later spiny sharks (e.g., *Acanthodes*) lack any toothlike structures. The phylogenetic position of the spiny sharks is still a debatable point. Some zoologists place them near cartilaginous fishes, while others put them near teleost fishes. The fact that they have characteristics of both groups may mean that spiny sharks are the ancestors of both these jawed fish classes. Free-swimming organisms, spiny sharks fed on plankton. Incidentally, the group should not be confused with the sharks we know today.

PLACODERMS (class Placodermi) were confined almost exclusively to the Devonian period. These were fishes in which the head and trunk were surrounded by armor, with only the pectoral fins and the rear of the trunk protruding from this covering. This is the reason some placoderms have even been described (erroneously) as turtles. In spite of the great degree of armor development, placoderms are not close relatives of teleost (= bony) fishes, for on the basis of most of their features they are closer to cartilaginous fishes. Like cartilaginous fishes, to cite one example, they lack gill covers. In fact, placoderms are sometimes placed as a subclass of the cartilaginous fishes. Two placoderm subclasses are distinguished on the basis of armor differences:

1. ARTHRODIRA; L from less than 10 cm to 5 m; they have a linkage between the head and the trunk armor; the hind part of the body is usually naked; there are "free" pectoral fins, one dorsal fin, and a heterocercal caudal fin. Examples of Arthrodira are: *Arctolepis, Coccosteus* (see Color plates, pp. 149 and 169), *Dunkleosteus* (see Fig. 10-10B), and *Kujdanowiaspis*. 2. ANTIARCHS (Antiarchi); L up to 30 cm; the head and trunk armor are firmly bound to each other; the eyes and the pineal opening and nasal openings between them form an eyeglasslike shape on the upper side of the head, which is armor-covered; the hind body portion is either scale-covered or naked; the pectoral fin is surrounded by a bony envelope (a spine-shaped lateral organ); there can be one or two dorsal fins, and the caudal fin is heterocercal. Antiarchs occurred from the Lower to the Upper Devonian period and perhaps into the Lower Carboniferous. One representative genus is *Bothriolepis* (see Figs. 10-11 and 10-12).

In the heavily armored Arthrodira, the trunk armor degenerates from the Lower to the Upper Devonian. The early arthrodirans have only a small opening for the pectoral fin throughout the armor covering. In Upper Devonian forms, on the other hand, the lower front part of the trunk armor is reduced to a clasplike structure, and the base of the pectoral fin is completely free. Therefore, the base surface of that fin is elongated. These armored arthrodirans had an ossified skull. Using the serial cutting method we described earlier, Erik Stensiö was able to study the structure of the brain and the pathways of the nerves and blood vessels

---

Ammonite Evolution. Ammonites descended from Lower Paleozoic orthocerates, ancestors of the modern nautilus. They proliferated in the Devonian into several groups, and, after several crises from which only a few groups survived, again in the Carboniferous-Permian, Triassic, and Jurassic-Cretaceous. Deviant forms appeared in the Upper Triassic, Middle Jurassic, and Cretaceous. Undulating suture lines began in *Bactrites* with one suture over the lateral siphon (E: exterior suture), and this trend increases within the individual ammonite groups until a series of twists in the shell are formed. The primary suture (white in the diagram) is trilobate in coiled Paleozoic ammonites (E: external suture; L: lateral suture; J: internal suture; L is paired), quadrilobate in the Triassic, and typically quinquelobate in the Jurassic and Cretaceous. *Crioceras* species returned to the tetralobate configuration of the primary suture, while the "Cretaceous ceratites" returned to the ceratite suture line pattern. In the clymenians of the Devonian, the siphon appeared on the inner aspect. U: Umbilical suture; A: sutures appeared only in the Paleozoic) between the outer suture E and the lateral suture L.

in armored arthrodirans (see Color plate, p. 172). He concluded that the brain of these fishes was similar to that in teleost fishes.

In addition to the armored arthrodirans, there were also fishes in which the armor was reduced to a few bony plates, as is the case in the PTYCTO-DONTIDS (Ptyctodontida), or those in which the armor degenerated into numerous small platelets as in the skatelike genus *Gemuendina* (see Color plate, p. 189), a Lower Devonian form. Male ptyctodontids possessed modified pelvic fins used as copulatory organs, a pair of cartilaginous processes at the front of the body, large tooth plates, and a whip-shaped tail. These features all resemble similar ones in modern Holocephali. In fact, their resemblance to chimaeras is so great that Tor Ørvig claims that they are ancestral to chimaeras. However, in both cases it seems we are dealing with two separate end products, for there are Mesozoic fossil chimaeras lacking all these features.

The early arthrodirans with the powerful, ventrally flattened trunk armor were bottom dwellers. Skatelike forms such as *Gemuendina* and the ptyctodontids were probably bottom-dwelling organisms as well, and their anatomic similarities with chimaeras are probably the result of adapting to similar ecosystems. In armored arthrodirans, the mobility of the head-trunk joint increased as the trunk armor degenerated. The eyes became considerably larger, and the jaws became more powerful. Some arthrodirans were true giants (e.g., the North American genus *Titanichthys*).

The ANTIARCHS were the most highly specialized placoderms. Their pectoral fin was surrounded by bony plates and developed into a spine-shaped lateral organ attached by a linkage to the breast armor and sometimes having a linkage within itself. This lateral organ was only used to anchor the animal in the floor; it had lost all locomotion functions. The belly of antiarchs is flat, and the mouth is terminal but on the underside. The armor is very heavy. All these features suggest that antiarchs were bottom dwellers.

CARTILAGINOUS FISHES (class Chondrichthyes) appeared much later than teleost fishes (to be described), for they did not arise until the late Lower Devonian or early Middle Devonian period. The fact that no cartilaginous fishes are known from the Silurian is not a result of poor preservation of cartilaginous fish fossils remains, for their teeth and scales are very resistant structures. Even cartilage fossilizes readily, especially in its calcified form. The lack of cartilaginous fish fossils is due instead to the fact that there were none of these fishes around during the Silurian. In complete contrast to the viewpoint of 19th-Century biologists, we know now that cartilaginous fishes are not the original fish group, for they developed rather late. While in ontogeny (individual development of the organism) the development of cartilage precedes the development of bones, the opposite occurred in fish phylogeny. Bones appear as external armor in early forms of various fish groups, even in jawless fishes, and

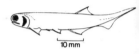

Fig. 10-7. *Triazeugacanthus,* an Upper Devonian acanthodian.

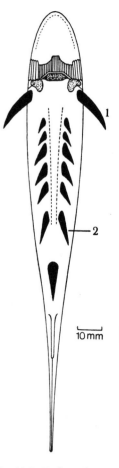

Fig. 10-8. *Euthacanthus:* ventral view of an early acanthodian showing the numerous spine pairs between the pectoral fins (1) and the pelvic fins (2).

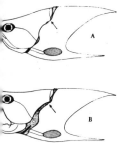

Fig. 10-9. Mobility of the head and the head-trunk joint in the arthrodiran *Coccosteus*. A=closed mouth, B=opened mouth.

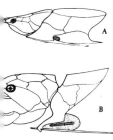

Fig. 10-10. Degeneration of the thoracic armor from an early arthrodiran (A, *Coccostolepis*) to a later larger form (B, *Dunkleosteus*).

Fig. 10-11. Head and thoracic armor of an antiarch (*Bothriolepis*), seen from above.

fishes with cartilaginous or completely vestigial skeletons did not arise until considerably later in geological time. Only the vertebral column has undergone a steady increase in ossification.

In the oldest cartilaginous fishes we can find the remains of true bony external skeletal elements, although these are in no way comparable to those in teleost fishes or the armor of agnathans or placoderms (placoderms being their closest relatives). As Tor Ørvig and Erik Stensiö showed, even the simple "dermal teeth" (the scales of modern sharks) are not primitive, simple structures but vestiges, the remains of ones from an earlier geological period. Many Paleozoic teleost fishes had composite scales (see Fig. 10-13A), in which new layers were added during life. The scales of modern sharks (see Fig. 10-13B) are also composite structures, and at their base are bony cells (true bone), which would not be there if cartilaginous fishes preceeded teleost fishes in evolution.

Fossil cartilaginous fishes are not well understood, because the teeth and scales they have left behind in the fossil record tell so small a story of their total nature. Uniform teeth occur in the most diverse cartilaginous fishes. The finds that are needed to explore fossil cartilaginous fishes in depth are fully preserved specimens, and once in a while paleontologists receive such treasures. An entire order of cartilaginous fishes, the INIO-PTERYGIANS (Iniopterygii), which are closely related to chimaeras, were discovered intact in Upper Carboniferous strata in the U.S.A.

In ELASMOBRANCHS (subclass Elasmobranchii), which includes the sharks and rays of today, the "upper jaw" (palatoquadrate) is movable and is not fused with the skull. The teeth have one or more dentine crowns with a bony or bonelike base and an enamel-like cap. The gill openings lack covers; they open directly to the outside. We cite here the following orders:

1. PRIMITIVE SHARKS (Cladoselachida and Cladodontida); rather large sharks, with a L of over 2 m; unsegmented fin ray bearers in the fins; no anal fin; cladoselachids have two dorsal fins, each with one spine in front of it, while cladodontids have one spineless dorsal fin; distribution (geologically) is from the Middle Devonian to the Lower Permian. *Cladoselache* (see Color plate, p. 149) is representative of the group. 2. TRUE SHARKS (Selachii); fins have one or several subdivisions in the fin ray bearers; there are two dorsal fins, each with a spine; an anal fin is also present; true sharks occur beginning in the Upper Devonian and extending into the present. Examples are *Ctenacanthus* and *Bandringa* (see Color plate, p. 149). 3. FRESH-WATER SHARKS (Xenacanthida); the pectoral fins have fin ray bearers arranged around a central axis (or archiptery-gium); a spine is at the rear of the skull, and after an intermediate space there begins a fin seam extending around the tail to the two anal fins; the teeth are double-crowned; distribution is from the Middle Devonian to the Upper Triassic. *Xenacanthus* (see Color plate, p. 149) is an example of this group.

Primitive sharks are closest to the original placoderm group, from which the true sharks also evolved. *Cladoselache* is a well-known form from the Upper Devonian Cleveland slate formations, a deposit that has had superb retentive characteristics and has preserved such fine details as muscular tissue impressions. The fresh-water sharks were an early branch of the placoderms, and as their name implies they adapted to a non-oceanic life.

The placoderms also include the HELICOPRIONIDS (order Helicoprionida), which often have characteristic spiralling teeth (see Fig. 10-17). Helicoprionids did not cast their teeth away as other sharks do, but instead they coiled them into a spiral, pushing the larger teeth further and further into the spiralling coil. Sometimes the spiral pattern has led imaginative paleontologists to formulating grotesque reconstructions of these animals. For example, the tooth spiral has been placed, quite erroneously, in front of the dorsal fin or at the very end of the elongated upper jaw like a coiled elephant's trunk!

CHIMAERAS (subclass Holocephali) have short jaws; the upper jaw (=palatoquadrate) is fused with the skull. The teeth are composed of plates (hence the old term Bradyodonti for fossil chimaeras) or of individual teeth that form a slowly developing spiral. Specialized hard tissue reinforces the outer surface of the teeth and grows throughout life. The gills are pushed forward to a place beneath the skull. There is a membranous gill cover over the single common gill opening. There are three chimaeran orders:

1. CHIMAERAS (Chimaerida); the teeth are composed either of plates or fused teeth; caudal fins are usually whip-shaped but are sometimes heteroceral; relatively small forms prevail in the Paleozoic; the order occurs from the Lower Carboniferous to the present. 2. EDESTIDS (Edestida); tooth spiral, but unlike *Helicoprion* (see Fig. 10-17) the teeth are cast off; there is an elongated rostrum ("upper jaw"); distribution temporally is from the Lower Carboniferous to the Lower Triassic, and a representative genus is *Sarcoprion* (see Fig. 10-19). 3. INIOPTERYGIANS (Iniopterygii); large protruberances in front of the upper and lower jaws; the teeth form small spirals sitting beside each other in the jaw; the greatly modified, high-set pectoral fin has a reinforced first fin ray; the caudal fin is symmetrical; the pelvic fins are modified into copulatory organs, a phenomenon occurring otherwise only in Mesozoic or modern chimaerids; iniopterygians are found only in Upper Carboniferous strata in the U.S.A., and an example of the group is *Sibyrhynchus* (see Fig. 10-20).

The chimaerid (Chimaeridae) forms (e.g., *Helodus* (see Color plate, p. 149) and *Menaspis*)) differ from each other greatly. Little is actually known about them, and interpretations about them vary considerably. Different paleontologists develop entirely different reconstructions of these chimaerids.

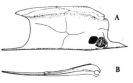

Fig. 10-12. A. Joint; B. Pectoral spine; both of an antiarch, *Bothriolepis*.

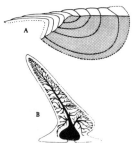

Fig. 10-13. A. Composite scale of a cartilaginous fish of the Paleozoic, seen in cross-section. B. Placoid scale of a modern shark, seen in cross-section.

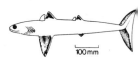

Fig. 10-14. Primitive shark (*Claudoselache*).

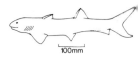

Fig. 10-15. A true shark (*Ctenacanthus*) from the Paleozoic era.

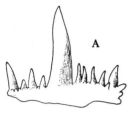

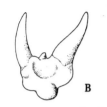

Fig. 10-16. A. Tooth of a primitive shark; B. Tooth of a fresh-water shark.

Fig. 10-17. Lower jaw fore end of a helicoprionid with tooth spiral.

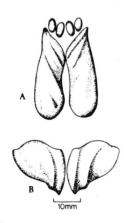

Fig. 10-18. Teeth of a Holocephali (*Deltoptychius*) from the Paleozoic. A. Upper jaw; B. Lower jaw.

There are still other Paleozoic cartilaginous fishes besides the placoderms and chimaeras. However, we cannot satisfactorily classify them, because too little is known about them or else they have extremely specialized features. Knowledge about fossil cartilaginous fishes will undoubtedly be increased in the future as more finds are uncovered.

Paleozoic placoderms led lives not unlike those of our modern sharks: they were predators. With the exception of primitive sharks, placoderms had pelvic fins modified into copulatory organs, and thus they utilized internal fertilization. Chimaeras inhabited the same ecosystem they do today; they inhabited ocean floors and lived there on hard-shelled prey (the edestids were an exception in this regard, for they occurred elsewhere). Iniopterygians occurred in shallow water covered by a nectonic, closed plant layer. In the fossil record, chimaeras appear after placoderms. These two fish groups are close relatives, but their exact phylogenetic origin and the relationship between chimaeras and placoderms is still unknown.

In the Upper Silurian—long before the first cartilaginous fishes appeared and only shortly after the spiny sharks made their debut—the first TELEOST FISHES (class Osteichthyes) appeared. In these fishes, the head and body have an external skeleton composed of true bones, and the internal head skeleton is at least partially ossified. The earliest teleost fishes have rhomboid scales, which surround the body as an external supporting, movable skeletal armor. The scales and head bones have three layers: bony substance, dentine, and an enamel layer (going from inside to outside). Primitive teleost fishes have heterocercal caudal fins. The biting jaws were formed from the upper and lower jaw bones, to which the teeth are either firmly attached or fused. There are three teleost fish subclasses: A. HIGHER BONY FISHES (Actinopterygii); B. LUNGFISHES (Dipnoi); and C. CROSSOPTERYGIANS or LOBEFIN FISHES (Crossopterygii).

Upper Silurian strata have delivered only the scales of teleost fishes, and these probably belonged to higher bony fishes, even though they lack an outer enamel layer. True ganoid scales (see Fig. 10-22) have been found in Lower Devonian Canada and Australia. The first intact actinopterygian finds are from the Middle Devonian. In ganoid scales new layers of ganoin (an enamellike substance) are added to the scale as the fish grows.

Three higher bony fish superorders are distinguished: 1. STURGEONS and PADDLEFISHES (Chondrostei), so called from their cartilaginous axial skeleton; the elongated caudal fin is heterocercal; chondrosteins occur from the Upper Silurian to the present, and typical examples of them are *Cheirolepis*, *Rhadinichthys* (see Color plate, p. 149), *Palaeoniscum* and *Chirodus*. 2. BOWFINS and GARS (Holostei), characterized by the beginnings of ossification of the axial skeleton and degeneration of the heterocercal caudal fin; temporal distribution is from the Upper Permian to the present; the only Paleozoic genus is *Acentrophorus*. 3. TRUE BONY FISHES or TELEOSTS (Teleostei); Triassic to the present.

Within these three major groups of sturgeons/paddlefishes, bowfins/ gars, and teleosts, the higher bony fishes (also called spiny-rayed fishes) have developed a wealth of diverse species. The spindly, typical fish shape (e.g., in *Palaeoniscum*) occurs, as do high-set (*Chirodus*) and elongated, eel-shaped bodies. The greatest degree of proliferative development is attained in the teleosts, which arose in the Triassic from the bowfin/gar superorder. The transformation from rhomboid scales to cycloid scales, the characteristic teleost scale type, or the loss of all scales can be seen in holosteins. Since phylogenetically only the Chondrostei and the Teleostei form a unit, we shall interpret the above three superorders as organizational stages. Chondrostei arose from primitive teleost fishes, and the true teleost fishes are from one family of the Holostei. These holosteins appear to embrace several phylogenetic lines of descent, which arose from other chondrosteins.

Some zoologists place the spiny-rayed fishes in one subclass and the other teleost groups into a second subclass, Sarcopterygii. Indeed, the lungfishes and lobefins are closer to each other than to the other teleost fishes, but there are so many differences between them that we shall retain a division into two subclasses as we did at the outset of this discussion. Lungfishes and lobefins share a similar scale structure. As in spiny-rayed fishes, the scales are basically composed of bone, dentine, and enamel, but within the dentine layer there is a porous canal system, which opens in small pores on the single-layer enamel coat. Thus, unlike the ganoid scales in spinyrayed fishes, the scales have a regular pattern of fine pores. Nothing yet is known about the function of this porous system; it may be somehow related to the lateral line system.

LUNGFISHES (subclass Dipnoi) have a unique arrangement of the bones in the roof of the skull. The chief lungfish characteristic are the tooth plates with their hardened tissue (see Fig. 10-25), as found in chimaeras. However, this major feature evolved in time and is not found in the oldest (Devonian) lungfishes nor in more recent long-snouted forms, in which small teeth line the jaws and roof of the mouth. The two oldest lungfishes are also the only ones with rhomboid scales; all others have cycloid scales, but with their porous canal system they retain this aspect of rhomboid scale structure. The fin ray supports are in the long, fleshy lobes of the paired fins along a central rod or archipterygium. The originally heterocercal caudal fin becomes distended and finally fuses with the dorsal fins, then with the anal fin, and this results in the formation of a continuous fin seam. The tendency seen in primitive fishes of degeneration of the ossification of the outer and inner skeleton can also be seen in lungfishes. During ontogenesis, the cartilage is not replaced by bone as in the early species.

After they flourished in the Upper Devonian, lungfishes have survived into the present without undergoing any major anatomical modifications. The two lines leading to modern lungfishes were present in the

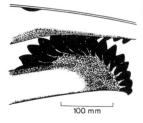

Fig. 10-19. Mouth of an edestid (*Sarcoprion*), with tooth spiral.

Fig. 10-20. *Sibyrhynchus*, an iniopterygian.

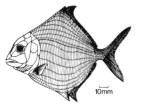

Fig. 10-21. A high-set chondrosteid.

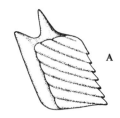

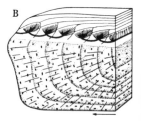

Fig. 10-22. Ganoid scale. A. Seen from above (outside); B. In cross section.

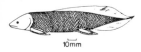

Fig. 10-23. Lungfish with fin seam (*Uronemus*) from the Permian period.

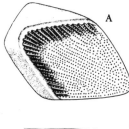

A

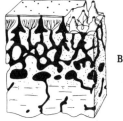

B

Fig. 10-24. Scale of a crossopterygian. A. Seen from above (outside); B. Cross-section.

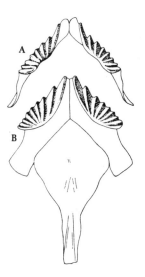

A

B

Fig. 10-25. Teeth of a lungfish (*Sagenodus*). A. Lower jaw; B. Upper jaw.

Carboniferous period. Furthermore, the Carboniferous ancestors of modern African and South American lungfishes also had the ability to survive drought periods by immersing themselves in mud. This is indirect evidence of the presence of lungs, which probably characterized all primitive teleost fishes. The lungs do not mean that lungfishes were related to tetrapods; in fact, lungfishes lack the nasal passage found in all tetrapods, and the nasal openings in lungfishes are in the roof of the mouth. The other developmental line is found in most Devonian lungfishes, which lived in the sea and differ little from the Australian lungfish of today. Examples include: *Uranolophus* (with rhomboid scales), *Dipterus* (see Color plate, p. 149), *Uronemus* (see Fig. 10-23), and *Conchopoma* (from the Permian period).

LOBEFIN FISHES or CROSSOPTERYGIANS (subclass Crossopterygii), with the exception of a few coelacanths (superorder Actinistia), are only known from the Paleozoic, and until 1938 even coelacanths were only found as fossils. The astounding find of the coelacanth *Latimeria chalumnae* showed that this group had not died out toward the end of the Cretaceous, as everyone thought it had! The major lobefin characteristic is the two-part internal skeleton of the head (see Fig. 10-26), which is composed of a nose-eye fore section and a rear earoccipital section. The two parts can also be followed in the skull roof. Some mobility exists where the two bone sections meet. Lobefins are externally characterized by the fleshy lobes on all the fins.

We divide crossopterygians into three superorders 1. COELACANTHS (Actinistia or Coelacanthida), which have the following features: two external nasal openings; no nose-throat passage (choane); no tooth-bearing maxillary bone; cycloid scales lacking a porous canal system; distinctive symmetrical tail shape; they have existed since the Middle Devonian. A typical genus is *Rhabdoderma*. 2. RHIPIDISTIANS (Rhipidistia), with the following characteristics: choane present; teeth with convoluted dentine; development from rhomboid to cycloid scales; porous canal system. a) Order HOLOPTYCHIIDS (Holoptychiida); two external nasal openings; paired fins with elongated, fleshy lobes and presumably with an archipterygium; distribution from the Lower to Upper Devonian period; examples are *Holoptychius* (see Color plates, pp. 149 and 169) and *Porolepis*. b) Order OSTEOLEPIDIDS (Osteolepidida); one external nasal opening, with the second modified into a nasal tear duct; internal skeleton of the paired fins of the same type found in tetrapods; distribution from the Middle Devonian to the Permian; examples are *Osteolepis*, *Eusthenopteron* (see Color plate, p. 169), and *Panderichthys*. 3. ONYCHODONTIDS (Onychodontida); skull division not evident; arrangement of the head bones similar to that in spiny-rayed fishes; tooth spiral at the fore end of the lower jaw; no fleshy fin lobes; distribution from the Lower Devonian to the Upper Carboniferous. An example of this group is *Strunius* (see Fig. 10-27).

COELACANTHS are such a conservative group phylogenetically that the coelacanths of today are hardly any different from the Devonian genera, the greatest difference being the regression of some head bones in the modern forms. Until *L. chalumnae* was discovered, coelacanth anatomy was only known from fossil evidence. Readers may be interested to learn that even today we only have direct information on adult coelacanth anatomy, for only adults have ever been caught! Here the fossil record has been helpful, for there are numerous juvenile coelacanth finds from the Upper Carboniferous; they belong to the genus *Rhabdoderma*. The amazing preservative qualities of the Illinois Upper Carboniferous deposit from which we have a juvenile has retained even the yolk sac (see Color plate, p. 244), the very tender organ that provides nutrition for the freshly hatched fish. During the yolk sac stage and after consumption of the yolk sac, the middle process on the caudal fin develops very rapidly in proportion to the growth of the rest of the body, and this kind of development is an example of allometric growth. Finally, the process becomes a long, whiplike appendage. During further development this part grows less rapidly than the rest of the body, and in the adult coelacanth it is only a stump.

The ONYCHODONTIDS are a distinct group. Their characteristic teeth have been found over a long period throughout the Paleozoic, but the only intact specimens are from the Upper Devonian. With their short snout and large eyes they look like chondrosteins, but they appear to be crossopterygians.

RHIPIDISTIANS are the only fish with a nose-throat passage, which joins the nasal capsule with the mouth and opens in the roof of the mouth. For this reason the group is thought to be ancestral to tetrapods (we recall that it was because of the lack of this nose-throat passage that lungfishes were eliminated as the forefathers of tetrapods). Rhipidistians definitely possessed lungs, but the presence of lungs has only been substantiated indirectly. Erik Jarvik's work with *Eusthenopteron*, an osteolepidid with cycloid scales, has made it the most thoroughly investigated fossil fish. We know about details of the head and trunk skeleton, the brain, the blood vessel anatomy, and the muscle attachment points in this genus. Jarvik used the serial cutting method pioneered by Stensiö.

The Color plate on p. 172 shows (below) a rhipidistian, and this is evidence that the effusion technique can be as precise as the serial cutting method in showing the details of the brain, nerves, etc., and this technique has the advantage that the specimen is not destroyed. It is left intact after a mold is made of the inside.

The symmetrical caudal fin of *Eusthenopteron* developed from a heterocercal caudal fin. However, the original, heterocercal, condition can be seen in recent Carboniferous and Permian osteolepidids. Surprisingly, they have retained the rhomboid scales as well, while rhomboid scales can only be found among holoptychiids in the earliest Lower Devonian genus

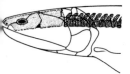

Fig. 10-26. Head of an osteolepidid crossopterygian (*Eusthenopteron*); the external and internal skeletons are shaded.

Fig. 10-27. Onychodontid crossopterygian (*Strunius*).

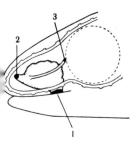

Fig. 10-28. Nasal capsule with nose-throat passage (choane) (1), external nasal opening (2=originally the fore nasal opening) and nose/tear duct passage 3=originally the rear nasal opening).

be *Porolepis*. The spindle-shaped body shows that lobefins were good swimmers, and their powerful jaws are probably a sign that they were capable hunters as well. As in all other Paleozoic fishes, they developed no flat, bottom-dwelling species.

Fishes are usually contrasted with the tetrapods (the four-footed vertebrates), and it is not difficult for us to distinguish our modern fishes from tetrapods. However, if we look back through the vast fossil record, we will find that we suddenly have tetrapods with fish features, namely the Ichthyostegalia, and fishes (rhipidistian crossopterygians) with features characteristic of tetrapods (see Color plate, p. 169).

Above all, fishes are distinguished from tetrapods in having fins. However, if we examine the paired fins of lobefin fishes, we find that their skeletal structure corresponds to that of a tetrapod. In fact, the humerus (upper arm bone) of a crossopterygian corresponds to that in a tetrapod in very fine details. The humerus is attached to the radius and ulna, just as we find in tetrapods. The five digits are also present in the lobe fin. Thus, it is not difficult to imagine that lobefin fishes used their paired fins to support the body as they moved along the floor or even on land in search of a new body of water.

It is not so easy to understand why fishes would leave their original habitat, the water with its rich food supply and natural means of body support without the elaborations we find in terrestrial animals. In this regard it is very enlightening to learn that the ancestors of tetrapods lived in regions subject to drought, and they were forced to leave one body of water and find another one from time so time. In this situation, the advantage of feet versus paired fins is obvious.

The ICHTHYOSTEGALIA (order Ichthyostegalia) already have made this great step. Their hind limbs are supported at the vertebral column above the pelvis, and the pectoral girdle, coming from the head, supports the fore limbs. In early tetrapods, as in crossopterygians, the limbs were used solely as supporting structures, assisting in undulating movement or simply to get the animal onto land. As contemporaries of crossopterygians, ichthyostegalians have numerous fish features, including a fin seam with true rays, a scale-covered body, a vestige of the gill cover, and the lateral line system.

Tetrapods correspond to rhipidistian crossopterygians in the structure of the skull roof, the arrangement of the bones in the cheek region, upper and lower jaws, in the presence of lungs, and in the internal openings of the nose-throat passage. Furthermore, these structural similarities are shared only with rhipidistian crossopterygians and not with lungfishes or coelecanths. Rhipidistians are divided into two orders, which differ in a few respects. If we examine only the nasal region, it is clear that only osteolepidids, with their external nasal opening and their nose-tear duct, are the closer relatives of tetrapods, while holoptychiids with their two external nasal openings are not as closely related.

The same relationship is found when we examine dental structure. All rhipidistian crossopterygians have teeth with convoluted dentine, but the folded structure in osteolepidids is so similar to that in the earliest tetrapods that the holoptychiids with their much more complex tooth structure could not even be considered as closer to tetrapods in this regard. This dental similarity is particularly striking in the lobefin genus *Panderichthys* and the ichthyostegalian genus *Ichthyostega* (see Color plate, p. 169). These are also features that show that all tetrapods could only have originated from osteolepidids or similar lobefin fishes.

The Ichthyostegalia have been found in strata belonging to the Uppermost Devonian or the Lowest Carboniferous, and they are the first, simplest tetrapods. The next tetrapods to appear are a series of amphibians in the Lower Carboniferous (see Color plate, p. 201). The LEPOSPONDYLI (a subclass) appeared very early; they are characterized by a unified vertebral column, although the group is quite heterogeneous in other respects. They are organized into four orders: AISTOPODS (Aistopoda), NECTRIDS (Nectridea), MICROSAURS (Microsauria) and PRIMITIVE SALAMANDERS (Lysorophia). The phylogeny of these groups is unknown, however, so we do not really know if these four orders are closely related to the Ichthyostegalia (as Alfred S. Romer assumes they are) or whether they arose from as yet undiscovered osteolepidids. We do not even know if these four orders form a single taxon (a phylogenetic entity). However, it is certain that they all became extinct in the Permian and left no descendants.

At any rate, there is no close relationship between the aistopods and the caecilians (order Gymnophiona) of today, for although both groups are characterized by a snake-shaped body, this body form was attained in both groups independently. Likewise, modern urodeles (order Caudata) are not closely related to the Lepospondyli, even though both groups have unified vertebral columns. As urodeles develop, their vertebrae ossify without displaying any signs of preliminary stages in the phylogeny of the vertebrae, so we have no information on how their ossified vertebrae came about phylogenetically. We cannot say anything about the original, primitive structure of the vertebrae. In the Lepospondyli and urodeles we are probably dealing once again with parallel evolution (the independent acquisition of similar structures).

Most Paleozoic amphibians are LABYRINTHODONTS (subclass Labyrinthodontia). They have retained the dental structure of osteolepidid lobefin fishes, and this dental structure has in some respects been complicated in that the entire dental cavity is filled with dentine. Small labyrinthodonts lack the dentine foldings, as we also found in Lepospondyli. The structure of the vertebrae is used to divide labyrinthodonts into several groups. They took from osteolepidid crossopterygians the ability to produce two vertebral disks in each body segment, a fore disk and a hind disk. However, these amphibians were evidently under selective pressure to form

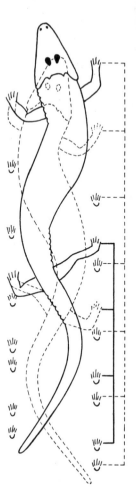

Fig. 10-29. Fishlike undulating movement in an early tetrapod (here an authracosaur).

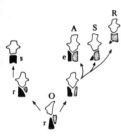

Fig. 10-30. Development of the vertebrae from osteolepidid crossopterygians (O) to labyrinthodonts on one side (r=rhachitome vertebrae, s=stereospondyle vertebrae) and batrachosaurs (e=embolomere vertebrae), reptiles (R), birds (A), and mammals (S) in other lines of phyletic descent. Black: front disc; stippled: rear disc; white: upper arch.

Fig. 10-31. *Branchiosaurus*, a neotonous (reaching sexual maturity in the larval stage) amphibian from the Paleozoic era.

just one vertebra per body segment, and this was accomplished in several ways: in some labyrinthodonts the rear disk became smaller (the rhachitome vertebrae of the Rhachitomi) or even regressed entirely (stereospondyle vertebrae of the Stereospondyli); in others the fore disk was reduced in size (the embolomere vertebrae of proto-reptiles) or disappeared entirely (in reptiles, birds, and mammals). It is conceivable that the two disks could grow together and fuse, and some theorists maintain that this was indeed the case in the Lepospondyli.

Among the many Paleozoic amphibians, we shall only cite those that were most prevalent in central Europe. These all belonged to the genus *Branchiosaurus*, a small form with external gills. These amphibians had many characteristics normally associated with amphibian larvae, and the genus is often cited as a fossil example of neoteny. Neotenous animals reach sexual maturity in the larval stage. The external gills are a clear sign that *Branchiosaurus* lived exclusively in water.

We shall discuss the Lissamphibia (fossil and living amphibians as well as reptiles that descended from Paleozoic amphibians) in more detail. The three modern amphibian orders (frogs and toads, urodeles {=tailed amphibians} and caecilians) are closer to each other than to any Paleozoic amphibian. One of the advanced features they all share are jointed teeth, which also appear in true teleost fishes. In Lissamphibia, however, that joint or link is in the dentine between the lower and upper tooth segment. Thus it appears that this feature appeared only once in evolution, so the three modern orders are a phylogenetic unit.

In the fossil Mesozoic and Cenozoic ancestors of modern amphibians, the only part of the tooth that can be found is the lower stump, for the tip invariably breaks off at that above-mentioned joint. In 1969, John Bolt described similar teeth structures in a rhachitome labyrinthodont. He was able to recover some individual pieces in which the tooth tip was still on the stumps. The teeth of this rhachitome (*Doleserpeton*) have a second tip on the inside surface; this is another feature found within the three Lissamphibian orders. *Doleserpeton* differs from other rhachitome labyrinthodonts in another important regard and is likewise considered to be closely related to the Lissamphibia because of this: the vertebrae are composed almost entirely of the rear vertebral disk. The Lissamphibia have unified vertebrae, but whether these arose from the rear disk of the vertebrae is not known. However, all known evidence indicates that *Doleserpeton* is very close to the origin of all three modern amphibian orders.

Just as the labyrinthodonts are ancestral to amphibians, the BATRACHOSAURS (Batrachosauria) are closely related to reptiles. The oldest reptilian species are from the early Upper Carboniferous period. However, the same (or even younger) deposits have batrachosaurs, primarily GEPHYROSTEGIDS (Gephyrostegidea). Thus, batrachosaurs did not appear before reptiles, and we must look for their common ancestor in still earlier strata.

Some of the most important distinctions between amphibians and reptiles will be cited here. Modifications occurred at the rear end of the skull roof, and these were associated with changes in the ear region. The tabular horn, which surrounds the top of the ear cavity in amphibians, was lost in reptiles, so the eardrum only contacted the skull bones with its front edge. A more easily appreciated difference between the two groups is concerned with their reproduction. Reptiles lay amniotic eggs, which are eggs covered with a protective membrane as protection against dehydration. These envelopes are not found in amphibian eggs. Amphibian eggs (amphibians lay their eggs in water) are known from the Upper Carboniferous (in Illinois), but it is very unlikely that we will ever find the earliest amniotic reptilian eggs laid on land.

To depict the development of amniotic eggs, Robert S. Carroll makes a comparison between extant lizards and salamanders. As he envisions the process, the transition from amphibian to reptile occurred in small animals that layed anamniotic eggs (those lacking membranes and certain embryonic organs) as some frogs do today. Amniotic eggs gradually developed from the anamniotic eggs, and this change made tetrapods independent of an aquatic life. In these small species the labyrinth (inner ear) is relatively large, and there is a maximum possible size of this structure. The great size of the labyrinth forced a modification in the ear region and resulted in the regression of the tabular horn; thus, selective pressure was responsible for the origin of this second major reptilian feature.

The temporal openings in reptiles are significant phylogenetic characteristics, for they are used to make some major subdivisions within the reptiles. In the earliest reptiles (the cotylosaurs; see subsequent discussion), the skull roof is closed, and this is the same condition occurring in amphibians: the chewing muscles are attached (as in labyrinthodonts) to the inside of the skull roof. With the opening up of the skull roof and the formation of temporal openings (fenestrae), the muscles have a greater opportunity to develop. This trend was carried to its greatest extent in the mammallike reptiles (the synapsids; see below); the lower temporal opening of pelycosaurs (order Pelycosauria) enlarged so much in the therapsids (order Therapsida) that the lower border of the opening only consisted of the mandibular arch, which was a narrow structure. In the course of further evolution leading to mammals, the musculature becomes attached to the entire skull roof. These muscles are the ones that enable mammals to assume many different expressions in contrast to the apparent stare in reptiles. The skull capsule in mammals is a secondary development from bones in the palatine roof and downward oriented processes in the skull roof bones of reptiles.

In recent years, reptiles from ever older strata (as far back as the early Upper Carboniferous) have been found, but most Paleozoic reptiles are from the Permian period. Reptiles reached their zenith in the Mesozoic

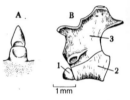

Fig. 10-32. *Doleserpeton*, a rhachitome labyrinthodont with characteristics of the Lissamphibia. A. Jointed tooth; B. Vertebra: 1. Fore disc; 2. Rear disc; 3. Upper arch.

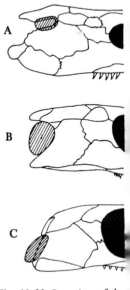

Fig. 10-33. Location of the eardrum (shown with diagonal lines) in: A. An anthracosaur; B. A gephyrostegid; C. A primitive reptile (Cotylosauria).

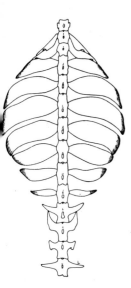

Fig. 10-34. *Eunotosaurus*, a primitive reptile with broadened ribs (a development parallel to a similar one in turtles).

(see Chapters 12 and 16). The COTYLOSAURS (order Cotylosauria), which appeared in the Upper Carboniferous, did not actually flourish until the Permian period. They are the only ANAPSIDS (subclass Anapsida) in the Paleozoic. The second anapsid order (turtles; Testudines) is not known until the Triassic, but at that time they were fully developed. *Eunotosaurus* (see Fig. 10-34), a Permian reptile from South Africa, is not an ancestor of turtles as originally thought but is now classified with the cotylosaurs. The rib spreading of this animal is simply a parallel development to the same phenomenon that occurred in turtles.

The sudden appearance of highly specialized animal forms, of which the turtles are an example, is a widespread phenomenon in reptiles. The process is known in the Paleozoic in MESOSAURS (Mesosauria or Proganosauria) and in the Mesozoic in ichthyosaurs (Ichthyopterygia), synaptosaurs (Euryapsida), and flying saurians (Pterosauria). The mesosaurs were the first reptiles to return to the water and to adapt to that habitat; their fore feet were very short, an adaptation to the "new" environment. Mesosaurs play an important role in Alfred Wegener's continental drift theory, since mesosaur finds are only known from South Africa and South America. Since these reptiles were fresh-water species and could not cross the ocean, it would indeed appear that these two continents were once part of the larger, hypothetical Gondwana. Further evidence of continental drift is the fact that the strata sequence in fresh-water deposits from these two continents are precisely the same, even in very fine details.

LEPIDOSAURS (subclass Lepidosauria), which comprise to majority of our modern reptiles, and ARCHOSAURS (subclass Archosauria), which include the famous dinosaurs of the Mesozoic era, first appeared in the Upper Permian period. One of the lepidosaurs was *Proterosaurus speneri*, which was one of the few tetrapods among the many fishes found in Permian deposits in central Germany.

The MAMMALLIKE REPTILES or SYNAPSIDS (subclass Synapsida) appeared even before the lepidosaurs, and they gave rise to the mammals (see Chapter 11). Synapsids occurred in the Upper Carboniferous and first flourished in the Lower Permian with the PELYCOSAURS (order Pelycosauria). Just as in amphibians, the more advanced forms leading to new body organization plans appeared earlier than those which stopped evolving once a certain phylogenetic stage was reached and survived into the present at that stage. The latter adapted so successfully to their habitats that they could freely proliferate. This is the reason, then, that within amphibians there were "proto-reptilians" alive in the Lower Carboniferous and which died out in the Permian, while modern amphibians (Lissamphibia) can only be followed back to the Triassic, a later period.

In reptiles, the first mammallike species are from the Upper Carboniferous, and the last are from the Jurassic, while all presently living orders (turtles, crocodiles, rhynchocephalians, and true reptiles) made their first appearance in the Triassic. Modern amphibians and reptiles display many different organization levels, each of which is the end of a long developmental series. An amphibian from which a reptile could arise or a reptile that could give rise to a mammal would have looked completely different from any modern members of these animal groups.

11   The Path to Warm-Bloodedness

By A. S. Brink

The mammallike reptiles or SYNAPSIDS (subclass Synapsida) are among the most interesting fossils of the Permian-Triassic period, for they bridge the gap between the classes of reptiles and mammals. Thousands of synapsid specimens from various orders, suborders, and families, animals which lived for over 100,000,000 years, are all evidence of the process of adapting to various habitats and worldwide climatic changes. Part of this process was the transformation from what is commonly called cold-bloodedness to warm-bloodedness. Cold-blooded animals are those whose body temperature changes to conform to ambient conditions, and such animals are technically known as poikilothermic (from the Greek ποικιλος = variegated); warm-blooded (homoiothermic) animals maintain a relatively constant body temperature.

Mammallike reptiles

Synapsids possess just one pair of temporal openings (fenestrae), and several synapsid skeletal features show that the group was approaching mammals. A distinction among synapsids is made between the more primitive pelycosaurs (order Pelycosauria) and the more advanced therapsids (order Therapsida). Therapsids, the predecessors of mammals, are arranged in three suborders of theriodonts (Theriodontia), dinocephalians (Dinocephalia), and anomodonts (Anomodontia). The theriodonts were the ones that gave rise to mammals. Their teeth are differentiated into incisors, canines, and molars, and within individual theriodont lines the skeleton takes on more and more mammalian features, as we shall see.

Fossil theriodonts are found everywhere in Permian and Triassic deposits, even in Antarctica. Their prevalence is yet another piece of evidence supporting the theory of continental drift. Comparative osteological studies on theriodonts have helped clarify the phylogenetic significance of this group. Theriodonts are intermediate between reptiles and mammals in terms of their dental formula, the differential development of teeth, and the jaw linkage, which acts as a sound transmitter. Of particular importance is the arrangement of the ribs, which shows that these animals had a diaphragm. Fossil evidence also shows that theriodonts had movable noses with tactile hairs, lips, and cheeks, just as mammals had.

History of their discovery

The history of the discovery of theriodonts long occupied a "back seat" to the more sensational finds of dinosaurs and other dramatic Mesozoic reptiles. In about 1860, the first pelycosaurs (order Pelycosauria) were found in North America; these animals were just the first step toward mammals. In South Africa, one of the major sites for finding mammallike reptiles, a specimen discovered in 1838 by Andrew Geddes Bain, a road engineer and amateur geologist, was later named *Dicynodon lacerticeps* (= two-toothed lizard skull) by the great paleontologist Richard Owen of the British Museum. A few years later, Bain sent additional specimens of mammallike reptiles to the British Museum, where they were described by Richard Lydekker and Govier Harry Seeley, who were impressed by the "advanced", mammalian qualities of these fossils. One of the most striking specimens in this regard was from the genus *Diademodon*, in which the teeth behind the canines had the broad surface resembling molars. Another one of interest was *Cynognathus crateronotus*, with its multi-crowned tearing teeth located behind the canines.

A Scottish physician and paleontologist, Robert Broom (1866–1951), made a great contribution to the study of mammallian predecessors. He had conducted studies of primitive marsupials in Australia for several years and thereby came upon the problem of the origin of mammals. When he heard in 1896 that fossils of mammallike reptiles had been found in South Africa, he made this continent his second homeland and devoted his life to studying mammal phylogeny. From that time on, South Africa was the center of this line of research, although mammallike reptiles had also been found in Permian-Triassic strata in North and South America, Europe, eastern Africa, southern and eastern Asia, and other places. The South African scientist James Kitching even demonstrated that mammalian predecessors once lived in Antarctica. In North America, famed paleontologists Edward Drinker Cope (1840–1897) and Othniel Charles Marsh (1831–1899) found a series of pelycosaurs, which belong to the earlier reptilian forerunners of mammals, but these two men were more interested in their sensational dinosaur finds and placed their emphasis with those creatures. The greatest collector during the time Broom worked in South Africa was Croonie Kitching. After World War II, his efforts were carried on by his son James Kitching, who made into a profession what had only been a hobby to his father.

From reptile to mammal

Before mammallike reptiles had been discovered, the ancestors of mammals could be traced to the Cretaceous, where they and other reptiles appeared just after the dinosaurs disappeared. As mammalian predecessors received more and more attention by paleontologists and were described more frequently in the literature, it became apparent that the first step from reptile to mammal was probably taken in the Carboniferous. This initial phylogenetic step was followed by a long, complex pathway through the

▷
Paleozoic and Mesozoic fossils.
Left, from top to bottom:
Skeleton of a small Carboniferous labyrinthodont (*Amphibamus lyelli*);
Fossilized group of trilobites (*Isotelus gigas*) that had drifted together;
A Jurassic *Limulus*.
Right, from top to bottom:
The oldest known jawless fish (Agnatha) from Ordovician North America;
An individual *Isotelus* trilobite;
Eurypterus lacustris, a sea scorpion from the Devonian, found in Buffalo, New York;
The graptolithine *Rastrites*, from Silurian Czechoslovakia.

▷▷
Two specimens of *Bothriolepis canadensis*, an Upper Devonian antiarch found in Canada.

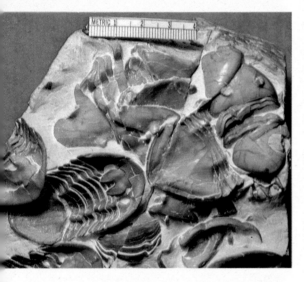

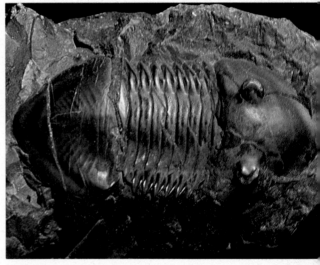

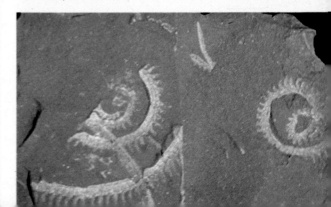

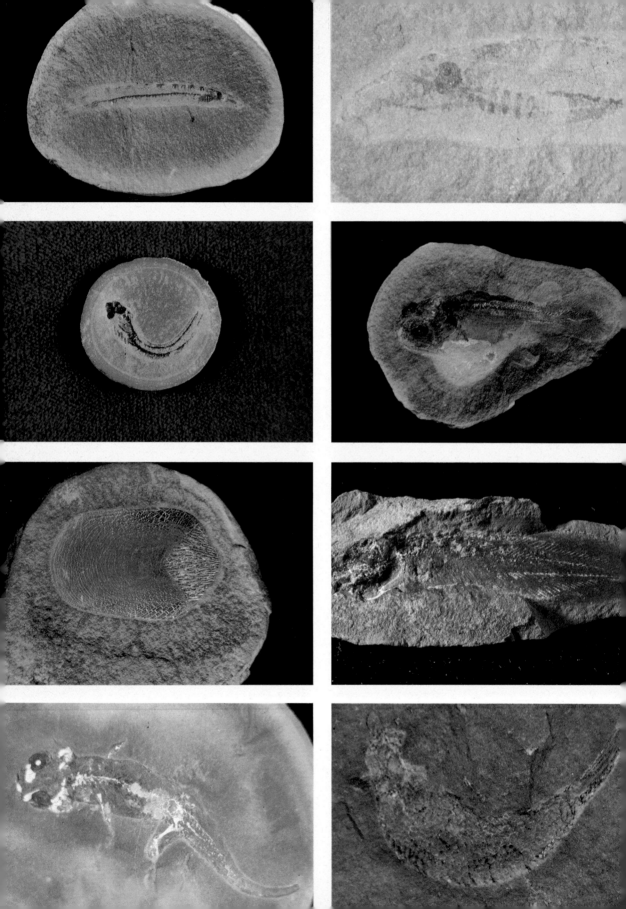

Permian and Triassic periods and into the Upper Triassic. Even after mammals appeared, it took a similarly long period through the Jurassic to the end of the Cretaceous period until they assumed the form they essentially have today.

Anatomical differences

The major anatomical differences between reptiles and mammals were known to 19th Century zoologists, of course, and even in earlier times. However, once the reptilian forerunners of mammals were identified, it was only then that zoologists realized that the transformation from the original ancestral reptile to a true mammal must have involved a tremendously profound adaptational change involving many different systems of the body. Anatomical studies revealed that livebearing and the suckling of young with mammary glands are intimately related to the presence of a diaphragm, warm-bloodedness, differential teeth, hair, the perception of sound, and the means of locomotion. All these changes occurred under the guidance of natural selection as part of a great adaptive process, which eventually led to in the appearance of mammals.

◁◁
A euomphalid Paleozoic gastropod (greatly magnified).

Displacement of the primary jaw joint

The most important change occurring at the level of the mammallian forerunners was the displacement of the original (primary) jaw joint, which in reptiles is formed by the quadrate bone and the articulate bone. These bones become less and less important in the mammal forerunners, and they are replaced by the squamosal and dentary bones. The quadrate and articulate bones become embedded in the middle ear and finally are transformed into the malleus (hammer) and incus (anvil) bones in the auditory apparatus of mammals.

◁
Upper Carboniferous vertebrates from Illinois. From left to right and from top to bottom: *Mayomyzon*, the only known cyclostome fossil genus (on the left an entire specimen; on the right the head and eye, showing the gill slits and digestive tract); juvenile acanthodian (*Acanthodes*), with large eyes; larval coelacanth (*Rhabdo-derma*), with the yolk sac (the bleached portion beneath the body); Scale of an adult coelacanth (*Actin-ista*); lungfish (*Conchopoma*), showing fin seam; Amphibian (*Amphibamus*, a rhachitome labyrintho-dont), with an impression of the body outline; A chondrosteid (*Elonich-thys*), bent as it settled into the stratum.

To what extent are the mammallike reptiles, especially the theriodonts, actually like mammals? And where is the line drawn between reptiles and mammals? In 1959, George Gaylord Simpson indicated the following possibilities: 1. The term "mammal" is confined to monophyletic groups (i.e., those of a single stock), and all others are classified as reptiles. 2. All groups that have reached the mammalian condition can be placed within one class. 3. All mammallike reptiles can be classified with the mammals, and the border between the two classes is drawn at the base of the mammallike reptiles.

One important question here is whether all theriodonts were warm-blooded. My research indicates that at least the higher theriodonts (e.g., *Diademodon* and *Cynognathus crateronotus*) were fully warm-blooded, breathed with the help of a diaphragm, bore hair all over their bodies, and suckled their young (whether incubated or born live) with mammary glands.

The changes in the jaw bones described above, as significant as they are for distinguishing reptiles and mammals, does not mean unconditionally that all the other mammalian features appeared simultaneously with jaw bone alterations. Even assuming the above-mentioned skeletal changes occurred, animals with the new skeletal anatomy could still be classified as reptiles if they were still cold-blooded, laid eggs, and lacked hair. On the other hand, if warm-bloodedness and other features appeared before the development of the malleus and incus bones, these animals could be designated as mammals. My studies indicate that the latter development occurred and that it is highly unlikely that all these changes happened simultaneously.

The best way for me to clarify these phylogenetic relationships in mammalian forerunners was to begin with the question of whether a

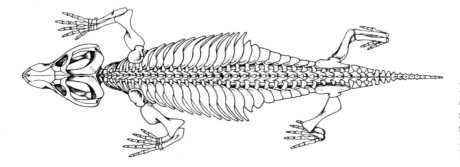

Fig. 11-1. *Thrinaxodon liorhinus.* Skeletal reconstruction based on a specimen in the National Museum, Bloemfontein, South Africa.

diaphragm was present or not. My research indicates that in all higher theriodonts (e.g., the genera *Thrinaxodon*, *Cynognathus* and *Diademodon*), there was a clear difference between pectoral ribs and lumbar ribs. A rib cage is present only in the chest region, and the pelvis is largely free of ribs. In this respect, these genera resemble mammals and deviate from reptilian anatomy, and rib differentiation of this type strongly suggests that these animals had diaphragms. The overlapping of ribs is probably

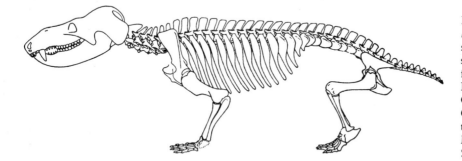

Fig. 11-2. *Diademodon laticeps.* Skeletal reconstruction from various specimens. The skull is from an animal in the D. M. S. Watson Collection, University College, London, and the trunk is from the National Museum, Bloemfontein, South Africa.

an adaptation to new rib functions probably caused by new muscle movement mechanisms. Overlapping and interlocking vertebral elements restrict the movement of the ribs along the longitudinal body axis in the pectoral region, and this would of course greatly influence the reptilian breathing pattern. In *Diademodon* the pectoral ribs are perpendicular to the vertebral column, so that rib movements do not greatly influence pectoral volume. In older animals, all lumber and pectoral ribs are fused with their vertebrae. These are all signs that a diaphragm was present.

Respiration via a
diaphragm

A diaphragm, which works partially involuntarily, facilitates uniform respiration, even under a heavy breathing load, so that when the animal is working hard it can still get enough oxygen to meet its needs. We can assume that the higher mammal predecessors had a greater activity level than earlier species due to their improved locomotor apparatus and to a higher metabolic level. These are likewise signs of a higher body temperature, and the differentiation of the teeth into incisors, canines, and molars in higher theriodonts supports the contention that these animals had a higher body temperature and greater metabolic rate. True mastication of food with high-crowned teeth invariably accelerates the rate of assimilation of food and hence would increase metabolic activity.

Secondary palate

Other evidence for the mammalian mode of breathing in the higher theriodonts is the hard (secondary) palate. The palate offers resistance to the tongue, and when the animal with a palate chews, it presses its food against the palate. Furthermore, the palate inhibits the passage of food down the nose-throat passage. Thus, and this is the actual function of the hard palate, a palate permits chewing and breathing to occur simultaneously. Mammals cannot stop breathing for any length of time while they eat, and certainly not for the duration of their eating time. The hard palate also facilitates sucking on mammary glands.

The structure of the nose-throat passage and the presence of additional large muscles in the nasal cavity, which resemble those found in mammals, also push the mammallike reptiles closer to the true mammals. These structures had the following functions: they cleaned, moistened and warmed inspired air, and they refined the olfactory sense (i.e., the sense of smell). Therefore, nasal muscles and other features of the nasal tract were basically an adaptation to a higher respiration rate. All these examples suggestive of a functioning diaphragm in higher theriodonts make it likely that these animals were warm-blooded. From the structure of the bones beside the nasal opening and their canals, one is led to the conclusion that higher theriodonts had a movable nose with tactile hairs and perhaps true lips as in mammals. The finding of a coiled *Thrinaxodon* (see Fig. 11-3) is further evidence for homoiothermy in this animal group. Only mammals coil like this to preserve body warmth under cold conditions; reptiles do not coil in this way.

Highly developed skin
glands

Just as hairs inhibit an unnecessary heat loss, sweat glands help prevent body temperature from rising too much. The strange depressions on both sides of the snout in *Diademodon* and the grooves leading from these depressions to the external nares might be signs of special skin glands used to transport fluid, perhaps to help moisten air before it is inspired. Whatever function they had, such highly developed glands must have arisen from simpler glands, which were not present in reptilian skin but which can be found in mammals in the form of sweat glands. Thus, the skin of higher theriodonts appears to have more mammalian qualities than reptilian qualities.

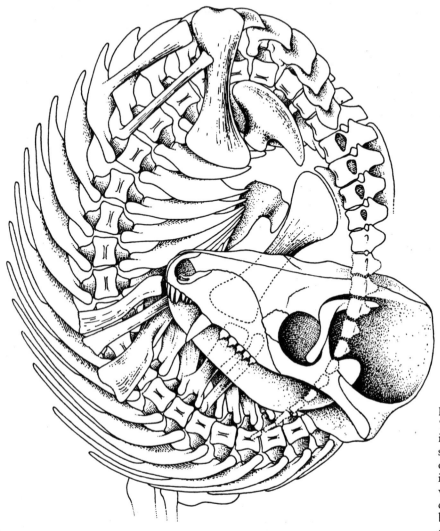

Fig. 11-3. Skeleton of *Thrinaxodon liorhinus* as it was found (natural size), seen from below. The coiled position of the body indicates the animal was warm-blooded, since body coiling serves to preserve body warmth under cold conditions.

Another important question is whether the higher theriodonts (at least in their molars) underwent a change of teeth, a phenomenon characteristic of mammals. Paleontologists have many different viewpoints on this topic. Some paleontologists found in young specimens of the genus *Diademodon* (probably newborn animals) that the teeth had the adult pattern. On the other hand, a number of paleontologists claim that several tooth changes occurred in genera such as *Thrinaxodon*, while tooth change in *Diademodon* was not completely mammallian in character. These paleontologists maintain that newborn *Diademodon* theriodonts had tapered molars and probably ate different food from the adults, with their flattened molars. This would suggest both that the young were suckled and the possibility that they changed teeth as they grew up. An animal with specialized skin glands in the snout region could also be capable of producing other specialized glands for milk production (probably by modifying simple sebaceous glands). Live-bearing seems more likely to have occurred in theriodonts than egg-laying and incubating the young. The rib-free pelvic region could also be mentioned here, since the absence of ribs in this region would permit the body to expand in the state of pregnancy.

It is very important to point out that the size of the skull at birth is not nearly as dependent on brain size as teeth size. The large head of theriodonts makes it more likely that they were born rather than hatched. In contrast with theriodont heads, the skulls of egg-laying and incubating reptiles and birds are always small, and newborn mammals have a much larger head in relation to the body than birds or reptiles. I find it difficult to imagine that theriodonts actually hatched from eggs.

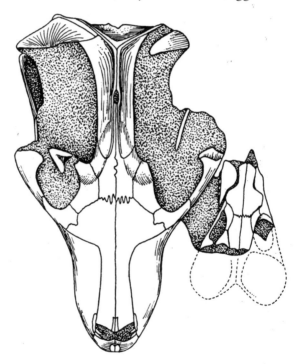

Fig. 11-4. The *Thrinaxodon* "mother and child" pair, shown as found (natural size) and seen from above.

The skull of a freshly hatched reptile is scarcely one-fifth the skull size of an adult, while the skull of a newborn mammal is proportionately much larger. One *Thrinaxodon* "mother and child" pair has been found (see Fig. 11-4) in which the skull of the presumably nearly newborn young is about one-third the size of the adult skull beside it. This again supports the hypothesis that these animals bore live young and suckled them. James A. Hopson takes a different view of these same theriodonts and claims that they laid eggs just as modern monotremes do. However, it happens that these monotremes (the echidna and the duck-billed platypus) are small animals with a short gestation period, and they

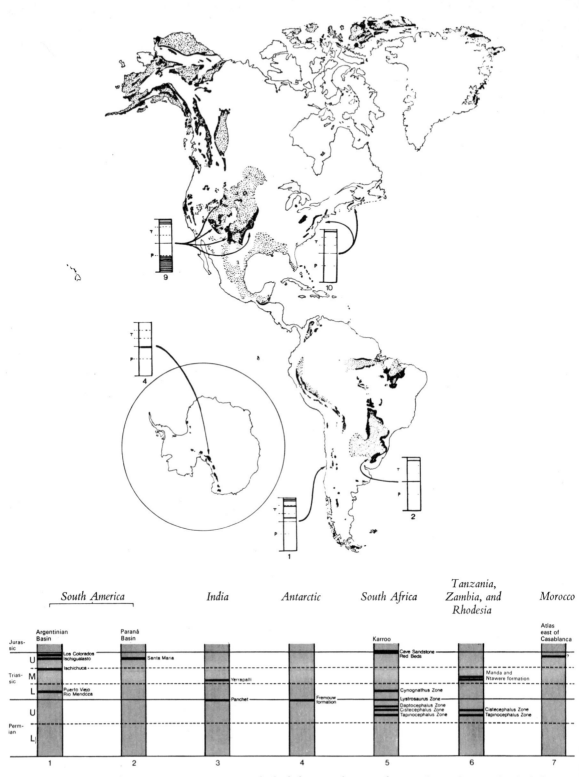

Fig. 11.5 Distibution of Permian-Triassic strata. Dark shaded areas: deposits close to the surface. Light shaded areas:

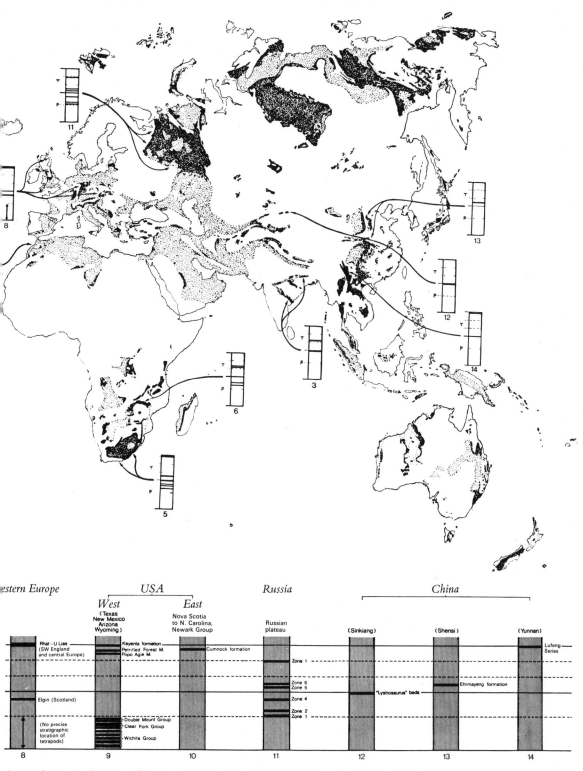

deeper deposits. The graphs beside deposits show the exact geographical and geological location of the various individual strata and fossil finds.

have relatively small skulls and vestigial dentition. Marsupials have solved the problem of embryonic development by giving birth to. their young while they are still in the embryo stage and nourishing them in the sack until they can live independently.

The previously mentioned modification of the lower jaw bones of reptiles into auditory ossicles is of great importance in evaluating the phylogenetic significance of mammalian precursors. As we mentioned, the jaw bones became incorporated into the middle ear as the malleus and incus, and they joined a third auditory ossicle, the stapes, to form a sound conducting system. Paleontologists have chiefly dealt with the questions of when and how this modification occurred, and only recently have they begun dealing with the question of why that occurred. Did this new ossicular auditory apparatus really improve sound perception?

There are essentially three opinions on this matter. According to the first, the three auditory ossicles form an amplifying system; according to the second the ossicles dampen excessively loud tones; and according to the third theory they weaken loud and very deep tones. None of these opinions is convincing, since amphibians, reptiles and birds, all of which have just one auditory ossicle, can distinguish sounds just as well as mammals, and they detect conspecifics, prey, and enemies. They are in no way slightly deafer versions of mammals. Experiments have shown that their ability to differentiate various tones is quite high indeed. However, the two additional ossicles, the incus and malleus, must have some important function, since they have been retained in every single mammalian species and never degenerated within any mammalian line.

Three-boned auditory apparatus

The following conclusion about these supplementary ossicles I find to be the sole tenable one: in reptiles and birds, sound is perceived with both ears, just as vision is perceived with both eyes. Mammals, on the other hand, perceive sound dimensionally ("stereophonically") just as they see dimensionally. Electronically it is impossible to measure precise directions and distances with just two microphones. At least four are needed to measure the three dimensions between each microphone pair. The ideal solution is to set up the four microphones at equal distances from each other. A similar effect is obtained if the microphones are placed in two pairs, *if* with the pairs one of the microphones has a delay mechanism so that four individual signals are received at four temporal instances and are then transmitted.

Stereophonic audition

From this one could postulate that the mammalian ear has a similar system with four "microphones" set up in two pairs. Sound does reach the cochlea of the inner ear, but it is delayed by the tympanic membrane and the auditory ossicles, which enable each ear to receive two signals, separated by just an instant in time. This tiny time difference is amplified in the ear on the other side of the head, where the amplitude of the sound is weakened by the head itself. This would indeed explain the advantage of the auditory ossicles. Furthermore, to produce even more effective results with stereophonic audition, all mammals have external ear muscles, which permit still finer sound orientation. Since mammalian predecessors lack external ear muscles, they were in this respect quite unlike the mammals.

The pineal organ: regulation of body temperature

The development of homoiothermy in mammallike reptiles was significant because it made them less dependent on external temperature changes than poikilothermic animals. The light-sensitive organ of reptiles (the pineal organ) assumed a new function in mammallike reptiles: it controlled body temperature. In mammals, the pineal organ disappeared and was modified into the epiphysis, and as such it functions in temperature regulation. Thus, with respect to temperature regulation, mammallike reptiles were basically different from reptiles and from mammals, and they formed an intermediate group.

The line of demarcation between reptiles and mammals is often seen as a true border and not as a zone of transition as it actually is. This transition zone, which extends from the mammalian precursors to the first appearance of early mammals, is not really a phylogenetic gap but a gap in knowledge on our part in following its development. This gap is being made continually smaller as more decisive fossils are being found, and some day it may be possible to eliminate the hypothetical line drawn between reptiles and mammals and to replace it with knowledge. One could classify those mammallike reptiles with hair and warm-bloodedness as an independent class between the reptiles and the mammals. But the higher systematic categories (families, orders, classes, etc.) are not physical entities but retrospective divisions made as we look back on the entire process of evolution. Among the mammalian predecessors there could easily be forms and even entire lines that remained at the reptilian stage throughout their development.

However the phylogenetic pathways led from the mammallike reptiles to mammals, there have been no demarcation lines or intermediate zones in the strictest sense between these two groups. There have only been a multiplicity of lines of descent, of which we have only found a few. In Chapter 13 we shall deal with these in more detail.

12 The Triassic, the Beginning of the Mesozoic Era

The Mesozoic Era comprises the time span from 70 to 230 million years ago, and it includes the three periods known as the Triassic, Jurassic, and Cretaceous. The dates chosen for the Mesozoic are a reflection of developments in animal life, not in plants, for the prevailing plants during the Mesozoic Era (numerous gymnosperms including seed and palm ferns, gingkos, and conifers) had appeared in the Permian (a Paleozoic period); "modern" plants, on the other hand, and we refer here to angiosperms, arose in the Lower Cretaceous. Thus, the floral "middle period" (known as the Mesophytic) extends from the Upper Permian to the Lower Cretaceous.

The Mesozoic has sharper boundaries in terms of its animal life. A great many animal groups died out at the turn from the Permian (Paleozoic) to the Triassic, and they were replaced by newer groups. Another, still greater "turnover", occurred at the end of the Mesozoic, and these are the criteria used to delineate the boundaries of this era. Naturally, only a portion of the entire animal world is affected by changes going on in certain groups at some particular time. The development of terrestrial mammals occurred across the Permian-Triassic boundary, and most marine mollusks and snails also developed to a great extent during the Cretaceous-Tertiary period. As in flora, there were thus few major interruptions in the continuity of faunal phylogeny during the Tertiary.

The name Triassic ("Three-stage period") stems from the division of Triassic deposits into sandstone, limestone and Keuper. Naturally, this is only true for central Europe and bordering regions. All three formations developed during the warm, dry climate in the so-called "Germanic basin", which sank down during the Triassic. A bleak lowland developed from central Europe to England during the Lower (sandstone) Triassic, and it filled with debris on its margins. The sea occasionally invaded Europe via Silesia and possibly from an existing North Sea, penetrating as far as central Germany. The sea moved even further southward toward the end of the Lower Triassic. This led initially

By H. Hölder and R. Schmidt-Effing

to the development of coastal flora and then to the formation of the Germanic calcareous (limestone) inland sea. In time this sea acquired more and more dissolved substances (minerals). As this calcareous sea gained more and more minerals it became ideal for fossilization, for the increasing mineral content killed off organisms living in the sea, and geological changes in the water permitted the formation of shell limestone deposits.

The sinking process of the Germanic basin slowly let up, and the sea retreated. This created new lowlands that extended into France. The Keuper lowland, as this one in Germany is called, was subject to flooding from time to time, and in time and with a new intrusion by the sea the Rhaelian deposits (see below) formed there. This was followed by a long period of flooding by the Jurassic sea.

The western Mediterranean region also shows signs of this triple division during the Mesozoic. In the east, from the eastern Alps to southeastern Asia, the great Tethys sea was the major geological feature. The triple division found in central Europe does not occur in Asia, and for this reason the Triassic is now divided into six subdivisions, recognized throughout the world. From bottom (oldest) to top (youngest), they are: Scythian, Anisian, Ladinian, Carnian, Norian, and Rhaelian.

The rapidly changing strata in the Tethys shows that this was not a deep sea; it was filled with numerous islands, bars, and coastal stretches. It included a rapidly sinking strip of the earth's crust, the Tethys geosyncline. Toward the east, the Tethys was connected to the old (and once larger) Pacific region. The Pacific at that time probably extended across what is now Central America, for the Atlantic did not arise until later. Thus, the Tethys was a broad stretch of water in which marine organisms could intermingle.

The Tethys was approximately in the equatorial girdle, and the climate at that time was much warmer than it is today. The continental regions to the north and south of the Tethys were deserts with occasional marginal seas. As the old Gondwana continent decomposed, lava formations formed great deposits in the south. Parts of the Gondwana continent drifted apart, and as they did so the Atlantic and Indian oceans formed. The Karoo basin of South Africa was particularly rich in saurian fauna and has been a great fossil source. The many tracks in the sandstone found in the valley of the Connecticut River show that this area also had a lot of terrestrial vertebrate life. These tracks also exist in Triassic formations in Europe.

There are just a few plants found only in the Triassic. Among the EQUISETUMS, *Equisetites* looks like many modern species in that group, but with a stem up to 15 cm thick and a height of 6 m, this Equisetum was much larger than any occurring today. *Neocalamites*, on the other hand, is joined with Paleozoic Equisetum having free leaves; *Schizoneura* (see Color plates pp. 204 and 286/287) has a thin shaft with numerous,

Fig. 12-1. *Rhaetavicula contorta*, a Rhaelian mollusk.

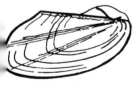

Fig. 12-2. The Triassic mollusk genus *Hoernesia*.

leaf-covered branches. *Pleuromeia* (see Fig. 12-3), an unusual plant with four root-bearing nodules at its base and a cone at the top, was probably the very last member of the tree-shaped lycopods, which had been so prevalent in the Paleozoic.

Of the old ferns, *Glossopteris* survived into the Triassic in the southern hemisphere. New fern families arose in the Upper Triassic. *Lepidopteris ottonis*, a fern, is a key Rhaelian fossil. It belonged to the seed ferns (which extended from the Paleozoic to the Jurassic) and therefore to the gymnosperms.

However, during the Triassic other gymnosperms, of fern or palm-like appearance, were the most important plants. These included the PALM FERNS (Cycadinae), GINGKOS (represented by the narrow-leafed *Baiera*; see Fig. 12-4), and CONIFERS (which existed in the Upper Permian and were represented in the Triassic primarily by *Voltzia* (see Color plate, p. 286/287) and *Yuccites* (see Fig. 12-5)). The conifers formed stands so vast they were probably comparable to modern evergreen forests.

Fig. 12-3. *Pleuromeia*, probably a lycopod.

Among the animal fossils of the Triassic, the UNICELLULAR ORGANISMS (subkingdom Protozoa) were not nearly as significant as they would prove to be in the Jurassic and later periods. Skeletons of flagellates and ciliates first appear in the Jurassic, and foraminifers (order Foraminifera) were not represented by many species in the Triassic. CONODONTS are among the important multicellular organisms from the Triassic (see Fig. 9-49 and the accompanying Chapter 9 text), and they are found in great variety during this period. Their systematic position is still a point of debate. Since we are forced by space limitations to restrict our discussion of Triassic fauna, we shall emphasize mollusks, arthropods, echinoderms, and vertebrates.

Fig. 12-4. Leaves of the gingko plant *Baiera*.

Only a few MOLLUSK (class Bivalvia) genera from the Paleozoic survived into the Triassic; one of these was *Myalina*. The Triassic is characterized primarily by new molluscan groups. The new genus *Claraia* is a close relative of the Carboniferous-Permian genus *Aviculo-pecten*. The modern wing oyster *Pteria hirundo* is closely related to the highly prevalent Triassic genera *Daonella*, *Halobia* and *Monotis*, which all had adhering fibers and butterflylike ribbed shells. In the Rhaelian there was a small mollusk named *Rhaetavicula contorta* (= the distorted Rhaelian bird), which as a key fossil was used to demonstrate (in the 19th Century) that the deposits in the Alpine and extraalpine region were comparable. Other important Triassic mollusks are those whose valve ligature is on an area transected by pits, as in *Hoernesia* (see Fig. 12-2). Triassic relatives of the scallops of today (genus *Pecten*) include *Pleuronectites*, *Entolium* and *Chlamys*. In the Triassic we first find mollusks in which one valve (in this case the right) becomes attached to a substrate (examples include *Plicatula* and *Placunopsis*). *Placunopsis* inhabited the shells of live ammonites and is also the first mollusk to build a reef (a small,

Fig. 12-5. *Yuccites*, a conifer.

Triassic mollusks:

Fig. 12-6. 1. *Enantiostreon.*

Fig. 12-7. 2. *Myophoria.*

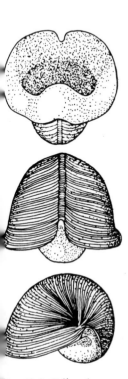

Fig. 12-8. *Bellerophon*, a primitive snail.

1-m structure!) Over a period of time, thousands of individuals grew on top of each other, each using the lower valve of their predecessor as a substrate. The genus *Enantiostreon* (see Fig. 12-6), which apparently could grow to the left or right, should probably be classified with the left-growing oysters (family Ostreidae). Members of the Limidae family are represented in the Triassic by the heavily ribbed *Lima striata* and the smoother *L. radiata* from shell limestone.

Important mollusks with two muscle impressions include the many *Myophoria* species, which arose from *Schizodus* of the Permian and which continue into the first trigoniids in the Rhaelian. The Triassic is also characterized by the first toothless mollusks, the pholadomyids (*Homomya, Pleuromya* and *Pholadomya*). Strangely, the thick-shelled megalodonts, which had apparently disappeared for a long time after the Devonian, are again found in Triassic deposits. Upper Triassic examples include *Megalodon* and *Conchodus.* Unless convergent evolution was occurring, we must assume that these mollusks had retreated to inaccessible areas during the intermediate time. Megalodonts often appear *en masse* in the Triassic, sometimes with life postures preserved. The first river mussel species appear in Triassic fresh-water strata, and they bear a considerable resemblance to the modern species *Unio pictorum.*

Unlike the mollusks and ammonites, GASTROPODS (class Gastropoda) of the Triassic are much more like older forms, even though some new species arise. The great diversity of forms found in modern gastropods, including those with long spines and respiratory tubes, do not occur in the Triassic.

The earliest snails were the BELLEROPHONTACEANS (class Bellerophontacea) from the Paleozoic. The very primitive genus *Bellerophon* (see Fig. 12-8), with a medial slit, has been found in Lower Triassic shallow water alpine deposits. A cross-section through one of these is practically identical to an ammonite cross-section, with the difference being that *Bellerophon* lacks septae. The planospiral shell, slit ligament, and the paired muscle impressions sometimes found on the specimens indicate that *Bellerophon* still had a partially segmented body plan. This places the genus phylogenetically between the monoplacophores and the true gastropods, in which the visceral sack has undergone a rotation of 180° and the adhering muscles have been reduced in number to just one.

The most primitive snails belong in the order Diotocardia, and the order includes a few "living fossils" such as *Pleurotomaria*, which are still found in certain deep parts of the Indopacific area. The order reached its zenith in the Mesozoic, for great numbers of diotocardians inhabited the Triassic seas. Like the primitive gastropod *Bellerophon, Pleurotomaria* and close relatives thereof (i.e., *Worthenia*) had a slit in the shell (see Fig. 12-9). The slit is open at the very front, but during growth the rear parts of the slit continually fill up with shell material, and the sculptured texture of this area differentiates it from the rest of the shell.

The LIMPET GASTROPODS, (Patellidae and related species), which are adapted particularly to rocky coastal regions, becomes much more important beginning in the Triassic (e.g., *Emarginula*; see Fig. 12-10). Others appearing in large numbers include NERITACEANS (e.g., *Neritaria* from Italy; see Fig. 12-11), TROCHACEANS (Trochacea), and the TURRITELLID GASTROPODS (e.g., *Undularia*), all of which are from the deposits of tropical to subtropical Triassic seas.

Fig. 12-9. The snail *Pleurotomaria*.

The OPISTHOBRANCHIA (Opisthobranchia) gastropods are represented in the Triassic by a single genus, *Actaeonina*. (see Fig. 12-12). Only doubtful finds exist for PTEROPODS (Pteropoda); today these snails are very important in marine planktonic communities. Likewise, the PULMONATA (Pulmonata) gastropods played a subordinate role in the Triassic, although the group has been known since the Carboniferous.

Thus, gastropods were far from their high point in the Triassic. Those present at that time were chiefly algae and seaweed feeders, and this meant that they could hardly inhabit the ecosystems where we find gastropods today. In spite of that, over 1000 Triassic snail species have been identified, especially from the shallow sea regions of the southern alps (St. Cassian, Italy) and from elsewhere in Italy, Peru, and Indonesia. Germanic deposits sometimes contain masses of Triassic gastropods.

The first SCAPHOPODS (class Scaphopoda) appeared in the Middle Triassic St. Cassian formation in the form of *Dentalium*, which occurs today by the billions in soft sea floors as deep as 4000 m. PLACOPHORES (class Placophora), a molluscan class whose distinct body division is reminiscent of the tryblidiaceans and hence the primitive mollusk *Neopilina*, have a shell consisting of eight plates that overlap at their edges and fall off after death. Thus, few fossil placophores have been found although the group made its appearance in the Cambrian period. One intact, 17 mm placophore (*Trachypleura*; see Fig. 12-14) was recovered from coastal shell limestone deposits near Berlin, Germany. It belongs to the order of simple-shelled placophores (Lepidopleurina), a group with few species which has since moved into the deep sea.

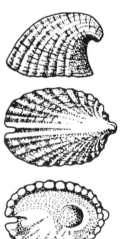

Fig. 12-10. The patellid snail genus *Emarginula*.

For the CEPHALOPODS (class Cephalopoda; see Color plate, p. 150/151), the Triassic period was a time of radical change. The great proliferation of nautiloids (subclass Tetrabranchiata) of the Paleozoic was a thing of the past by the Triassic, and the only survivors from this vast group were two ORTHOCERATE (order Orthocerata) genera, which had straight shells and survived into the Upper Triassic. A greater number of NAUTILIDS (family Nautlidae) were present; these are the coiled forms. They included some genera whose shells were highly sculptured and which did not live beyond the Triassic. *Encoiloceras*, with prominent ribs, had a large open hilum and thus was rather primitive. Of the "smooth" genera, we shall only mention *Syringonautilus*, which probably was the ancestor of post-Jurassic members of this subclass once most other Triassic nautiloids had died out. *Germanonautilus bidorsatus* (see Fig. 12-15) is a characteristic member from Germanic deposits.

Fig. 12-11. The neritacean snail *Neritaria*.

Fig. 12-12. *Actaeonina.*

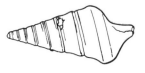

Fig. 12-13. The turritellid snail *Undularia.*

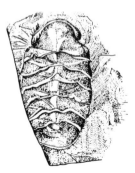

Fig. 12-14. The placophore snail *Trachypleura.*

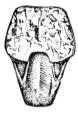

Fig. 12-15. *Germanonautilus bidorsatus.*

AMMONITES (subclass Ammonidea) are important key fossils for the late Paleozoic and the Mesozoic eras, and as such they are of extreme importance. A vocabulary has developed to describe all the facets of the complex ammonite evolutionary pathways, and we shall briefly deal with this evolution. Ammonite shells are usually coiled within one plane (=planospiral), but some are coiled in a snaillike way, and there are also straight, rod-shaped ammonite shells and those in which the convolutions are not planospiral throughout their length; the final loop can be straight for a certain distance and then be coiled.

The shells consist of a living chamber and air chambers, whose septae are penetrated by a tube called a siphuncle (see Fig. 12-16). In some phylogenetically old ammonites, the septae of the air chambers are bent concave toward the mouth, while in later ammonites the septae are convex. The connection between the septum and the inner shell wall is formed by a rather complex fused network called the suture line, a feature used to distinguish individual ammonite groups (see Fig. 12-17). Those parts of the suture line extending toward the mouth of the shell are called saddles, while the parts extending toward the beginning of the shell are called lobes. The suture line is simple in the early ammonites of the Devonian and Carboniferous. The lobes and saddles are smooth and usually have simple curves (they are called goniatitic sutures). In other ammonite groups only the saddles are smooth curves, for the lobes have a fine jagged structure (ceratitic sutures). Finally, in the great majority of ammonites, saddles and lobes are jagged (ammonitic sutures). Juvenile ammonites do not have suture lines as complex as those characterizing adults, and the distinctive features of the suture lines come with growth. Thus, even in ammonites that as adults have complex, jagged lobes and saddles, the initial sutures are goniatitic.

However, a regular suture change occurs throughout ammonite phylogeny (see Fig. 12-17). In the early ammonites up to Permian, the first suture has just four lobes (one external lobe, two lateral lobes and one internal lobe); since only one side of the animal is counted, this anatomical plan is known as trilobate ("three-lobed"). The first suture in most Triassic ammonites is quadrilobate, while post-Jurassic ammonites are typically quinquelobate. Thus, the lobe structure in ammonites (see Color plate, p. 224) becomes more complex as the group evolves.

The suture line (the line produced from the interface between a septum and the shell wall) can easily be seen on the inside of the shell. The outside of the shell must be carefully removed to expose the suture line. Suture lines should not be confused with growth lines, which are evident on the outside of the shell and are also systematically important. Furthermore, one must keep in mind that the growth of an individual shell is often interrupted by pauses of various lengths.

The sculpturing of the shell is just as significant a feature of the ammonite shell as the suture line. There are smooth and ribbed ammonite shells, and ribbed shells include those with rows of nodes, spines, thorns, and constrictions of various sorts. These differences are not evident in young ammonites, for they do not appear until a certain growth stage is reached, and in some species they are lost with even more advanced age.

Toward the end of the Permian period, ammonites underwent their first great reduction. Hardly any Permian genera survived into the Triassic. The last MEDLICOTTIAN (see Chapter 9) from the Permian was *Sageceras*, with numerous two-lipped lobes and two ridges; this species occurred in the Upper Triassic. A new, initially simple, ammonite group related to the medlicottians proliferated in the Lower Triassic. They had fewer lobes, but the lobes bore notches, and the saddles were initially smooth. These were the first ceratitic forms, which included the simple ribbed *Xenodiscus* (see Color plate, p. 224) from the Upper Permian and the smooth-shelled *Ophiceras* (Lower Triassic). These early Triassic ammonites were primarily distributed in the eastern Tethys, in the Pacific, and in the polar seas.

Then, in the Scythian (earliest) stage of the Triassic, there was a veritable explosion in ammonites, and several distinct families arose. There were broad-hilumed, narrow-hilumed, inflated, and high-mouthed ammonites, most of which had ceratitic suture lines. Some characteristic genera were the NORITACEANS (superfamily Noritaceae) *Gyronites*, *Flemingites*, *Meekoceras* and *Sibirites* as well as CERATITACEANS (superfamily Ceratitaceae), which appeared with several families in the Scythian but chiefly proliferated in the Middle Triassic. Ceratitaceans were more highly sculptured than noritaceans and included genera such as *Beyrichites*, whose high-mouthed shell bore faint sickle ribs; *Balatonites*, with ribs and rows of nodules; the inflated looking *Acrochordiceras*, with deeply cleft, bulging ribs and knots; *Ceratites* (see Color plate, p. 224 and Figs. 12-23 and 12-24), to which we shall return; and many others. A great number of other Triassic ammonites, which became particularly prolific in the Upper Triassic, often had much richer shell sculpture, and in these the suture lines are primarily ammonitic. One strikingly beautiful member of this group is *Trachyceras* (see Fig. 12-19), with its spiralling ridges. The thick-ridged *Tropites* (see Color plate, p. 224) has forward-arching forked ribs, while the ribs in *Juvavites* are forked but not arched, and they have constrictions.

A still greater degree of jaggedness in the suture line, producing a pattern that looks like delicate moss, is found in the thick disk-shaped or nearly spherical ARCESTACEANS (Arcestaceae), for example in *Arcestes* (see Fig. 12-20), *Joannites* and *Cladiscites* (see Color plate, p. 224). The latter has close-set ribs on an otherwise smooth shell. Jaggedness reaches its highest point of development in the ammonite genus *Pinacoceras* (see Fig. 12-21) from the Upper Triassic, whose narrow-hilumed, sharply tapered disk shell has high flanks. Presumably the degree of jaggedness is related to the stability of the shell, and as these mollusks anchored them-

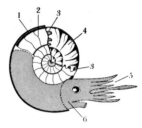

Fig. 12-16. Ammonite anatomy.
An ammonite shell (*Ceratites*), revealing some of the interior structure, with septae, siphuncle, and the soft body of the ammonite. On the above right a piece of the inside of the shell is exposed; it becomes filled with slime. 1. Siphuncle; 2. Septae; 3. Suture line; 4. Shell interior; 5. Tentacles; 6. Soft body with eye (upper surface is dotted).

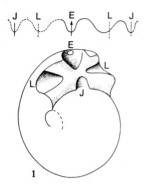

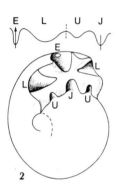

Fig. 12-17. Suture lines of ammonites, showing their phylogenetic development: 1. Devonian-Permian: trilobate primary suture with 3 (i.e., a total of 4) lobes; 2. Triassic: quadrilobate with 4 (6) lobes; 3. Jurassic-Cretaceous: quinquelobate with 5 (8) lobes.
E. External lobe; L. Lateral lobe; J. Internal lobe; U. Umbilical lobe.

selves in various depressions, the suture lines may have assisted them in gaining a firm hold on the substrate. It is impossible to assign each of the nearly uncountable folds to some specific function. It appears that nature permitted the elaboration of suture lines to that extent allowed by the limits of natural selection, and here it would seem that those limits were rather great.

Ammonites were most diverse at the end of the Triassic, and it is interesting to note that in the Upper Triassic ammonite species evolved that had not only simple suture lines but even lost the closed spiral of the shell. *Choristoceras* (see Fig. 12-22), for example, began uncoiling itself, and *Cochloceras* (see Color plate, p. 224) had a spiral in the shape of a turret. This is only the first of several times that we witness a return to an earlier state, for the same thing occurs several times in the Upper Jurassic and in the Cretaceous periods. Why should something like this occur? There are several opinions regarding this problem; according to one, the ammonites experienced a sort of biological crisis toward the end of the Triassic, and they degenerated, while another viewpoint holds the opinion that this was an adaptation or a re-adaptation to specific life habits. Ammonites may have remained in some area where it became advantageous to have an open or an unusual spiralling shell.

The Tethys preserved some 100 ammonite genera, while the Germanic basin, with its higher salinity, did not offer such favorable fossilization conditions, and the finds there are subsequently much more modest. As new connective routes to the Tethys opened up, a single Tethys ammonite (*Paraceratites binodosus*) found its way to the still over-saline marginal seas of central Europe. However, this species apparently met favorable conditions, for it proliferated in Europe and gave rise to a number of new species.

The ancestral form, with its forked ribs and two nodule rows, looked like "*Ceratites*" *atavus*. It was followed by species belonging to the genus *Progonoceratites* with swollen, knotted forked ribs. Initially the shell diameter was approximately 5 cm, but shell size increased somewhat in time. Furthermore, the sculpture of the living chamber became obliterated; *Acanthoceratites enodis* (see Color plate, p. 288) was nearly completely smooth. One side group developed, which had a much larger shell and on the living chamber in adults (or on a part of it) bore simple, usually narrow, but prominent ribs. In *Progonoceratites spinosus*, the ribs terminate in a sharp spine, and the forked rib structure is confined to the inner shell convolutions; this is a sign that rib growth ceases at a certain age. The same condition is found in species of *Ceratites*, although the youngest parts of the shell sculpture consists of distinct, swollen ribs. In *C. nodosus*, the most familiar species, the ribs terminate in a thickened structure but not in a spine.

In the uppermost shell limestone strata, *Ceratites* gives way to the very large, disk-shaped shells of *Discoceratites* (see Color plate, p. 288).

In the early Keuper period, the mollusk-bearing limestone seas retreated, and with that the development of Germanic ceratites ended.

The continuous phylogenetic development of ceratites makes distinction into discrete species a difficult task, but the changes occurring over time are significant enough to permit a certain amount of species differentiation in the shell limestone deposits. A number of these are important key fossils. The proliferation of ammonites in Germanic seas serves as a superb example of how possible evolutionary pathways are restricted, influenced, and channeled by the environment.

All specialized Triassic ammonites disappeared toward the end of the Triassic period. The reasons for their disappearance are unknown. Whenever a large animal group suddenly disappears, it is sometimes postulated that increased cosmic radiation influenced the rate of mutation, but if this were true we would expect all animal groups to change dramatically during that time, and this is not what happens. Of the ammonites, only two or three lines reached the Jurassic; they all arose from primitive ceratites of the Triassic/Permian border, and they are known as LYTOCERATES (Lytocerata), PSILOCERATES (Psilocerata), and PHYLLOCERATES (Phyllocerata). All three, but especially the psilocerates, proliferated. One small Upper Triassic form from northern Iran, *Phyllytoceras*, bore a striking resemblance to the Jurassic genus *Psiloceras*. The study of transition species, such as we seem to have here, is made more difficult by the lack of fossil specimens.

BELEMNITES (suborder Belemnoidea), whose chambered shell portion (= phragmocon) is surrounded by a calcite rostrum (see Chapter 14), were not uncommon in the Tethys, but there were few belemnite genera. *Aulacoceras* had an open, longitudinally channeled tapered shell. In *Ausseites* the rostrum was smooth and up to one meter long. The chambered inner part of the shell of these species have small angles, making them more slender than true belemnites. If such specimens are found and are lacking a rostrum, they can easily be confused with orthocerates. In both groups the septae are rather widely separated. The Californian Triassic *Metabelemnites* appears to be a true belemnite. However, belemnites did not significantly proliferate until the beginning of the Jurassic period.

Triassic TENTACULATES (phylum Tentaculata) are hardly represented in the Triassic by BRYOZOANS (class Bryozoa), but BRACHIOPODS (class Brachiopoda) were next to the mollusks in prevalence during this period. They were not very diverse, however, and the group began declining at the Permian/Triassic border. Brachiopods were continually being displaced by the newly proliferating mollusks, which occupied the same ecological niche.

The only survivors from the vast (subclass) Ecardines brachiopod group of the Paleozoic were long-lived forms such as *Orbiculoidea*, *Crania* and *Lingula*. All three genera have survived into the present, giving these

Fig. 12-18. *Ophiceras*, a medlicottian; the suture line is shown below.

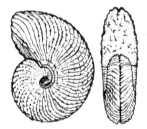

Fig. 12-19. *Trachyceras*, a ceratitacean.

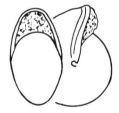

Fig. 12-20. *Arcestes*, an arcestacean.

Other Triassic ammonites:

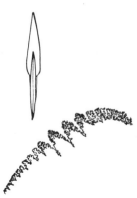

Fig. 12-21. *Pinacoceras.* Below is the suture line.

Fig. 12-22. *Choristoceras.*

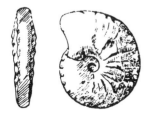

Fig. 12-23. *Ceratites atavus.*

Fig. 12-24. *Ceratites nodosus.*

hardy "living fossils" an extremely long phylogenetic lifetime of more than 400 million years. The (subclass) Testicardines suffered even more losses in the Triassic, and with just a few exceptions the only members left of this group were rhynchonellids (order Rhynchonellida), terebratulids (order Terebratulida), and spirifers (order Spirifera), which thrived for some time.

Rhynchonellids were represented by the numerous sharply ribbed species of genus *Rhynchonella* and by the smooth-shelled genera *Rhynchonellina* and *Norella.* Their stalk opening is usually closed off by calcite secretions. The arm structure is composed of two hooks attached to the base of the hinge. *Coenothyris* was an important cherry- to walnut-sized genus with a round stalk hole; it sometimes appeared in masses in Germanic shell limestone. A terebratulid, it possessed a loop-shaped arm structure. Spirifers, with arm structures something like goose feathers, were represented in the Triassic by the nearly halfmoon shaped *Spiriferina* (see Fig. 12-25) and *Tetractinella,* whose valves were supported by four sharp, radiating ribs. Unlike the impunctate *Rhynchonella,* the shells of *Spiriferina* and *Coenothyris* are penetrated by countless fine holes, which can be seen with a magnifying glass.

With the end of the Permian period, approximately 230 million years ago, many large ARTHROPOD (phylum Arthropoda) disappeared, primarily the TRILOBITES (subphylum Trilobita), SEA SCORPIONS (family Eurypteridae, class Merostomata; L up to 2 m), and of insects the PALEODICTYOPTERANS (order Palaeodictyoptera), whose members had flown through the Carboniferous and Permian forests. The ecological niches freed by the disappearance of these groups permitted new animals to occupy them, and many of these new species were ones we know today. One could even state that all modern arthropod orders have been present since the Triassic period, even though some of these were only represented by rather primitive species at that time. There are no post-Triassic fossil arthropod orders.

New crustacean species (excepting some rare Permian precursors) include the MALACOSTRACANS or HIGHER CRUSTACEANS (subclass Malacostraca) in the form of shrimp and lobsters, and new insects included the NEOPTERANS (Neoptera), which includes all modern FLYING INSECTS (subclass Pterygota), excepting ephemeral flies and dragonflies (for they had existed in the Paleozoic). Complete transformation as we know it in modern insects (=holometaboly) probably existed in the Permian, but it was in the Triassic that this phenomenon became widespread.

There are also some arthropod classes in which few new species arose. The MEROSTOMATES (Merostomata) of the Paleozoic, which had been so diverse in that era, were represented in the Triassic by the single genus *Limulus.* Externally, in spite of the great time span between the Triassic and the present, there are few differences between the Triassic species and the modern horseshoe crab, and *Limulus* is one of the finest examples of a so-called "living fossil". However, Triassic horseshoe crabs lived in all seas, and today there are just two, sharply separated distribution regions, one along the North American Atlantic coast with the king crab (*Limulus polyphemus*) and the south-east Asian region with the "Moluccan crabs" *Tachypheus* and *Carcinoscorpinus*.

Only a few scorpion finds represent Triassic arachnids, and pantopods and myriapods are hardly known from that period. The number of CRUSTACEANS (class Crustacea) was so much the greater. Some of the most interesting ones were long-lived forms such as certain PHYLLOPODS (sub-class Phyllopoda): *Triops* (see Fig. 12-32) and *Isaura* (see Fig. 12-31). *Triops cancriformis*, with a L of several centimeters, had an oval carapace that hid most of the body. Its rod-shaped, movable trunk is divided into segments and terminates in a forked tip (furca). This species still occurs in fresh water, and its eggs can remain viable in dry mud. *Isaura minuta*, which has existed since the Devonian, is another extremely old species. It formed entire fossil banks ("Estherian strata", since the former genus name was *Estheria*); its dorsal armor gives it a molluscan appearance (see Fig. 12-31). Because of this, the suborder to which it belongs is named Conchostraca (=mollusk-shelled). The dorsal armor is not shed with each molting, so it displays concentric growth lines.

In contrast, the OSTRACODS (Ostracoda), which has been such a sig-nificant subclass from the Cambrian to the present, are scarcely known from the Triassic and then in the form of some very resistant, long-lived forms such as *Bairdia* (see Fig. 12-34) and *Cythere*. *Bairdia* has been found in entire deposits from the sea, filling them like countless white dots. Unlike the externally similar *Isaura*, whose body has many limb pairs, ostracods have just three to five.

COPEPODS (subclass Copepoda), which include the cyclopids (family Cyclopidae) occurring today throughout the oceans of the world, are only represented by a single find in Germany. The oldest confirmed ISOPOD (Isopoda, an order) finds are from the Triassic (Germany and Tanzania).

With the DECAPOD order (Decapoda) we pass to the most interesting crustacean group, whose phylogenetic proliferation began in the Tri-assic. They include the shrimps and prawns (e.g., *Aeger* and *Penaeus*) that are of such commercial importance today. All the decapods can be easily recognized with their laterally compressed body, the thin, smooth shell, and the jagged rostrum. Good swimmers and jumpers, *Aeger* chiefly inhabited shallow seas during the Triassic, just as they do today.

ARMORED CRUSTACEANS developed from common ancestors at the Permian/Triassic border, and their modern representatives are classified in the four suborders of Reptantia, Astacura, Anomura, and Brachyura. They are characterized by a dorsally flattened body, and modern members of this order include such familiar species as lobsters and crabs, many of which have been widely distributed since the Triassic. *Pemphix sueuri*

Brachiopods, whose shells are penetrated by fine holes:

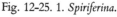

Fig. 12-25. 1. *Spiriferina*.

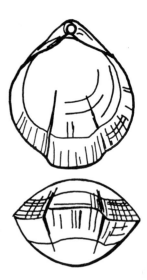

Fig. 12-26. 2. *Coenothyris*.

Other brachiopods:

Fig. 12-27. 1. *Lingula.*

Fig. 12-28. 2. *Rhynchonella.*

Fig. 12-29. 3. *Norella.*

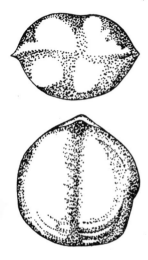

Fig. 12-30. 4. *Rhynchonellina.*

(see Fig. 12-33), which was the size of a hand and had a wart-covered dorsal armor and long antennae, is from central European shell limestone deposits. Anomurans, the group including the hermit crab, were probably widely distributed in the Triassic, although they have never been found in these deposits. However, the spoor known as *Rhizocorallium* (see Fig. 12-36) are probably their tracks. These U-shaped structures, produced from resting and feeding, are quite like those produced by the modern genus *Callianassa.* Crabs are not known from the Triassic.

With the development of homometaboly, the NEOPTERANS (Neoptera), the largest insect group, were able to proliferate in the Triassic. Of the PALEOPTERANS (Palaeoptera) with stiff wings, the only ones to survive into the Triassic were MAYFLIES (order Ephemeroptera; they spend most of their life in the larval stage) and DRAGONFLIES (order Odonata). Both have continued to flourish into the present. Primitive neopterans, which could only fold their wings backwards in the resting position and which were widely distributed in the Paleozoic, included COCKROACHES (suborder Blattaria) and ORTHOPTERANS (order Orthoptera). Those that were important in the Triassic included modern members of the cockroach suborder (cockroaches; family Blattidae) and the LACUSTOPSIDS (family Lacustopsidae), which are closely related to modern locusts.

With the exception of wasps, which do not appear until the Lower Jurassic, and fleas, which arise after the Cretaceous, all major insect groups characterized by holometaboly have existed since the Triassic. Only the wings have been found from the oldest butterfly, *Eoses triassica* (see Fig. 12-37), which is a Middle Triassic species found at the great Triassic insect site on Mt. Crosby in Australia. The butterfly probably had a chewing apparatus, since there were no angiosperms when this butterfly lived. The first FLIES (Diptera) are from central Europe and Upper Triassic Russia. These were thread-horns (suborder Nematocera), which according to Soviet zoologists gave rise to wasps and sawflies.

The vast BEETLE (order Coleoptera) group, which contains approximately 250,000 species, is first found in Permian formations but includes Triassic species found in Europe, South Africa, Kirghiz U.S.S.R., and the Antarctic. However, our knowledge about Triassic beetles is fragmentary. The thick wing covers have been well preserved, but they alone tell little about beetle phylogeny.

All three ALDER FLY orders (superorder Neuropteroidea), which includes megalopteran snake flies, rhaphidid snake flies, and lace-wings, have also been found in Permian formations. Triassic lace-wings have been recovered in the Kirghiz U.S.S.R. and Australia. Some of the Australian specimens are classified with the psychopsids (family Psychopsidae), an alder fly group occurring solely in Australia, where it is known as the silky lace-wing due to the brilliantly colored wings and the long, silky hair.

The (extant) 300 SCORPIONFLY (Mecoptera) species in this insignificant

order were represented in the Permian and Triassic. Individual scorpion
fly families were confined to certain regions, just as they are today. Thus,
the choristids and nannochoristids (families Choristidae and Nanno-
choristidae) occur only in Australia, both as fossils and today, and the
snow scorpionflies (families Bittacidae and Boreidae) are chiefly found in
the northern hemisphere.

Fig. 12-31. *Isaura*, a
phyllopod (enlarged).

During their long geological history, the ECHINODERMS (phylum
Echinodermata) have always been confined to the ocean, but they proliferated there and have come to occupy every marine ecosystem, from
rocky reefs to the darkest deep sea and from the open sea to coastal mud.
Prevalent echinoderms in the Triassic changed from the chiefly sessile
pelmatozoans (Pelmatozoa) to mobile forms, which are known as eleu-
therozoans and include holothuroids, sea urchins, starfishes, and brittle
stars.

The many pelmatozoan classes of the Paleozoic were reduced to just
one insignificant crinoid group in the Triassic. These INADUNATES (Inadunata) disappeared during the Triassic and gave way to a new group,
the ARTICULATES (Articulata), from which modern crinoids and feather
stars evolved. Many modern crinoids occur in deep seas, but during the
Triassic they were prevalent in shallow, temperate waters. Intact specimens
of *Encrinus carnalli* (see Fig. 12-38) and *Encrinus liliiformis* (see Color plate,
p. 365 and Fig. 12-39) are among the most valuable fossils available,
for the soft parts of the body are rarely retained. Other less well-
known genera include *Holocrinus*, *Isocrinus*, *Dadocrinus* (see Fig. 12-40)
and *Pentacrinus*.

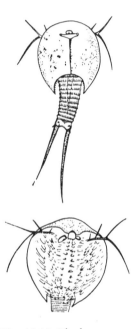

Fig. 12-32. The large
Triops cancriformis.

In *Encrinus*, the bowl-shaped cup at the base of the crown contains
the internal organs and is capped by the mouth/anal opening. It is surrounded by the arms, which with the help of the pinnulae pull food
toward the crinoid and into the mouth. The crown rests on a stalk up
to 2 m long, which is usually anchored at its base with an adhering disk.
The stalk is built up of round, coinlike elements called trochites
(trochus = tire), and trochites have been found in such numbers (as they
are in one southern Indiana site well-known to geologists and paleon
tologists, not to mention amateurs) that they form kilometers of beach!

Crinoid habitat is not usually the same as the deposition site, for many
crinoids became separated from the base and turned into free-swimming
organisms that move into an entirely different ecosystem. Sessile crinoids
probably produced veritable forests in shallow waters, and they could
only survive where the water was relatively clear. This explains the
sudden disappearance of crinoid colonies, for sudden shifts in sediment
and clouding of the water would kill them via "asphyxiation". The
incoming sediment also provided excellent fossilization conditions.

Although SEA URCHINS (class Echinoidea) were very uncommon in the
Triassic, some decisive steps in their evolution were taken during this
period. Of the numerous Paleozoic orders of the subclass Perischoechinoi-

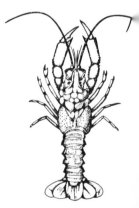

Fig. 12-33. *Pemphix sueuri*.

Fig. 12-34. *Bairdia*, an ostracod

dea, only the CIDAROID (order Cidaroida) sea urchins survived into the Mesozoic, and these animals are still found below depths of 50 m. The most well-known cidaroid is *Miocidaris* (see Fig. 12-41), which may have existed in the Carboniferous but certainly occurred in the Permian and survived into the Triassic. Its shell plates overlapped like roof tiles, with the ambulacral fields covering the interambulacral fields and affording the capsule some mobility. The Upper Triassic miocidarids gave rise to the modern cidaroids (*Balanocidaris*) with their stiff capsules. They became important in the Jurassic.

The EUECHINOID subclass (Euechinoidea), occurring from the late Mesozoic to the Cenozoic, began with "typical" forms in the Middle Triassic, with "irregular" species first appearing in the Lower Jurassic. Triassic euechinoids were relatives of the modern deep-sea echinothuriids (family Echinothuriidae), another group with movable shell plates as in *Miocidaris*. The ECHINACEAN superorder (Echinacea), to which the edible sea urchin (*Echinus esculentus*) belongs, is first found in Middle Triassic St. Cassian formations (genera *Triarechinus* and *Lyssechinus*).

Like sea urchins, STARFISHES (class Asteroidea) also crawl and dig through the floor, as do brittle stars as well. Only shallow seas species are known in the Triassic, and starfishes are quite uncommon among those. One of them, *Trichasteropsis* (see Fig. 12-45), is a modern-looking starfish whose arms are strengthened by a lower and upper row of marginal shields.

BRITTLE STARS (class Ophiuroidea) were chiefly represented by the still extant order of OPHIURIDS (Ophiurida), which includes *Aspidura* with its large mouth plates and grainy plates. Since the genus was discovered on the inside of the shells of mollusks, it was thought that *Aspidura* was a scavenger that consumed the meat of dead mollusks. However, it appears more likely that upon being suddenly covered with sedimentation, the brittle stars attempted to free themselves and accidentally got into the waiting open mollusk shells (see Fig. 12-43), where they suffocated. Later the sediment hardened, and the aragonite mollusk shell was dissolved, leaving the open inner cavity with the dead brittle star. The calcite elements of aspidurans, unlike the aragonite of the mollusk shells, did not dissolve or decompose under pressure or temperature changes (diagenesis).

Of the SEA CUCUMBERS (class Holothuroidea), we find the microscopic, anchor- or wheel-shaped ossicles (see Fig. 12-44), which were embedded in the skin. Ossicles occur in alpine Triassic deposits.

The great calcite deposits of the northern and southern Alps, formed from Triassic marine life, are a massive witness to the life of that period. These formations are from marine reefs and they are examples of particularly interesting fossil communities. The alpine mountain formation processes have caused the fossil deposits to become very confused and quite unlike their original disposition, but these fossils have recently been researched with the help of modern techniques.

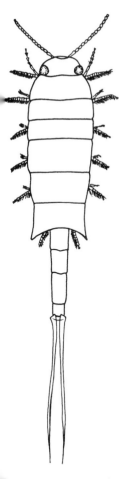

Fig. 12-35. A copepod from the subclass Copepoda.

Numerous reef systems comparable to the Great Barrier Reef of today formed in the Middle and Upper Triassic on the Tethys shelfs. The Göll system (see Fig. 12-46) in the Bavarian Alps near Berchtesgaden is an example of these reefs. The complex is divided into a central region approximately one kilometer wide, a front section that once led to the open sea, and a rear section where the true reef-building activity occurred.

Fig. 12-36. A fossil crustacean spoor designated "*Rhizocorallium*".

The reef coral lived in a symbiotic relationship with green algae, and the reef must have sunk continually over millions of years, since the coral acted to build up the reef and the algae tore it down. What is left today must be appreciated in terms of the community of reef builders and reef inhabitants, the latter being the destructive organisms.

Divided and undivided sponges (Sphinctozoa and Inozoa) as well as branched coral colonies, low-growing coral and high individual coral stands were the most important reef builders. A "change in scenery" occurred at the Permian/Triassic border: while the Paleozoic tabulate corals died out, the modern, radially symmetric corals (known to zoologists as Madreporaria and to paleontologists as Cyclocorallia) took their place. They were chiefly represented by the bilaterally symmetrical pterocoral. Calcareous sponges first made a significant contribution to reef-building activities in the Permian of Texas (in the classic Capitan Reef). They also participated in reef building in the Triassic central alpine region, and in later parts of the Mesozoic they were displaced by corals and siliceous sponges. Apparently the calcareous sponges were the first to invade the ecological "hole" left by the widespread extinction of Paleozoic reef-building coral.

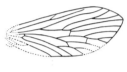

Fig. 12-37. *Eoses triassica*, the oldest butterfly.

Other important reef builders included calcium-producing red and blue algae, hydrozoans, and bryozoans. Sessile foraminifers also participated. They formed a crust on top of the work of the other reef builders, and the clumps they made could sometimes be of impressive dimensions. Other foraminifers were characterized by thick, porcelainlike shells.

Crinoids:

The reef builders are contrasted with reef inhabitants, which through their destructive activity and their own hard parts contributed to the formation of reef sedimentation. All boring organisms were destructive to reefs; this includes certain algae, porifers (sponges), bryozoans, gastropods (snails and others), mussels, and worms. Parts of the coral were actually consumed by holothuroids, crustaceans, and fishes. Hard parts of reef sedimentation were produced by bottom-dwelling foraminifers, gastropods, mollusks, brachiopods, crinoids, holothuroids, and some algae. Crustacean parts, stumps of fish teeth, and the bones of marine reptiles have also been found in this sediment. Many of the reef inhabitants had no hard parts, and we have to use their tracks to make educated guesses about their habits.

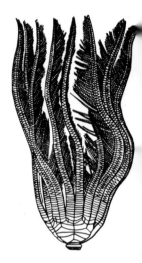

These two groups of organisms—those that build and those that destroy—lived in equilibrium (the latter being helped by the surf), and as

Fig. 12-38. 1. *Encrinus carnalli*.

the reef grew it was also being torn down. Mighty fragmentary blocks, which could be as large as boulders, formed the pre-reef section in the open sea, where they joined the fine-grained basin deposition, which was often rich in ammonites. Basin depth probably did not exceed 200 m.

The rear side of the reef was characterized by shallow, quiet inlets where great deposits were built by thick-shelled mollusks (*Megalodon*), and their activity produced the vast deposits we see today.

This habitat has produced many *Magalodon* specimens, usually in life postures, as well as numerous echinoderms, gastropods, corals, and calcareous algae. Among the calcareous algae, the dasycladaceans were the most important Middle Triassic reef builders in what are now the Dolomite mountains of Italy. The sea inlets, which were usually several kilometers long, extended to the west and north into the Alps and the highly saline lagoons, and it was here that the main Dolomite was deposited which today forms, among other things, the mountains of Höllengebirge, the Lechtaler Alps and those of the Allgäu.

As we previously discussed (Chapter 7), vertebrates during the Paleozoic had entered all water habitats as cartilaginous and teleost fishes, went on land as amphibians, and as reptiles became reproductively independent of water. Evolution of fishes and amphibians came to a standstill in the Triassic, while the reptiles continued to develop (as saurians).

The only Triassic CARTILAGINOUS FISHES (class Chondrichthyes) are members of the genera *Hybodus* and *Acrodus*, and all we have of them are their teeth and fin spines. They reappear in the Jurassic (see Chapter 14). These fishes are prevalent in deposits from seas and marginal sea bodies in limestone, clay, and reed and Rhaelian sandstone. The so-called bonebeds, which are bone aggregations in shallow water, are particularly important sites for cartilagionous fishes.

g. 12-39. 2. *Encrinus iiformis.*

On the other hand, the STURGEONS (superorder Chondrostei) and GARS (superorder Holostei) were quite prevalent. Some of them had asymmetrical caudal fins. *Colobodus* had ribbed scales and crushing placodont teeth. Other representatives were *Dollopterus* (with large fins), which like *Thoracopterus* of the Upper Triassic (see Fig. 12-25) probably leaped much like modern flying fishes.

In contrast to these species, the 1-m pikelike *Saurichthys* (see Fig. 12-53) had a symmetrical caudal fin. The conical teeth in the tapering snout have grooves and often are found by themselves. Scales have been reduced to just four longitudinal rows. The small Upper Triassic *Semionotus* (see Fig. 12-54) is covered with thick ganoid scales, and large keeled scales are found along the dorsal line in front of the dorsal fin. The species occurred in huge swarms in shallow water. Some Triassic genera (e.g., *Pholidophorus* and *Caturus*) were closely related to true teleost fishes; the latter is related to the modern North American bowfin (*Amia calva*).

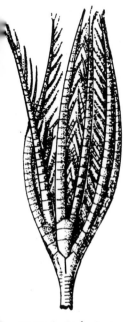

g. 12-40. 3. *Dadocrinus.*

One of the most famous Triassic fossils of all is the lungfish *Ceratodus*, which lived until the Upper Jurassic and which was related to the Australian lungfish (*Neoceratodus forsteri*), a modern species which was discovered much later. LUNGFISHES, which were much more prevalent in the Paleozioc than in the Mesozoic, had many cartilaginous skeletal elements as in Chondrostei and Holostei. This means that the skeleton decays rapidly, and the decomposition rate is perhaps the reason that no intact Triassic lungfish skeletons have been recovered. Usually all that we find are the convoluted teeth covered with a thin enamel layer. The teeth were used for masticating small organisms and plants. The bony lungfish skull consists of a few large plates. The Australian lungfish has a "lung" bladder, which corresponds to the swim bladder occurring in many other fishes, to survive drought periods in oxygen-poor pools. It is possible that the Triassic precursors of the modern lungfish had a similar structure and could therefore exist in similar unfavorable conditions.

Among the LOBEFINS or CROSSOPTERYGIANS, the RHIPIDISTIANS (superorder Rhipidistia), which have inner nasal openings and from which the first amphibians and thus all tetrapods evolved in the Devonian, did not survive beyond the Paleozoic Era. However, the COELACANTHS (superorder Coelacanthida), to which the modern coelacanth (*Latimeria chalumnae*) belongs, have been found in Triassic deposits in central Europe and elsewhere. All of them resemble the modern coelacanth, especially in the distinctive three-part caudal fin. Finds such as those from Triassic alpine strata suggest that during this period the coelacanths left the inland waters and began migrating into the sea, where the modern coelacanth is found.

As we have seen, the Triassic was a time of new development and proliferation for many invertebrates, but the situation was quite different for amphibians, because many of the Permian groups became extinct during this time, and the only new ones to appear are very late LABYRINTHODONTS (sublcass Labyrinthodontia). The dentine of the slender, sharp teeth of these animals has a complex, labyrinthine pattern of convolutions (see Fig. 12-56). The skull roof of labyrinthodonts is composed of large, ridged plates (see Fig. 12-55), and therefore another name for the group is Stegocephalia (=roof-skulled). The roof may be semi-oval, triangular or semicircular and is completely closed save for the nose and ear openings and the tiny parietal eye opening. The ear slits form a triangular depression at the rear of the head; this is for the tympanic membrane. The occipital joint consists of two conelike structures (double condyles).

The eye openings in the narrow, triangular skull of *Trematosaurus* are located well laterally, while they are closer to the body mideline in other genera (e.g., *Parotosaurus*) and are separated by the frontal bones. The most well-known amphibian is probably *Mastodonsaurus*, with a skull 1.25 m long (see Color plate, p. 286/287). In front of the nasal openings there are two other openings to accommodate the large fangs of the lower jaw. In *Cyclotosaurus* the ear slits are closed in back, and there are

Fig. 12-41. The cidaroid *Miocidaris*.

Fig. 12-42. *Triarechinus*, a Triassic euechinoid.

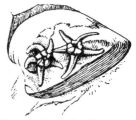

Fig. 12-43. A fossilized brittle star (*Ophiura scutellata*; 2 specimens) in a mollusk valve.

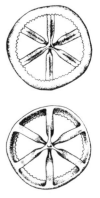

Fig. 12-44. Ossicles of holothuroids (enlarged).

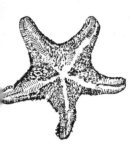

Fig. 12-45. The "modern" looking starfish *Trichasteropsis*, seen from below.

two auditory openings in the roof of the skull. Since the skull length of these two amphibians is approximately two-fifths the body length, *Mastodonsaurus* would have an overall length of approximately 3 m. *Metoposaurus* (see Fig. 12-58) has particularly prominent breast plates of sutured bone; the plates correspond to the sternum and clavicle. In the vertebrae of these amphibians the intermediate vertebral elements play a more significant role than in the more uniform vertebrae of primitive reptiles related to labyrinthodonts. Labyrinthodonts had weak limbs; small bones were replaced by cartilage. These organisms became more and more aquatic, as the increasingly flattened skull indicates.

PLAGIOSAURS (*Plagiosaurus*; see Fig. 12-57), with their short, broad skull roof and large, round eye openings, deviate from other labyrinthodonts. The skull roof is particularly broad in *Gerrothorax* (Rhaelian); like

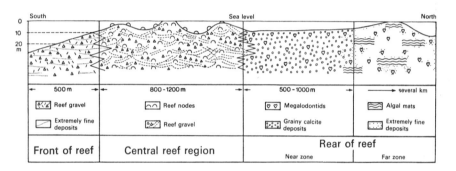

Fig. 12-46. North-south cross-section through the reef of the upper Göll. Drawn after H. Zankl, 1969.

the modern axolotl, *Gerrothorax* had gills and bony gill arches as an adult and therefore spent its entire life in water. As in other plagiosaurs, its back was covered by small, knotty dermal bones. A thick network of ventral ribs and scales protected the belly. One amphibian group, whose skulls were not quite as wide as in plagiosaurs, was confined to the southern hemisphere in the Triassic; examples are the genera *Brachyops* and *Batrachosuchus*.

The most primitive Triassic saurians were small, late COTYLOSAURS (order Cotylosauria), the reptile ancestral group. The skull roof was closed as in amphibians and lacked the temporal openings behind the eyes as in more advanced reptiles. These openings were used to provide points of attachment for lower jaw muscles. One representative cotylosaur is the small *Koiloskiosaurus*, which had a broad, triangular skull 7 cm long that was tapered in front; external ear depressions; a large parietal eye; palatine teeth; thick vertebrae; and five-digit limbs. These are all primitive features which indicate that *Koiloskiosaurus* and its still smaller relatives *Procolophon* (South African Lower Triassic) and *Telerpeton* (Scottish Upper Triassic), as the last cotylosaurs, were not members of a major phylogenetic line.

TURTLES (order Testudines) likewise had a closed skull roof. Turtle tracks have been found in sandstone but good skeletal finds date from

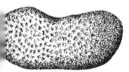

Fig. 12-47. The massive reef coral *Palaeastraea*.

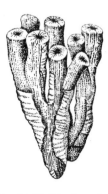

Fig. 12-48. The branched reef coral "*Montlivaltia*".

the Keuper formation in Germany (*Proganochelys*; see Fig. 12-59; and *Triassochelys*; see Color plate, p. 286/287). The few skull bones, the unusually extensive armor of vertebrae and ribs, and the absence of teeth and replacement by horn all show that turtles have been highly specialized animals from their very beginning. We would expect that the fossil record would be able to show us how all these changes came about, but this is not true at all, for the Keuper turtles are characteristic members of the Testudines order, and only a very close examination would reveal that they are actually primitive species. The arched dorsal armor and the ventral armor, of which only the bones (not the horn) have been retained in fossils, completely surround the trunk, and the number of skull bones is reduced. However, the deep depressions characterizing the skulls of later turtles are only beginning to form in Keuper turtles. That is, the spaces permitting the turtles to retract its head were not present in these primitive turtles.

What is most interesting is that *Triassochelys* is the only known turtle genus with palatine teeth and hidden, non-erupting teeth buried in the externally toothless jaw bones. This is evidence of toothed ancestors; these animals have not yet been found. In *Eunotosaurus watsoni*, a fragmentary turtle-like fossil from Permian South Africa, what may have been parallel evolution has led to a toothed skull and spoonlike broadened ribs, which some paleontologists feel may have surrounded the body in turtle fashion. The two Keuper turtles probably inhabited water shores in the Keuper geosyncline.

The THERAPSIDS (order Therapsida), the mammallike reptiles we discussed previously (see Chapter 10), occurred with numerous species in the Triassic but were replaced by the first mammals at the turn to the Jurassic (see Chapter 13). The ICHTHYOPTERYGIANS (subclass Ichthyopterygia), which would become so important in the Jurassic, were quite diverse in the Triassic. As in turtles, ichthyopterygians combined primitive and specialized features in their skeletal structure. The primitive characteristics are the structure of the vertebrae, ribs, and skull. Their streamlined shape, however, is an advanced feature reflecting adaptation to life in the water. Other specializations (also adaptations for this habitat) include the paddlelike modified limbs and the indented caudal fin. The paddle structure came about as a result of shortening the upper and lower arm, the upper and lower leg bones, and the digits (which increased in number, an unusual development in the animal kingdom). Ichthyopterygians also had a long snout (composed of intermaxillary bones) for catching prey and a large number of short vertebrae that looked like checkers and were connected only by connective tissue and cartilage. They had loose thorny vertebral processes.

The primitive features of ichthyopterygians led zoologists for some time to believe that these animals may actually have been amphibians that descended directly from fishes. However, recent work has shown that

Fig. 12-49. A colony of the calcareous sponge *Peronidella*.

Fig. 12-50. The calcium-producing red algae *Solenopora*.

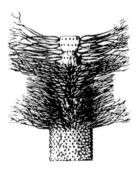

Fig. 12-51. Reconstruction of the algae *Diplopora*. Below with calcium shell, above without, and on the very bottom with only the. calcium shell. At the very top are the whorled branches; they are removed in the middle to show the internal structure.

Fig. 12-52. The flying-fishlike *Thoracopterus*.

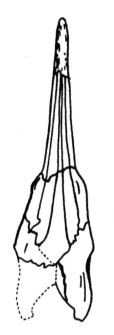

Fig. 12-53. *Saurichthys* skull.

Fig. 12-54. *Semionotus*, a ganoid fish.

these are indeed reptiles. Triassic ichthyopterygians, in comparison with their Jurassic descendants, were completely adapted to water. They descended from terrestrial ancestors and are the best known example of a return by reptiles into the water, the habitat from which Devonian vertebrates had come onto land.

The most primitive ichthyopterygians, the OMPHALOSAURS (*Omphalosaurus* and *Grippia*) are from Triassic North America and Spitsbergen. They have a relatively short snout with dome-shaped (i.e., specialized) cracking teeth and hind limbs that have not yet been modified into paddles. These are all signs of descent from terrestrial or at least coastal animals *Mixosaurus* (see Fig. 12-60) is from another ichthyopterygian group. Barely one meter long, it had sharp teeth. Fossil specimens of this genus have been found in German limestone and well-preserved skeletons have been recovered in Lower Triassic slate deposits of Tessin. The broad forelimbs (now larger than the hind limbs) bear numerous, almost square, digits arranged in fives. The upper and lower arm bones are much shorter in contrast to omphalosaurs. The slightly downward arching caudal vertebrae bore a low caudal fin as the chief propulsion organ. The pair of high-set temporal grooves characterizing ichthyopterygians first appear in the skull roof of mixosaurs.

Shell limestone sometimes contains the flat, gleaming black teeth of the placodont *Placodus* (see Color plate, p. 285), the most familiar member of the order Placodontia (formerly Sauropterygia). Today the PLACODONTS are classified as an individual order restricted to the Triassic. Like ichthyopterygians and plesiosaurs they have a pair of high-set temporal openings (fossae), but they are more restricted by other bones than in the ichthyopterygians. The earliest placodonts are from Middle Triassic-Tessin. *Helveticosaurus*, in spite of its sharp teeth, is placed with the placodonts. With a slight increase in digit number, the genus displays features of the early stages of paddle formation and hence an adaptation to life in the water. In the somewhat later *Placodus* the digit number appears to be "normal", but the weak bones of the limbs indicate that the animal was a swimmer.

Placodonts perhaps moved with the aid of the limbs and the long tail. The teeth in the rather high skull also show signs of an aquatic life: *Placodus* has ten pencil-shaped grabbing teeth in front and twenty square, placodont teeth in the sides of the jaws and in the dentine bones. The animal probably fed on mollusks and brachiopods, pulling them up from the floor and crushing their shells. A network of powerful pectoral ribs protected the belly, while the dorsal vertebrae had high thorny processes and above them a row of round dermal bones forming a dorsal ridge.

Another placodont, *Cyamodus* (see Fig. 12-62) appeared simultaneously with *Helveticosaurus*. Its skull, which was extended to the rear had a tapered snout and teeth of different sizes but all of which were flat and pea-shaped. In place of the single-rowed dorsal ridge of *Placodus*,

Cyamodus possessed a dorsal armor of small, knobby bony plates. Similar armor covering is found in later genera with fewer teeth: *Psephosaurus*, *Placochelys* and *Psephoderma* from the Keuper period; they all look like turtles. *Henodus chelyops* (see Color plate, p. 286/287 and Fig. 12-63) looked even more turtle-like (and its scientific name in fact means turtle-like). The rectangular, very flat skull has just one bean-shaped tooth in each jaw (the entire scientific name means "turtlelike single-toothed animal"). This therefore represents the extreme in tooth reduction, a process that occurs in numerous vertebrate groups and which in turtles has led to a complete loss of teeth in the modern species. Humans have degenerated last molars (the so-called wisdom teeth); they correspond to the last teeth remaining in *Henodus*. The diet of *Henodus* consisted of the shells of small mollusks and crustaceans, organisms that can be found in the same deposits with these placodonts. It appears that *Henodus* scooped the shells (and mud) up from the floor, and after pressing the water out of its mouth with a special filtration apparatus, the prey was chewed with the four last teeth.

We now come to those Triassic reptiles with two temporal fossae behind the eyes. They arose from the Permian EOSUCHIANS (order Eosuchia), which split in the Triassic into three daughter groups: RHYNCHOCEPHALIANS (order Rhynchocephalia), SQUAMATANS (order Squamata), and THECODONTS (Thecodontia). The latter are now extinct but gave rise rise to crocodiles, flying saurians, birds, and what are popularly called dinosaurs.

Rhynchocephalians have survived into the present with just one species, the tuatara (*Sphenodon punctatus*). Two bony bridges are located beneath the temporal fossae, as in the crocodile skull. The scientific name of the tuatara (= wedge-toothed) stems from the flat, triangular teeth, which fuse in adulthood. The name of the order (= beak-headed) refers to the hooked bend in the fore jaw in some extinct rhynchocephalians (e.g., *Stenaulorhynchus* from the east African Triassic). This hook-shaped structure was presumably used for digging. The five claws on the fore legs likewise suggest that this animal was a digger. The claws were the size of those found on a medium-sized dog. Further possible evidence for digging behavior is the strong pectoral girdle with its powerful muscles. The diet probably consisted of roots, for which the small, blunt teeth would be appropriate. The teeth are not only on the jaws but also (and this is a primitive feature) on the palate (the hard roof of the mouth). The parietal eye of the tuatara (which is covered with skin) is found in some of its fossil relatives, but not in the above-mentioned genus, which in this respect was more specialized than the tuatara of today.

The SQUAMATANS (order Squamata), which includes lizards and snakes, are characterized by a quadrate bone that does joins not only the lower jaw but also the upper part of the skull (by a movable joint). Furthermore, the lower temporal fossa has the shape of a sinus with a

Labyrinthodonts:

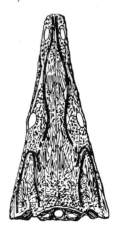

Fig. 12-55. *Trematosaurus*. Entire (reconstructed) skull, seen from above.

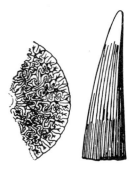

Fig. 12-56. Tooth of *Mastodonsaurus*. Left: in cross-section.

Fig. 12-57. *Plagiosaurus*. Skull with lower jaw fragment and neck-breast armor.

Fig. 12-58. *Metoposaurus*.

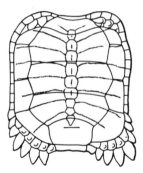

Fig. 12-59. Reconstruction of the dorsal armor of the Keuper turtle *Progano-chelys*.

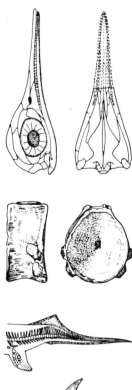

Fig. 12-60. The ichthyop-terygian *Mixosaurus*. Skull, vertebra, and tail.

lower opening, since the lower bony bridge is vestigial in squamatans. This order is well represented in the fossil record. In *Prolacerta* of Lower Triassic South Africa the bridge is interrupted by just a small hole; in later squamatans this gap becomes larger and larger as the phylogenetic line leads to the true lizards (suborder Sauria).

In the Triassic there was already a lizard line in which the members were amphibious. This included *Askeptosaurus* with its long catching snout and the unusual, up to 6 m long squamatan *Tanystropheus longo-bardicus* (see Color plate, p. 286/287 and Fig. 12-64). Its small skull was attached to a neck with nine rodshaped vertebrae; the neck was half the length of the body, the second half being composed of the extremely abbreviated trunk section and the tail. It is thought that this animal lived half in the water and half on land, using its long neck on land as a terrestrial fishing rod (just as we depict in our Color plate). The great marine saurian *Thalattosaurus*, which had more typical body proportions, was further developed for aquatic life. The genus is from Upper Triassic California. From its appearance and presumed life habits the genus might seem to be a predecessor to the mosasaurids of the Upper Cretaceous (see Chapter 16). However, mosasaurids adapted to water independently of Triassic species, and there are no linking fossils from the Jurassic that would join the mosasaurids with *Thalattosaurus*.

Terrestrial thecodonts (order Thecodontia) are called PSEUDO-SUCHIANS (suborder Pseudosuchia). The eye openings and temporal fossae in the high, narrow skull (see Fig. 12-66) are joined by another (preorbital) fossa. The limbs in these animals had five digits, which increased in length from the first to fourth digit, the fifth one being shorter. Some primitive genera (e.g., *Erythrosuchus*, from Middle Triassic South Africa and shown in Fig. 12-66) lacked armor, while others (e.g., the 3 m long *Stagonolepis* from German and English Keuper and the smaller *Aëtosaurus*; see Color plate, p. 286/287) had dorsal and ventral armor composed of rectangular dermal bones.

Prestosuchus and *Ticinosuchus* (see Color plate, p. 286/287) were pseudosuchians several meters long with an armor covering restricted to narrow rows of plates. The former genus is from Brazil and the latter from southern Tessin. Both genera are considered to be responsible for the *Chirotherium* spoor (see Fig. 12-68), which are often found in sandstone. These spoor have been found in Triassic deposits from Spain, France, England, Germany, and North America. The animals made these tracks in moist soil, which solidified in that shape when the soil dried up. To see them in proper perspective we have to view these spoor from the bottom in mirror image. Thus, what seem to be the left feet are actually the right feet, and vice-versa. The name *Chirotherium* (=hand animal) was originally chosen because of the prominent fifth digit on the hand or foot. In 1925, Wolfgang Soergel, using the distance between footprints and the size of the prints, deduced the weight distribution of the body and thus gained an approximate idea of what the animal looked like before it was even discovered as a fossil. The "hand animal" had an uneven gait, for its fore feet were smaller than the hind feet.

There are also tracks lacking fore feet impressions. This tells us that they were made by animals that at least sometimes ran on just two legs. Later it was found that two-legged running occurred in pseudosuchians, since the group had a tendency to develop short fore limbs. Fore limb reduction led to the birdlike appearance in several genera, e.g., *Ornithosuchus* from Scotland and, even more so, *Saltoposuchus* (see Fig. 12-69) from German Keuper. We shall return to two-legged locomotion patterns in later discussions of the dinosaurs (see Fig. 12-72 and text thereto) and flying saurians and birds (see Chapter 15).

A Triassic group of crocodilelike river inhabitants with long, low skulls, were descendants of pseudosuchians or perhaps even eosuchians. These were the PHYTOSAURS (suborder Phytosauria; more recently, Parasuchia). Unlike crocodiles, the long snout in phytosaurs is composed of lower jaw bones, not those of the upper jaw. In view of the unequally long prey-seizing teeth, the name phytosaur (=plant-eating saurian) is completely inappropriate. It is in fact based on an error. The snout fragments found in 1828 in the Neckar Valley near Tübingen, with their blunt teeth, suggested that this was indeed an herbivore. Actually, what was found in 1828 was only the impression of the lower jaw with empty tooth depressions.

In most phytosaurs the front of the snout is broadened into a spoonlike shape and is arched. The nares are well to the rear on a tuberosity; thus, they could breach the water surface when the snout was underwater. No particular modification occurs in the limbs, and they were used for locomotion on land as well as swimming (with the help of the tail in the latter case). Phytosaurs probably lived in habitat not unlike that of modern crocodiles. Their armored covering is similar to that of crocodiles: it has thick, angular or round bony plates with a pitted texture, on which (unpreserved) horny shields lay. The shields were in longitudinal rows on the back, with smaller shields on the belly.

Most phytosaurs are from Europe and North America. The oldest genus (*Mesorhinus*) has a skull structure most similar to pseudosuchians: the snout is not as long as in other phytosaurs, and the nares are not as far back as in young genera. The later forms include *Nicrosaurus* (see Color plate, p. 286/287 and Fig. 12-70) and *Mystriosuchus*, both from Europe. *Nicrosaurus kapffi* has a high snout whose ridge extends forward nearly at the height of the skull roof. In contrast, the snout of *Mystriosuchus planirostris* and the closely related *Belodon* (see Fig. 12-71) is low and narrow. It is possible that the high- and low-snouted forms were merely different sexes, since injuries are sometimes found on the high snouts, and these could have resulted from rivalry fights between males. Advanced phytosaurs have temporal fossae open toward the rear (i.e. lacking a septum).

Placodonts:

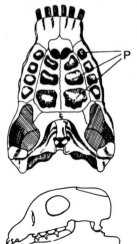

Fig. 12-61. 1. *Placodus.* P=teeth.

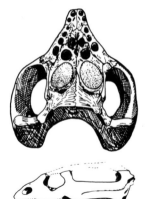

Fig. 12-62. 2. *Cyamodus.*

Fig. 12-63. 3. *Henodus.* (Scale is 1:30 of actual size).

The resemblance between phytosaurs and CROCODILES (order Croco-dylia) is only due to convergent evolution. However, both groups share common ancestors. They probably arose from eosuchians and then

Fig. 12-64. The extremely long-necked lizard *Tanystropheus longobardicus*.

Fig. 12-65. Skull of *T. longobardicus*.

developed further via pseudosuchians. Like rhynchocephalians, crocodiles possess two temporal fossae pairs. The additional fossa found in front of the eye in pseudosuchians occurs only in primitive crocodiles and is then lost. The dorsal armor, and often the ventral armor as well, remains (if we except a few Jurassic marine crocodiles). A new feature appears in the palate, which receives a "double floor" created by a bend in the

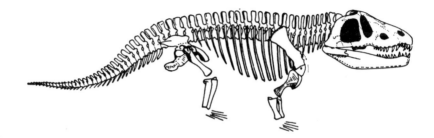

Fig. 12-66. The pseudo-suchian *Erythrosuchus*.

jaw and dentine bones. In the intermediate space created by this bending, air coming in from the nares is led to the rear, and the more "modern" the crocodile, the more rearward the air enters the mouth cavity. This is the same mechanism that permits in mammals respiration without disrup-tion of chewing activity. The earliest crocodiles were the PROTOSUCHIANS (suborder Protosuchia) from Upper Triassic South Africa. The North

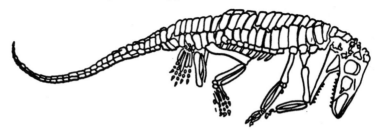

Fig. 12-67. Aetosaurian (*Aëtosaurus*) skeleton as it was found.

American *Protosuchus* stratum is from the Lower Jurassic. The pseudo-suchian heritage is evident in the relatively short snout, the slender limbs (which are longer in the rear) and the two rows of plates on the back.

Descendants or at least "cousins" of the protosuchians were the Triassic DINOSAURS, which in the Jurassic and Cretaceous periods reached the immense sizes most of us associate with dinosaurs. In contrast with pseudosuchians, the thigh bone does not join a panlike, closed pit but an open depression in the pelvic bone (an open acetabulum), an adaptation that increases mobility of the leg. Dinosaurs diverged into several lines during the Triassic. Only a few finds are available of the early Triassic beginnings of the bipedal, herbivorous ORNITHISCHIANS (order Ornithischia), in whose pelvis the pubis bone (similar to birds) is parallel to the ischial bone and is directed rearward. The three middle toes are the longest. The SAURISCHIANS (order Saurischia) were much more diverse. They have a typical saurian pelvic structure, and they included forms that ran on all four legs, which apparently descended from quadripedal eosuchians or pseudosuchians (e.g., *Melanosaurus*, an Upper Triassic saurischian from South Africa).

In other saurischians, such as the 5–7 m long *Plateosaurus* (see Color plate, p. 286/287 and Fig. 12-72), the fore limbs are so short the animals must have moved bipedally. The small, light skull on the swanlike, long

Fig. 12-68. Tracks—presumably made by *Chirotherium*, transected by the tracks of some smaller, unidentified animal.

neck bore flat, sharp-edged teeth. This suggests that *Plateosaurus* fed on small animals. The fore legs have sharp claws. *Plateosaurus* and its relatives were apparently a blind side branch. However, nothing is known about what drove these animals into extinction. They apparently lived along oasislike shores.

Fig. 12-69. A pseudo-suchian that ran on two legs, *Saltoposuchus*.

Another saurian group, which also belongs to the saurischians, continued the bipedal, predatory behavior of the pseudosuchians. These COELUROSAURS (infraorder Coelurosauria) included the delicate *Procompsognathus* (see Color plate, p. 286/287), which had particularly long metatarsal bones, and the large *Halticosaurus* (one specimen of which is 5.5 m long). The related CARNOSAURS (infraorder Carnosauria, genera *Carnosaurus* and *Pachysaurus*), which were sturdy predators that arose in the Upper Triassic and developed into huge bipedal forms in the Jurassic and Cretaceous periods.

The last saurian group we shall mention were the aquatic SAUROPTERYGIANS (order Sauropterygia). They were once placed close to the placodonts (see Color plate, p. 285) since they, too, have a pair of upper temporal fossae. However, a depression on the lower edge of the skull indicates that, like lizards, they must have descended from ancestors with

Phytosaurs:

Fig. 12-70. 1. *Nicrosaurus*; skull and lower jaw.

Fig. 12-71. 2. *Belodon*; skull and lower jaw.

two pairs of temporal fossae and therefore are an independent line. The oldest finds are the NOTHOSAURS (suborder Nothosauria, genus *Nothosaurus*; see Color plate, p. 285). As in all sauropterygians, the lower arm and lower leg in nothosaurs are very short. The eyes on the top of the

g. 12-72. *Plateosaurus*, saurischian dinosaur at ran on two legs.

flat skull are suggestive of swimming. This is supported by the tapering snout (adapted for seizing prey), the hooklike elongated neck (produced from an increased number of vertebrae), and the still longer tail. The number of phalanages (the small bones comprising each of the five digits on the limbs) are not quite the same as in terrestrial saurians (2.3.4.5.3 vs.

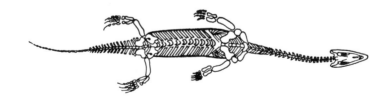

g. 12-73. *Ceresiosaurus*, nothosaurian sauropygian (see text).

2.3.4.5.4). Thus, nothosaurs are sorts of ecological "bastards" or intermediate forms, hanging between an aquatic and a terrestrial life. In fact, their German vernacular name (in the original German edition of this encyclopedia) is BASTARD SAURIANS. The phalanges of the fingers and toes begin increasing in the Middle Triassic *Ceresiosaurus* (on the feet at least 2.3.5.6.6). We have here the transformation of limbs from members used for terrestrial locomotion to paddles; at this point the process is in its very earliest stages. However, the short lower arms of *Nothosaurus* are already a clear indication that this process is occurring.

Simosaurus has a broader, shorter skull. The Swiss paleontologist Emil Kuhn-Schnyder discovered an ear slit on the hind edge of this skull, something found otherwise only in amphibians and very primitive reptiles. This has led to the yet unresolved question of whether sauropterygians and other saurian orders arose independently of each other from amphibians. The issue of polyphyletic descent is still being argued.

Another nothosaur was the small *Pachypleurosaurus*, found *en masse* in Tessin. Its temporal fossae are small, and the genus has the typical number of phalanges. Its bones are thickened, a condition known as pachyostosis. This is found in saurians occurring in water and in marine mammals (e.g., sirenians). Originally it was thought that pachyostosis was a result of metabolic difficulties caused in the transition to the new life habits of saurians. However, today it is felt that pachyostosis has positive adaptive value, since the thick bones are useful in diving and staying on the floor in shallow waters. A related, likewise small form, *Neusticosaurus*, also has pachyostosis.

Cymatosaurus (see Fig. 12-74) and the rare, narrow-snouted *Pistosaurus* led to the PLESIOSAURS (suborder Plesiosauria). In both genera the snout-forming intermaxillary bones extended posteriorly to the frontal bones, and the nasal bones (which were medial in nothosaurs) were pushed apart and regressed. This evolutionary process strengthened the seizing jaw. The process continued further in Jurassic plesiosaurs, and in some genera the intermaxillary bones even pushed the frontal bones apart.

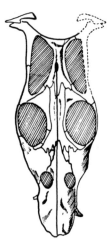

Fig. 12-74. Skull of the plesiosaur *Cymatosaurus* with the long snout and large, long temporal depressions.

13   The Origin of Mammals

By G. G. Simpson

Mammals developed from the once worldwide therapsids (see Chapter 11), of which fossils are known from the Middle Permian (250 million years ago) to the Middle Jurassic (160 million years ago). During this time, therapsids developed into numerous lines, and during the 90 million years in which therapsids occurred, each line developed into more and more mammallike animals, but they did so to different extents and in different ways. The mammals arose from one of the most progressive lines or perhaps from just a small group.

Differences from reptiles

Modern mammals differ from reptiles in the following ways: 1. They suckle their young; 2. They do not lay eggs (with the exception of monotremes) but bear live young; 3. Mammals have hair; 4. They are "warm-blooded" (i.e., homoiothermic) and therefore have much greater control over internal body temperatures; 5. Mammals have four-chambered hearts. However, if we compare modern mammals with modern reptiles, we should understand that this is not equivalent to comparing their ancestors to each other. The mammallike reptiles all became extinct, save for those that gave rise to true mammals. They differed considerably from the ancestors of modern reptiles and were a phylogenetically independent line toward the end of the Carboniferous Period (approximately 300 million years ago).

Therefore, mammals and their reptilian predecessors can only be compared realistically by using the therapsids from the fossil record. None of the above listed features can be directly observed in fossils. However, direct evidence suggests that some of these features were partially or fully developed in the most advanced mammalian precursors. We would expect that to be the case, for anatomical modifications only take place over very long time spans. Also, it would not be necessary for live-bearing and suckling to develop simultaneously and at the same evolutionary rate. Obviously, a comparison of modern monotremes with other mammals shows that these features do not develop at the same rate in all mammal lines.

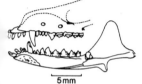

5mm

Fig. 13-1. *Eozostrodon.* Reconstruction of the left upper jaw (seen from the outside) and the right lower jaw (seen from the inside).

The border between advanced therapsids and primitive mammals is a very fine one and is somewhat artificial. It can only be drawn by dental and skeletal features preserved in the fossil record. As we would expect, the mammal state is sometimes achieved by advanced reptiles, whose advanced state is determined by the sum total of their anatomical features. Examples for that are the presence of a complete, ossified (or so-called secondary) palate and the differentiation of the teeth into incisors, canines, and molars. In some respects, mammalian precursors continually approached the mammalian condition and reptile-mammal intermediate species attained the complete mammal structure (e.g., in the transformation of the link between the upper skull and the lower jaw into the middle ear ossicle). Other features were not found until they occurred in pure mammals, and suckling is an example of one such feature.

Due to the rapid rate of evolution and the gaps in the fossil record, it is at present not possible to define the exact line of descent leading to mammals or to pinpoint the time from which true mammals existed in this phylogenetic continuum. However, fossil mammals first occur in the Upper Triassic and in the Triassic-Jurassic transition region from the Rhaelian to the Hellangian (i.e., approximately 190 million years ago).

The first of these early mammals was discovered in 1847 in Rhaelian strata in Württemberg, Germany. For a century thereafter, paleontologists had nothing but a few individual teeth with some doubtful features testifying to this group of animals. Since 1947, however, many new finds have been unearthed, and some of the most important of these have not been fully described yet. Furthermore, there is a great gap between these early mammals and the Middle to Upper Jurassic mammals. In consequence thereof, there is little agreement at present among paleontologists regarding the phylogenetic relationships between Rhaelian-Hellangian mammals. Most authorities distinguish four families of these early mammals: 1. HARAMIYIDS (Haramiyidae; formerly Microlestidae or Microleptidae); 2. SINOCONODONTIDS (Sinoconodontidae); 3. EOZOSTRODONTIDS (Eozostrodontidae); and 4. KUEHNEOTHERIIDS (Kuehneotheriidae or, better, Kuehneontidae).

The only records we have of HARAMIYIDS are individual teeth from Germany, England, and Switzerland. Each tooth has two separate roots, something very common in mammals but extremely unusual in reptiles, and an oval, deep crown whose edge has an irregular number of ridges. It is possible (although very unlikely) that these animals were related to the multituberculates that appeared later (see subsequent discussion) or were ancestral to them. On the other hand, there is no certainty as to whether haramiyids were actually mammals.

SINOCONODONTS are represented by incomplete skull and lower jaw fragments from China (*Sinoconodon*, which may be related to the next family) and a skull from a partially retained skeleton (*Megazostrodon*) from Lesotho, South Africa. Their molars bear three differentiated major

Features: skeletal and dental structure

Fossil sites from the Rhaelian-Hellangian and Jurassic:

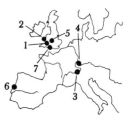

Fig. 13-2. 1–4. Rhaelian-Hellangian. 1. Somerset, England; 2. Glamorgan, Wales; 3. Hallau, Switzerland; 4. Württemberg, Germany; 5. Stonesfield (Middle Jurassic), England; 6. Leiria (Upper Jurassic), Portugal; 7. Purbeck (Upper Jurassic), England.

Fig. 13-3. 1. Rhaelian-Hellangian: Quthing, Lesotho 2. Upper Jurassic; Tendaguru, Tanzania.

Fig. 13-4. Rhaelian-Hellangian: Lu-feng, Yunnan, China.

crowns arranged in a row. Although thorough studies have not yet been made on sinoconodonts, it appears that these animals are probably related to the Middle Jurassic to Upper Cretaceous triconodonts.

Many traces of EOZOSTRODONTIDS have been found in England and Wales, while others have been recovered in Switzerland, Lesotho, and China. However, studies of some of the most important finds have not yet been made. The molars resemble those of sinoconodonts, but they are more complex since they have a line row of crowns. Many opinions are voiced over the relationship between eozostrodontids and subsequent groups. Some paleontologists group eozostrodentids with triconodonts, along with sinoconodontids, while others claim they are ancestors of docodonts and should be classified within that order. The presently most tenable theory combines both these viewpoints insofar as the eozostrodontids are grouped with triconodonts, and the docodonts are thought to be a separate group but closely related to eozostrodontids.

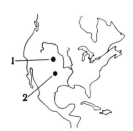

Fig. 13-5. Upper Jurassic: 1. Como Bluff, Wyoming, 2. Canyon City, Colorado.

We shall discuss KUEHNEOTHERIIDS, which are perhaps of special importance for mammalian evolution, in the description of the symmetrodonts.

According to our present knowledge, mammals arose at the beginning of the Jurassic Period approximately 190 millions years ago. Even then, there were several mammalian lines. As we describe below in more detail only two of them (multituberculates and pantotheres) were ancestral to modern mammals. Mammals represent a major evolutionary step forward in contrast to reptiles and therapsids, for they progressed in terms of obtaining and utilizing food, in metabolism, in adapting to the environment, reproduction, raising the young, and perhaps in locomotion as well. These advantages were not well developed in the early stages of mammalian evolution, however. It took some 125 million years until mammals replaced reptiles as the predominant terrestrial animals.

Molars of various Mesozoic primitive mammals, each with a lateral view and a top view below it.

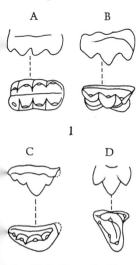

Fig. 13-6. Group 1: Upper molars. (Continued on p. 289.)

Even today, the fossil record of early mammals is a very modest one. There is virtually nothing in the 30 million year gap between the Rhaelian-Hellangian forms we have been discussing and the next "younger" early mammals from the Middle Jurassic (more specifically, from the Bathonian stratum). The first fossil early mammal was found in 1764, but Middle Jurassic mammals even today are known from just a single site (Stonesfield, England). Upper Jurassic mammals from the Kimmeridgian and Purbeckian deposits have been found in several localities, and they are thus better understood. Late Jurassic mammals have been located in Swanage (Dorset, England; in 1854), Wyoming (1878), Tanzania (1910–1912), and Leiria, Portugal (1959). In Africa only a single toothless lower jaw has been found, but the other sites listed above have housed great collections of fossil fragments. The Portuguese finds have not yet been fully described. Nevertheless we know that there were early mammals during the entire Jurassic Period, which lasted approximately 55 million years, and mammals were undoubtedly present on every continent. However,

▷▷
A Triassic landscape and its life:
Plants: Cycadeans (a), Equisetum (d), and a conifer forest (b) with *Voltzia* and a growth of *Pleuromeia* (c), a relative of Carboniferous sigillaria-ceans.
Animals: the lizard *Tanystropheus* (1), Keuper turtle *Triassochelys* (2), the armored aetosaurid *Aetosaurus* (3), *Henodus* (5), the skull of *Nicrosaurus*, a pseudosuchian (6), *Ticinosuchus* (7), *Mastodonsaurus* (8), and several plateosaurs (*Plateosaurus*; 9).

in spite of their wide distribution and long time span, the small fossil record of these organisms gives us only a small amount of insight into these animals. An intact skull has never been found, and there are only a few skeletal fragments available (not to even mention an entire skeleton).

The known Middle and Upper Jurassic early mammals are divided into five distinct groups, which are designated as order:

1. The MULTITUBERCULATES (Multituberculata) as such are known from the Upper Jurassic. As we mentioned in the discussion of haramiyids, it is doubtful that multituberculates appeared in the Rhaelian and Hellangian. Multituberculates lived throughout the Cretaceous and even survived into the early part of the Tertiary (from the Paleocene to the Upper Eocene). They may have been ancestral to the modern monotremes (class Monotremata), but this is not known with certainty.

2. TRICONODONTS (Triconodonta) lived in the Middle and Upper Jurassic. They include the sinoconodontids and perhaps the eozostrodontids as well in the Rhaelian/Hellangian. Their Triassic early members were perhaps the ancestors of other early mammal orders.

3. DOCODONTS (Docodonta) are only known from the Upper Jurassic. They may have evolved from Rhaelian-Hellangian eozostrodontids or similar tricondonts. No post-Jurassic descendants of this group have been found.

4. SYMMETRODONTS (Symmetrodonta) are documented from the Upper Jurassic to the Middle Cretaceous. Some paleontologists include the Rhaelian-Hellangian kuehneotheriids with these. Early symmetrodonts resemble kuehneotheriids and may have belonged to this group. Symmetrodonts were presumably the ancestors of pantotheres and thus all

▷
Life in a Triassic sea:
On land is the amphibian *Nothosaurus*; swimming are primitive ganoid fishes (right: *Ptycholepsis*); *Placodus gigas* has a mollusk meal on the floor, and an ammonite swims by on the right; behind it is a crinoid colony (*Encrinus liliiformis*).

▷▷▷▷
The evolution of the ammonite genus *Ceratites* (in shell limestone) from fork-ribbed to simple forms and repeatedly to relatively smooth forms, always with an increase in size. Subgenera: *Progono-ceratites Acanthoceratites, Ceratites, Discoceratites*.
—The shell of *Cer-dorsoplanus* contains the shells of three adult mollusks of the genus *Placunopsis*. Drawing after H. Schmidt (1928) and R. Wenger (1957).

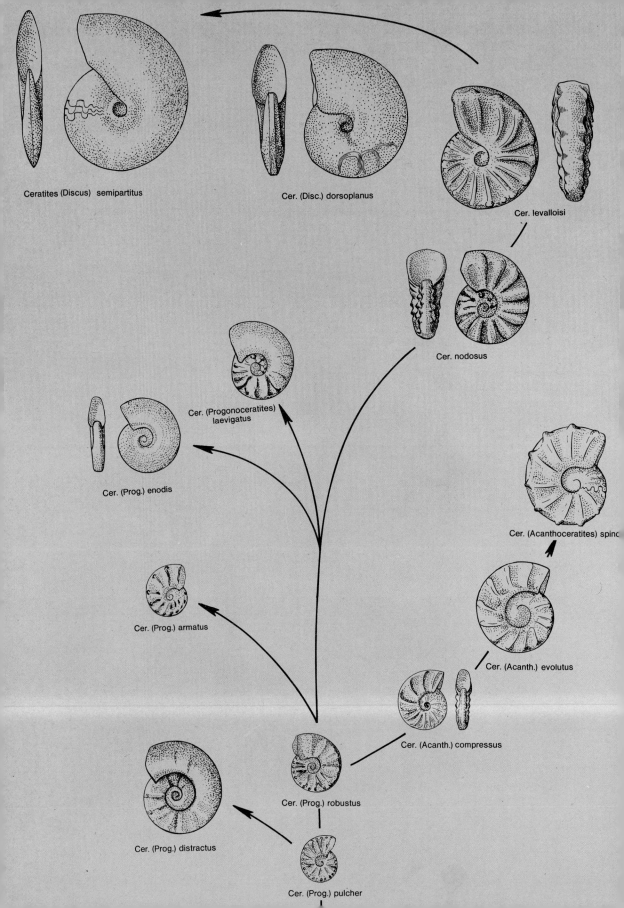

Ceratites (Discus) semipartitus

Cer. (Disc.) dorsoplanus

Cer. levalloisi

Cer. nodosus

Cer. (Progonoceratites) laevigatus

Cer. (Prog.) enodis

Cer. (Acanthoceratites) spino

Cer. (Prog.) armatus

Cer. (Acanth.) evolutus

Cer. (Acanth.) compressus

Cer. (Prog.) distractus

Cer. (Prog.) robustus

Cer. (Prog.) pulcher

modern mammals except the monotremes. Younger symmetrodonts, however, are specialized forms with no known descendants.

5. PANTOTHERES (Pantotheria), sometimes called Eupantotheria, lived from the Middle Jurassic to the Lower Cretaceous. They presumably evolved from Rhaelian-Hellangian forms resembling kuehneotheriids. Pantotheres include various extinct lines, and today it is thought that pantotheres were the ancestors leading to all modern mammals (excluding monotremes; see Color plate, p. 221).

Even if multituberculates are considered separately from their possible ancestors (haramiyids) and descendants (monotremes), they maintained themselves for one of the longest periods known (nearly 90 million years) for any mammalian order. During this long period they changed considerably, but only within specific adaptive borders. They had large incisor teeth, which were cylindrical and used for seizing, or needlelike teeth (in some later species these developed into gnawing teeth), and simple upper premolars for holding prey. They also had shearing lower premolars, whose number (as in upper premolars) decreased during evolution, and finally there were two upper and lower molars with crowns arranged in rows. The molars gradually increased in number and adapted to the front-rear chewing motions of the lower jaw. True canines were absent. The skull, not well known in Jurassic forms but well described in Upper Cretaceous and Lower Tertiary species, differs considerably from the skull structure of true mammals (i.e., marsupials and placentals).

Recently it has been shown that the portion of the skull surrounding the brain in multituberculates is similar to that of modern monotremes (the platypus and echidna). This leads us to be believe that monotremes developed from earlier multituberculates, even though there are skull skull differences (particularly in the face region) between the two groups. Furthermore, the teeth of young platypuses (echidnas lack teeth) are highly differentiated. Some teeth features of the earliest multituberculates (from Kimmeridgian Portugal) resemble those of the still older haramiyids and may reflect a phylogenetic relationship between these groups. Until more complete haramiyid finds are located, however, there is little convincing evidence supporting this sort of relationship.

Multituberculates deserve our attention because excepting triconodonts they outlived all other mammal orders. Although no multituberculates are alive today which might reveal their adaptations, it appears likely that multituberculates were fruit and grain eaters. If that were true, it would mean that they had very little competition for food until the rodents appeared in the Lower Tertiary. In terms of feeding patterns, multituberculates would be quite unlike modern monotremes.

TRICONODONTS even outlived the multituberculates. They probably lived from the Uppermost Triassic (Rhaelian China and Lesothos) to the Late (but not Latest) Cretaceous (Companian Canada), a time span of more than 100 million years. Yet triconodonts evolved and diverged very

Between haramiyids and egg-layers

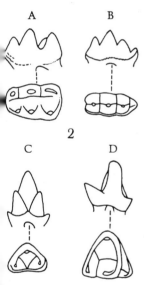

Fig. 13-6. (Continued.) Group 2: Lower molars. A. Multituberculates Jurassic to Eocene, here Upper Jurassic). B. Triconodonts (Jurassic to Cretaceous, here Upper Jurassic). C. Symmetrodonts (Jurassic to Cretaceous, here Upper Jurassic). D. Pantotheres (Jurassic to Cretaceous, here Upper Jurassic).

little during this enormous period of time. It is of course possible that triconodonts did indeed evolve substantially over their 100 million year lifespan and that the meager fossil record does not bear witness to these changes. All we have of triconodonts are fragments, and the only nearly intact finds (from Lesotho) have not been thoroughly analyzed.

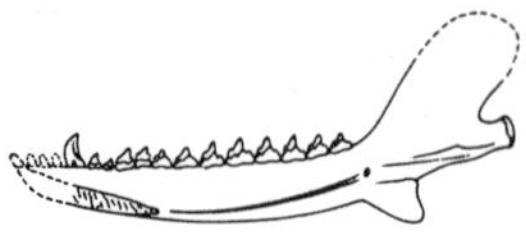

Triconodonts had narrow incisors, large cutting canines, simple seizing premolars and usually three to four molars with three major crowns and one or two subsidiary crowns arranged longitudinally. The jaws were operated with powerful cheek and temporal muscles. The lower and upper molars created a scissors action for cutting off food and also for sharpening the edges of the canines. This is the characteristic dentition of predatory carnivores. Like most Mesozoic mammals, triconodonts were small; they were no larger than a modern house cat. The last triconodonts from the Middle and Upper Cretaceous had a distinct row of molars, which formed a surface like a pair of long, jagged scissors.

DOCODONTS are represented only by a few upper and lower jaws and some teeth fragments. They are from the Upper Jurassic period of Europe and North America. Their laterally broadened molars bore a crown pattern of blunt ridges with depressions between them, which indicates that the teeth were used for mashing food. These animals may have developed from triconodonts or similar forms (e.g., eozostrodontids, which some paleontologists classify as docodonts). However, there are important differences in the shape (anatomy) of these two groups. Docodonts appear to have died out relatively early. They were an unsuccessful developmental branch.

SYMMETRODONTS are also almost exclusively represented by upper jaws, lower jaws, and parts of teeth. The chief characteristics of symmetrodont teeth are found in kuehneotheriids from the Rhaelian and Hellangian. We must comment here on the sole kuehneotheriid specimen, *Kuehneon duchvense* from Rhaelian Wales, of which all we have are individual teeth. These few teeth, however, are of great phylogenetic significance. They bear a great resemblance to the teeth of Upper Jurassic and Cretaceous symmetrodonts, although the external anatomy of *Kuehneon* indicates that it was a direct ancestor of pantotheres and is thus classed with all marsupials and placental animals (including our own species, *Homo sapiens*). It would be fantastic if this sole genus were the true ancestor of all mammals excepting the monotremes, but even though this is unlikely, *Kuehneon* surely gives us an idea of this mammal ancestor. The crowns of the upper and lower molars (see Figs. 13-9 and 13-10) consisted chiefly of interlaced major crowns, which were triangular, and subsidiary crowns at the base of the major crowns. This crown pattern was more specialized in Upper Jurassic and Cretaceous symmetrodonts, which became a side branch and then died out.

True PANTOTHERES, which differ considerably from their symmetrodont precursors, are known from the Middle Jurassic to the Upper

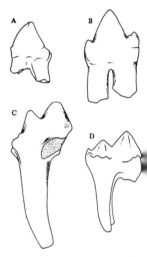

Fig. 13-7. A docodont. Inside of the right lower jaw, seen approximately life size.

Fig. 13-8. Teeth of Rhaelian triconodonts. A. A specimen from Hallau, Switzerland. B. *Eozostrodon parvus* Parrington. C. *Eozostrodon problematicus* Parrington. D. Second Hallau specimen.

Cretaceous. Although we only have jaws and incomplete dentition from them, pantotheres are known to be a very significant group. They possessed small incisors, large canines, grabbing premolars and complex molars. Molars existed in different numbers among various pantothere genera. The lower molars had three major crowns as in symmetrodonts, but they were in an irregular, slanted triangular arrangement. They were followed by a low crown surface (or talonid) with one (sometimes up to three) small ridges. A large number of pantotheres were present toward the end of the Jurassic Period. The few known examples are divided into three families.

The DRYOLESTIDS (Dryolestidae) were the most prevalent, but these were probably just another of the side branches that led to extinction instead of descendants. Another family, the AMPHITHERIIDS (Amphitheriidae), is probably closer to most later mammals. The lower jaw dentition resembles the original dentition of marsupials and placentals. Even the upper molars (which are not known in genus *Amphitherium*) retain this similarity in their outline and functional features. However, amphitheriids differ so much in other respects that little can be said about their phylogeny until more transition forms have been found. These small pantotheres were insectivorous. Their chewing apparatus had many developmental potentials, which were subsequently realized in the various Cenozoic mammals (excepting monotremes).

In summary, the current state of the art of paleontology permits us to state that mammals developed in the Triassic from mammallike reptiles, probably from the theriodonts (suborder Theriodontia; see Chapter 11). We do not yet know which of these mammal predecessors formed the bridge to the early mammals. Early mammals diverged at the beginning of the Jurassic Period. One group was composed of the primitive symmetrodonts, which gave rise in the Middle Jurassic to pantotheres. Then, during the Cretaceous, various pantothere lines gave rise (perhaps simultaneously, but more probably over a period of time) to the ancestors of marsupials or methatheres (Marsupialia or Metatheria) and the placental mammals (Placentalia or Eutheria) and hence all modern mammals with the exception of the monotremes, which today occur only in Australia. Three other groups evolved during the Jurassic in addition to this major line, and these were the tricondononts, docodonts, and multituberculates. The phylogenetic relationship between these groups and to that line leading from the symmetrodonts via the pantotheres to marsupial and placental mammals is still quite unclear. It it possible that a side branch of multituberculates led to the monotremes. The docodonts and triconodonts died out and left no descendants.

Dryolestids

Amphitheriids

Molar-shaped kuehneotheriid teeth:

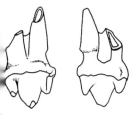

Fig. 13-9. Left upper jaw, inside (left) and outside.

Fig. 13-10. Left lower jaw, inside (left) and outside.

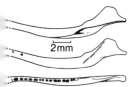

2mm

Fig. 13-11. *Kuehneon*; reconstruction of the lower jaw: right jaw inner view; left jaw, outside and top view.

14 The Jurassic

The second period of the Mesozoic Era began 195 million years ago and ended 135 millions years ago, according to radiometric data. The above-ground and subterranean deposits belonging to the Jurassic system and bordering on the (older) Triassic and (younger) Cretaceous strata were only identified after a long period of research, often utilizing key fossils (see Chapter 3). The name Jurassic is derived from the Jura Mountains of Switzerland (Jura = forested mountains in Celtic). Jurassic deposits arose in the sea and continue in central Europe as the Swabian and Franconian Alb.

Jurassic formations began with the rapid transgression of the sea onto the land, a process that occurred in parts of central and western Europe in the beginning of the Jurassic and later in eastern Europe as well. The upper border of the Jurassic is marked by the retreat of the sea, which led to brackish and fresh-water deposits (such as the Purbecks and Wealdon formations in England). In southern Europe the sea did not retreat and covered the land into the Cretaceous period, which has made stratigraphical distinctions difficult to determine in this part of the world.

The Jurassic is divided into several parts, which are designated, from upper (more recent) to lower (older) as follows: A. Upper Jurassic with the stages: 1. Purbeckian; 2. Kimeridgian; 3. Oxfordian; 4. Callovian; B. Middle Jurassic with the stages: 1. Bathonian; 2. Bajocian; C. Lower Jurassic with the stages: 1. Toarcian; 2. Pliensbachian; 3. Sinemurian; 4. Hellangian. The exact placement of some of these stages within the three major parts of the Jurassic is still a point of disagreement among various authorities. The system used here is from *Paleobiology of the Invertebrates* by Paul Tasch (1973).

The great Tethys sea, just as in the Triassic, was greatly branched during the Jurassic. It extended from southeast Asia westward to what is now the Mediterranean and from there presumably to Central America and in some places joined the east and west coasts of the Pacific Ocean. The Atlantic, on the other hand, arose later when the American continent

By H. Hölder

Deposit-forming small Jurassic fossils (greatly enlarged):

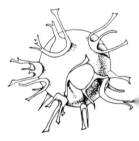

Fig. 14-1 A flagellate algae, *Hystrichospaera systematophora*.

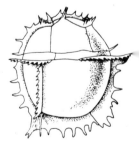

Fig. 14-2. A flagellate algae still found today, *Goniaulax ornata*.

Jurassic foraminifers
(greatly enlarged):

Fig. 14-3. 1. *Lenticulina
quenstedti.*

Fig. 14-4. 2. *Citharina
implicata.*

Fig. 14-5. 3. *Spirillina
polygyrata.*

migrated westward. The Tethys tended to homogenize animal life since it was a continuous body of water, but at the same time it was so extensive it could not prevent the development of zoogeographical differentiation from east to west. Zoogeographical differences were even more pronounced, however, between the northern (boreal) region and the Tethys. The central European region between these two areas was alternately influenced by one or the other.

The modern continents had largely become entities by the Jurassic, although in some places they had a different appearance and location than they do today. Shelf seas extended to the edges of the continents and only temporarily penetrated inland. The deposits from these shelf seas have given us most of the fossil clues to the geological and zoological picture of the Jurassic earth. Some innercontinental basins sunk down during the Jurassic, and they were filled with great continental deposits. Thus, they also contain numerous terrestrial animals and have contributed significantly to the terrestrial fossil record from this period. An example of this are the Morrison formations of North America, an Upper Jurassic formation with reptiles. Key fossils from both land and sea are found in border regions between continental and marine deposits.

Climatic differences were no longer as distinct throughout the world as they were in the Triassic. There were no glacial periods during the Jurassic, but Europe experienced a temperature decline in the Lower Jurassic. Great salt deposits in Germany formed during the Upper Jurassic, are evidence of a warm dry belt north of the Tethys, a belt known since the Triassic. Coral reefs in the Jurassic extended as far north as Great Britain, but there were many more coral species further south. Biologically, the Jurassic period is significant for the diversification of ammonites, the reef-building siliceous sponges, the rise of the dinosaurs, and the first flying saurians and birds. Like the Triassic, the Jurassic was an intermediate epoch regarding plant life.

Of the Jurassic flora, the GYMNOSPERMS (Gymnospermae (see Chapter 8) predominated. Upper Triassic (i.e., Rhaelian) plants are distinguished from Jurassic plants by the nature of the ammonites occurring in the same strata, for Triassic and Jurassic ammonites differ. The dipteridacean fern *Thaumatopteris* was an important Lower Jurassic key fossil. Its delicate foliage suggests that it occurred in a moist, warm climate, which at that time prevailed as far north as Greenland. Middle Jurassic sandstone depositions at river mouths in northeast England and Scotland have preserved a rich flora, including Equisetales thickets in life position. One plant group, Caytoniales, is named from Cape Cayton on the coast of Yorksire.

The large inner Asian land mass of the Jurassic was covered with conifers and ginkgo forests, which are indicative of a moist, temperate climate. Ferns were predominant along the warm, moist coastal stretches of the Tethys in what is now the high mountainous region of southern

Asia and further eastward in the old Pacific region. The flora of the inner Asian continent has been delivered to us in great, intercontinental geosynclines where plant materials from higher terrain mixed with the local flora of the lowlands and swamps, producing the richness of plant life that led eventually to the production of commercially important coal regions.

Coastal forests spread at the turn of the Jurassic to the Cretaceous, and their plant life has been retained in the sandstone, clay, and coal deposits of Wealden. They resemble the plants from the Lower Jurassic. One of the ginkos (*Ginkyoites*) had broad leaf tips, and it led to the modern kinkgo tree (*Ginkyo biloba*). Among the bennettitaceans (see Chapter 8) of this period, *Cycaeoidea*, with its spherical "trunk" covared with leaves and large flowers, was a most impressive plant. It had tufts of long, pennate fronds.

Fig. 14-6. 4. *Guttulina jurassica.*

The last members of the fern genus *Glossopteris* (see Chapter 8), a characteristic genus in the southern hemisphere since the Paleozoic and one that greatly influenced the nature of flora during the Paleozoic, are found in the Lower Jurassic in southern China and Mexico. Flora from Greenland to South Africa were rather uniform. Climatologically, the Jurassic was a very stable period in the earth's history, a fact that can also be appreciated in the fauna of the period.

Ever since there have been animals with hard body parts, deposits have formed consisting in part or completely of these animal parts. The smallest fossil-producing organisms of the Jurassic, the dwarf or nanno-fossils whose structure was only closely determined with the aid of electron microscopes, were primarily the COCCOLITHOPHORIDS (Coccolithophoridae), from Tethys calcite. These were a group of unicellular PHYTOFLAGELLATE ALGAE (Phytoflagellata) whose gelatinous shell was surrounded by a capsule of loose, round plates or coccoliths and which were widely distributed in plankton since the Jurassic. One cubic centimeter of Jurassic limestone can contain millions of coccoliths, which themselves are composed of still smaller "micelles". In 1970 the amazing discovery was made that coccolithophorids also occurred in fresh water.

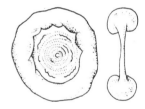

Fig. 14-7. 5. *Bullopora rostrata* on belemnite.

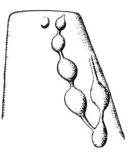

Fig. 14-8. 6. *Orbitopsella.*

Phytoflagellate algae also include the larger, but still microscopic, DINOFLAGELLATES (order Dinoflagellata) and perhaps the *Hystrichosphaera* (see Fig. 14-1). Dinoflagellates, like coccolithiphorids, are first found in the Lower Jurassic. Their girdled capsules consts of tightly bonded plates composed of a chemically very resistant organic substance. The Jurassic record of dinoflagellates contains primitive and very advanced species, including genera still found today (e.g., *Goniaulax*; see Fig. 14-2), and this suggests with near certainty that there were pre-Jurassic dinoflagellates that have not yet been discovered. Hystrichosphaera were distributed in the Paleozoic, and after what appears to be their absence in the Permian and Triassic they flourished anew in the Jurassic period. They began with simple latticed-shell forms (*Leiosphaera*, which lacked spinelike processes

Fig. 14-9. 7. *Globigerina cf. helvetojurassica* (from above).

Fig. 14-10. A well preserved radiolarian sphere with round porous openings; diameter 0.08 mm.

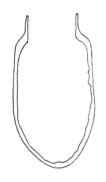

Fig. 14-11. Longitudional section through the skeleton of *Calpionella elliptica*, a ciliate (height 0.1 mm).

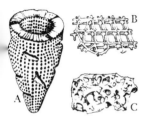

Fig. 14-12. The cup-shaped dictyonid *Craticularia paradoxa*. A=shell; B=lattice skeleton; C=thickened upper surface.

(hyrstricho = spine), and these forms reappeared in the Upper Jurassic, a period much richer in planktonic organisms. The spineless varieties also included those that lived as cysts on the ocean floor (*Tasmanites*), which were formerly and erroneously thought to be plant spores. Dinoflagellates and Hystrichosphaera, even as fossils, have a hole and a cover used during hatching.

The Jurassic FORAMINIFERS (Foraminifera), which belong to the RHIZOPOD class (Rhizopoda), compared to earlier periods, developed quite rapidly. Agglutinizing forms had predominated in the Paleozoic (see Chapter 9), but in this group only a few new genera were added in the Jurassic (e.g., *Hormosina*, *Haplosticha*), while shelled species multiplied considerably. This is particularly true of the NODOSARIDS (Nodosaridae), whose still extant genera *Lenticulina* (see Fig. 14-3), *Lagena* and *Citharina* (see Fig. 14-4) began in the Jurassic. Other calcite-shelled, bottom-dwelling foraminifers also arose during the Jurassic (e.g., *Bulimina*, *Spirillina* (see Fig. 14-5), *Discorbis*, *Guttulina* (see Fig. 14-6) and *Epistomina*). Some foraminifers (e.g., *Bullopora*; see Fig. 14-7) form small chains which grow on the shells of mollusks and similar animals as irregularly spiralling nodules (e.g., *Nubecularia*). Large, complex LITUOLIDS (Lituolidae) are found in the Jurassic Mediterranean region. One of them, *Orbitopsella* (see Fig. 14-8), had a disk-shaped shell with a diameter exceeding 2 cm. Such large size, which also occurs in the Carboniferous and Permian in fusulines (see Chapter 9) and in the Eocene in nummulites (see Chapter 17), was probably related to the higher water temperatures of the sea in what at that time was a tropical zone. It was in these seas that during the Jurassic, after beginnings in the Triassic, the first non-bottom dwelling foraminifers developed. These were the GLOBIGERINS (*Globigerina* and others; see Fig. 14-9), which were planktonic foraminifers with needle-like processes on their spherical bodies. These features permitted them to drift in the ocean and leave the ground, and they were the first foraminifers to do so.

The delicate siliceous skeletons of RADIOLARIANS (order Radiolaria; see Fig. 14-10) first appear *en masse* in deposits from substantial depths. These organisms occur not only in limestone but also in the red, green or gray radiolarite, in which the radiolarians are visible as tiny white dots. The silicic acid found in these rocks (e.g., from Upper Jurassic alps and the Balkans) from deep sea water did not always arise from the radiolarians, and in cases where it had other origins the radiolarians usually form a small component of the entire deposit. In the course of evolution, more primitive radiolarians with spherical latticed skeletons gave way to others with a heteropolar (=with differentiated poles) skeleton on a longitudinal axis.

The highly differentiated class of CILIATES (Ciliata) was represented in the Jurassic by the CALPIONELLENS, whose cup-shaped skeletons have been found in masses with the much smaller coccolithophorids in the Upper

Jurassic of the alpine-Mediterranean region. Calpionellens have been used as important key fossils. They are often classified with the modern tintinnids (Tintinnidae), whose skeletons are composed of organic materials. Thus it is suggested that the organic materials in the fossil members were replaced by calcium after the death of the animal. However, there are some studies indicating that Jurassic and Cretaceous calpionellens had a primitive calcareous skeleton.

Different structural materials occurring within one animal class, as we find in the calpionellens, is not at all unusual. SPONGES (phylum Spongia), for example, contain horn, calcite, and silicic acid as skeletal material. These primitive multicellular animals have been known since the Cambria, but they did not form a significant portion of marine fauna until the Jurassic period. Deposits from the Lower Jurassic Alps contain spiculite, which to a large extent is composed of single-axled silicon needles from disintegrating sponges. A mass proliferation of siliceous sponges begins in the Upper Jurassic. Their needles penetrate the body through a firm skeletal structure. These sponges have often retained their plate-shaped, cuplike or cylindrical external form in the rocks. These are of course only "mummies" of calcite, which penetrated the softy body through the pores between the needles after the animal died, and in many cases the calcium deposits replaced the original silicic acid component. Sometimes the skeleton can be removed from rock intact with the use of diluted hydrochloric acid.

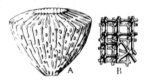

Fig. 14-13. The thick-walled cup of *Pachytei-chisma carteri*. A=shell; B=skeletal structure with complex intercrossings.

Fig. 14-14. Profile of a sponge (porifer) reef from the Upper Jurassic. It begins with isolated sponge growths, which grow and spread across neighboring facies. Later the reef dies and is covered by depositions. The length of the section shown here is 1 km.

The GLASS SPONGES (class Hexactinellida), which are also known as Triaxonia, have a regular, cubical skeleton built up from a plan in which three needles always meet at right angles, and their six rays join these needles with neighboring ones. Differences among various species occur in the overall skeletal configuration and in differences in arrangement and size of the canals, which conducted food-bearing water into the stomach with the help of collar cells. Furthermore, many genera are characterized by a covering layer with its own needle pattern.

In DICTYONIDS (order Dictyonida) the needles intersect in a simple pattern. *Casearia* is divided by constrictions and also has a very characteristic external appearance, something atypical of sponges. *Ramispongia*, as its name implies, is ramified or branched; *Tremadyction* has a dense pattern of alternating canal openings; the cup-shaped *Craticularia* (see Fig. 14-12) has non-alternating canal openings. *Stauroderma* has a re-

Fig. 14-15. Shell of the dictyonid *Cypellia dolosa*.

Jurassic reef-producing corals:

Fig. 14-16. *Thamnasteria.*

Fig. 14-17. *Stylina labechi.*

Fig. 14-18. *Lithodendron.*

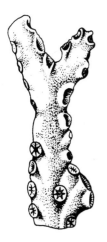

Fig. 14-19. *Enalohelia compressa.*

inforced skeleton on the outside of the body, the source of reinforcement being star-shaped needles called stauractines. LYCHNISCARIANS (Lychniscaria) often look like dictyonids, but their skeletal structure differs in that the intersections are not simple but are surrounded by eight diagonal struts and hence look like lanterns (The Greek *lychnis* = lantern). The group includes the thick-walled cup-shaped genera *Cypellia*, *Pachyteichisma* (see Fig. 14-13) and other genera.

Dictyonids are opposed to LITHISTIDS or stone sponges (order Lithistida), whose still thicker skeleton often has many tiny, root-shaped, gnarled parts called desmones, which are interconnected. If you permit your eye to wander across enlarged pictures of lithistids, they take on the appearance of the foamy water in a waterfall. The underlying needle type in lithistid skeletons is one- or four-axled. Typical genera are *Cnemidiastrum* and *Platychonia*.

Sponges live in groups or lawnlike growths. During the Jurassic, and only during that period, they formed colonies and even built reefs along with calcareous algae, and the reefs raised above the surface of the water in a cup shape (see Fig. 14-14). The reef began with a colony along the floor whose skeletons offered a favorable and propagatable habitat for newer generations. Since the decomposing soft bodies continually produced calcium, which precipitated in the sea water and which of course was not present in such great concentrations in sponge-free areas, the colony was built up over the course of many generations and finally protruded beyond the surface, sometimes attaining a height of more than 100 m. Neighboring reefs could merge and form great systems along the floor of the ocean. A large amount of evidence indicates that reef building was associated with moderate depths and quiet, warm water with a great amount of calcium precipitation. The calcium supply "mummified" the sponge skeletons and prevented them from decomposing rapidly.

The largest known sponge reef girdle extended from the Tethys in what is now the Mediterranean and the European marginal sea from the eastern Swiss Jura to the Franconian Alb. It was also present around Cracow, Poland. The sponge reefs produced highly resistant, massive calcite rocks, but the incessant action of erosion left great clefts in these deposits, and they contribute to the beauty of the landscape in modern Jura regions in southern Germany and Poland. Erosion has often bared the delicate structural features of these sponges, and the fine skeletons can be seen under a magnifying glass. The massive boulders produced by the fossil sponges seem completely unlike the delicate fine structure. On high parts of the rocks one can see the old relief of the sponge-covered sea floor, which appears in places after the rocks have eroded sufficiently. Thus, the sponge structures from the Jurassic are an impressive example of how geological and zoological activities in the past have influenced the modern apperance of the surface of th eearth.

CALCAREOUS SPONGES (class Calcarea), whose skeleton consists of calcite needles, are typically smaller than siliceous sponges and are restricted to shallower water.

When we speak of reefs, we usually are discussing CORAL. Thus it is not surprising that the sponge reefs were originally thought to be coral reefs. Actually, true coral reefs were as widely distributed as sponge reefs in the warm, shallow Jurassic seas. In fact, sponge reefs are often covered on top by corals, because as sponge reefs grew higher they reached a water level that was too shallow and too agitated for them to continue growth. This shallower region, however, was suitable for corals, which need water where sunlight can penetrate and which is agitated and hence oxygen-rich.

Coral skeletons, which are much more resistant to wear than sponge skeletons, depend neither on calcium production (via wearing) or quiet water for the growth of reefs. They form vertical walls with closely set individual elements, and these offer protection from the pounding surf; furthermore, coral reefs could grow by budding. The highly branched coral reefs we find in modern quiet lagoons also existed in the Jurassic under similar conditions. Intact retention of fossil coral reefs is an indication of a rapid sinking of the ocean floor in that particular area. Reefs like this are rarely found, however, and more commonly we encounter the debris from eroded reefs, since the shallower the water the more rapidly the reef is destroyed. The recesses of reefs (and the bits and pieces of reef debris) are accompanied by a rich assortment of fauna, consisting especially of mollusks, brachiopods, crustaceans, sea urchins and crinoids.

As in sponges, corals are still an important part of the animal kingdom. Distinguishing Jurassic and modern corals requires very specialized knowledge, since the Jurassic species all belong to the modern stony corals, which have predominated among corals since the Triassic. Jurassic coral is best fossilized when the original calcareous (aragonite) skeleton has been subsequently transformed into silicic acid, the acid often stemming from decomposing siliceous sponges.

Although the BRACHIOPODS (class Brachiopoda) were long past their flourishing period of the Paleozoic, these animals were much more significant in the Jurassic than they are today. Only a few SPIRIFER (order Spirifera) species survived into the Jurassic. *Spiriferina* (see Fig. 14-20), which had a porous shell, was the last spirifer with a spiral loop structure. The small *Cadomella* (see Fig. 14-21), a late member of the otherwise unlooped STROPHOMENIDS (order Strophomenida), also had a small loop but probably developed it independently. The strophomenids probably gave rise to the small THECIDEIDS (family Thecideidae), which adhered to mollusk shells and corals in the Jurassic and which in the Cretaceous developed a curious double loop at the base of the valve.

Most Jurassic brachiopods were RHYNCHONELLIDS (order Rhynchonellida) and TEREBRATULIDS (order Terebratulida), both of which existed

Brachiopods (Brachiopoda):

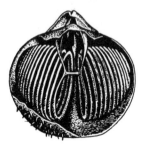

Fig. 14-20. *Spiriferina rostrata* (exposed).

Fig. 14-21. *Cadomella quenstedti* (width 8 mm).

Fig. 14-22. *Rhynchonella boueti.*

Mollusks (1-11):

Fig. 14-23. 1. *Gryphaea arcuata.*

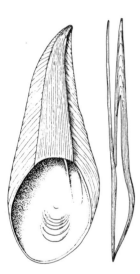

Fig. 14-24. 2. *Lithiotis problematica.*

Fig. 14-25. 3. *Parvamussium pumilus.*

since the Paleozoic. They were represented by numerous families and genera. Rhynchonellids (see Fig. 14-22) typically have a ribbed, thick, fuzzy to shiny shell. The top of the shell has a sharp tip (hence the name rhynchonellid; from the Greek ρυγχος = beak), beneath which there is a small hole. The corresponding hole in terebratulids is larger, and the smooth shell is covered with a regular arrangement of fine pores. There are also internal differences between these two brachiopod groups. In rhynchonellids, the fleshy ciliated apparatus used for respiration and feeding is attached to two short hooks, while in terebratulids this apparatus is borne by a rather extensive, complex calcite loop. The members of both orders have an indentation along the midline of the pedicle valve and a corresponding protuberance on the brachial valve which influences the configuration of the shell margin. This simplifies the separation of the two major water currents passing through the ciliated apparatus; one current, with fresh water, comes in through the side while the water bearing waste material is expelled in the middle.

Rhynconellids and terebratulids were most prevalent in the Jurassic in shallow seas and in reef facies. *Pygope*, an Upper Jurassic genus from the Alpine-Mediterranean region, has an extremely unusual shape; the shell indentation becomes transformed into a round hole passing through both valves during the course of growth. Since water being passed out now goes through this hole (it would otherwise pass through the opened valves) separation of fresh water and waste-bearing water is even more efficient than in other brachiopods, and this higher efficiency was apparently valuable for this genus, which occurred in greater depths where the ocean floor had less food and less oxygen than at the depths where other brachiopods were found.

Of the INARTICULATE BRACHIOPODS (class Inarticulata), the Jurassic (and still extant) representative was the small lingula (*Lingula*), with its calciphosphatic shell; *Discina* species, with horn shell; and the calcareous (calcium carbonate) shelled *Crania*. Brachiopods sometimes appear in masses in Jurassic deposits. *Rhynchonella lacunosa* was a colonial form occurring on the edges of sponge reefs. Sandy terrain, in contrast, could be entirely free of brachiopods.

The BRYOZOANS (class Bryozoa), which were related to brachiopods, and SERPULIMORPHS (Serpulimorpha), existed in many different species and were prevalent organisms often settling on fossil sponges, coral, and mollusk shells.

BIVALVE MOLLUSKS (class Bivalvia), which played a much more modest role in the Paleozoic than brachiopods, began to overtake them in the Mesozoic, a process which took a decisive turn in the Jurassic. Jurassic molluscan fauna includes some forty families (compared to fiftyfour at present). Nineteen of these had survived from the Paleozoic, and sixteen of these families are still alive today. Ten families have been identified since the Triassic, and twelve more appear in the Jurassic period. OYSTERS (superfamily Ostreacea) particularly flourished at this time, after beginning in the Triassic. In these oysters the left shell fused either during the juvenile period or throughout life. Jurassic oysters look very modern, although the genera so prevalent then (e.g., *Gryphaea*, *Exogyra* and *Lopha*) are hardly significant at all any more. Monomyarians (Monomyaria) included the order Filibranchia, with the families of scallops (Pectinidae), file shells (Limidae), wing oysters (Pteriidae), and spiny oysters (Spondylidae), all of which sometimes appeared in great masses. Two Lower Jurassic Tethys genera, *Lithiotis* and *Cochlearites*, are of uncertain systematic position. One of the pectinids was *Parvamussium pumilus* (*pumilus* = dwarf) from the Middle Jurassic (see Fig. 14-25), which had radiated internal ribs but was smooth on the outside. The species is also known as *Pecten personatus*.

Fig. 14-26. 4. *Pholadomya*.

Of the DIMYARIANS (Dimyaria), some of the families beginning in the Jurassic included Veneridae, Cyprinidae, Isocardiidae, Cyrenidae, Corbulidae, Thraciidae and Pholadomyidae. Of special importance were the cardiniids (Cardiniidae) and the trim trigoniids (Trigoniidae), which appeared in western America during the Jurassic and only later migrated into European Jurassic seas (see Color plate, p. 329). *Diceras* (see Fig. 14-30), with its snaillike coiled shell, was a characteristic organism from coral reef edges. It occupies a transition position between the Triassic *Neomegalodon* and the Cretaceous rudistids (see Chapter 16).

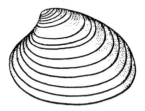

Fig. 14-27. 5. *Cardinia*.

Of the aforementioned mollusks, some lay freely on the ocean floor; others had an adhering fiber known as a byssus fused to their shell; and still others, such as pholadomyids, burrowed into the floor (see Color plate, p. 329) and have been recovered in life postures. We do not know whether some pectinid mollusks could flee through opening and closing the shell or whether some limids (*Lima*) built nests. Nevertheless, the life style of mollusks living in self-bored tubes becomes significant in the Jurassic. We encounter such living recesses in thick mollusk shells, large belemnite pieces, coral skeletons, and calcaerous banks, which even in Jurassic seas were hard surfaces. With some luck one can also find the inhabitant of these recesses, and various mollusk families of this group include the chemically or mechanically boring genera *Hiatella*, *Lithodomus*, *Gastrochaena*, *Martesia* and *Xylophaga*. Not all these mollusks bore through substrates throughout their life, and they are not always dependent on their own ability to dig a home. Sometimes the adults and larvae inhabit empty cavities together with members of non-digging genera and other animals.

Fig. 14-28. 6. *Trigonia navis*.

The order Protobranchia includes the mud-dwelling NUCULID MOLLUSKS (family Nuculidae), which were also prevalent in the Jurassic. They possess a primitive toothed valve, which also characterizes the filibranchiate taxodonts (suborder Taxodonta), a group including the ARK SHELLS (Arcidae). They lead to *Parallelodon* (see Fig. 14-31), which survived into the Jurassic, while *Arca*, *Barbatia*, *Cucullaea*, *Grammatodon*

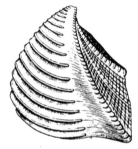

Fig. 14-29. 7. *Trigonia costata*.

Fig. 14-30. 8. *Diceras arietinum.*

Fig. 14-31. 9. *Parallelodon.*

Fig. 14-32. 10. *Discohelix dictyota.*

Fig. 14-33. 11. *Harpagodes oceani.*

and *Isoarca* arose in the Jurassic. The last two genera did not reach the Cretaceous.

SNAILS or GASTROPODS (class Gastropoda) were also distributed in the Jurassic, but since they had few prominent shell characteristics they have been difficult to classify. Most of these were marine snails, which were formerly classified as a single group (Prosobranchiata). The oldest ones, the order Diotocardia, include the beautiful SLIT SHELLS (Pleurotomariidae) and the genus *Discohelix* (see Fig. 14-32), a flat-shelled gastropod found only in the Jurassic and Cretaceous. The LIMPETS (Patellidae) appeared in the Jurassic, and their form was probably a subsequent modification to their sessile life. Examples of patellid genera are *Scurria* and *Helcion*, as well as the curious slipper shell (*Crepidula*), a member of the calyptraeid family (Calyptraeidae). It is the only extant snail that lives in small, chainlike colonies. One small genus, *Spinigera*, lives in clay soil, and its long spinelike processes are used to prevent the animal from sinking in mud. The modern giant conch (*Strombus*) is not unlike the Upper Jurassic genus *Harpagodes* (see Fig. 14-33 and Color plate, p. 329). The slender, highly convoluted genus *Nerinea* often appears *en masse* in coral reef facies and the vicinity thereof. This gastropod has ridges on the inner shell wall, and they can be seen on the outside as grooves that follow the shell spiral.

Jurassic brackish and fresh-water deposits contain several "foregill" migrants from the sea such as *Valvata*, *Bithynia* and *Hydrobia*. What were formerly called "hind gill" gastropods (Opisthobranchia) can also be found in the Jurassic. The ampullariid snails and related species (e.g., *Bulla*) arose in the Jurassic period. The Basommatophora were represented by such genera as *Lymnaea* and *Physa* (Lymnaeidae), *Planorbis* (Planorbidae), and other fresh-water gastropods from the Purbeckian stage at the turn of the Jurassic to the Cretaceous. These gastropods were presumably phylogenetic end links on the evolutionary pathway from the sea to the land and then back to fresh water.

Of the shelled CEPHALOPODS (class Cephalopoda), whose only present members are the NAUTILIDS (family Nautilidae), the AMMONITES (subclass Ammonoidea) flourished and reached their zenith in the Jurassic (see Color plate, p. 224). With some 550 genera, not all of which have been distinguished with certainty, they were more diverse than at any time in the Triassic and in the adjoining Cretaceous. Their structural plan and the significance of the suture line in ammonites was thoroughly discussed in the chapter on the Triassic. At the beginning of the Jurassic, after a threatening crisis (see Chapter 12), ammonites underwent a new proliferative phase that reached its height in the Upper Jurassic. This new phase led to the development of many shell styles and sculpturing that occurred in Triassic ammonites (see the chapter on the Triassic), but there were also distinct Jurassic features, and it is not difficult for paleontologists to distinguish ammonites from the two periods. All Jurassic ammonites differ from most Triassic species in an early juvenile characteristic that becomes apparent after careful preparation of the specimen and when viewed under a low power magnification: Jurassic ammonites have a greater indentation of the seams of the first two septae (the pro- and quinquelobate septae; see Color plate, p. 224). This characteristic becomes more and more complex throughout ammonite evolution beginning in the Devonian. The Jurassic style of the septal seam is reached in the ancestral forms of Jurassic ammonites (lytocerates and phyllocerates) in the Triassic, and they are the starting point for the new stage of proliferation attained in the Jurassic. This feature was thus a more advanced one, and one that permitted greater evolutionary development of the group than other adaptations. In terms of external appearance of ammonites, the new proliferative phase began with very simple forms; the smooth *Psiloceras* (see Fig. 11-40) was the ancestor of the majority of Jurassic ammonites (suborder Ammonitina), with the exception of the independently developing suborders of LYTOCERATES (Lytoceratina) and PHYLLOCERATES (Phylloceratina).

Ammonites have been particularly important in subdividing the Jurassic into various stages. The quinquelobate primary suture in all Jurassic ammonites does not mean that these ammonites reached a higher stage of suture line complexity than their Triassic forerunners, for there were some Triassic species in which the entire suture line was highly branched and more so than in any of the Jurassic forms.

Psiloceras, which is found in the deepest Jurassic bank known in central Europe, led to *Caloceras*, a form with simple ribs, a development carried still further in *Curviceras* (see Fig. 14-40 for a summary). A groovelike interruption of the ribs on the outside occurs in *Schlotheimia*, while a ridge forms in *Alsatites*, *Arietites*, and related species. These two characteristics —a groove and a ridge—first appear only on the inner convolutions (outer groove in *Saxoceras*, ridge in *Caloceras*) and do not extend across the entire shell until later forms. In these instances, therefore, a juvenile characteristic (the groove or ridge) does not reflect ancestral features but actually precedes a future development. It does not act repetitiously (palingenetically) but rather as a predisposition (proterogenetically); assuming, of course, that we are correctly interpreting the evolution of these features. Evolution occurs in many different ways, sometimes very rapidly instead of in a slow, regular progression, and one must not forget these different possibilities.

The ridge in *Arietites* and other related species can occur between two deep furrows or appear from the sides of the shell. This group consists of small species to those the size of wagon wheels (a similar size difference occurs in *Schlotheimia*). In the round-mouthed descendants of *Arietites* ("aegocerates": *Promicroceras* and *Echioceras*), the ridge is suppressed by invading ribs from the outside, while in the high-mouthed, narrow-hilumed forms it develops into a cuneate process leading from the narrow

Characteristic Jurassic ammonites:

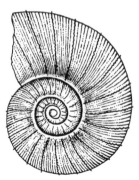

Fig. 14-34. *Lytoceras fimbriatum.*

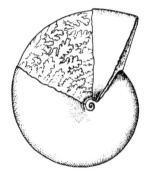

Fig. 14-35. *Phylloceras.*

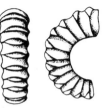

Fig. 14-36. *Promicroceras planicosta.*

shell cross-section (genus *Oxynoticeras*; see Fig. 14-40). This prominent feature, the ridge, has apparently been subjected to selective pressure during the course of evolution leading to its development and alternating recession. *Microderoceras* appeared while *Arietites* and relatives were still alive. Its flank ribs are bordered internally and externally by a row of nodules, a feature developed much more prominently by the spiny *Eoderoceras*.

All the aforementioned ammonite genera are from the beginning of the Jurassic. Processes like the formation of simple ribs, nodules and ridges continued through the Lower Jurassic, with increasing complexity of configuration (e.g., in genera *Polymorphites*, *Acanthopleuroceras* and *Androgynoceras*). The evolutionary relationships between all these genera have not yet been fully clarified. Other genera evolved ribs that took on a sickle shape (*Tropidoceras*), while still others had forked ribs (*Coeloceras*). *Liparoceras* has a rib and nodule sculpture reminiscent of the later kosmocerates. In about the Pliensbachian stage, the ammonites become less diverse, particularly because of the predominance of a new genus, *Amaltheus*. It has a jagged ridge and spiral stripes (see Fig. 14-41), with flat, slightly sickle-shaped rib folds. The generic name is a reference to the goat Amalthea from Greek mythology, since the shell of this ammonite is somewhat like the coiled horns of certain ungulates. The ridge looks something like the earlier genus *Oistoceras*, which also had outward oriented ribs. However, the spiral sculpturing and a specific feature of the inner suture line (the double division of the umbilical lobe U_1; see Fig. 12-17 for an explanation) are suggestive of lipocerate ancestry, which share these two features. However, much to paleontologists' dismay, the lipocerates have different sculpturing on other parts of the shell and a ridgeless shell as well!

The predominance of amaltheans is true only for northern regions and central Europe. Amaltheans are less common in the Mediterranean than other ammonites with simple sickle ribs and a smooth ridge (*Arieticeras* et al.) and those with radiating forked ribs (*Dactylioceras*). Thus we have evidence of zoogeographical differences within an animal group, a topic to which we shall return.

The amaltheans disappeared before the Toarcian as rapidly as they had appeared, while the now large-sized HARPOCERATES (with sickle ribs) and DACTYLIOCERATES (with fingerlike ribs) propagated. In many harpocerates the ridge is separated from the shell chambers by a floor. Later, and for a short time, sickle ribs with lateral furrows appear (*Hildoceras*, named after the Abbess Hilda from the Whitby monastery above the ammonite-rich Yorkshire coast of northeastern England).

Hammatoceras appeared in about the Bajocian; it has deeply cleft ribs and a ridge that evolved from ridgeless ancestors (probably *Dactylioceras*). In related species the ribs assumed a sickle shape and thus there was a second line of sickle-ribbed ammonites. These related forms included

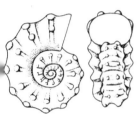

Fig. 14-37. *Microderoceras*.

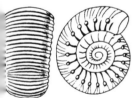

Fig. 14-38. *Coeloceras*.

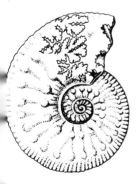

Fig. 14-39. *Liparoceras*.

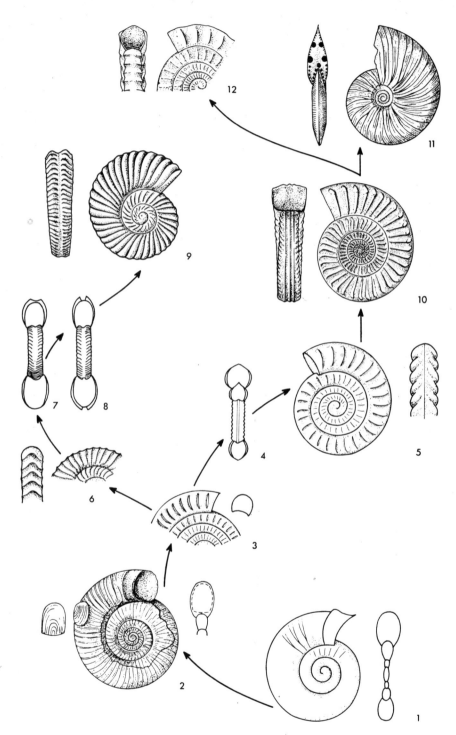

Fig. 14-40. Evolution of Hellangian ammonites (psiloceratelike) in phylogenetic representation.
1. *Neoptychites antededens*
2. *Psiloceras planorbis* with lower jaw (anaptychu
3. *Caloceras*
4 and 5. *Alsatites*
6. *Curviceras*
7. *Saxoceras*
8 and 9. *Schlotheimia angulata*
10. *Arietites bucklandi*
11. *Oxynoticeras oxynotum*
12. *Echioceras raricostatum*

Fig. 14–41. *Amaltheus margaritatus.*

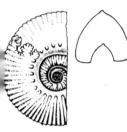

Fig. 14–42. *Hammatoceras.*

Fig. 14–43. *Pleydellia.*

Fig. 14–44. *Sphaeroceras brongniarti.*

Pleydellia (see Fig. 14–43) in the Lower Middle Jurassic and *Leioceras* and *Ludwigia* (see Color plate, p. 375) with their partly thin, partly thickened shells and high mouth position on these diverse shell styles. Some of the shells are "sharp-backed" (oxyconic). In adults the shells have shallow slits but suture lines with a great many lobes. This was once thought to be indicative of a phyllocerate heritage, but there is no confirmation for this assumption. This group also contains the Bathonian SONNINIANS, which sometimes had nodules on the shells, and the just subsequent, sickle-ribbed OPPELIANS.

Other, presumably likewise hammatocerate, lines led to renewed disappearance of the ridge to species with thick convolutions and ribs surrounding them: the OTOITES, SPHAEROCERATES, MACROCEPHALITES, and STEPHANOCERATES. The last named, as well as the GARANTIANS and PARKINSONIANS, had an outer furrow, which like the ridge came and went in ammonite evolution.

After *Strenoceras*, a relative of *Garantiana*, there surprisingly came forms with an open spiral (*Spiroceras* and *Parapatoceras*) and even some with rodshaped shells (e.g., *Acuariceras*; see Fig. 14–46). Since the very first coiled ammonites arose from straight ones, these later straight ammonites, which were rare in the Triassic and Jurassic but more prevalent in the Cretaceous, were thought to be aberrant, "atavistic" (i.e., reverting to a primitive type) organisms. Actually, in purely morphological terms, this was only an apparent atavistic development. However, the development of uncoiled ammonites occurred for a fairly long period of time during the Middle Jurassic. It is not an example of so-called retrograde evolution but probably the result of adapting to some ecological niche with new functions for the straight-shelled ammonites. One of the most unusual forms to appear in this line of development is the gnomelike genus *Oecoptychius* (see Fig. 14–49).

Macrocephalites then led, in the Callovian, to the KOSMOCERATES (in which the outside of the shell is flattened), the CADOCERATES (with sharper exteriors), and the QUENSTEDTOCERATES, whose ribbed peak led to the jagged ridge in Upper Jurassic CARDIOCERATES. Cardiocerates thus returned to an amalthean feature! PERISPHINCTS (see Color plate, p. 329) arose about the same time as stephanocerates, which they resembled. Their name is derived from the constrictions found at intervals along the ribbed convolutions of the shell. The singly or multiple cleft ribs are usually on the outside.

Perisphincts became quite prevalent in the Upper Middle Jurassic, and in the Upper Jurassic they became the predominant ammonite family. They are difficult to distinguish from each other, but some genera or subfamilies have been useful key fossils. One feature that repeatedly appeared in this line was a medial rib interruption or groove, as in *Reineckeia* with lateral spines (Bathonian), *Idoceras* in the Callovian, *Nebrodites* in the Kimeridgian and *Berriasella* in the Alpine-Mediterranean

Purbeckian and Infravalangian ·(the lowermost Cretaceous period). The more recent species in the small genus *Sutneria*, with mouth processes on the shell, had an external medial furrow. A striking Kimeridgian genus, *Hybonoticeras*, has a medial furrow between two outer rows of spines. The PELTOCERATES (Bathonian stage), had typical inner convolutions but outer ones with strong, spiny, simple ribs that were flattened on the outside. This group led to the EUASPIDOCERATES, which had nodules in the juvenile stage and often had a quadratic shape in cross-section. Euaspidocerates differ from the typically spined but otherwise sculptureless ASPIDOCERATES of the Upper Jurassic. They had a round or oval cross-section.

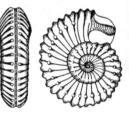

Fig. 14–45. *Strenoceras subfurcatium.*

During the Kimeridgian, with the genus *Ataxioceras* (from the Greek αταξςα = disarranged), a perisphinct group arose with an irregular rib interval and ribb clefts as well as particularly deep constrictions. *Rasenia* is characterized by a bundling of the ribs at nodules near the sutures, and this genus gave rise to the related genus *Aulacostephanus* (aulux = furrow), which also had a medial groove. It in turn led to *Gravesia* (no furrow !), which has a cross-section similar to many stephanocerates (e.g., *Teloceras*). Thus, a repetition of certain characteristics leads again and again to externally similar features and often confusingly similar species at different points in time, and these species can only be distinguished by specialists using rather esoteric characteristics like the juvenile suture line. Naturally, a collector is usually assisted by the association of a particular perisphinct with other organisms that give a key to the origin of the particular perisphinct. However, it is always possible for an unaltered genus to reappear among some animal community, and this could have a confusing effect for the systematist. Zoological classification of these organisms can only be made utilizing morphological features. And of course it is only thus possible to evaluate morphologically different fossils as key fossils.

Fig. 14–46. *Acuariceras acuarium.*

The Jurassic ammonites, which, as we saw, began in the early part of the Jurassic with *Psiloceras* and were descended from the lytocerates, continue into the Cretaceous ammonites as a rather broad animal group undergoing constant evolution. The lytocerates themselves, and the phyllocerates as well, hardly changed from the Jurassic into the Cretaceous. Toward the end of the Jurassic, an uncoiled (but not an ammonite) form broke off from the main lytocerate line. This was *Protancyloceras* (see Fig. 14–52). With that genus began the numerous "aberrant" Cretacous forms, whose shells were variously open, spiralled, hooked, snail-shaped, or rod-shaped.

It has long been known that the ammonite fauna of different Jura regions are dissimilar. The ammonite populations of northern, eastern, and central Europe were all different. The phyllocerates and lytocerates were the most prevalent species; they were probably adapted to greater ocean depths and thus became rarer in the shallow seas of the continental marginal zones (in some cases they are entirely absent). Other differences among ammonites, such as the number of species, were probably related

Fig. 14–47. *Quenstedtoceras lamberti.*

Fig. 14-48. *Cardioceras cordatum.*

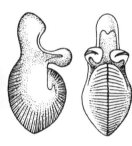

Fig. 14-49. *Oecoptychius refractus.*

Fig. 14-50. *Perisphinctes.*

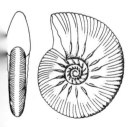

Fig. 14-51. *Aulacostephanus.*

to differences in water temperature. The Tethys was characterized by a tropical climate, for example. Assays of oxygen isotopes in the calcium of animal skeletons are used to determine prevailing temperatures. These measurements have shown that one part of central Europe was sub-tropical to tropical, while Scotland and northern Siberia were cooler, with prevailing temperatures between 13 and 25° C. (certainly higher temperatures than they experience today, however). The climatic differences during the Jurassic were not as great as they are today, as our discussion of flora also indicated. One test of the above temperature measuring technique was made by comparing animals from deep ocean floors with the results obtained from planktonic organisms, which would obviously have lived in warmer temperature surroundings. The test results for the temperatures were confirmed by the organisms used. This test has recently been used to demonstrate that the ammonite underwent rhythmic temperature changes as it grew, and these corresponded to annual temperature changes.

Amaltheus was for some time the sole ammonite genus between northern Canada and northern Siberia, and it was generally the predominant genus in western and central Europe. In the Mediterranean, however, the genus was not as prevalent as the sickle-ribbed ammonites (the hildoceratids, with the genera *Arieticeras, Protogrammoceras,* etc.) and dactyliocerates. Mixed fauna occurred in the Alps, southern France, Spain, and on the Japanese island of Honshu. Like today, we had during the Jurassic a climate-dependent decreasing number of species moving from south to north. Somewhat later, the genus *Dactylioceras* (see Color plate, p. 224) was almost as prominent in the north as the amaltheans had been before. Further to the south, this genus lived with numerous harpocerates, which then penetrated to the Arctic at the turn to the Middle Jurassic. In the later part of the Middle Jurassic, the difference between the Tethys fauna and the more northerly region increased in that the cadocerates and kosmocerates developed in the north and, like the amaltheans before them, did not penetrate further southward. The same is true in the Upper Jurassic for the *Quenstedtoceras* cardiocerates arising from the genus *Cadoceras.* In central Europe a mixed fauna prevailed between more southerly perisphincts and oppelians on one hand and the northerly kosmocerates, cadocerates, and their descendants on the other.

The parkinsonians, macrocephalites, and reineckeians were typical ammonite families of the Tethys and its bordering areas in the upper Middle Jurassic, while the perisphincts and oppelians were characteristic groups in the Upper Jurassic. Perisphincts inhabited the northern regions as well, but only with a few genera (*Dorsoplanites* and others). These northern forms include the genus *Virgatites* (see Fig. 14-54), which was limited to the inner Russian Upper Jurassic Portlandian. *Craspedites,* another northern group, was even more widespread. An arm of the ocean extended from the southern Asian Tethys to the east coat of Africa as far as Madagascar, probably bordered in the east by a land mass where we now find the Indian Ocean. Ammonites resembling those from more northerly regions propagated here, and they developed into independent forms. One of them, the Lower Jurassic genus *Bouleiceras* (see Fig. 14-56), extended as far west as Spain.

A shallow bay, extending from the nothern part of this arm of the sea to the southwest, contained the most unusual Toarcian and Middle Jurassic ammonites known. They only occurred in this central Arabian Jura region of the Dschebel Tuwaiq (where this bay was located). They are characterized by unusually simple suture lines, which is probably related to the muscles used for moving upward. This simplification may also be an adaptation to living in shallow water. The first to describe these peculiar fauna, William J. Arkell, distinguished numerous genera and species, while the respected evoultionary scientist Ernst Mayr claims that a closed population like that one would predictably lead to a great deal of diversity within just a few species.

Fig. 14-52. *Protancyloceras garaense.*

The Pacific Ocean, which existed by the Jurassic, was not surrounded by the marginal mountains it is today but by geosyncline zones with turbulent floors and volcanic activity. The ammonites from this zone are not uniform, for some arose from the Tethys, which stood between southeast Asia and Central America and some from more northerly regions. Climatic differences permitted only a few genera to propagate around the edge of the Pacific. Thus, this great ocean was as much an isolator with its size and depth as it was a connector between distant land masses.

Fig. 14-53. *Dorsoplanites dorsoplanus.*

The zoogeographic differences among fossil ammonites show that the empty shells did not drift aimlessly through the oceans. They tended to become embedded in the floor near their native habitat. In some cases there is even evidence of association of specific ammonites with a specific facies. The use of ammonites as key fossils is only valuable for relatively limited regions. Evaluating them stratigraphically over greater distances requires knowledge of other fauna overlapping with ammonites.

The soft ammonite body was so perishable that there is not a single known case of a fossil specimen. However, the shape of the living chamber suggests that the body could assume various shapes from a compact structure to a wormlike body. Impressions on the inner walls give clues about body musculature, as do (perhaps) the rear suture lines. Otherwise we must use the modern nautilus (*Nautilus*) for making conclusions about the ammonite body. Only recently has it been shown that ammonites had an upper and lower jaw, with a radula (or rasping tongue). What was once thought to be an operculum (or closing cover) is now understood as a lower jaw, at least in Lower Jurassic ammonites. If that should be confirmed by additional evidence, then the differences between the "dentition" types in ammonites, in adaptation to the diet, were rather substantial.

Fig. 14-54. *Virgatites virgatus.*

Fig. 14-55. *Craspedites fulgens.*

BELEMNITES (order Belemnoidea) do not appear externally to be

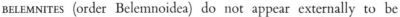

Fig. 14-56. *Bouleiceras arabicum.*

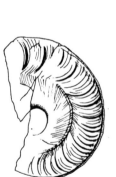

Fig. 14-57. The oppelian ammonites *Ochetoceras palissyanum irregulare* and *Streblites tenuilobatus.*

cephalopod descendants and relatives of ammonites. The chambered shell (phragmocon) does not become visible until the rostrum is broken open. It looks like a tapered bag with simple, hourglass-shaped septae. On the edges, as in ammonites, they are penetrated by the siphuncle, which is inflated between septae. Unlike the condition in ammonites, a rearward arched portion of the mantle, which secreted calcium, surrounds the chambered shell from the outside. Thus, belemnites (like the squid of today) possessed an internal skeleton, which only rarely is retained in the fossil. Most finds lack the rostral part of the chambered shell, which protruded well beyond the tip of the mantle. The calcite pen is usually missing as well. A few finds do demonstrate that belemnites possessed an ink bag; this part is retained in the fossil as a black mass. The weight of the mantle tip was probably useful in maintaining a horizontal position as the animal swam through the water with a blast of water through its funnel.

Belemnites are known since the Devonian, but they are not among the most prevalent animals until the Jurassic and the Cretaceous. They sometimes appear *en masse* in banks. The individual animals are often all facing in the same direction, which reflects the prevailing water current at that time. This often assists in replacing belemnite fragments that have fallen from decomposing formations. Although they were prevalent, belemnites were not nearly as diverse as ammonites. However, they are key fossils for certain time periods and zoogeographic parts of the Jurassic seas.

The Jurassic-Cretaceous belemnites, like ammonites, began in the Lower Jurassic with simple, small forms like those from Hellangian Germany; after that they proliferated considerably and led to numerous belemnite lines. By the Sinemurian stage, the compact mantle tips and relatively long chambered inner portion were several centimeters long (*Nannobelus acutus*). In the subsequent Jurassic, the length of the mantle tip increased in relation to the internal part. There were cigar-shaped forms (*Passaloteuthis paxillosa*), slender, shank-shaped belemnites (*Hastites hastatus*), short-fingered genera (*Dactyloteuthis digitata*), and spitlike thin ones (*Salpingoteuthis acuaria*). In some *Salpingoteuthis* species the mantle tip is divided in two parts, a compact portion and, on top, a loose or hollow part called the epirostrum. The significance of this division is unknown. The large belemnite rostrums have a loose lamellar structure along the line from the tip of the shell to the tip of the rostrum (the apical line). The rostrums in these belemnites (*Megateuthis gigantae*) are over $\frac{1}{2}$ m long, suggesting that the animals had a L of 1–2 m.

A first developmental progression closed with these giants. The development extended from the beginning to the middle of the Jurassic and was characterized by a continual increase in size. All members of this group have two or more short furrows at the tip (sometimes appearing incompletely closed), features probably arising from the fusion of the tip with the surrounding mantle envelope. A new belemnite group then appeared in the Middle Jurassic of Europe, after the rod-shaped *Rhabdobelus* arose as its predecessor in the Toarcian. These belemnites had a smooth tip and a well-developed ventral furrow extending posteriorly to the middle or even the end of the rostrum (*Belemnopsis*, *Hibolithes*). Another in this group was *Duvalia*, with a flatly pressed tip (Jurassic/Cretaceous Tethys), the predecessor to *Belemnites pressulus* (a tiny, drop-shaped and variable belemnite from the early Upper Jurassic).

As was the case with ammonites, belemnites since the Middle Jurassic also included characteristic northern genera (*Cylindroteuthis*, *Pachyteuthis* and *Acroteuthis*), which occasionally penetrated to the south. Upper Jurassic Germany, on the other hand, contains almost exclusively belemnites from one genus (*Hibolithes hastatus*), which is in great contrast to the variety of ammonites found in this Tethys region. Within the Tethys, there are differences between the eastern (Indopacific) region and the western (Mediterranean) region regarding belemnites.

A comparison between Jurassic cephalopods and those of today will show that at that time there were roughly equal numbers of cephalopods with external skeletons and those with internal skeletons. Today, the only members of the former group are the nautilooids (genus *Nautilus*), the last members of an old and once very diverse subclass. The same subclass gave rise to the ammonites in the Paleozoic. Cephalopods with internal skeletons are flourishing today, more than they ever have before. Their great advantage over the (better protected) internal skeleton cephalopods is their greater mobility. Poorer body protection has been offset by possessing the ink sac for fleeing. Belemnites are, of course, no longer alive. In the modern genus *Spirula*, only the chambered portion of the belemnite type body remains, and that is highly coiled. All other presentday cephalopods (the Dibranchiata), excepting octopuses, have a calcareous or horny cuttlebone (or pen) in the back, which in the Teuthoidei is comparable to the belemnite calcite pen. The more complex pen of squids (suborder Sepioidei) is of questionable phylogenetic origin. All these cuttlebone-bearing animals may not have descended from belemnites; they could have evolved their structures independently (in parallel evolution), and it could be that these organisms were greatly inferior to belemnites for a long time and therefore have not shown up in the fossil record. The first true squid cuttlebones are from slate and calcite plates from the Jurassic, materials which are good for fossilizing animals that are rare in other deposits. A number of different Jurassic squid genera have been described (*Geoteuthis*, *Loligosepia*, *Pleisioteuthis* and *Trachyteuthis*). The ink sac is sometimes retained. The thickness of one cuttlebone found in Upper Jurassic Cuba indicates that the animal concerned was a true squid.

Modern cephalopods are divided into the "four-gilled" forms (Tetrabranchiata) and the "two-gilled" forms (Dibranchiata). On the other

Cephalopods of the Jurassic and the present

hand, it is more appropriate to divide fossil cephalopods into those with an external shell (Ectocochleata) and those with an internal skeleton (Endocochleata). In fossils it is impossible to determine how many gills there were in each species, and we cannot arbitrarily draw analogies between modern forms and extinct ones, since some modern cephalopods may have regressed or developed even further in regard to their gill number.

The only living tetrabranchiate genus is the one we have already mentioned several times, the nautilus (*Nautilus*). A similar situation occurred at the beginning of the Jurassic, when *Cenoceras* was the only genus left from the diverse Triassic nautiloids. The evolutionary development of this early Jurassic genus was modest in comparison to the phylogeny of the many ammonite genera, and during the course of the Jurassic *Cenoceras* was replaced by several other nautiloid genera. Some had spiral stripes, marginal ridges, and septae, which were much more arched than in modern nautiloids: these Jurassic forms, therefore, had more highly curved suture lines as a result of their septal configuration. The common biological crisis that faced ammonites and nautiloids at the turn of the Triassic to the Jurassic was probably a consequence of physical forces that threatened the survival of several animal groups.

Decapods

After the last trilobites disappeared (see Chapter 7 on the Paleozoic), the post-Triassic seas were increasingly populated by DECAPODS (order Decapoda). Their skeletons have frequently remained in fine-grained, thin plates of Jurassic rocks (often only as exuviates). *Aeger* (see Color plate, p. 382) was a swimming decapod with a slender shape and swimming feet on the hind body. Walking forms, which were often flattened from the back to the belly and had sharp scissorlike members, included genera such as *Eryma*, *Eryon*, *Uncina* (Jurassic only), *Glyphea*, *Mecochirus* and *Palinurus* (spiny lobsters, a group still found today). In one of these bottom-dwelling walking groups, in a post-Lower Jurassic development, the abdomen was reduced in size and still later was bent under. This modification led to the origin of short-tailed TRUE CRABS (suborder Brachyura), which are so prevalent on our coasts today. Their anatomy permits these animals to burrow into the ground faster than other crab species. These crabs appeared in the Lower Jurassic (*Eocarcinus* and *Goniodromites*) and the Upper Jurassic (*Prosopon*). Some of the living relatives of these primitive crabs have migrated into the deep sea. In suborder Anomura, genus *Magila*, the hind part of the body had fewer hard parts and had to be buried in the mud of the ocean floor, and this behavior led to the rise of the hermit crab, which must carry a snail shell about to protect the soft hind body. The striking size difference in the claws of these crabs was established after the Jurassic. Differentially sized claws have been seen in Jurassic decapods, however, but these were most likely the result of regeneration of a claw subsequent to prior loss or injury of that claw.

Of the primitive crustaceans, the small, bivalved OSTRACODS (subclass Ostracoda) were important, prevalent key fossils. Analysis of ostracods can tell whether the water in which they lived was of normal or subnormal salinity. The calcareous plates of barnacles (subclass Cirripedia, genera *Archaeolepas* and *Pycnolepas*) have been found, and the results of their efforts are often seen as slit-shaped depressions in thick mollusk shells and belemnite rostrums. These were caused by boring barnacles belonging to the order Acrothoracica.

ISOPODS, which have been found since the Triassic, are particularly prevalent in certain Upper Jurassic limestone from former reefs and in lagoons. PARASITIC ISOPODS (suborder Epicaridae) caused diseased swellings on the thoracic armor above the gill region in short-tailed crabs, just as they do today.

The KING CRAB or LIMULUS (*Limulus*; see Color plate, p. 241 and Fig. 14-62) is one of the most famous Jurassic fossils. It is a merostomate (class Merostomata) within the subphylum Chelicerata. The Jurassic king crab is sometimes classified as a separate genus (*Mesolimulus*), distinct from the extant species of horseshoe crab (*Limulus polyphemus*) occurring on the Atlantic coast of North America. However, some authorities place them within the same genus. This is one of those relatively rare cases of an animal group hardly changing its morphology at all through millions of years. Sometimes fossil king crab tracks are found in huge numbers, for these walking organisms moved across soft floors and left their tracks behind them. A single track occasionally leads to a fossil king crab, and what we have on record are the last steps that animal ever took. In some cases it appears that king crabs swam into lagoons and were then killed by the excessive salinity trapped there.

The first primitive butterflies, which began with *Eoses* in the Triassic period, proliferated during the Jurassic along with flowering plants (the latter since the Middle Cretaceous), on whose juices and pollen they fed. Beetles were prevalent in the Jurassic; one green lacewing (*Kalligramma*) found in Upper Jurassic Germany had a wing span of 25 cm.

As we mentioned in Chapter 9, the stalked ECHINODERMS (phylum Echinodermata)—namely the CRINOIDS (subphylum Crinozoa or Pelmatozoa—were much more significant in the Paleozoic than, for example, mobile forms such as the ECHINOIDS or SEA URCHIN (subphylum Echinozoa). The situation is now the reverse of that, and the Jurassic is a transition period between the Paleozoic and modern conditions, for at that time there were about as many sessile echinoderms as mobile ones. Indeed, the crinoids in the Jurassic were long past their Paleozoic zenith, but in the Jurassic they produced their largest representatives, which are among the most magnificent fossils available of any animal group. Sea urchins flourished for the first time in the Jurassic, and both sea urchins and crinoids evolved a modern adaptation: crinoids assumed a floating life manner (with no adhering stalk), and sea urchins began digging into the ground. The STARFISHES and BRITTLE STARS (the latter also known as SERPENT STARS; both are in the subphylum Asterozoa; see Color plate, p. 382) were not

Jurassic belemnites:

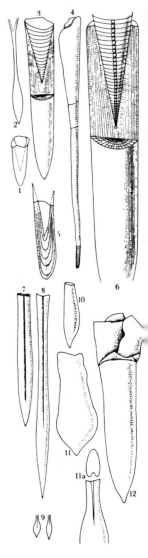

Fig. 14-58. Older Jurassic belemnites: 1. *Nannobelus acutus*; 2. *Hastites clavatus*; 3. *Passaloteuthis paxillosa*; 4. *Salpingoteuthis acuaria*; 5. *Dactyloteuthis irregularis* (sections); 6. *Megateuthis gigantea* (shell and tube). Younger Jurassic belemnites: 7. *Belemnopsis fusiformis*; 8. *Hibolithes hastatus*; 9. *Belemnites pressulus*; 10. *Duvalia tithonica*; 11. *Duvalia*

infrequent in the Jurassic, but they did not differ substantially from asterozoans in previous and subseqent periods.

CRINOIDS (class Crinoidea), with their stalk and their arms at the crown, have an entirely different structure from the free-moving starfishes, sea urchins, and related forms. Their mouth is on top of the body, not on the bottom. Jurassic crinoids were highly developed forms (having had a long phylogenetic history by that time) and had fewer plate elements in the cup than most Paleozoic forms. Since the cup cover, which was built of very small platelets, was highly mobile, these Mesozoic crinoids and those of the Cenozoic era as well are grouped together in the ARTICULATE subclass (Articulata). The Jurassic was a period of renewed proliferation for crinoids, and in adapting to new life habits they assumed greater variety than they had before.

The most frequently found parts of crinoids are the star- or coin-shaped trochites of the stalk with their delicate pentaradial structure and their central canal, but other individual parts have been known to form long banks composed of vast masses of crinoid elements. The calcite deposits cleave readily and have a sparlike sheen. Intact crinoids have been found in Toarcian Germany. One of these, *Seirocrinus* (formerly *Pentacrinus*) *subangularis* (see Color plate, p. 366), was the largest crinoid that ever lived. The stalk sometimes reached a length of 18 m, while the cups were always very small and had five arms. The arms were branched and thus had a delicate, flowerlike appearance. Even as fossils, it looks almost like the tiny arm elements of these crinoids are moving about just as they did in the water of their natural habitat in the Jurassic. Unlike most sessile crinoids, the end of the stalk was not anchored to the floor, but to drifting wood by means of small adhering disks. The wood has been retained in the fossil record as black bituminous coal. The largest continuous find of fossil invertebrates is currently the 18 m long plate with a piece of driftwood on which there is a colony of hundreds of individual *Seirocrinus* crinoids; it is at the Hauff Museum in Holzmaden, Germany.

The impetus responsible for crinoids to assume a quasi-planktonic life drifting about was the fact that the sea floor in the Lower Jurassic was oxygen-poor. Bitumen, hydrogen sulfide, and ferrous sulfide collected there, and they tended to kill off benthonic life. The black rock coloration and the presence of sulfur springs are indicative of this. *Seirocrinus* escaped these threatening conditions by adapting to a quasi-planktonic life style in the ventilated upper water levels. If the wood on which the colony floated sank to the floor, and it would do so under the weight of all those crinoids, the soft crinoid body parts decayed slowly because this part of the water had little oxygen and almost no scavengers. Thus, crinoids from these kinds of deposits have been marvelously retained. Another crinoid that floated on wood, *Pentacrinus fossilis*, is much smaller and has a short stalk with close-set, long cirrae.

Stalkless crinoids are first found from the Middle Jurassic (*Palaeoco-*

dilatata (lower-most Cretaceous); 12. *Pachyteuthis excentralis.*

master, Solanocrinus) and in the limestone of the Upper Jurassic (the latter resembling the modern genus *Antedon*). *Saccocoma*, a member of another family, is a small crinoid with small capsules and branching arms equipped with paddlelike structures. These stalkless, free-swimming crinoids, which embody the currently prevailing type, appear suddenly in the Jurassic and probably arose from various lines. One Swiss Bathonian genus known since 1950 (*Paracomatula helvetica*), although it apparently was a free swimmer, retained the stalk as an adult, and the stalk had five cirrae-equipped arms. In the modern crinoid genus *Antedon*, a stalk only occurs in the juvenile stage; it is later cast off. Therefore, *Paracomatula* is a true transition form that probably arose from the *Pentacrinus* crinoids but for which no intermediate links have been identified.

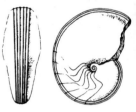

Fig. 14-59. *Aulaconautilus sexacarinatus* (Upper Jurassic).

Another adaptation occurring in crinoids was a reinforcing of the stalk and transformation thereof into a trunk, which was anchored on knobby "roots" to mollusk shells or a hard substrate of a reef. The crown also consisted of a few, but strong, elements with small, powerful seizing arms. These kinds of crinoids were adapted to agitated water and were particularly prevalent in coral reef habitat (typical genera being *Apiocrinus*, *Eugeniacrinites* and *Millericrinus*). Yet even this life mode could lead to a loss of the stalk, as in the Lower Jurassic crinoid genus *Cotylederma*, whose cups were in the shape of small, calcareous bowls with a simple structure; the cups could adhere directly to ammonites.

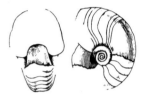

Fig. 14-60. *Cenoceras intermedium*—the last of the Triassic nautiloids in the Lower Jurassic.

Of the Paleozoic SEA URCHINS (subclass Perischoechinoidea), only the CIDAROIDS (order Cidaroidea) reached the present. Their geometrical capsules with large spiny warts in the interambulacral fields gave these sea urchins a striking appearance among the marine fauna of fossil and modern seas. After the disappearance of one of the earliest cidaroid families, they were represented (from the Permian on) by the MIOCIDARIDS (family Miocidaridae) in the small genus *Miocidaris*, a group that survived into the Jurassic. The last miocidarid, *Pachycidaris*, lived in the Oxfordian stage. Miocidarids retained the primitive feature of mutual mobility of the plates, which partially overlapped onto each other. This body plan led to rapid decomposition upon the death of the animal and few intact finds exist.

From the Rhaelian onward (i.e., shortly before the beginning of the Jurassic), and into the present, the stiff-capsuled CIDARIDS (Cidaridae) have survived and are currently represented by a few species. They were very rare in the Lower Jurassic, but they proliferated after that time so quickly that well-preserved specimens from the Upper Jurassic are common. The spines (large major spines and small ones called milliar spines) are often found by themselves, and intact capsules with spine systems are rarities. *Plegiocidaris coronata*, an inhabitant of the Upper Jurassic mud and coral reefs, is found in practically every paleontological collection. Unusually large, fan-shaped flat spines stem from the genus *Rhabdocidaris* (see Fig. 14-68). The stiffening of the capsule offers greater protection and, as unusual as it seems, enhances mobility, since motile muscles can be attached

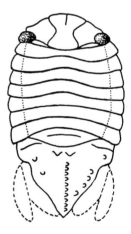

Fig. 14-61. The fossil isopod *Cyclosphaeroma woodwardi* from Upper Jurassic England; it is three-lobed and has facetted eyes (here the animal is enlarged).

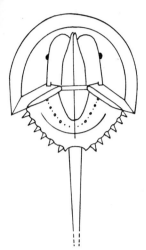

Fig. 14-62. The limulus *Limulus walchi* (reduced in size).

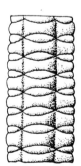

Fig. 14-63. Cross-section and part of the stalk of the crinoid *Seirocrinus subangularis* (enlarged (2:1).

to a stiff structure better than to a pliable one. The powerful dentition of these forms, has ridged (aulodont) composite teeth.

Another feature which enhanced the proliferation of sea urchins appeared in the late Triassic: the arrangement of the ambulacral plates was modified. In cidaroids they run in parallel rows, but in later forms the plates divide and coalesce and thereby form a much more complex system whose function has not yet been explained. This modification was responsible for the rise of the superorders Diadematacea and Echinacea. The differential development of the plate mosaic, the stiffness or mobility of the plate connections, and the nature of the teeth are used to distinguish different sea urchin groups. Jurassic genera include *Eodiadema*, *Pedina*, *Diademopsis*, *Hemicidaris*, *Pseudodiadema* and *Phymosoma*. Some of these and related genera can only be differentiated from cidaroids by closely examining the ambulacral fields. Others have broader fields and sometimes large spines and spine warts as well, and in these sea urchins the spine covering is more uniform and often more delicate as well. There are deposits absolutely filled with sea urchins spines (such as from the pin-thin spines of the genus *Diademopsis*).

All the aforementioned sea urchins, with their radial geometrical structure, lived on fairly firm sea floor, as we can deduce from the adjacent rocks and from modern sea urchin habitats. They took up food with the ventrally located mouth and released wastes through the dorsal anus; the mouth and anus were both medial. Wastes were carried away at once by the water current. In forms with a delicate spine covering, such as *Pedina*, the Jurassic was a period of change to a new life mode: living and burrowing in soft mud, in which large spines would of course be a hindrance. Thus, as the spines became smaller, the sea urchins adapted to this new way of life. And a second modification occurred simultaneously, also in relation to this new life mode: the anus was moved either toward the rear (as in *Pygaster*) or to the underside of the body (*Holectypus*).

What is actually happening here is the early evolution of bilaterally symmetrical, so-called "irregular" sea urchins. By the Cretaceous, the mouth had moved to the front of the capsule. The adaptive value of these changes is clear; if the sea urchin digs in the mud, a medial, dorsal anus would be disadvantageous because it would cover the animal with its own wastes. Also, the animal must have a way of feeding in fresh water through the mouth (the water providing both food and new oxygen), just as the modern sea urchin genus *Echinocardium* does with ambulacral feet. The optimal positions for the mouth and anus to perform these functions in a burrowing sea urchin are in front and in back, respectively. The sea urchin would dig little depressions in front and in back of it for creating intake and output spaces, and when the wastes had filled the depression behind the animal, the sea urchin moved on somewhat further and built new depressions. This way of living, which began in the Jurassic, was apparently so successful that the "irregular" sea urchins became the prevailing groups and radially symmetrical sea urchins have by now become far less signifi-cant, although they are still important on hard reef and rock substrates. The rise of irregular sea urchins does not mean that all of them burrow in the mud, incidentally; there are also forms that have taken up life on the open floor as well.

Not nearly as many vertebrates have been retained in the fossil state as invertebrates (with their calcareous shells and skeletons) have. The skeletal structure in vertebrates has been such that it enhances decay of body parts after the death of the animal. An exception to this is found in what are called bonebeds, which occur in certain areas at the lower border of the European Jurassic. These arose from places where shallow, active water between land and sea collected the bones of animals from both habitats. These deposits also contain fossil feces (coproliths = fecal rocks), which have been retained because of their calcium phosphate content. Deposits like these have been valuable sources for vertebrates, and the first teeth of those small animals we now know were the first mammals were found in such a place. Unfortunately, there are still no intact specimens of these vital transition forms that took the great steps between reptiles and mam-mals. Fishes and reptiles (the latter also called saurians), on the other hand have been recovered in fairly plentiful quantities and have often been pre-served in astonishing detail in bituminous deposits, slate, and lithographic calcite.

The bonebeds often contain the teeth and fin spines of CARTILAGINOUS FISHES (class Chondrichthyes), particularly of small and large sharks(*Acro-dus* and *Hybodus*). The rest of the skeleton is only preserved when there is an ample supply of calcium salts in the cartilage and very quiet waters for proper embedding of the animal. Once in a while these conditions are met, and they have delivered some magnificent fossil fishes. Whole examples of *Hybodus hauffianus*, with a L of 1.5 m, have been found, and they only lacked pure cartilaginous vertebrae; these were unusually well-preserved specimens and are rarities. One of these skeletons contained 250 belemnite rostrums as undigested material in the stomach. Unusually large fin spines with star-shaped protuberances (*Asteracanthus*) probably belong to rectan-gular or rugose teeth plates, which are designated *Bdellodus*.

While SHARKS have been found in three orders ever since the Devonian, RAYS and SKATES (order Rajiformes) are only known since the Jurassic. Angel sharks (suborder Squatinoidei) have a ray- or skatelike shape, but the squatinoids are placed with the sharks due to their anatomical features (e.g., the fact that the fore edge of the pectoral fins do not fuse with the head). The group includes the angel shark (*Squatina*), which is still found today and whose beautiful skeletons have been recovered from Upper Jurassic Europe. Another genus found there, *Rhinobatis*, is a true skate 160 cm long. In contrast to the rounder form of the angel shark, this ray has a rhombic outline, which leads in front into a toothless skull process and in the rear into a whip tail (see Fig. 14-73).

Fig. 14-64. *Pentracrinus fossilis.*

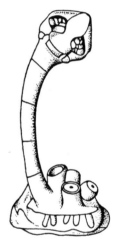

Fig. 14-65. *Eugeniacrinites.*

Fig. 14-66. *Plegiocidaris coronata.*

Fig. 14-67. *Plegiocidaris coronata* spine.

Other sea urchins:

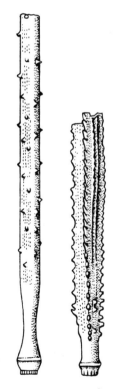

Fig. 14-68. Spines of *Rhabdocidaris maxima* and *Rhabdocidaris nobilis.*

CHIMAERAS (subclass Holocephali) are found since the Paleozoic era, and according to some paleontologists they arose from the old armored fishes. Members of the chimaera family Chimaeridae have only been found since the Jurassic, however. They fed on prey with hard shells or skeletons and could so feed because they had a stiff linkage between the cartilaginous upper jaw and the skull and also had a few powerful teeth plates. *Ischyodus*, a chimaera with a spined dorsal fin and one of several Jurassic genera, preyed upon belemnites but rejected the hind end with the rostrum (unlike sharks). Not surprisingly belemnites have been found together with *Ischyodus* skeletons, the former characteristically chewed up. Sharks and Jurassic rays and chimaeras have been astonishingly durable, relatively unchanging animals, and the world picture of cartilaginous fishes, with the sole exception of the giant sharks from the Tertiary, has been about the same from the Jurassic to the present.

The situation was quite different in TELEOST (BONY) FISHES (class Osteichthyes). The prevailing Jurassic species, the sturgeon/paddlefish super-order (Chondrostei) and the bowfin/gar superorder (Holostei), exist today only in remnants. These were actinopterygians (commonly called higher bony fishes or spiny-rayed fishes; subclass Actinopterygii) with a partially ossified skeleton. The vertebrae were partially or completely cartilaginous and therefore have not been well preserved in the fossil record. Some forms were almost scaleless, such as the 3 m long sturgeon *Chondrosteus*. The slender, narrow-snouted *Saurichthys*, whose teeth have often been found in bonebeds, was followed by the Lower Jurassic *Acidorhynchus* (which had regressed scales), a genus not to be confused with the scaled, larger genus *Aspidorhynchus* from the Upper Jurassic. Most ganoid fishes, in contrast, possess a characteristic covering of rhomboid scales. These black, shiny scales are covered by a dentinelike layer of ganoin. This group includes the high-set *Dapedius* (see Color plate, p. 376 and Fig. 14-72) and the thicker *Lepidotus* (see also Fig. 14-72), with its round, black, gleaming placodont teeth. Skeletons of both genera have been recovered, fully intact, in Toarcian slate. The high, narrow PYCNODONTIDS *Gyrodus* and *Microdon* from the Upper Jurassic had a chewing dentition well-supplied with quite a few teeth.

The trend toward degeneration of the ganoin layer and a rounding of the originally rhombic scales led to the cycloid scales, which are composed almost entirely of skeletal material. In *Pholidophorus* this replacement of the older scale type by the newer can be seen on one and the same individual. This is one of the clues suggesting that holosteins gave rise to the TRUE TELEOST FISHES (superorder Teleostei) with their fully ossified vertebral column and cycloid scales. The concomitant reinforcement of the skeleton enhanced mobility and permitted the armored scale covering to become weaker. *Leptolepis*, *Thrissops* and *Aethalion* are some of the earliest Jurassic true teleost fishes, and they probably arose from several ganoid ancestors. They were the introduction to the vast proliferation of modern teleost fishes, which now inhabit the seas and fresh water of the earth and number between 10 and 20 thousand species. Strictly speaking, if teleost fishes did arise polyphyletically (i.e., from various lines) as we suggest here, they should be classified from the very beginning in different lines. However, basic research in teleost fishes is not so extensive that polyphyletic descent can be incontrovertibly proven, and for the present it is useful to classify all the teleost fishes within one systematic taxon (category), although we keep in mind that a feature like increasing ossification of the bones could occur via parallel evolution.

The CROSSOPTERYGIANS or COELACANTHS (order Crossopterygii), which gave rise in the Devonian to terrestrial animals, were still rare in the Jurassic period. *Holophagus* (?= *Trachymetopon*) is very similar to the coelacanth *Latimeria chalumnae* discovered in 1938 (see Color plate, p. 356). Folded teeth plates, which are particularly known from Lower Jurassic bonebeds, are from LUNGFISHES (order Dipnoi), namely from the genus *Ceratodus*, which is closely related to the modern Australian lungfish (*Neoceratodus*).

AMPHIBIANS (class Amphibia) are extremely rare during the Jurassic. This is not only due to the fact that amphibian remains from marine deposits would not be present in great quantities, but also because the Triassic LABYRINTHODONTS (subclass Labyrinthodontia, also called Stegocephalia) hardly survived into the Jurassic. The habitat was there for the growth of the modern ANURANS or FROGS and TOADS (order Anura), but there are few finds of these creatures. The first known frog is from the Jurassic. Its hind vertebrae have fused into a kind of coccyx bone, and these enabled the frog (*Notobatrachus* from Middle Jurassic Argentina with an uncertain predecessor, lacking a coccyx, from the Triassic of Madagascar) to jump better. Confirmed URODELE (order Caudata) species have been found in the Jurassic but are sparse. A relationship between urodeles and the Paleozoic Lepospondyli is still uncertain.

Among the vertebrates of the Jurassic, the SAURIANS or REPTILES had the most striking shapes. The earth was warmer than it is now, and hence those reptiles were much more widely distributed than is the case today. The word reptile means crawling animal, but the term is not very appropriate, especially if we consider the diverse reptilian fauna of the Jurassic, which included very long-legged reptiles and marine species as well. Only modern reptiles are truly "reptilian", in the strictest sense.

The THERAPSIDS (order Therapsida), members of the order that in the Triassic gave rise to mammals, were present in the Jurassic but only in the form of a single, disappearing family, the ICTIDOSAURS (Ictidosauridae). All other therapsid families had died out by the Jurassic, and none of them left descendants. The first small mammals may have posed a danger for reptilian egg survival. Whatever the causes may have been for their decline as a group, the great change of terrestrial predominance from reptiles to mammals began in the Jurassic.

We must make a few comments on this great change. The study of

Sea urchins:

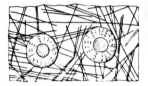

Fig. 14-69. *Mesodiadema crinifera.*

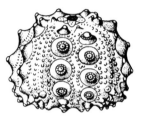

Fig. 14-70. *Hemicidaris crenularis.*

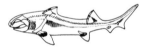

Fig. 14-71. *Hybodus hauffianus*, a fully intact shark 1.5 m long from Lower Jurassic Holzmaden (Germany)

Two ganoid fishes (reconstructed):

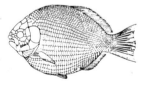

Fig. 14-72. 1. *Dapedius pholidotus.* 2. *Lepidotus elvensis.*

Transitional forms between reptiles and mammals

Triassic mammalian predecessors has shed a great deal of light on mammalian evolution. During the Triassic we find that the rear lower jaw bones degenerate in favor of the tooth-bearing (dentary) bone. As soon as the lower jaw consists solely of the dentary bone and the quadrate and articulate bones have been incorporated into the auditory apparatus as inner ear ossicles (see Chapter 11), we are speaking of mammals. This is simply the joint occurrence of some features found in all modern mammals however, and although there is no doubt about classifying any living animal as a mammal or a reptile, the only Jurassic finds that can be so classified are those in which the lower jawbone has been well preserved. The Jurassic species correspond to the transition forms between the pure reptilian condition and the mammalian condition. We do not yet know to what extent the internal organs, and above all the auditory organs, became modified in the Jurassic and moved toward the configuration found in mammals. One animal from Upper Triassic South Africa has been found in which the mammalian type jaw linkage between the dentary bone and the squamosal bone exists along with the reptilian quadrate-articulate linkage. The selective advantage of the mammalian lower jaw was simply that it permitted more efficient mastication of food, assisted by the multiple crowned teeth in place of the single-tipped reptilian tooth.

The reptilian ancestors of mammals died out during the Jurassic, but other saurian groups continued to evolve, including turtles (known since the Triassic) and lizards (the latter with just a few known Jurassic genera), while snakes apparently are still absent in this period. The most significant terrestrial reptiles, however, were the ARCHOSAURS (subclass Archosauria; from the Greek αρχων = ruler) on land and the ichthyosaurs and plesiosaurs in water. The latter two groups had existed since the Triassic.

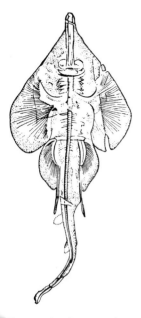

Fig. 14-73. The 1.6 m long true skate *Rhinobatis*.

The first marine TURTLES also appeared in the Jurassic, while terrestrial Jurassic turtles are relatively rare. One of the most astonishing processes in nature is the phenomenon of changing a life mode, which had originated in the sea and through many modifications adapted to terrestrial life, back to the sea again. Turtles are an example of just such a process, and it appears that the sea offered new survival possibilities to turtles during the millions of years they slowly adapted to terrestrial life! In Upper Jurassic thalassemydidae turtles (family Thalassemydidae; see Color plate p. 330/331), the shell had openings on the side and in the middle of the belly. This made the animal lighter, which was an advantage in the water, and this trend of developing a light shell with many openings continued into the Cretaceous. The clawed limbs may have born webbing. Since these turtles have been recovered from areas formerly near land or island, it seems likely that they led an amphibious life.

There are many Jurassic specimens of ICHTHYOSAURS or ICHTHYOPTERYGIANS (subclass Ichthyopterygia), a group confined to the Mesozoic. Many of these finds have even included the skin and other soft parts (see

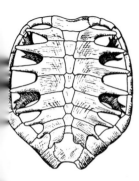

Fig. 14-74. The marine turtle *Eurysternum wagleri* from the Eichstätt calcite of Solnhofen, Germany.

Color plate, p. 367). The ichthyopterygian body outline is very dolphin-like, but it differs from delphinids in having a dorsal fin and a vertical caudal fin. Jurassic ichthyopterygians were even better adapted to marine life than their Triassic predecessors.

Fully intact skeletons of the ichthyopterygian genus *Stenopterygius* (see Color plate, p. 367) are superb examples of how well some of these animals have been preserved. The skull has large lateral optical fossae with a ring of protective plates around the eyes, and an elongated snout with close-set teeth, which in cross-section have labyrinthine foldings of the dentine as in labyrinthodonts. The vertebral column consists of a large number of short amphicoelous (biconcave) vertebrae, whose vertebral processes are loosely attached. The vertebrae and their spinal processes were connected solely by cartilage, not by bone as in terrestrial animals. The caudal portion of the vertebral column is longer than the thoracic portion and is bent downward at a 45° angle. The terminal elements thereafter support the lower lobe of the propeller-shaped caudal fin. This probably helped maintain the body at an acute angle to the water surface, facilitating breathing through the nares. Fishes are built differently; any bend in the vertebral column is always upward and extends into the upper lobe. Delphinid vertebrae, on the other hand, are straight and terminate in the middle of the horizontal caudal fin. The caudal fin was the chief propulsion organ in ichthyopterygians, and it was much larger than the other fins, which were used for guidance. However, these other fins (which correspond to the limbs of terrestrial vertebrates) were superbly adapted to life in the water. The arm, leg, hand, and foot bones are all short and form a mosaic of polygonal plates with a rudder function. The great number of phalanges made the limbs quite flexible. The pelvic bones are very small; long ribs formed a large rib cage.

Ichthyopterygians probably had the same adaptations for diving we find in modern whales. Unborn young have sometimes been found beneath the ribs (see Color plate, p. 367: *Stenopterygius crassicostatus*). The young were live-born and were fully developed at birth, as is sometimes the case in modern fishes, amphibians, and reptiles. Live-bearing in ichthyopterygians was one of many adaptations made to living in water. A gynecologist, Leipmann, investigated birth in ichthyopterygians in 1926 by comparing numerous maternal skeletons. He found that the normal birth position for the young ichthyopterygian was to come out with the posterior end of the body first, but other transverse positions also occurred in abnormal births, and some of these were fatal to the mother. After the death of the mother, an unborn young could still be ejected by the force of the gases building up inside the mother's decaying body.

Some ichthyosaur skeletons have been preserved with their skin or with transformed fatty substances, and the latter are possible evidence of warm-bloodedness. Although Richard Owen described his in 1839, the

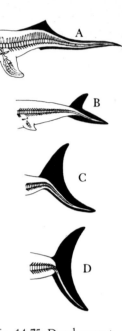

Fig. 14-75. Development of the caudal fin in ichthyosaurs:
A. *Mixosaurus nordenskjöldi* (Triassic) from Spitsbergen; B. *Stenopterygius quadriscissus*, a young animal (Lower Jurassic) from Germany; C. Adult *S. quadriscissus*; D. *Ichthyosaurus?* (Upper Jurassic) from Solnhofen, Germany.

retention of soft parts in fossils was virtually unknown throughout most of the rest of the 19th Century, and some or most paleontologists thought that these parts could not possibly be retained. Owen's claims were incontrovertibly substantiated in 1892 when Bernhard Hauff prepared a 1.2 m ichthyosaur skeleton in which the entire outline of the skin had also been preserved. These specimens have shown us how remarkably similar ichthyosaurs were to modern dolphins, with their propellor-shaped caudal fin and triangular dorsal fin. Ichthyosaurs fed on fishes and, most importantly, on cephalopods, whose hooks have been found in huge numbers inside ichthyopterygian stomachs.

Since intact skeletons with some soft parts have been found, it appears that ichthyosaurs lived in oxygen-poor seas and were buried in mud. Similar geological conditions exist today in relatively closed off bodies of water such as the Black Sea and the Baltic Sea. Upper water levels always have many fauna, and bodies of animals such as ichthyosaurs falling into the depths escape being eaten by these surface animals and also escape scavenging activity because of the absence of scavengers in low water levels where there is scarcely any oxygen for respiration. The exact extent to which a particular sea floor would offer favorable fossilization conditions would vary, of course, from place to place. The dark color of the rocks from these deposits (produced by the presence of ferrous sulfide and bitumen) gives us an idea of the prevailing geophysical conditions in those seas at that time.

Two Jurassic ichthyosaur lines have been identified, and they probably diverged in the Triassic: Longipinnati and Latipinatti (the names mean narrow-finned and broad-finned, respectively). *Stenopterygius* is a major member of the first group and is characterized by relatively narrow fins. This group also includes *Leptopterygius* (over 10 m long and with powerful teeth) and the unusual *Eurhinosaurus*, whose toothed upper jaw protrudes well beyond the lower jaw and probably reflects a specialized means of feeding. In the other ichthyopterygian lines, those with broader fins, there is an increased number of phalanges in the hands and a greater number of fingers and toes. Two typical representatives are *Ichthyosaurus* (=*Eurypterygius*; see Fig. 14-76) and the Upper Jurassic genus *Ophthalmosaurus*.

The rarer PLESIOSAURS (suborder Plesiosauria; see Color plate, p. 356) were another group of Jurassic marine saurians. They belong to the sauropterygian order (Sauropterygia). Plesiosaurs resemble ichthyosaurs in that their limbs have been transformed into paddle-like structures and are short and broad. There is also a greater number of fingers and toes, but the individual finger and toe elements are not polygonal; they are whorled. There are no other anatomical similarities between plesiosaurs and ichthyosaurs. The neck is long, but the tail is short and lacks a fin. Plesiosaurs moved only with the use of the limbs, and these structures were much longer and more powerful than the limbs of ichthyosaurs.

Progressive adaptation of fore limbs of saurians to aquatic life:

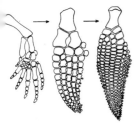

Fig. 14–76. Left: terrestrial reptile; middle: *Mixosaurus* (Triassic); right: ichthyosaur (*Eurypterygius*) from the Jurassic.

Fig. 14-77. Hind limbs of sauropterygian: left: *Lariosaurus* (Triassic); middle; *Geresiosaurus* (Triassic); right: the plesiosaur *Trinacromerum* (Upper Cretaceous).

The long neck probably stretched out like a fishing pole when the plesiosaur swam through the water as it hunted. Arm and leg paddles require large, flat bones to perform properly, to which joints and muscles were attached, and plesiosaurs were so equipped. The ribs were powerful, and the roof of the gastric cavity was reinforced by stable ribs. Pectoral ribs also occur in many reptiles of today and are the fine structures we see in our fishes; they were fine structures in ichthyosaurs as well. No plesiosaur embryos have been recovered, which may suggest that these animals temporarily left the water from time to time, just as marine tortoises do, to lay their eggs on land.

True plesiosaurs, with complete fins; flat, broad and relatively short arm and upper leg bones; and an increased number of fingers and toes, first occur in the Jurassic. There are two major plesiosaur groups, the plesiosaurs (in the narrower sense; family Plesiosauridae), with their small head and long neck consisting of 30 to over 40 cervical vertebrae (*Plesiosaurus* in Lower Jurassic Europe and *Muraenosaurus* in Oxfordian England and the pliosaurs (family Pliosauridae), with a relatively shorter neck consisting of just 20–30 vertebrae and a rather large head (including the 3 m *Peloneustes* and the gigantic *Pliosaurus*). The increasing size of the rear bony plates of the pectoral girdle at the expense of the front ones indicates a strengthening of the retracting muscles at the cost of the extensor muscles, and this adaptation gave these cumbersome animals considerable speed in the water in spite of their size. Strong retractor muscles would enable the fins to strike the water that much harder. In plesiosaurid plesiosaurs, the front and rear bony elements were about equal in size, which indicates that the front and rear stroke were about of equal strength. This means they could not swim as fast as those with larger hind plates, but the long neck probably enabled them to fish through a wider area.

Numerous CROCODILE (order Crocodylia) groups also adapted in interesting ways to marine life. One of these groups were the teleosaurs (family Teleosauridae, members of suborder Mesosuchia). The Lower Jurassic genus *Steneosaurus* (see Fig. 14–78) contained members several meters long. Their long, slender snouts had a spoonlike broadened tip. Their armor covering, which consisted of numerous pitted bony plates, gave them an appearance quite like the Indian gavial. They were surely good swimmers. Their slender hind limbs and long tail are indicative of efficient swimming behavior, and the short forelimbs were undoubtedly unsuited for terrestrial movement. Some splendidly preserved teleosaur skeletons have been recovered from slate deposits stemming from the Toarcian stage (see Color plate, p. 368, for one of these magnificent fossils). Sometimes the stomach region contains up to a handful of gravel, which were used just like chickens use gravel to assist in masticating food. Since the kind of gravel that has been found could only have originated from the coastal surf gravel, teleosaurs must have lived on the coast.

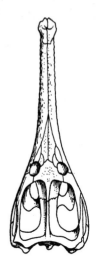

Fig. 14-78. Skull of the teleosaurid crocodile *Steneosaurus*.

Fig. 14-79. Skull of the brachiosaur *Brachiosaurus*.

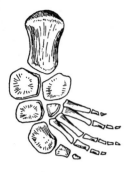

Fig. 14-80. Hand of the marine crocodile *Geosaurus*. The flat members of the first finger reflect the phylogenetic modification, already underway, of the fore limbs into paddles.

Fig. 14-81. In saurischians the pelvis was structured with the ischium facing backward and the pubis downward or forward.

Fig. 14-82. Skull of the large, bipedal saurischian *Antrodemus*.

One contemporary of *Steneosaurus* was the much smaller, more delicate *Pelagosaurus*. Large teleosaurs have also been found in Upper Jurassic strata, and calcareous slate has yielded small, armored terrestrial crocodiles (family Atoposauridae). One very interesting group were the armorless marine crocodiles (suborder Thalattosuchia), which were just 1–2 m long and whose tails had a bend in them (*Geosaurus*). Elongated spinal processes show that the caudal fin musculature attached to this bend point. Thus, the thalattosuchians were even more efficiently adapted to water than the Lower Jurassic teleosaurs. Further evidence of this are the very small fore limbs, in which the upper and lower arm bones and the members of the first finger are short and flattened while all other hand bones have the typical anatomy. Here, then, in this completely different reptilian group, began the transformation of the limbs into paddles, a process that reached completion in the ichthyosaurs. Marine crocodiles died out (for unknown reasons) shortly after the beginning of the Jurassic, and they never reached the paddle stage of development.

Like the crocodiles, the DINOSAURS evolved from the phylogenetically important PSEUDOSUCHIANS (suborder Pseudosuchia). The word dinosaur means "giant" or "terrifying" saurian (from the Greek δειν = fearful and σαυρα = lizard). The group did contain some very small animals, although it also contained the largest terrestrial animals that ever lived (contrary to widespread popular opinion, the largest baleen whales of today are much bigger than any dinosaur ever was). Particularly rich Jurassic dinosaur finds are in what were formerly coastal stretches, shallow seas, and inland bodies of water. Some famous dinosaur sites include the Tendaguru in eastern Africa, sites in Algeria and Morocco, and particularly the Morrison formation in the western U.S.A. Jurassic dinosaur fossils have also been found in Europe, Asia and Australia.

Some Jurassic SAURISCHIANS (order Saurischia) were fleet predators. They had a very delicate skeleton composed of slender, light bones, and with their short fore limbs and long tails they looked something like kangaroos. One small genus, *Compsognathus* (see Color plate, p. 286/287), had three, slender clawed fingers and birdlike toes. Its skull has two prominent openings (or fossae) in front of the eyes. *Ceratosaurus* from Colorado was much larger and had a higher skull and a horn on its nose. The heavily built, up to 8 m long *Megalosaurus* (see Color plate, p. 330/331), which had strong dagger-shaped teeth, has been found as early as the Lower Jurassic. Like *Antrodemus* (see Fig. 14-82) from Upper Jurassic North America, it looked quite like its mightier relative, *Tyrannosaurus*, the famous Cretaceous dinosaur.

In addition to the carnivorous bipedal dinosaurs there were tetrapedal, herbivorous saurischians that reached lengths unequalled by any land animals before or since that time. These all belonged to the suborder Sauropodomorpha. They had a massive trunk and a long neck and tail (see Fig. 14-83 for an example). Their fore legs, as in the primitive four-legged pseudosuchians from the Triassic, were somewhat shorter than the hind legs, in which the upper leg bone was much longer than the lower leg bone. The small size of the skull and the vertebrae, quite unlike the bones of the rest of the body, suggest strongly that these giant animals spent most of their time in bodies of water like ponds and lakes. The water would have helped support the massive body. This group contained, among many other genera, *Cetiosaurus* and *Bothriospondylus* from Middle Jurassic England, 23-m *Brachiosaurus* from eastern Africa, and *Camarosaurus, Brontosaurus* and *Diplodocus* (see Color plate, p. 354/355) from North America. In *Brachiosaurus* (see Color plate, p. 338), the large, high-set nasal openings are set on a low, flat snout. In other genera named above the skull is much higher. In *Diplodocus* and *Dicraeosaurus* the skull is horselike. These last two genera received their names (from the Greek δοκός = beams, διπλόος = double, δικραεος = bifurcate) because of the deep cleft of the spinal processes in the front of the vertebral column, an adaptation that accommodated the muscles supporting the neck. *Elosaurus* (from North American Morrison strata) was a smaller dinosaur just a few meters long.

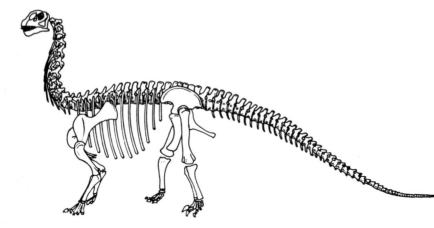

Fig. 14-83. *Camarosaurus,* an Upper Jurassic (Morrison) North American brachiosaur 12 m long.

In ORNITHISCHIANS (order Ornithischia), the ventral pelvic bones (the ischium and the pubis) are parallel and directed rearward (compare with Fig. 14-81 of a saurischian pelvic configuration). When the muscles attached to these bones are reconstructed, it becomes evident that this configuration enabled ornithischians to run rapidly on land on their hind legs, even though this pelvic arrangement is retained in some tetrapodal relatives. All the ornithischians were herbivores. One of the small ones was *Nannosaurus* from Upper Jurassic North America. Most of its relatives reached lengths of several or many meters: *Camptosaurus* (see Color plate, p. 330/331), *Laosaurus*, and *Dysalotosaurus*. The generic name of the last of these is from the Greek δυσάλοτος (hard to catch). These animals had five fingers and four toes, and they were relatives of the Cretaceous iguanodonts (family Iguanodontidae), who only had three toes touching the ground.

Besides the bipedal, short fore-limbed ORNITHOPODS (suborder Ornithopoda), others of this suborder in the Jurassic had longer fore limbs, and they ran on all for legs, bending somewhat while doing so. They had hardened parts of the skin, which in their earliest representatives consisted of a row of narrow ridged plates above the vertebral column. These animals were from the very early Jurassic of England. The fact that their plates were small contradicts the theory that the weight of the plates caused these animals to assume a four-legged posture when they moved. One of the heavy, most impressive dinosaurs was one with a double row of armored plates, the Upper Jurassic genus *Stegosaurus* (the Greek στέγε = plate; see Color plate, p. 356). The rear part of the tail bore long, sharp spines instead of plates, and in the closely related east African genus *Kentrurosaurus* (κέντρος = spine; ουρά = tail) all the plates are replaced by spines. The spine-covered tails of these unusual, heavy STEGOSAURS (suborder Stegosauria), in which, as in some others,

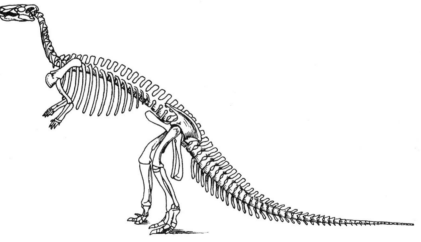

Fig. 14-84. *Camptosaurus*, a 5 m long ornithischian from the Morrison formation in North America (the ischium and pubis are directed backwards, and the latter has a forward-directed projection called the prepubis).

the pelvic vertebrae had a nerve knot larger than the brain, must have been formidable defensive weapons. This kind of excessive development of a feature is sometimes found in animal groups reaching the end of their phyletic development. Other developments in ornithischians took place in the Cretaceous and led to still other forms, which often had horns on the skull and to us seem most impressive.

The RHYNCHOCEPHALIANS (order Rhynchocephalia), which still survive today but only in the form of the New Zealand tuatara, arose from Permian eosuchians (Eosuchia). Even in the Jurassic they did not play a significant role. *Homoeosaurus* was a small, terrestrial rhynchocephalian, while the very lizardlike and likewise small genera *Pleurosaurus* and *Acrosaurus* from the Upper Jurassic were adapted to coastal waters. Another small genus, *Ardeosaurus*, was one of the few true Jurassic lizards. We shall report on the most "modern" creatures of this period, the flying saurians and primitive birds, in a separate chapter.

The slate deposits at the foot of the Swabian Alb near the German villages of Holzmaden, Ohmden, Zell and Bad Boll have been quarried for centuries (see Color plate, p. 366). The slate collected there has been used for floor covering and for lining ovens and other purposes. A young chemist, Alwin Hauff, came to these deposits in the middle 19th Century, and he was attempting to find a way to obtain oil by carbonization from the slate. His plans were a failure, but he continued in the quarry business, settled in Holzmaden, and brought his son Bernhard into the business. Two paleontologists convinced Bernhard to devote his time to paleontology and to permit working these slate deposits for their fossil value. It had long been known among scientists that Swabian slate contained some of the earliest rocks ever recognized as bearing fossils. One of the fossils that Bernhard found was the first one ever to have the soft parts preserved (an ichthyosaur), and after his discovery was made the region was declared protected for paleontological research. A Hauff Muesum has now been built at this site and contains some of the magnificent specimens that have been recovered at this classic site.

The Holzmaden fossils: an example of the exploration of a find,
by B. Hauff

The procedures used at Holzmaden to work the deposits are similar to those used elsewhere and will be described here briefly. Stone breakers, under the constant supervision of paleontologists, are used to expose parts of the slate in the hopes of finding fossils, and any areas that have a brownish color or look different from slate are evaluated by a paleontologist and possible exploited further during the search for fossils. Once any piece of slate has been removed and has exposed a fossil, the removed slate must often be examined for hours to find some trace of that fossil in it.

Fossil-bearing slate is then brought to the shop for closer examination and preparation. Small skeletal parts are exposed and evaluated for the extent to which they have been preserved and in what kind of condition they exist. Preparing specimens requires hours of exacting work with chisels of various sorts. Fossils do not simply cleave from the rock in which they rest, at least not from slate. The entire process of freeing a fossil from rock can take months, but the fruits of this kind of work can be the splendid specimens such as those illustrated on the Color plates, pp. 365–368.

The so-called Posidonian slate is named after the fossil genus found there, *Posidonia bronni*, a Toarcian genus. Composed of a fine bituminous sediment, this deposit has offered excellent preservation conditions for fossils. The bitumen and pyrite content of Posidonian slate give it a very dark color.

The most frequently found organisms in these deposits are ICHTHYOPTERYGIANS (subclass Ichthyopterygia), from the prevailing Jurassic animal

Fig. 14-85. *Posidonia bronni.*

group adapted to living in water. Their origin is still uncertain. They had attained marine life by the Triassic, and to adapt to life in water their bodies had to undergo considerable modification. It assumed a streamlined shape, and any source of resistance caused by armor or protruding organs was eliminated during the course of evolution. Naturally, this all occurred as part of the process or natural selection, whereby those animals lacking armor survived better of reproduced in higher numbers and hence survived better than any with armor. Over a long period of time, those animals more efficiently adapted to the conditions in water prevailed over all the others and eliminated them. The skull protruded forward and was tapered toward the front. The head passed into the body without a distinct neck section. A powerful bony ring formed about the eyes. The large dorsal fin was used to stabilize the body along the longitudinal axis. The vertebral column bent downward at the tail, and a symmetrical, vertical and caudal fin arose. The new caudal fin was a new organ for propulsion, which could push the body forward by beating up and down. The pectoral and, particularly, the pelvic girdles degenerated after they lost their supporting function. There was no longer any selective pressure to retain these structures. There may have been selection pressure against them. The limbs, which had become modified into multi-membered paddles, were used for steering and stabilization. The originally more powerful hind limbs played the subordinate role.

Ichthyopterygian reproduction was extremely interesting. Since these animals lived exclusively in water, they could not lay their eggs on land. So they bore live young. We have already discussed how females have been found with unborn young. We also know about their diet, since the stomach contents have often been preserved with the rest of the animal. The contents consist of the remains of cephalopod tentacular hooks. The development of these reptiles had ended by the time the Posidonian slate was being deposited, for ichthyopterygians became extinct in the Cretaceous.

Ichthyopterygians were the most prevalent reptiles in Jurassic seas. For no other fossil reptile are there so many specimens available to paleontologists, and some of the finds include soft parts of the body. There are enough fossil ichthyopterygian specimens to give rather complete phyletic information on many species. The most commonly found genus is *Stenopterygius* (see Color plate p. 367), of which there are many species. Less frequent ones include *Eurhinosaurus* and *Leptopterygius*.

The SAUROPTERYGIANS (order Sauropterygia) took an entirely different route as they adapted to living in water. Their ancestors, the Triasic nothosaurs (suborder Nothosauria), had already become marine organisms. Unlike the slender, high oval shape of ichthyosaurs, the PLESIOSAUR (suborder Plesiosauria; see Color plate, p. 356) body was like a broad rowboat. The tiny head was at the end of a long, powerful and very mobile neck. Long sharp teeth lined the jaws, which could be opened very wide.

▷▷
Coastal landscape from the Upper Jurassic. In the rear is the huge *Brachiosaurus* (7); in front the carnivorous bipedal *Megalosaurus* (1) and two ornithischians: *Camptosaurus* (2; lying) and the spiny *Polacanthus* (3; from Jurassic/Cretaceous deposits in Wealdon, England); the short-tailed flying saurian *Pterodactylus* (5); a pair of *Archaeopteryx* (6) climb a cycadacean tree; on the beach is the top of the shell of the marine turtle *Thalassemys* (4).

The limbs, which were supported by massive girdles, had developed into elongate paddles with many separate elements. They were used for propulsion, while the tail was used for steering (note that the opposite was true in ichthyopterygians). These animals lived on the high seas and thus are very rare at the Holzmaden site, which was formerly an inlet of the ocean.

MESOSUCHIANS (suborder Mesosuchia) have also been found at Holzmaden, although not in as great numbers as ichthyosaurs. *Steneosaurus* (see Color plate, p. 368) has been one of the genera in that slate. Steneosaurs retained some of the characteristics of terrestrial animals even though they lived in shallow water near the coast. They evolved from pseudosuchians (suborder Pseudosuchia), which also gave rise to dinosaurs and the precursors of birds. The steneosaur body in cross-section is broader than it is high. The long, slender snout is broadened at its tip into a spoonlike structure. There are large temporal fossae (openings). The fore limbs are very delicate and are smaller than the hind limbs. The tail is as long as the body from the snout top to the pelvis. Like all crocodiles, steneosaurs had an armorlike covering. The back was covered by a double row of rectangular (sometimes ridged) plates with many small depressions, while the armor on the belly from the chest to the pelvis consisted of oval plates firmly bound to each other. Rocks are sometimes found inside steneosaur stomachs; they are gravel the animals ingested on land. Young steneosaurs (they were born on land) moved into the sea gradually, and therefore finds of juveniles are extremely rare.

The FLYING SAURIANS (order Pterosauria; see Chapter 15) were the first reptiles to begin conquering the air (and the term "conquering" is

▷
An Upper Jurassic sea floor:
On top are shiny-scaled fishes, with a belemnite at the top left; below on the ground are two ammonites (a perisphinct and the tightly coiled *Taramelliceras trachinotum*; beside them are three examples of the mollusk *Trigonia* and a snail shell (*Harpagodes*); the burrowing mollusk *Pholadomya* digs in the mud (the siphuncle is shown). In the middle and to the right is the edge of a porifer reef are various cup- and bowl-shaped siliceous sponges; brachiopods (ridged rhynchonellids and smooth terebratulids) are at the foot of the reef; mollusks are in the very front. In the far rear there is a coral reef.

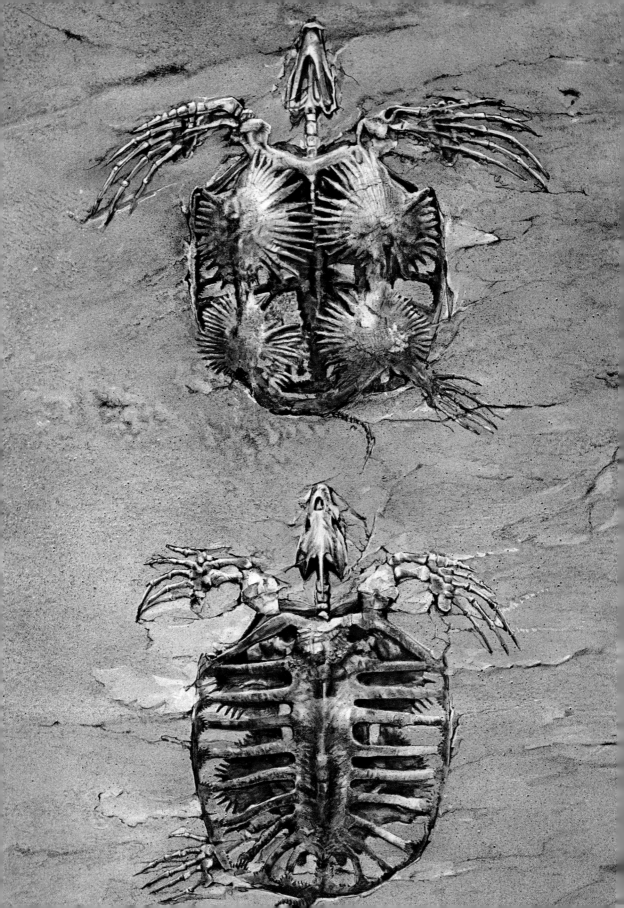

Flying saurians

◁

Skeleton of the giant marine turtle *Archelon ischyros* from Cretaceous North America. Above: ventral (bellyside) view; below: dorsal (back) view. The original is in the Yale Peabody Museum of Natural History.

Bony fishes

Crinoids on drifting wood

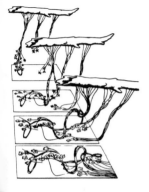

Fig. 14-86. Crinoid deposits in Holzmaden slate (after A. Seilacher).

used advisedly, for it often has inappropriate, unbiological implications !) Like bats, they had a flight membrane between the body and the greatly elongated fourth finger. The skeleton was very light, as we find in modern birds. Jurassic flying saurians were usually the size of sparrows or doves and had a maximum wing span of 1 m.

Fishes are of course an important part of any marine deposit, such as the Posidonian slate. Finds from Posidonian slate indicate that during the Jurassic the superorders of ganoid fishes (Chondrostei and Holostei) were replaced more and more by teleost fishes (superorder Teleostei), the latter being ancestral to most modern fishes. Teleost fishes have not been the only fishes to survive into the present, of course, and we still have cartilaginous fishes, a few ganoid fishes, and one coelacanth.

The heavy ganoid fishes with their dark, gleaming scales look particularly "primitive," a typical example being *Lepidotus*, which grew to a L of 1m. Another, *Dapedius* (see Color plate, p. 376), was more prevalent. Unlike these two, *Pachycormus* bore thin scales. However, its much stronger internal skeleton (which still lacked an ossified vertebral column) could support a more powerful musculature, something needed to provide speed and agility to a predatory fish like this one. The small, sprat-shaped *Leptolepis* was a true teleost with an ossified vertebral column.

Holzmaden deposits also contain the decapod *Uncina* (an arthropod) and magnificent crinoid colonies (see Color plate, p. 366). They often settled on drifting wood with the mollusk *Inoceramus*. Each individual animal, which as an adult could have a stalk several meters long, had its organ hidden within the cup structure. From the base of the cup, a five-pointed star, arose the seizing arms in strict geometrical arrangement. They were composed of thousands of calcareous plates. The plates were smaller the further they were from the body. The size and life expectancy of a crinoid colony was determined by the load capacity of the piece of wood supporting the colony. Small colonies are found on small pieces of wood, with larger colonies on the more substantial pieces of wood. Once the growth of the crinoid/mollusk colony exceeded the supportive capacity of the wood, the entire community sank to the floor of the sea.

The great majority of all the fossils recovered at Holzmaden are CEPHALOPODS (class Cephalopoda), especially AMMONITES and BELEMNITES, which we described previously. Their presence or absence in any particular area is a result of the prevailing ecological conditions there and to which every animal group responded according to its own sensitivities. We have found that the original habitat was generally the same as the place where the ammonites and belemnites became embedded. Some strata bear great masses of these cephalopods (and mollusks, too). Local concentrations like this must have been caused by water currents.

Geoteuthis squid lacking a pen and with an intact sack full of ink are great rarities and are evidence of extremely favorable fossilization conditions. Some strata contain only the pen (=cuttlebone) of *Beloteuthis* and none of the soft body parts. All that we have of ammonites are their shells, and no trace of the soft parts have ever been found. Ammonites are present as millimeter-thick, flat shells, and they are particularly suited for fossilizing. The closed shell is so heavy that it lies on the sea floor after the animal dies or sinks down after gases of decomposition raise the decaying body. The Posidonian slate contains many individual ammonite but not many different species. The ammonites found there include *Lytoceras*, *Phylloceras*, *Dactylioceras* and *Harpoceras*. Belemnites are represented by several species found throughout the deposit.

Of the MOLLUSKS, Holzmaden slate contains just two families, but a member of one of them (*Posidonia bronni*) is a key fossil of Posidonian slate. *Inoceramus*, the mollusk occurring with crinoids, is another key fossil from this deposit. Other mollusks, such as *Pseudomonotis* and *Pecten*, occur only in certain parts of the slate. BRACHIOPODS (class Brachiopoda) are only found in the lowermost and upper strata of this slate. Three orders of them—discinids, rhynchonellids and spirifers—occur here. The fossils of the Posidonian slates are completed by microfossils, the tubes created by scavenger feeding activity, and a few extremely rare plants as well.

Mollusks

A close examination of the strata in this slate in association with the paleontological studies has given insights into the origin of this deposit. Parts of the Black Sea and Sea of Asov have been implicated in the development of Holzmaden slate. The Jurassic sea near Holzmaden was a broad bay several hundred meters deep with a well articulated coastline in a subtropical climate. Its depositions stem from what is called the Vindelician continent to the south and southeast, and a great deal was also contributed by microorganisms. As water moved off and up from the deposition and more and more was deposited, these contributing materials left up to 1/20th of their original mass behind in the form of a deposit. Deformations of various materials deposited within this system varied according to their physical qualities; thus, bones and skulls of large saurians or thickwalled shells were not pressed as flat as the thin shells of some of the ammonites, the skeletons of delicate fishes, or the crinoid cups. The individual hard, round calcareous plates composing crinoid cups have hardly been altered at all, and comparable structures in belemnites have also been well retained. Dissolved minerals in the water around the settling fossils behaved differently (in a chemical sense) in different parts of the water, depending on concentration differences and other factors. The various strata within the slate and some of the calcareous intermediate strata owe their existence in part to an exchange of materials during the first settling period of the fossils. Purely physical forces have also played a role in stratification.

The Jurassic sea near Holzmaden

Deposition and settling can only occur where there are no currents on the ocean floor, since these would disturb the strata. In fossils we repeatedly find that the first part of the body to enter the sediment is in better condition than the upper side of the body. This means that there

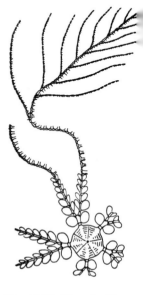

Fig. 14-87. The crinoid *Saccocoma tenella* (enlarged 2x).

must have been some water movement, although it was not strong
enough to transport the animals away. The relative absence of floor
currents has a favorable effect on fossilization since there is little oxygen in
such areas, and this enhances the preservation of organic substances in all
fossils, including their soft body parts. These conditions were all met in the
Posidonian slate, and this is the reason it has been such a gold mine for
fossil fauna.

The fossil contents stem from numerous habitats. Some were driven
into sediments from or near land or from rivers (e.g., terrestrial plants
drifting wood, flying saurians, young steneosaurs), but all of these kinds
of fossils (excepting driftwood) are very rare. Others came from the open
sea (e.g., some of the large fishes, plesiosaurs and certain ichthyosaurs like
Eurhinosaurus and *Leptopterygius*). In most of these cases the site of embedding is the same as the habitat of the living animal. We must emphasize
that the fossil record represents just a small fraction of all the organisms
that lived here and in this vicinity during the time the fossils we have were
embedded. The Holzmaden fossils are found in all strata of the deposits,
and this is a sign that the great prevalence of preserved animals at this site
is not the result of some catastrophe that killed off huge numbers of
animals all at once. Each stratum has its characteristic fauna, and these
fauna were laid down over a long period of time.

The following climatological and geophysical conditions probably
prevailed during the time the organisms in the Posidonian slate existed:
a sea a few hundred meters deep extended far from the coast and was
not subject to regular incurrents from the ocean. The deeper waters in
this arm of the ocean were low in oxygen content and were poisoned in
fact by hydrogen sulfide. Above these deadly depths was a zone rich in
life. Prevailing winds and convection currents in the water caused movement about the vertical axis of this system and acted like a sort of great
funnel. A similar situation, incidentally, is found today in the Black
Sea. Everything that was dead or could not swim on its own was eventually pulled into this circulating zone and sank down in the place where the
water current was least, where they were embedded in the sea floor.
This would explain the localization of fossils at Holzmaden. Whether a
circular current actually existed there is not known with certainty, but
it is likely that a system like the one described above did exist.

Germany contains two other classic fossil sites, Eichstätt and Solnhofen. These sites have been systematically dug since the 19th Century,
and some of the scientists that have evaluated their fossils included such
great names as Georges Cuvier and Louis Agassiz. Solnhofen had long
been exploited by the lithographic industry, since its slab limestone was
ideally suited for lithography. The limestone there was also exploited for
flooring and other commercial uses, and even today the site is worked
for commercial purposes. Naturally, this activity has been at the expense
of the fossils buried within the limestone slabs.

The limestone at Solnhofen was deposited along the edge of the great Tethys sea, the body of water extending from Europe to Asia. The continental deposition occurred some 130,000,000 years ago. A reef composed primarily of algae and sponges was built up parallel to the shore. Today these reefs are in the form of massive limestone and dolomite cliffs. They existed in water 50-100 m deep and died off when the depositions forming there made the water shallower.

The reefs produced a strong relief as they grew and coalesced with each other. Sometimes they formed isolated, lagoonlike bodies of water, and these were the places where the Solnhofen fossils were deposited.

Thus the Solnhofen strata were laid down in great pools in back of living reefs. All indications are that the fossils were deposited in quiet, largely standing water. The formation of limestone occurred through pure chemical decomposition of the reefs or by the action of algae or bacteria (and possibly both contributed to the limestone). Reefs like these must have existed in tropical conditions with relatively high water temperatures. Parts of the water would have had a temperature of 25° C.

Each pool in the Solnhofen system had its own microclimate and characteristic biota. Individual sections of the limestone deposits differ accordingly. The swimming crinoid species *Comatula pinnata* is prevalent around Solnhofen but absent in Eichstätt. On the other hand, another swimming crinoid (*Saccocoma pectinata*) is found in huge amounts in Eichstätt. The serpent star *Geocoma carinata* is found almost exclusively near another town in this system, Zandt. The crab *Eryon arctiformis* is known only from Solnhofen, while *Cycleryon propinquus* is restricted to Eichstätt. And so it goes, one species after the other.

All the above organisms lived in the pools in this Solnhofen/Eichstätt/Zandt system. Other characteristic (and also pool) fauna include the small, spratlike teleost fish *Leptolepis sprattiformis*, the long-armed crab *Mecochirus longimanus* (see Fig. 14-89), ammonites, and still other fishes and crustaceans. Microfossils such as foraminifers and ostracods have been found as well.

Many of the fossils animals and plants originally lived elsewhere. Most of the insects at Solnhofen flew there from the continent, as did flying and terrestrial saurians. Terrestrial plants, naturally, likewise came from the continental region, and some (like cypresses) occurred in substantial amounts. The surf brought in seaweed, and sea urchins swam in from the sea. Reef-inhabiting fishes like *Gyrodus* and *Gyronchus* swam in and cracked open the shelled animals with their semicircular placodont teeth. Once in a while large animals like very large fishes, marine crocodiles and ichthyosaurs also swam accidentally into the Solnhofen pools.

The Solnhofen limestone is lacking in large bottom-dwelling organisms, particularly in mud burrowers. This is probably because the habitat was relatively unfavorable for them. The warmth of the shallow water was responsible for the low oxygen content thereof (since oxygen solu-

The origin of the slab limestone

Fig. 14-88. A swarm of small teleost fishes (*Leptolepis sprattiformis*), which have been particularly prevalent in Solnhofen (Germany) deposits. The direction of the water current can be determined from the position of the fishes!

Fig. 14-89. The long-armed crab *Mecochirus longimanus* (L 20 cm).

▷

Early (Paleozoic) amphibians and reptiles. Above: The armored amphibian *Eryops megacephalus*. The skeleton is from Lower Permian Texas.
Below: Primitive reptile *Bradysaurus baini*. It is from Permian South Africa. Both specimens are in the Bavarian State Collection for Paleontology and Historical Geology in Munich.

bility decreases with increasing temperature). This would mean that less oxygen was present to contribute to decay of dead animals, which would enhance their fossilization. Some evidence indicates that the upper levels of the water were better ventilated and contained more drifting and swimming organisms. It is also possible that the mass appearance of small organisms depleted the oxygen supply and killed off the larger animals.

There is some evidence that the water retreated, at least temporarily, for some fishes in death throes have been recovered, and the marks they left appear to be those of a stranded fish when the water has washed away. Tracks of crustaceans such as *Mesolimulus walchi* and *Mecochirus longimanus* (see Fig. 14-89) with the dead animal at the end of them have also been found. These animals may also have been subject to a rapid decrease in the water supply, and they crawled about in vain searching for their vital water. They might also have crawled about underwater and died from a lack of oxygen. There are also signs of crystallized sodium chloride on the limestone, something we would expect when sea water evaporates. Yet, in spite of all this evidence, some doubt exists about whether water actually retreated or not. There are no clefts that are customarily caused when land dries up. The origin of the limestone slabs has never been fully explained, either. All these questions await further research before they can be more adequately resolved.

The animal world of the Upper Jurassic

The Eichstätt and Solnhofen fossils give us a typical picture of life in the Upper Jurassic. Nearly all the large animal groups are represented in these deposits. The most numerous crustaceans were DECAPODS (order Decapoda) in all their various forms. *Phyllosoma* larvae also occur. INSECTS are present in particular variety, and approximately 180 insect species have been described here. They include locusts, roaches, dragonflies, water bugs, ephemeral flies, beetles, horntail wasps, butterflies, and flies. Some of the dragonflies had a wing span of 19 cm, which is more than modern tropical forms have.

The fishes were also well represented. CARTILAGINOUS FISHES found at Solnhofen include sharks, rays, and chimaeras. Of the SPINY-RAYED or HIGHER BONY FISHES, the ganoid species were the prevailing ones. Teleost fishes are only represented by a few species (e.g., *Leptolepsis*, *Anaethalion* and *Thrissops*), but there are many individual fossils fishes (especially of *Leptolepis*). Entire strata have been covered with *Leptolepis sprattiformis* (see Fig. 14-88), and these fishes were clearly oriented with the water current.

REPTILES, which reached their zenith during the Jurassic, are represented by turtled, ichthyosaurs, rhynchocephalians (of which only the tuatara of New Zealand has survived into the present), lizards, crocodiles, flying saurians, and a small (50–60 cm L) dinosaur, *Compsognathus*. The reptiles were the first animals to enter the air, and this was an advanced feature of these reptiles. They were advanced in other ways, too. One fossil flying saurian, *Rhamphorhynchus* (see Color plate, p. 415), was covered

◁
Skeleton of the brachiosaur *Brachiosaurus brancai* from Tendaguru strata of Upper Jurassic east Africa. This specimen is in the Paleontology Museum of Humboldt University, East Berlin.

with hair; hair today is found only in mammals. However, it is certain that flying saurians were not related to mammals. We must assume that a hair covering (and presumably warm-bloodedness associated with that hair covering) evolved separately in those reptiles having it.

Perhaps the most interesting fossil of all at Solnhofen (and hence from the Jurassic) is the well-known primitive bird *Archaeopteryx lithographica* whose discovery and evolutionary significance we discuss in the following chapter. The species was named after a single feather found at Solnhofen in 1860. In 1857 the animal was thought to be a flying saurian, and it was not until 1970 that paleontologists realized that this flying animal was *Archaeopteryx*, a bird and not a reptile.

15 The Conquest of the Air

By H. Wendt

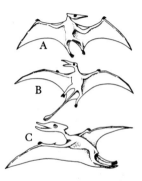

Fig. 15-1. Flying saurians (order Pterosauria):
A. *Nyctosaurus*, an Upper Cretaceous blunt-tailed form lacking the skull projection; its wing span measures 2.10 m.
B *Rhamphorynchus*, An Upper Jurassic long-tailed flying saurian; wing span 90 cm; C. *Dsungaripterus*, a short-tailed Upper Cretaceous giant form.

From leaps to gliding

The air is a third medium besides land and water that offers living possibilities to organisms. However, vertebrates did not enter this niche until very late (the Mesozoic and Cenozoic eras). INSECTS had invaded the air much earlier, for in the Carboniferous there were already giant dragonflies, roaches, beetles, and others of the insect group that could fly. A second insect proliferation occurred in the Permian (see Chapter 7), when insects with complete metamorphosis arose; during the Mesozoic they evolved into the members of present-day orders. A third flourishing period for insects came with the development of flowering plants after the beginning of the Upper Cretaceous (see Chapter 16) and during the Tertiary (see Chapter 17). Flowers were exploited as food sources.

The first vertebrates to move for short distances in the air were apparently flying fishlike species from the Triassic, such as genus *Thoracopterus*, one of the ganoid fishes and not related to the flying fishes of today, which belong to suborder Exocoetoidei. Their flying ability, of course, was not well developed and was not very effective. FLYING FISHES do not actually fly, either, for their pectoral fins are only used like aircraft wings. They glide, like some amphibians, reptiles, and mammals. Even the BUTTERFLY FISH (*Pantodon buchholzi*), with its greatly enlarged winglike pectoral fins, can only make leaps through the air. The only modern fishes that actually beat their pectoral fins are HATCHET FISHES (Gasteropelecidae), and even these can only move a few meters through the air as they fly.

In terrestrial animals, movement through the air is chiefly accomplished by gliding from one tree to another. This is a valuable adaptation for feeding and for fleeing enemies, since it means those animals do not have to first go down to the ground to get to the next tree. Selective pressure for gliding is particularly strong in areas where trees are not too close to each other and where the animal could not simply jump from one to another. Nonetheless, gliding by means of a flight membrane occurs relatively infrequently in the animal kingdom.

The sole gliding amphibians are the southeast Asian rhacophorid frogs (*Rhacophorus*). The Borneo flying frog (*R. pardalis*) and *R. nigropunctatus* have particularly large flight skins on their hands and feet, and these can brake the animals in very high leaps. FLYING FROGS starting from a height of 5 m can glide for a distance of 7 m. Among reptiles, the so-called FLYING SNAKES (genus *Chrysopelea*) can simply drop from considerable heights. FLYING DRAGONS (*Draco*) are better flyers, since they possess winglike skin lobes supported by ribs on the sides of the body. When they jump, they spread their lobes and undergo a nearly vertical "flight", steering slightly upward just before they land. They can cover stretches of up to 50 m or more in unusual instances.

Fig. 15-2. Skeleton of the long-tailed flying saurian *Campylognathus zitelli* from the Lower Jurassic (Holmzaden, Germany). The wing span is approximately 1 m.

Gliding mammals use essentially the same technique employed by flying dragons. Gliding flight is developed in various marsupials. The GLIDING POSSUM (*Petaurus*) and the small PYGMY FLYING PHALANGER (*Acrobates*) have flight skins on the sides of their bodies, and the fore and hind limbs are used to spread the skin. The body becomes rectangular when the flight membrane is outspread, and these animals glide forward toward their goal and can even steer away from their original path to reach another goal. One of the largest marsupial glideflyers is the GREATER GLIDING POSSUM (*Schoinobates volans*), a relative of the koala. Greater gliding possums can weigh as much as 1½ kg. Since the flight membrane does not begin until the elbow, greater gliding possums form a triangle when they spread their "wings".

Some arboreal rodents also glide. The FLYING SQUIRREL (subfamily Pteromyinae) has a hair-covered flight membrane on the sides of its body, and the membrane is outspread like a parachute when the animal glides. Their flights can be as long as 50 m, and like gliding possums, flying squirrels can change direction as they "fly". In the unrelated SCALY-TAILED SQUIRRELS (Anomaluroidea), the parachute is not outspread from a bone on the wrist but from the elbow. This flight skin can be 8 cm long. The AFRICAN SMALL FLYING SQUIRRELS (*Idiurus*), members of the scaly-tailed squirrel group, can glide as far as 100 m.

Fig. 15-3. One of the finest preserved flying saurian skeletons (*Rhamphorhynchus phyllurus*) from the Upper Jurassic. The impressions of the flight membrane and the rhombic skin 'sack at the tail have been clearly retained. *Rhamphorhynchus* was the most familiar, prevalent, and geologically youngest long-tailed flying saurian.

Finally we shall mention the FLYING LEMURS. These are two species within their own mammalian order (Dermoptera). They evolved from insectivores and developed independently of and parallel with primates and bats. The oldest fossil flying lemur is the Paleozoic (Lower Tertiary) *Planetetherium*. Their flight membranes are not only between the limbs; they practically envelope the entire body. A fore flight skin (the propatagium) extends from the head to the arms; an extensive lateral flight skin (the plagiopatagium) is spanned between the arms and legs; and a caudal flight membrane (the uropatagium) runs from the hind limbs to the tail. Flying lemurs are the only mammals in which the tail is part of the flight apparatus. However, these animals do not truly fly. Flying lemurs are herbivores, and they glide up to 70 m from one tree to another seeking food.

In none of these cases can we speak of "conquest" of the air since none of the above animals flies to any great extent, and most of them can only glide. They support their leaps with built-in parachutes, and they are no more flying than a parachutist is when making a drop. True active flight (albeit in various ways) has only developed within three vertebrate groups: flying saurians, birds, and bats. It is quite possible that the ancestors of these animals began with parachutelike flight skins and in time developed more complete flying equipment. The largest flying animal of all time was a fossil reptile, the pteranodon (see Chapter 16 and Color plate, p. 406/407). Bats have evolved true echolocation with their ultrasonic emissions, and no other fliers are as well equipped to fly at night. However, flight *per se* is more highly developed in birds than in any other vertebrates.

Flying saurians and birds evolved from the reptile order of thecodonts (Thecodontia), but they are no more closely related than that, for they have developed independently along their own lines ever since their earliest beginnings. Thecodonts were widely distributed from the Upper Permian to the Upper Triassic. They were carnivores and walked on all four limbs in a crocodile manner and to some extent on two legs in a somewhat kangaroolike way. Some species were arboreal and may have been glide fliers. Upper Triassic England has yielded a small saurian, *Scleromochlus taylori*, whose very delicate skeleton and greatly elongated hind limbs give the impression that it could glide from branch to branch or tree to tree. It is not known whether this saurian was in the flying saurian line or ancestral to birds.

Besides the archeopteryx (to be described), the FLYING SAURIANS (order Pterosauria) are the most "modern" Jurassic animals. The oldest skeletons are from the Lower Jurassic and are fully developed at that time. Undoubtedly, flying saurians have a longer history, but the intermediate forms between thecodonts and flying saurians have not yet been identified. All we can say at the present is that flying saurians evolved from pseudosuchians (suborder Pseudosuchia). Pseudosuchians were an extremely important reptile group from which flying saurians, crocodiles, dinosaurs and birds arose.

The new feature in flying saurians is that they have a greatly elongated fourth finger. The flight membrane extends from this long finger toward the body and hind limbs. The flight skin probably acted like a parachute initially and was simply a modified segment of folded skin. In birds and bats the other fingers are part of the flight apparatus, but this was not true in flying saurians. Their three other fingers were very short and were probably used for dangling from trees. Anatomically it was possible for them to hang either head down, like bats, or head up. The legs were short and had five digits. The very light, pneumatic skeleton bore a sternum with a medial ridge. The clavicles disappeared as flying saurians adapted to leaping off the ground to begin flight.

Fig. 15–4. Life posture of of the stump-tailed flying saurian or pterodactyl *Pterodactylus fraasi* after erroneous drawing of the English paleontologist H. G. Seeley at the end of the 19th Century. He shows the pterodactyl standing on all four legs! The first report of this reptile is from 1784, when the Italian naturalist C. A. Collini found near Eichstätt an "unknown amphibian marine animal of doubtful zoological position". O. Abel then introduced the notion that pterodactyls did not move across the ground but could fly like bats and could rest by hanging from trees.

The older flying saurian suborder are the RHAMPHORHYNCHIDS (Rhamphorhynchida; since the Lower Jurassic), whose members have a long tail reinforced by needlelike cords. The group includes such genera as *Dimorphodon* and *Dorygnathus*; the Upper Jurassic genus *Rhamphorhynchus* had a rhombic rudder at the end of its tail. The second suborder, the PTERODACTYLIDS (Pterodactylida), are first found in the Upper Jurassic. Their tail has regressed to just a stump. These stump-tailed flying saurians lived until the end of the Cretaceous, when they developed into some giant species. Some of the smaller pterodactylids include *Pterodactylus*, with its powerful teeth and *Ctenochasma*, with its peculiar straining beak. Its long, narrow jaws had fine, brushlike teeth, which were used when the animal glided just over the water surface and grabbed up food.

Flying saurians were widely distributed during the Jurassic. They have been found in Purbeckian England, in a limestone deposit near Tashkent, and in Solnhofen, the classic site of which we spoke in the previous chapter. Jurassic flying saurians varied in length from the size of a sparrow to that of a buzzard, and only the giant pterodactyls of the Upper Cretaceous were larger. Yet by the Cretaceous the flying saurians were on the decline. Rhamphorhynchids were practically extinct, and the pterodactylids were quite rare. The only ones left were giant forms (genus *Dsungaripterus*) with a wing span of 2 m or the highly specialized types like the one with the filtration beak we mentioned earlier.

During the Upper Cretaceous, however, flying saurians experienced another growth spurt, with the PTERANODONS (family Ornithocheiridae with the genera *Pteranodon*, *Ornithocheirus* and *Titanopteryx*). Since these were giants, it seems likely that only flying saurians with a very one-sided life style could compete with birds, which were becoming quite prevalent during the Cretaceous. The smaller Jurassic flying saurians fed chiefly on free-swimming crinoids, insects and small fishes, grabbing them out of the water or from the air; the large pteranodons were primarily fish-eaters that preyed out on the open sea (where there was no competition from birds). *Pteranodon ingens* is believed to have had a wing span of up to 7.5 m. It has been found in Upper Cretaceous North America and Eurasia. These giant pteranodons had toothless, pelicanlike jaws, a long bony skull outgrowth and what is called a notarium near the vertebral column. The notarium consisted of several front dorsal vertebrae that were nearly fused together, with an acetabulum or cup-shaped structure for the shoulder blade.

One of the first classical students of the Jurassic, Friedrich August Quenstedt, found hairs on a fossil flying saurian. In 1908 the German paleontologist Wanderer claimed that in all probability, flying saurians bore hair like mammals and thus were not poikilothermic ("cold-blooded") but homoiothermic ("warm-blooded"). Twenty years later, a flying saurian with hair on its neck, back and on the flight skin was found. That find demonstrated that flying saurians were the third verte-

Fig. 15-5. The pteranodon *Pteranodon ingens*. In 1872, the American paleontologist O. C. Marsh first found fragments of this flying saurian in North American Cretaceous deposits. Marsh's find eventually turned out to contain some 600 pteranodon skeletal fragments. He calculated that these animals had a wing span of up to 7.5 m, and he described these, the largest winged animals that ever lived, as follows: "This was a truly gigantic dragon, even for this country where there are so many large things."

Fig. 15-6. Skeletal remains (the skull is absent of the archaeopteryx *Archaeopteryx*, which was found in 1861 in Solnhofen, Germany. Today this specimen rests in the British Museum in London.

brate group (along with birds and mammals) to attain homoiothermism, and for this reason they had to be classified in an independent class. It is not unusual that they should have been warm-blooded, for flying saurians had to be active for long periods of time. It would be difficult for a poikilothermic animal to live like that, for their internal temperature more or less matches that of their surroundings. Body temperature is kept uniform in "warm-blooded" animals.

Equally surprising as the discovery of hair was the finding that the brains of flying saurians were more highly developed than those of other reptiles. It was a true "bird brain" (we mean in the phylogenetic sense, of course!!), although flying saurians and birds are not closely related. The flight ability of flying saurians contributes to the high brain development. As is also the case in their toothless jaws (in the Cretaceous giant pteranodons), this development was an example of parallel evolution. No one has established yet whether flying saurians gave birth to live young or laid eggs. The separation by two bones along the midline of the connected pelvic halves suggests that they may have given birth to live young.

Since flying saurians achieved such a high degree of organization (a higher one than all other saurians), the questions arises why the long-tailed forms died out during the Jurassic and why the giant pteranodons also died out, quite suddenly, at the end of the Cretaceous. It is likely that these grand flyers were so specialized they were unable to walk on land. They were an evolutionary dead end. Furthermore, studies have shown that the flight membranes of these saurians, with the elongated fourth finger, was ultimately inferior to the feathered bird wing.

One of the most exciting chapters in the history of paleontology is the fact that the oldest known bird, the archeopteryx (*Archaeopteryx lithographica*), was found in the very same Jurassic slate in Solnhofen that had contained the first pterodactyl (*Pterodactylus*) ever discovered (which C. A. Collini found in 1784 but erroneously described as a marine animal). Before we describe the archeopteryx in detail and discuss its phylogenetic significance, we shall briefly trace the story of the discovery of this animal. The generic name *Archaeopteryx* stems from the Greek αρχαιο (= ancient) and πτερόν (= feather).

Fig. 15-7. Thomas H. Huxley saw in this "proto-bird" the ideal link between reptiles (below) and birds. He postulated the existence of a proto-bird (above) with reptilian teeth, claws on the wings, scales on the body, and a long lizard tail. Soon thereafter an impression of a small feather was found in Solnhofen slate!

When Darwin's great supporter, Thomas H. Huxley, attempted to postulate how birds arose from reptiles, he described a hypothetical proto-bird. It was a feathered creature with reptilian teeth, claws on the wings, scales, and a long lizardlike tail (see Fig. 15-7). Huxley never suspected that soon after his model was drawn, a primitive bird would be discovered that corresponded to his postulated animal, even in many of the details of its anatomy. The only fossil evidence of a proto-bird prior to the discovery of the archeopteryx was an impression of a single feather collected in 1860. This impression did not yield a great deal of information other than the fact that it was from the Jurassic. Until the feather imprint was discovered, it was thought that birds had arisen during the Cenozoic, but this feather was three times as old as that. The bird from which it came existed during the age of reptiles. Hermann von Meyer, the man who had the feather, named this as yet undiscovered bird *Archaeopteryx lithographica* after the lithographic slate in Solnhofen where it was found. One year later, workers at Solnhofen came across a slate section with a nearly intact skeleton of this primitive bird (lacking only the skull).

A physician who collected fossils obtained the slate section at once and began offering it to various museums and scientific institutes, but under some rather unusual conditions. The man allowed scientists to take a brief glance at the specimen but not to make notes on it or to draw it. The result of this childishness was widespread belief among many naturalists (who had not seen the specimen) that the entire matter was a hoax. One paleontologist who looked at the specimen outsmarted the odd physician, for this scientist had an excellent memory and was a skilled drawer! Working for the Munich natural history museum, this paleontologist visited the physician several times to peek at the fossil. After each visit, the paleontologist rushed back to his hotel and drew what he could remember. Eventually he made a precise drawing of the specimen that was as informative as the fossil itself!

The animal thus depicted was the size of a pigeon and had long flight feathers on its fore limbs. The long tail also had feathers. Since this animal had a long tail (and birds of today all have very short tails), the zoologist who first classified the animal placed it with the reptiles and called it *Gryphosaurus problematicus*.

Famed English paleontologist Richard Owen read the description of the animal, and he quickly managed to buy the specimen from the physician who still owned it, as German museums were still bickering with the physician to obtain it for their own collections!

When Huxley saw the specimen, he realized it was the missing link between reptiles and birds he was certain existed. Darwin was also excited over the find, because the archeopteryx supported his theory of evolution which at that time was still controversial. One problem was the missing skull of the archeopteryx specimen. If *Archaeopteryx* had a toothless beak, there would be no doubt at all that it was a bird, albeit a very primitive one. If it had teeth, it would have to be classified as a transition form from the reptiles to birds.

Another archeopteryx was found near Solnhofen sixteen years later, and this one included the head. The head and upper leg showed traces of scales, while feathers covered the rest of the body. The feathers on the long tail were particularly large. The animal had a beak-shaped mouth indeed, but there were reptilian teeth inside resembling those in thecodonts.

Amazingly, this second specimen got into the hands of the family of that same physician and again was the subject of a great deal of speculation

Fig. 15-8. Life depiction of *Archaeopteryx lithographica*.

Fig. 15-9. In 1956, this (the third) archeopteryx was found near Solnhofen.

▷
A skeleton of the well-known dinosaur, *Tyrannosaurus rex*, from Upper Cretaceous Montana (American Museum of Natural History).

DINOSAURS WITH HORNS

Fig. 15-10. Skeleton of the archeopteryx (*Archae-opteryx*), compared to a modern dove; with compared parts shown in black. In the modern bird the occipital portion of the skull is extended, and the flight bones are partially fused. Also, the pelvis has developed into a firm structure. The sternum has greatly enlarged and is the point of attachment for the flight muscles. The ribs are broader, and the bony tail has become much shorter.

◁
Two skeletons of hadrosaurs (*Anatosaurus nnectens*) from Upper Cretaceous North America (American Museum of Natural History).

and bargaining. Finally the great German industrial firm Siemens bought the specimen, ensuring that this one would *not* be lost to Germany, and to this day it is on display at the mineralogical museum of the Berlin University.

These two birds were the subject of different phylogenetic interpretations. The Berlin bird was called *Archaeornis siemensi* owing to differences in the coracoid bones and the pelvis. It was thought to be ancestral to flying birds, while the London *Archaeopteryx* was allegedly the ancestor of running birds. On the basis of more recent, thorough studies it is generally agreed that the two are one and the same genus and species.

A third archeopteryx was not found for another 60 years (1956), when a specimen was dug up just 200 m from the place where the first one had been found. This bird (see Fig. 15-9) also lacks a skull, but the limb bones and feathers have been particularly well preserved. A fourth *Archaeopteryx* was recently found in a collection where it had been unrecognized as such for a long time. That completed the current fossil record of these primitive birds.

No transition forms between saurians and the archeopteryx are known. It seems to have burst forth fully developed. However, what *seems* to be true, as we often find in biology, is not actually the case! The structure of the feathers, which arise from the same anatomical origin as reptilian scales, show signs of a long phylogenetic heritage. Paleontologists have simply been unable to find the intermediate forms between archeopteryx and its saurian precursors. Actually, since these precursors would have been terrestrial animals with delicate skeletons, we would not expect them to fossilize very readily, and it is not surprising at all that they have not been found. We can only hope that there will be an extremely lucky find some time in which one of these precursors was indeed preserved as a fossil. This case is yet another example of how fragmentary the fossil record is. What so often appears to be the sudden flourishing of an animal group is almost always a simple lack of the relevant fossil specimens demonstrating the gradual growth of these groups. Uninformed critics of evolution who claim that the theory does not explain the sudden appearance of many animal groups rarely appreciate this fact.

In spite of the great gaps in the fossil record, the archeopteryx shows beautifully how arboreal, delicately built saurians, who perhaps were gliders as well, could have made the transformation to flight (and hence the first birds). The toothed jaws of this bird are reptilian, as are the brain, the free metacarpals and metatarsals, and the tail. In addition to the feathers, a major avian feature is the fusing of the clavicles into a furcula. Studies on the third archeopteryx specimen have shown that the vertebrae and long hollow bones are definitely avian in nature. The sternum has no ridge (unlike all modern birds), and in this respect *Archaeopteryx* is less birdlike than the flying saurians.

If archeopteryx had no feathers, which force us to classify the animal as a bird, there could be a serious debate on whether the avian or the reptilian features predominate in this species. Furthermore, if *Archaeopteryx* died out without leaving descendants, it would have to be classified as a member of an unusual saurian order. Actually, the archeopteryx was the first member of a new animal class. They inhabited the forests of the Upper Jurassic and moved about by climbing as well as flying. Their flight ability, which we can evaluate from the structure of the fore limbs and the brain, was not nearly as well developed as in flying saurians. This leads us to ask now this species could survive and what sort of selective advantage would enable it to compete successfully against pteranodons and others.

Like the hand of a modern bird, the archeopteryx hand has just the first three fingers. In modern birds these fingers are incorporated into the flight apparatus. This, too, is true of *Archaeopteryx*, although the fingers bore sharp claws. The ancestors of the archeopteryx must have been climbing reptiles, not runners who lifted themselves into the air after a running start. Evolution toward avian species probably began in the Triassic, if not in the Upper Permian. The ancestor of birds was, as one American paleontologist has put it, a "climbing and jumping Pro-Avis". As feathers developed from the reptilian scales of these "pro-avids", the decisive distinction between flying reptiles and birds was made. The selective advantage of feathered flyers which opened up the future to birds in contrast to flying saurians was the better flying technique that can be achieved with feathers. Their flight technique was basically a superior one, and even though *Archaeopteryx* could not move through the air nearly as efficiently as pterodactyls, *Archaeopteryx* had the anatomical predisposition to a much more successful means of aerial locomotion. It was this that gave them the selective edge over all the flying saurians and which ultimately led to the extinction of flying saurians and proliferation of avian species.

Another great advantage of feathers was the insulation they offered, which was certainly more efficient than the hair covering of flying saurians. Anyone who has slept under a down comforter or worn a down jacket in the middle of winter knows how superb feathers are as insulation. The archeopteryx' primary feathers were very loosely attached, and the resultant flapping prevented them from being the efficient flyers modern birds are with their feathers firmly attached to a substrate. During evolution, one of the three archeopteryx fingers was modified into a base for the feathers, and this development was subject to such selective pressure and was hence so advantageous it occurred in every avian line. The flight membrane of flying saurians, which was attached to a single finger, could not compete with this avian adaptation of true wings. Recent studies have shown, incidentally, that *Archaeopteryx* was a very active flyer, even though it was only capable of rather awkward wing movements.

Fig. 15-11. Wing of *Archaeopteryx lithographica* with 9 primaries and 14 secondary feathers (reconstruction). The feathers are more efficient for flight than the flight membrane of the flying saurians and mammals (i.e., bats). In most accidents in which the bones are uninjured only one or two feathers are lost, and the hole created by the absence of that (those) feather(s) is filled in by the adjacent ones.

Another great gap exists between *Archaeopteryx* and the toothed Cretaceous birds (see Chapter 16). These birds invaded numerous new habitats, and some of them even lost the ability to fly. The first modern birds began appearing toward the end of the Cretaceous.

The evolution of flying mammals (i.e., BATS; order Chiroptera) began with as yet unidentified insectivores in the Cretaceous, and the first fully formed bats appear in the Lower Eocene. We shall mention them briefly for comparative purposes. Although bats are geologically a very old group, there have been and still are many bat species, a sign that they have been a highly successful group. They largely avoided competition with birds by being active at night and at dusk. Some birds, such as owls, are indeed active in dim conditions, but bats can forego light completely because of their echolocation: no bird can do this.

No gliding mammalian species is known that would serve as a model for the development of bat flight, although we assume that the characteristic flight membrane as a parachute was an earlier development. The fore limbs of bats, however, have been modified into true wings and are not for gliding alone. The upper and lower arm bones, and particularly the metacarpals and all the fingers except the thumb, are extremely elongated and support the elastic flight skin (or chiropatagium), which surround the fingertips, the tarsals, and the tail (either partially or completely). The role of the hand has given bats their scientific name, for Chiroptera literally means "hand fliers" (from the Greek χειρο = hand and πτερόν = feather).

It is inaccurate to characterize bat flight as a fluttering, for bats engage in true flight, and bats can execute vertical and circular flight patterns just as birds do. Some bats are very speedy fliers; they have long, narrow wings. Slower bats have broad wings, and some can even hover like hummingbirds. In technical terms they are no less able fliers than birds, and their prevalence over millions of years is itself evidence of the success of their locomotive pattern.

Although it is thought that the ancestors of bats appeared at the end of the Mesozoic, transition forms between climbing insectivore ancestors and fliers have not been found. However, there is a fossil bat record, of which one superb example is the intact skeleton of the Lower Eocene *Icaronycteris index* from Wyoming. It has features of the two modern bat suborders, fruit bats and insectivorous bats, which is evidence that these two lines did not develop independently.

The great diversity of flying insects and modern birds and the variety of their adaptations to various habitats, as well as the number of species of modern bats, are all evidence of the significance of the conquest of air. Oceans and high mountains were no longer distribution boundaries for flying terrestrial animals. The impressive distances covered by some of our modern migratory birds and the annual migrations undertaken by some butterflies and bats show that animal species can traverse vast dis-

ig. 15-12. Comparison f the wings of a ptero-actyl and a bat. The ourth finger in the ptero-actyl was greatly elon-ated and supported the ight membrane, which xtended to the body and ne hind limbs. This skin cts as a sort of parachute. he other fingers, unlike a birds, were not part of ne flight apparatus but vere probably used for anging from trees.

tances not only in water but also in the air. Some of these extend halfway around the earth. The development of flight is comparable in its far-reaching significance to the first steps taken on land by Paleozoic coelacanths.

▷
Skeleton of the 4 m high iguanodon (*Iguanodon bernissartensis*), a bipedal Lower Cretaceous ornitho-pod found in a coal mine in Bernissart, Belgium (Senckenberg, Museum, Frankfurt, Germany).
▷▷
Skeleton of *Diplodocus carnegii* from the Morrison deposit of Upper Jurassic Utah. In the background is the skeleton of one of the most familiar dinosaurs, the triceratops (*Triceratops prorsus*), this one from Upper Cretaceous Wyoming (Smithsonian Institution).

By E. Thenius

Above: Complete skeleton of a small, short-necked plesiosaur (L about 3 m) from the Niobrara deposits in Cretacean Kansas.
Middle: A stegosaur skeleton (*Stegosaurus stenops*) from Morrison Upper Jurassic North America (American Museum of Natural History).
Below: The coelacanth *Latimeria chalumae*, a species still found today; it was discovered in 1938 in the western Indian Ocean (Senckenberg Museum, Frankfurt, Germany).

16   The Cretaceous

The Cretaceous is the most recent period in the Mesozoic Era. The animal and plant groups arising in this period characterize not only the Tertiary and glacial biota but also include those found in the present. New animal groups included the higher, placental mammals (subclass Eutheria), modern birds (subclass Neornithes), modern teleost or bony fishes (superorder Teleostei) and colonial insects, while the Cretaceous flora included flowering plants or angiosperms (division Angiospermae).

Dating

According to radiometric data, the Cretaceous period began about 135 million years ago and ended 65 million years ago; thus, it lasted about 70 million years. Small differences in various dating techniques make it difficult to draw precise borders between the Jurassic and Cretaceous and between the Cretaceous and the Tertiary. The Cretaceous is divided into several stages, and again using the system employed by Tasch in his *Paleobiology of the Invertebrates* (1973) as we did in the Jurassic chapter, these stages are the following, from oldest to youngest: Infravalanginian, Valanginian, Hauterivian, Barremian, Aptian, Albian, Cenomanian, Turonian, Coniacian, Santonian, Companian, and Maestrichtian. The border between the Lower and Upper Cretaceous is drawn at the Albian-Cenomanian boundary. Differences of opinion exist among paleontologists about the geochronology of the Cretaceous, and the appearance and extinction of some ammonite groups, to cite just one example, may be listed by some authors as occurring in the Jurassic and by others as taking place in the Cretaceous.

Origin of the name

The origin of the word Cretaceous is the Latin *creta* = chalk and *aceus* = of the nature of. White, chalky deposits, consisting almost entirely of calcium carbonate (i.e., limestone), are characteristic of the Cretaceous. Many of the deposits are from flinty concretions arising in part from the fossil skeletons of siliceous sponges that produced shelf sea deposits, which were in time enclosed by calcium carbonate. Tiny (microscopic) shells of unicellular coccolithinean algae, which were Cretaceous planktonic organisms, contributed to Cretaceous formations. After they died, these algae sank to the ocean floor, where their bodies decayed and their shells remained behind to produce vast deposits in the form of limestone. Naturally, limestone deposits do not characterize the entire Cretaceous period. As in previous periods, the Cretaceous system also includes various sedimentary (stratified) rocks originating from deep-sea bituminous deposits.

The Upper Cretaceous in central Europe is particularly characterized by stratified lime marl and sandstone. These were originally deep-sea deposits, but during the mountain formation period they were tectonically modified and were shifted about and dragged onto land. Thus, these deposits do not occur at the site of their original habitat nor are they by any means at their original depth.

The Upper Cretaceous experienced vigorous mountain building activity. The largest, highest mountains of today were formed during this period and as a result of later tectonic activity in the Lower and Upper Tertiary. The Alps, Carpathians, Caucasus, Himalaya system, the Rocky Mountains, and parts of the Andes were all formed during this time. The deep-sea origin of stratified lime marl deposits is of great importance in understanding the paleogeographic situation and for appreciating the evolution of Cretaceous fauna and flora.

Mountain formation

During the Cretaceous, as geophysical, oceanographic, geological, and paleontological evidence indicate, the great Gondwana continent split into individual continental masses. South America and Africa were part of one land mass only during the Lower Cretaceous. The fact that they were joined has been deduced from evidence such as the resemblance between fresh-water ostracods from deposits in northeastern Brazil and western Africa. Transgression by the sea, which led to the cleavage of South America and Africa, occurred later in the Cretaceous. Also during the Lower Cretaceous, the sea invaded the region between eastern Antarctica and southern Australia, and during the subsequent Lower Cretaceous South America separated from Africa and from the Australo-antarctic continental block.

Paleogeographic situation

This means that only during the Lower (in fact the Lowest) Cretaceous could animals and plants have passed between Africa and South America and between Australia and the Antarctic. Since there may have been an island chain linking South America and the Antactic, fauna and flora could conceivably have migrated through this route.

Evidence for such migrations and hence for the existence of a unified southern continent exists in rather substantial amounts. There are remarkable similarities between the fishes, amphibians, and reptiles of Africa and South America, similarities known to be based on phylogenetic affinities. Examples occur in the osteoglossids (family Osteoglossidae), lungfishes (Lepidosirenidae), tongueless frogs from the suborder Aglossa (Surinam toads in South America and clawed toads in Africa), and the pelomedusid turtles (Pelomedusidae). The disjunct (i.e., discontinuous)

Zoogeography and the former southern continent

Several Cretaceous gymnosperms (Gymnospermae):

Fig. 16-1. *Araucaria toucasi.*

Fig. 16-2. *Baiera brauniana.*

Fig. 16-3. *Cycadeoidea marshiana.*

distribution of modern marsupials (opossums and rat opossums in the Americas and all the other families in Australia) is likewise explained by postulating the existence of a unified southern continent that was subjected to subsequent cleavage. The distribution of the above marsupials does not conform to a pattern extending across southeastern Asia but one extending through Antarctica. The possibility of an antarctic route is also supported by the astonishing relationships among various insect groups (e.g., beetles, cicadas, non-biting midges, stoneflies, ephermeral flies and scorpion flies). These insects occur in southern South America and in Tasmania, Australia and New Zealand, and in spite of the vast separation involved, these insects are more closely related to each other than to other insects found in their own continents. The distribution of many plants (e.g., *Araucaria, Podocarpus* and *Nothofagus*) can similarly only be explained by assuming the former existence of Gondwana (see Color plate, p. 76).

Another important factor in appreciating Cretaceous fossils was the climate of that time. Evidence for climatological conditions has been gained from various fossils indicative of specific climatic conditions (various plants, coral reefs, ground formations, etc.), and these fossils act as a kind of "fossil thermometer". They yield information about marine temperatures and other prevailing weather conditions. The plentiful flora, the presence of reef-building coral and large, often thick-shelled and reef-building mollusks (like rudistids from genus *Hippurites*), snails and many species of reptiles are all indications that Upper Cretaceous central Europe was primarily tropical in character. One could interpret this climate as a result of a migration in the earth's poles, but it is more likely that there were no glacial periods in the Cretaceous (or in the Jurassic and Triassic either, for that matter). The evidence suggests that there were no polar caps during the Cretaceous and therefore no cold periods anywhere. Glacial periods will be given extensive treatment in Chapter 19 of this volume.

Numerous Cretaceous fossil plants—in the form of leaves, fruit, stems spores, and pollen—are signs of the proliferation that occurred in flora during this period. The Lower Cretaceous was the scene of one of the most decisive changes in the world's plants. It was a time when the prevailing plants were conifers, such as araucarians and others, and ginkgos (*Ginkyoites* and *Baiera*; see Figs. 16-1 and 16-2). Bennettitean palm ferns (*Cycadeoidea* and others; see Fig. 16-3) and gymnosperms belong to the Mesophytic, but they first appeared in the Upper Cretaceous in the form of angiosperms (division Angiospermae), and the great diversity of these plants is a strong indication that the botanical transformation into the Cenophytic era (the floral equivalent of the Cenozoic, or recent era, in animals) was underway in the Cretaceous and was in fact realized during that period (see Color plate, p. 74/75).

The "sudden" and also prevalent appearance of angiosperms in the

Lower Cretaceous has long been a problem to botanists. There are two major interpretations of this phenomenon. According to one, angiosperms developed during the Lower Cretaceous from as yet unknown predecessors. According to the other viewpoint, angiosperms arose either in the Lower Mesozoic or even in the Upper Paleozoic. The absence of fossil angiosperms in pre-Cretaceous deposits is explained by the fact that they appeared at higher elevations or in arid regions, and in neither situation would they have fossilized. The alleged Jurassic angiosperm fossils (*Homoxylon, Sanmiguelia, Sassendorfites* and *Eucommiidites*) are not convincing proof that this plant group existed prior to the Cretaceous, for all these plants survive in the form of wood or leaf fragments or as pollen grains with angiosperm*like* features.

Interestingly, plant evolution shows that several groups of plants evolved toward the angiospermate condition (angiosperms being plants whose seeds are enclosed in an ovary; from the Greek $\alpha\gamma\gamma\epsilon\iota\sigma\nu$ = vessel and $\sigma\pi\epsilon\rho\mu\alpha$ = seed). Plants often developed means to have sessile large female spores (macrospores) to protect the germinating seeds. Receptacles were sometimes formed in the wood, and even leaf structure (in terms of clefts and venation) are signs of achieving a slight degree of independence from the external environment. All these adaptations were subject to favorable selective pressure and led to the proliferation of plants possessing them. Similar trends have been found in some palm fern species (bennettiteans and Caytoniales) from the Jurassic, and this has led to designating the bennettiteans as hemiangiosperms.

Upper Cretaceous angiosperms include plants that are now extinct (e.g., genera *Credneria, Dryophyllum* and *Dewalquea*; see Fig. 16-4) but also some still alive today, such as sassafras (see Fig. 16-5), figs, ash, viburnum (*Viburnum*), grapevines, poplars, *Pandanus, Pistia*, wax myrtle (*Myrica*), arrowroot (*Canna*), willows and others. Since all we have of these are leaf fragments, we do not have a basis for assigning these plants to present-day genera with absolute certainty. All of them are woody plants that are relatively primitive and are near the base of the plant world. It was formerly assumed that the eucalyptus and other Australian plants lived in Europe during the Upper Cretaceous, but so far this has not been confirmed with fossil evidence.

To evaluate the phylogenetic development of Cenozoic plants (which occurred in the Upper Cretaceous), paleobotanists use pollen grains because they not only arise from swamps and fresh-water deposits but also from marine deposits and arid regions (saline lakes). It is very difficult to identify isolated fossil spores and pollen (i.e., those not within spore cases). Flora of Cretaceous seas included the prevalent red algae lithothamnians (*Archaeolithothamnium*) and the coccolithineans, which were just beginning to proliferate. These calcareous ostracods were planktonic organisms that contributed to limestone deposits, as we mentioned earlier in this chapter. Their skeletons are microscopic and are composed of calcite

Several Cretaceous angiosperms (Angiospermae):

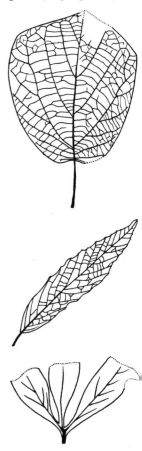

Fig. 16-4. The extinct species *Credneria triacuminata, Dryphyllum subfalcatum* and *Dewalquea trifoliata*.

Fig. 16-5. The still extant sassafras laurel (*Sassafras progenitor*).

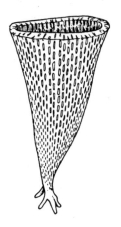

Fig. 16-6. *Ventriculites striatus*, a porifer.

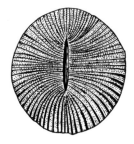

Fig. 16-7. The anthozoan *Cyclolites*.

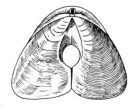

Fig. 16-8. *Pygope junitor*, a Lower Cretaceous brachiopod, relatives of which have survived into the present.

layers, but the massive prevalence of these ostracods made them a significant contributor to Cretaceous limestone formation.

These were the botanical conditions prevailing during the Cretaceous, and they are important since only by also considering plants can one understand Cretaceous animal evolution. We shall now discuss the animal groups that arose and how they developed during the Cretaceous period, following the same zoological order we have used in previous chapters.

The most prevalent UNICELLULAR ORGANISMS OR PROTOZOA (subkingdom Protozoa) were foraminifers (order Foraminifera). There were small and large foraminifers and the large ones played some role in rock formations. Various Upper Cretaceous deposits were formed from the fossil skeletons of large foraminifers. Some of these deposits (e.g., the orbitoid sandstone from the late Upper Cretaceous) are named from the foraminifer group of which they are composed. Orbitoids and alveolines were both active foraminifer groups in deposition formation. The large size of these protozoa and the diversity of protozoan species indicate that near-tropical conditions prevailed when they lived. Some of the small planktonic foraminifers globotruncanes, (globorotalians and globigerines) were important key fossils (see Chapter 3 for an explanation of the term key fossils). They have been used, as have the large foraminifers, to trace the course of individual foraminifer groups throughout the Cretaceous. In the Lower Cretaceous Tethys sea the calpionellens (see Chapter 14) were evolving rapidly and therefore have been used as key fossils in that habitat. These microfossils had a cup-shaped calcite shell, and since calpionellens lived at the same time as the tintinnids (e.g., *Tintinnopsella*), it is thought that the two groups are related.

PORIFERS OR SPONGES (phylum Spongia) were not at all uncommon in shallow marine deposits (including such genera as *Ventriculites*, *Coeloptychium* and *Barroisia*; see Fig. 16-6), but they were not as significant as they had been in Jurassic seas. *Barroisia* and *Cryptocoelia* were the last sphinctozoans (order Sphinctozoa).

Among COELENTERATES (subkingdom Coeleneterata, phyla cnidarians and comb jellyfishes), the only Cretaceous species we have are of course those that fossilized. It would be impossible for a great many coelenterates (such as polyps and medusas) to be preserved, and these have been lost. The most commonly encountered Cretaceous coelenterates are anthozoans (class Anthozoa), of which the most prevalent were madreporarians (order Madreporaria), which occurred is isolated forms (*Diploctenium* and *Cyclolites*; see Fig. 16-7) and colonial coral forms (*Actinastraea* and *Balanophyllia*). Some of these were descendants of Jurassic forms, while others were members of new phyla of skeletal corals, of which some of the most important are the siderastreids, cyclolithids, fungiids and the dendrophyllids. Vast reefs comparable to those from the Triassic and Jurassic are rare in the Cretaceous. Of the hydroids (class Hydrozoa), the hydroid polyps (Hydroidea) replaced the stromatoporoideans (Stromatoporoidea) and ellipsactinoideans (Ellipsactinoidea).

BRACHIOPODS (class Brachiopoda), which had been so prevalent during the Jurassic, were receding in the Cretaceous. The rhynchonellids (family Rhynchonellidae, subclass Testicardines or Articulata), which had been so prevalent during the Jurassic, became rare in the Cretaceous period (*Peregrinella* and *Cyclothyris*). The terebratulaceans (*Terebratulina* and *Pygope*; see Fig. 16-8) and terebratatellaceans (*Kingena* and *Magas*) were sparsely represented. Members of these last three families have survived into the present, however. One genus in the subclass Ecardines, *Crania*, is still extant. The only prevalent brachiopod group during the Cretaceous period were the thecidiids (family Thecidiidae), a Testicardines group found in Upper Cretaceous seas.

BRYOZOANS (class Bryozoa) became numerous and diverse during the Upper Cretaceous, and cheilostomates (order Cheilostomata) were particularly prolific. The order includes the suborder Anasca and the genera *Flustrellaria*, *Lunulites*, *Ctenopora*, *Lagynopora* and others. Ascophorans (suborder Ascophora) appeared for the first time; these genera (e.g., *Beisselina*, *Dacryoporella*, *Fusicellaria*) had water sacks. In terms of number of genera, the most prolific bryozoans in the Cretaceous were the cyclostomes (Cyclostomata), which reached their zenith during the Cretaceous.

MOLLUSKS (phylum Mollusca) of this period represent further evolution of those groups present in the Jurassic. This was the period of large inocerames (*Inoceramus*), "reef-building" rudistids and various ammonite forms. These were the three major Cretaceous molluscan groups. Gastropods or snails (class Gastropoda) were almost exclusively represented by prosobranchiates (*Prosobranchiata*= "fore-gilled animals"). The euthyneuran (subclass Euthyneura) opisthobranchiates (Opisthobranchia= "rear-gilled animals") were chiefly in the form of actaeonellans (*Actaeonella* and *Trochacteon*); in parts of the Alps they occur in great masses. For the first time we also find members of the thecosomes (order Thecosomata), which are marine snails that drift in the water and include the modern genus *Clio*. Thecosomes are highly specialized opisthobranchiates in which the shell, the creeping sole and the lateral lobes of the foot have regressed, while the head lobes have been modified into winglike fins.

The most important streptoneuran (sublcass Streptoneura) snails were the taenioglossids (suborder Taenioglossa), with the cerithiid gastropods (Cerithiidae; genus *Cerithium*), the littorinoidean gastropods (superfamily Littorinoidea), the cypraeoidean gastropods (superfamily Cypraeoidea), the naticoideans (superfamily Naticoidea with the genus *Natica*) and finally the rostellarians (Rostellaria). Neogastropods (suborder Stenoglossa or Neogastropoda) also proliferated, and they produced during the Cretaceous the first muricid snails (family Muricidae with genus *Ocenebra*), buccinids (Buccinidae with genus *Sipho*) and pleurotomids (genus *Turris*). Some of the most striking gastropods were the

Bryozoans

Mollusks

Streptoneuran gastropods

nerineans (*Nerinea* and *Itieria*), which arose during the Jurassic and whose shells look somewhat like those of the cerithioideans (superfamily Cerithioidea). However, their massive shafts bear spindle folds, and the shell cavity is relatively small in comparison with the shell. Nerineans died out in the Upper Cretaceous. Diotocardian gastropods (suborder Diotocardia) included pleurotomarioideans (*Pleurotomaria*, *Bathrotomaria* and *Conotomaria*), which were still worldwide inhabitants of shallow seas; trochoid snails (*Trochus* and *Turbo*) and neritoid snails (*Neritopsis*) were common; and patellids (*Patella*) and others (e.g., *Haliotis*) are found for the first time with certainty.

Molluscan fauna are chiefly characterized by the large inocerames (*Inoceramus*), which arose from floating organisms. They evolved rapidly and have been used as key fossils for Upper Cretaceous marine deposits. The rudistids, however, were particularly diverse; and they were the most prevalent Cretaceous mollusks. Expecting a few species with a snaillike coiled shell (*Requienia* and *Caprina*), these were sessile mollusks appearing in such great numbers that they formed large banks. Their shell looks very corallike, with its firmly attached, cup- or horn-shaped ground valve and the free rooflike valve. Openings for the siphuncle (breathing tube) or pores and canals in the roof valve enabled these cosmopolitan rudistids to respire without opening their shells. Two representative rudistid genera were *Hippurites* (see Fig. 16-9) and *Radiolites*. The variability and diverse species of these rudistids are indications that these were highly successful sessile marine organisms. The rapid evolution of the shell, the development of septallike structures and the canal system in the roof valve, all of which developed in the Upper Cretaceous, can be interpreted as modifications facilitating respiration. It is somewhat surprising then, that this particularly diverse group should have become extinct in the Upper Cretaceous.

The first teredinids (family Teredinidae) appeared during the Cretaceous period. They were the other extreme in Cretaceous mollusks teredinids bored into drifting wood. They did not really proliferate until the Cenozoic era. Of the many other Cretaceous mollusks, we cite only the trigonians (*Trigonia*), which were distributed in every ocean in the world but which declined toward the end of that period. Among the taxodonts (suborder Taxodonta), the first glycymerids (*Glycymeris*) appeared, but they did not flourish until the Tertiary, and they are still found today. Freshwater mollusks (superfamily Unionoidea) included river mussels (*Unio*) and the first members of the genus *Margaritifera*. Others found today in brackish water (e.g., *Dreissena* and *Congeria*) did not arise until the Tertiary.

The most significant Cretaceous mollusks were the still highly prevalent CEPHALOPODS (class Cephalopoda), which we encountered previously in the nautiloids (subclass Tetrabranchiata), the ammonoids (order Ammonoidea) and the belemites (suborder Belemnoidea). Nautiloids were still found in warm seas (e.g., the genus *Hercoglossa*), but they were now

Floating and sessile mollusks

Fig. 16-9. The mollusk *Hippurites gosaviensis*.

First teredinids and other mollusk groups

rare. In contrast, the number of ammonoid species and the diversity thereof was much greater than in the Jurassic. The heteromorphic ammonites, whose acquaintance we made in the Upper Triassic and the Jurassic, became much more prevalent. The coiling of their shell is a secondary development from straight-shelled forms. In some of these the shell convolutions did not make contact with each other (*Crioceratites*); in others the shell spiral was replaced by hook-shaped arches (*Hamites* and *Ancyloceras*) or by a straight shell (*Baculites*; see Fig. 16-10). Some heteromorphic ammonites had snaillike coiled shells (*Heteroceras* and *Turrilites*; see Fig. 16-11) or a completely irregular conformation (*Nipponites*). Some of these modifications regressed partially or completely during the course of later evolution. The factors that gave rise to these heteromorphic shapes have been the subject of a great deal of discussion. Functional aspects of any causative factors for these shell forms have not been identified. Ammonites become extinct at the end of the Cretaceous.

Evolution altered not only the shape of the ammonite shell but its sculpturing as well, including such important systematic features as the suture line, the mouth, the cross-sectional configuration, and the size of the shell. Since these changes occurred relatively quickly in an animal group composed of swimming marine organisms that were widely distributed, ammonites have been superb key fossils for the Cretaceous. Some of the more significant among the many of them are the ancylocerates (*Crioceratites, Ancyloceras, Heteroceras*), scaphites (*Eoscaphites, Scaphites, Hoploscaphites*), desmocerates (*Desmoceras, Puzosia, Holcodiscus, Pachydiscus*), holplitaceans (*Hoplites, Douvilleiceras, Leymerella, Sonneratia, Schloenbachia*) and the acanthocerates (*Ancanthoceras, Mortoniceras, Mantelliceras, Tissotia, Barroisiceras*). In general the shell sculpturing becomes more complex, but in addition it also acquires various nodule and spine structures (in *Douvilleiceras, Mammites, Acanthoceras, Collignoniceras, Ancyloceras*). Sometimes a simplification in structure occurs, something which in a few lines (e.g., *Tissotia*) also happens to the suture line, which can be modified secondarily from a complex ammonitic pattern to what is called a pseudoceratitic pattern. Giant Upper Cretaceous ammonites with a shell diameter of nearly 2 m have been found (e.g., *Pachydiscus seppenradensis*). While Jurassic ammonites displayed climatically controlled zoogeographic differentiation, evidence for similar faunal differentiation has not been found for the Cretaceous, although it is entirely possible that such differences did indeed exist.

In the cephalopods lacking an external skeleton (subclass Endocochlia or Coleoidea), which are also known as dibranchiates (Dibranchiata), the most important phylogenetic event during the Cretaceous is the single find of an Upper Cretaceous octopus (*Palaeoctopus*). Belemnites continued to be the most significant dibranchiate group, and there were many different belemnites (e.g., *Duvalia, Belemnitella, Actinocamax, Belemnella*). Practically all we have of them are their massive rostrums. Belemnites are important

▷
Crown of the crinoid *Encrinus liliiformis*. The ten-armed cup has a stalk composed of coin-shaped trochites. The stalks attained lengths of up to 1.5 m and were anchored to the floor or to a mollusk shell with rootlike adhering disks. However, the stalks could lose their hold on the substrate. Since the skeletons of crinoids usually decay, this nearly intact example of especially rare and hence valuable. It is from Middle Triassic Germany (Paleontological Collection of the University of Tübingen).

▷▷
Fossils from the foot of the Swabian Alb, Germany.
Above left: one of the Posidonian slate deposits in Württemberg;
Above right: Colony of the crinoid species *Seirocrinus subangularis* from Lower Jurassic Germany. The crowns of this species reach a diameter of nearly 1 m, while the stalks were up to 15 m and usually adhered to drifting tree trunks (Museum Hauff, Holzmaden, Germany);
Below: A Hauff museum worker prepares a crinoid specimen using a hammer, chisel, and a graver.

Key fossils: belemnites

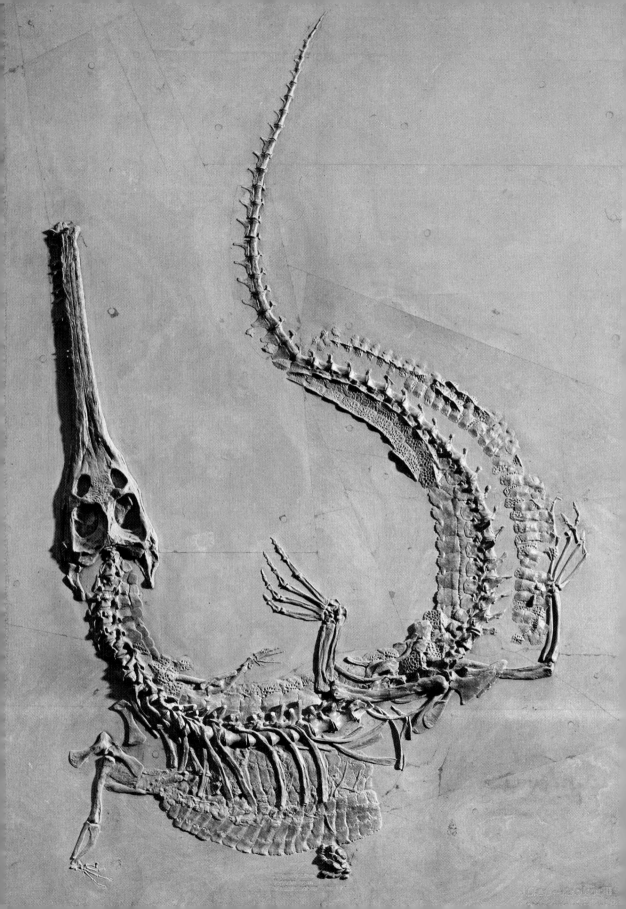

key fossils for the Upper Cretaceous due to their rapid evolution at that time. A few true squids also lived at that time (suborders Sepioidei and Teuthoidei), but we know less of these than we do of the Jurassic forms.

The collective term "WORMS" incorporates several different phyla and almost exclusively refers to animals lacking any hard parts and which therefore do not normally fossilize. The only signs of their existence in earlier periods (such as the Cretaceous) are their living chambers, tracks, and other indirect evidence. Various "worm" chambers have been described as chondrites and fucoids, while their tracks are known as helminthoideans. Helminthoideans are meandering spoor. U-shaped passages, known as *Rhizocorallium*, are found in shallow sea deposits, but they were probably created by crablike arthoropods.

Articulata

We come next to the ARTICULATA (phyletic group Articulata), the articulated animals. Evidence of annelid worms exists in the form of tracks, living chambers and similar spoor. Crustaceans comprise most of the Articulata finds. The ostracods (sublcass Ostracoda), which are classified as "lower" crustaceans, are extremely important microfossils. They are particularly prevalent in marine deposits dating from the Cretaceous period. Important ostracods were members of the orders Myodocopida and Podocopida, the latter found almost exclusively in fresh-water deposits.

Cirripedia or barnacles (subclass Cirripedia) are represented during the Cretaceous almost solely by the relatively primitive goose barnacles (suborder Lepadomorpha; *Archaeolepas, Euscalpellum*). Encrusting barnacles (suborder Balanomorpha) make their first appearance with the chthamalids (family Chthamalidae, genus *Catophragmus*). Verrucomorphs also existed then (they include the modern genus *Verruca*). Decapods (order Decapoda) were the most prevalent "higher" crustaceans. Other relatively common crustaceans included spiny lobsters (*Palinurus, Glyphea, Eryon*), isopods (family Sphaeromatidae), anomurans (suborder Anomura) in the form of *Callianassa*, hermit crabs (*Paguristes*), the crab genus *Galathea*, and the astacurans (suborder Astacura) with the long-tailed crabs (*Eryma, Palaeastacus* and *Astacus*), and especially the great crab suborder Brachyura (with the genera *Prosopon, Pithonoton, Homolopsis, Cenomanocarcinus* and *Raninella*). Although there were many higher crustaceans (all in subclass Malacostraca) in the Cretaceous, their actual flourishing occurred in the Cenozoic.

The remains of arachnids, myriapods and insects from the Cretaceous are only sparsely available. This means that our understanding of these three groups is still quite meager. This lack of fossils is chiefly due to the fact that all the above groups are terrestrial animals and would hardly be represented in marine deposits. Amber and fossil-bearing coal have been found at only a few Cretaceous sites, and in some cases the fossils in these materials have not yet been closely examined. We might mention that the earliest flea known was dug up in Lower Cretaceous Australia. Prior to this find, the oldest fleas were found in Tertiary strata. A few remains of

◁◁
Ichthyosaurs from Holzmaden, Germany.
Top: *Stenopterygius quadriscissus*;
Middle: A member of the same genus, with fossilized skin;
Bottom: *S. crassicostatus*. This is a female with five embryos beneath the ribs; just above and to her right is a newborn juvenile (Museum Hauff, Holzmaden, Germany).

◁
The Lower Jurassic crocodile *Steneosaurus bollensis* (Holmzaden, Germany), whose skeleton has been preserved in amazing condition, even showing very fine features. The animal was pressed flat, moving the dorsal and ventral armor. Other distinct features include the two large temporal fossae (openings) behind the eyes and stomach stones (below to the left of the body) used for digestion. L=3 m. (Paleontological Institute, Tübingen University, Germany).

king crabs (order Xiphosura; also called horseshoe crabs) are evidence of the existence of these chelicerates during the Cretaceous. Anatomically they are intermediate between the Jurassic genus *Mesolimulus* and the modern genus *Limulus*. They are also proof that only one line now exists of what during the Paleozoic was such a vast animal group.

All modern ECHINODERM (phylum Echinodermata) classes were represented in the Cretaceous. Among the crinoids (class Crinoidea), there were both sessile, stalked forms (*Metacrinus*, *Rhizocrinus*) and stalkless planktonic crinoids (*Uintacrinus* and *Marsupites*; see Figs. 16-12 and 16-13), whose fossilized remains often occur *en masse* in shallow sea deposits. They were not members of the feather stars (order Comatulida), for feather stars are the only planktonic crinoid group in the Cenozoic (albeit a very diverse group). Sea urchins (class Echinoidea) are present with a wealth of different species, of which the so-called irregular sea urchins were not uncommon. The "regular" sea urchins of the Cretaceous included *Echinothuria*, *Cidaris*, *Heterodiadema* and *Salenia*, while the "irregular" forms included *Galerites*, *Archiacia*, *Cassidulus*, *Hemipneustes*, *Echinocorys*, *Toxaster* and *Micraster*. These "irregular" sea urchins burrow in soft substrates and have developed some remarkable adaptations for this kind of life, including degeneration of the spine covering and the chewing apparatus, the translocation of the anus and bilateral shell symmetry. These features distinguished them from the "regular" sea urchins. There are few remains of starfishes and serpent stars, and all that remain of Cretaceous holothuroids (class Holothuroidea) are their tiny calcite sclerites.

Other invertebrates do not occur in Cretaceous strata, nor do jawless vertebrates (superclass Agnatha). However, all other vertebrates are well represented. These include the cartilaginous fishes, the teleost or bony fishes; amphibians, reptiles, birds and mammals. The Cretaceous fossil record of these vertebrates has cast considerable light on the evolution of these different lines.

A kind of modernization took place among CARTILAGINOUS FISHES (class Chondrichthyes): rays developed further, and primitive sharks such as the old hybodontids (Hybodontidei) were displaced by modern genera such as mackerel sharks (*Lamna*), mako sharks (*Isurus*), great white sharks (*Carcharodon*) and others. A number of primitive sharks still held on, although some became extinct later; those that held on included such genera as *Hexanchus*, *Scapanorhynchus* and *Squalicorax*. Their teeth show that these were true predators. Whale sharks (family Rhincodontidae), which are filter feeders that consume algae, are still absent. The ptychodontid rays (*Ptychodus*) were distributed throughout the world during the Upper Cretaceous. Their placodonlike teeth are adapted for mashing food, just as we find in the modern eagle rays (family Myliobatidae), although there is no direct relationship between the two groups. Ptychodontid dentition is composed of individual teeth, while eagle rays have tooth plates. Specialized rays, like devilfishes and electric rays, are not known in Cretaceous

"Heteromorphic ammonites"—Triassic and Jurassic cephalopods that proliferated greatly during the Cretaceous. Their planospiral coiling was altered during the Cretaceous:

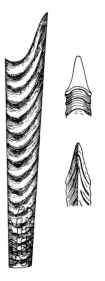

Fig. 16-10. *Baculites anceps*, an elongated ammonite. Left: the longest portion of the shell is straight (the coiled juvenile shell is rarely found); right: inner and ventral view of the head end.

Fig. 16-11. A snail-shaped ammonite: *Turrilites costatus*.

strata. The first sawfishes (family Pristidae) appear in this period, and different genera within the sawfish family (e.g., *Onchopristis* and *Sclerorhynchus*; see Fig. 16-14) are distinguished by the development of the sawlike structure.

The most significant BONY OR TELEOST FISHES (class Osteichthyes) are the spiny-rayed fishes or actinopterygians (subclass Actinopterygii) with their ganoid fishes and "true" teleost fishes. Ganoid fishes (super-orders Chondrostei and Holostei) of course were giving way to the more advanced teleost fishes, but they also became more "modernized" and have managed to survive into the present. The modernization process began in the Cretaceous and was completed in the Cenozoic. The cosmopolitan Lower Cretaceous ganoid fishes included sturgeon (genera *Acipenser* and *Paleosephurus*) along with the more primitive genera *Indaginilepis* and *Coccolepis*; Cretaceous ganoid fishes occurred not only as the primitive genera *Lepidotes*, *Belonostomus* and *Pycnodus* but also as the modern gar (*Lepisosteus*) and the bowfin (*Amia*). Numerous teleost fish groups make their first appearance in the Cretaceous, including the tarpons (Elopiformes), herring (Clupeiformes), osteoglossids (Osteoglossiformes), salmonids (Salmoniformes), codfishes (Gadiformes) and primitive perchlike species (Perciformes).

Flying fishes

Two interesting fish groups found in the Cretaceous were the flying fishes (*Chirothrix*), which evolved from the lanternfish group, and the marine giant osteoglossids (*Xiphactinus* = *Portheus*). These highly specialized forms are exceptions to the rule that most Cretaceous teleost fishes were primitive spiny-rayed fishes. There are practically no finds of carp or catfishes (Ostariophysi). Only additional studies can show whether these specialized fishes underwent adaptive radiation at a later time. Fossils of polypterids (Polypteriformes) dating from Cretaceous Africa vary little from modern forms. Lungfishes belonging to the modern genus *Neoceratodus* have been found in Australian Cretaceous strata.

The COELACANTHS, LOBE-FINNED FISHES or CROSSOPTERYGIANS (order Crossopterygii), which have proven to be of such great interest because of their significance in the evolution of terrestrial vertebrates, are uncommon in the Cretaceous but have been repeatedly found in shallow sea deposits (e.g., *Macropoma* and *Mawsonia*). These are members of the suborder Coelacanthini that were found before the modern coelacanth, *Latimeria chalumae* (see Color plate, p. 356), was discovered in Africa in 1938. Until the modern coelacanth was found, of course, it was universally believed that the group had become extinct! The absence of coelacanths in the rich Tertiary formations is understood as a result of these fishes migrating into greater depths during that period. The geophysical conditions at these depths are not amenable to fossilization. Stalked crinoids and various brachiopods and crustaceans were subject to similar conditions.

Urodeles or tailed amphibians

The Cretaceous AMPHIBIAN (class Amphibia) fossil record is quite fragmentary, although numerous skeletons from fresh-water and continental deposits have been discovered since the 1960s. These recent finds have not been thoroughly investigated, and analysis of them may add to our knowledge of this animal group. Urodeles (order Urodela) have been found repeatedly in Cretaceous strata. They first appeared in the Upper Jurassic. Cretaceous urodeles include Asian giant salamanders (family Cryptobranchidae; genus *Scapherpeton*), sirens (suborder Sirenoidea; genera *Ramonellus*, *Habrosaurus*) and salamanders (suborder Salamandroidea; genera *Opisthotriton* and *Prodesmodon*). The two salamander genera are from the woodland or plethodon family (Plethodontidae). The Cretaceous urodele record shows that they had diverged into the major modern groups during that period.

ANURANS (order Anura or Salientia) included "old frogs" (Archaeobatrachia) such as tailed frogs (Ascaphidae), pipids (Pipidae; genera *Cordicephalus* and *Saltenia*) and discoglossids (Discoglossidae). "New frogs" (Neobatrachia) found in the Cretaceous include the spadefoot toads (Pelobatidae), which first appeared in the Jurassic period. "Modern" species do not yet occur, and it appears that toads and frogs did not flourish until the Cenozoic era. Interestingly, the pipid frogs (family Pipidae) were nearly cosmopolitan during the Cretaceous (with *Cordicephalus* in Asia and *Saltenia* in South America). The precursors to what are now pipid toads (*Pipa*) and African clawed toads (*Xenopus*) existed in the Cretaceous as two separate lines. Pipids had diverged from the other anurans as an aquatic line in the Jurassic. During the Tertiary they were still found in the northern hemisphere (e.g., *Palaeobatrachus*). The spadefoot genus *Megophrys* diverged from the other spadefoot toads at the end of the Cretaceous. Caecilians (order Gymnophiona), which occupy a special position among amphibians, have not been found in Cretaceous strata.

As interesting as these vertebrate finds are, they are hardly comparable to the fossil REPTILES (class Reptilia) of the Cretaceous. Reptiles were the most prevalent vertebrate class in the Cretaceous period, and they had conquered every major habitat and had even produced an entirely new group of marine species. Birds were successfully competing with reptiles in the air, of course, but they did not prevent the development of the giant pteranodons (*Pteranodon*, a member of the flying saurian order Pterosauria) during this period. The Cretaceous, the last Mesozoic period, can truly be called the *Age of Reptiles*. Cotylosaurs (order Cotylosauria) and therapsids (order Therapsida) had either disappeared or were completely displaced by mammals.

We now turn to the TURTLES (order Testudines), which have a very long phylogenetic history. (We became acquainted with a great many Jurassic marine species in the chapter on that period). Some of these turtles gave rise to Cretaceous descendants (family Thalassemyidae; genera *Tropidemys* and *Cimochelys*). At the same time, some new and distantly related turtle groups arose (genera *Baena* and *Eubaena*, both fresh-water turtles), and they survived until the Ice Age. Some "modern" species also

Anurans

Turtles

Stalkless crinoids:

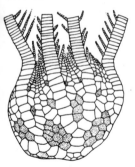

Fig. 16-12. *Uintacrinus socialis.*

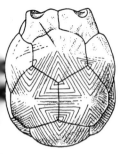

Fig. 16-13. *Marsupites testudinarius.*

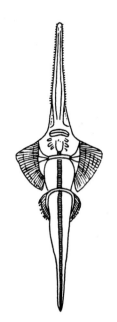

Fig. 16-14. *Sclerorhynchus.* The first sawfishes occur in the Cretaceous.

existed, such as the many genera of dermatemydid turtles (e.g., *Compsemys, Tretosternon* and *Tsaotanemys*), side-necked turtles (family Pelomedusidae; genus *Podocnemis*), soft-shelled turtles (family Trionychidae; genera *Trionyx* and *Aspideretes*) and marine turtles (family Cheloniidae), the marine turtles being present in a great many species. They included green turtles (*Chelonia*) and loggerheads (*Caretta*), which are still extant, and a number of extinct genera (*Corsochelys, Desmatochelys* and *Glaucochelone*). Some marine turtles were giants (families Toxochelyidae and Protostegidae; genera *Toxochelys, Archelon* and *Protostega*), with lengths of up to 6 m. Their limbs were modified into paddles, and their shell extended as far as the lateral plates (marginalia), which were joined with the ribs. Marine turtles reached their zenith with the giant marine turtle (*Archelon ischyros*; see Color plates, pp. 332 and 406/407) of the Upper Cretaceous. Toxochelyids and protostegids disappeared during the Lower Tertiary and left few descendants.

Whether terrestrial turtles (family Testudinidae; also called true tortoises) existed in the Cretaceous is questionable, but the genus *Gyremys* may have been a member of this family. At any rate, Cretaceous turtle fossils show that a number of genera still occurring today (loggerheads, green turtles, side-necked turtles and soft-shell turtles) lived in that period and that numerous other lines (e.g., leatherback turtles and pitted-shelled turtles) were in the early stages of development. Of ecological interest is the fact that during the Cretaceous period, side-necked turtles (which today are found only in the southern hemisphere) lived in Europe and North America. New Guinea pitted-shelled turtles (*Carettochelys*) were also found in Cretaceous Europe.

Among the LEPIDOSAURS (subclass Lepidosauria), the rhynchocephalians (order Rhynchocephalia) of Jurassic Europe have not been found in Cretaceous deposits. It appears that they had become confined to New Zealand, where today their last representative, the tuatara (*Sphenodon punctatus*), is still found. Eosuchians, which had been prevalent in the Triassic, were reduced to an amphibious reptile in the Cretaceous (*Champsosaurus*). Squamata included Upper Cretaceous lizards and snakes. The latter consisted entirely of hemophidians (Hemophidia), namely pipe snakes (family Aniliidae; genus *Coniophis*) and boids (family Boidae; genus *Madtsoia*). They are phylogenetically close to monitors (infraorder Varanomorpha) and simoliophids (family Simoliophidae). The simoliophid genera *Pachyophis* and *Simoliophis* from the Lower Cretaceous have vestigial limbs and primitive skull features and were originally classified as snakes because of these.

The close relationship between monitors and snakes is clearly displayed in the modern Bornean lanthanotid *Lanthanotus borneensis*, a monitor with some characteristics found otherwise only in snakes. Relationships between the aigialosaurs (*Aigialosaurus*) and dolichosaurs (*Dolichosaurus*) from the Upper Cretaceous also exist. These groups are close to the ancestral snake group. They were marine reptiles with shortened limbs, and their vertebrae possess the supplementary articulations also found in snakes. This raises the question, as yet unanswered satisfactorily, of whether the precursors of snakes are from a water habitat.

The most well known fossil monitor species are undoubtedly the MOSASAUR, (Mosasauridae). Found only in the Cretaceous, some of the mosasaurs attained gigantic dimensions. They had a rudder tail and paddle-like fins (see Fig. 16-17), both of which were effective adaptations to marine life. The limbs were not suited for terrestrial locomotion. In the more highly specialized forms (e.g., *Plotosaurus* and *Tylosaurus*; see Color plate, p. 406/407), the phalanges were increased in number (a condition known as hyperphalangia, which also occurs in ichthyosaurs and cetaceans) and this was related to the paddle modification of the fins. However, mosasaurs probably moved through the water chiefly by using an undulating motion. They assumed the kind of body plan that the Jurassic marine crocodiles (family Teleosauridae) had. These crocodiles were nearly extinct by the Lower Cretaceous. Some mosasaurs (*Clidastes, Platecarpus, Globidens, Plotosaurus, Mosasaurus* and *Tylosaurus*; see Fig. 16-17) attained a L exceeding 10 m. They were distributed nearly throughout the world. They probably preyed upon various free-swimming animals, and scratch marks on ammonite shells suggest they also ate these shelled cephalopods. A joint in the middle of the lower jaw enabled mosasaurs to open their mouth very widely and thus swallow large prey.

True monitors have been found in the Upper Cretaceous. Members of the family Varanidae included the genera *Palaeosaniva, Pachyvaranus* and *Telmasaurus*. Genus *Parasaniwa* was a pseudo-monitor (family Parasaniwidea). Others from the Cretaceous include lateral fold lizards (family Anguidae; genera *Peltosaurus* and *Gerrhonotus*), skinks (family Scincidae; genus *Sauriscus*), whiptails (Teiidae; genera *Chamops* and *Meniscognathus*), agamids (Agamidae; *Macrocephalosaurus*) and chameleons (Chamaeleontidae; *Mimeosaurus*). *Gerrhonotus* is a member of the modern alligator lizard family. True lizards (Lacertidae) and geckos do not yet appear.

ICHTHYOSAURS (order Ichthyosauria) and PLESIOSAURS (order Plesiosauria; see Color plate, p. 356) were also distributed throughout the Cretacean seas. However, ichthyosaurs were definitely declining, and the family Ichthyosauridae contained just a few Cretaceous genera (e.g., *Macropterygius, Myopterygius* and *Ophthalmosaurus*). The stenopterygiid family Stenopterygiidae) contained the single genus *Platypterygius*. This recession continued throughout the Cretaceous, and the last ichtyosaurs were the genera *Ophthalmosaurus* and *Myopterygius*. Plesiosaurs, on the other hand, reached a final flourishing point and were represented by the small-headed, long-necked elasmosaurs (*Elasmosaurus* (see Color plate, p. 406/407), *Hydrotherosaurus, Styxosaurus* and *Morenosaurus*) and the long headed, short-necked polycotylids (*Cimoliasaurus, Polycotylus* and *Trinacromerum*). The plesiosaurs did not survive the Cretaceous, however, and they died out

▷
Rich ammonite collection with several species in the genus *Ludwigia* (Jurassic). The specimen is in the Paleontological Institute of the University of Tübingen, Germany.

Plesiosaurs

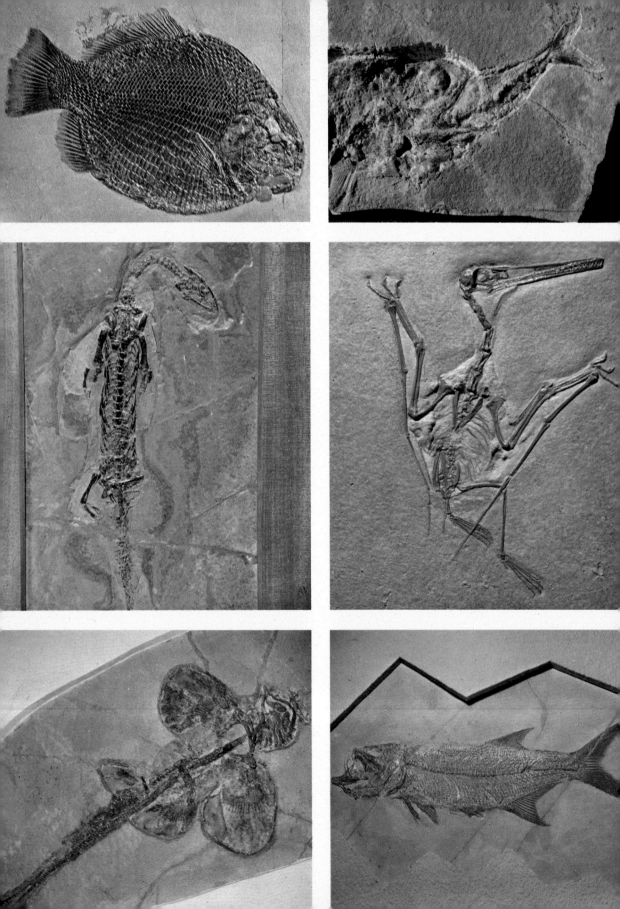

at the very end of the period. They last appear in the Companian and Maestrichtian stages. Pliosaurs, which were short-necked reptiles that were quite diverse in the Jurassic, reached their peak of development in the last giant form (*Kronosaurus*) from Lower Cretaceous Australia. It had a L of approximately 12 m and a skull nearly 3 m long. Plesiosaurs lived on fishes and cephalopods. The way in which they adapted to marine life, which was quite different from the adaptations found in ichthyosaurs and mosasaurs, was described in our chapter on the Jurassic.

The last and most important large reptile group is designated as the ARCHOSAURS (subclass Archosauria). Their sole survivors are the crocodiles, which existed in several lines during the Cretaceous as well. Some of the large reptiles, such as the thecodonts (order Thecodontia) were long extinct by the Cretaceous. However, the so-called dinosaurs proliferated during this period and flourished as they never had before (especially the saurischians (order Saurischia) and the ornithischians (Ornithischia)). CROCODILES were primarily represented by the old mesosuchians (suborder Mesosuchia), but they underwent a noteworthy change from their Jurassic predecessors. The widely distributed Jurassic marine crocodiles belonging to the teleosaur family (Teleosauridae) disappeared in the Lower Cretaceous except for the sole genus *Heterosaurus*. Many of them were displaced by thalattosuchians (family Metriorhynchidae; genus *Geosaurus*). These long-snouted marine crocodiles were well adapted for marine living, for they had a rudder tail and hind limbs modified into elongated paddles. Other Cretaceous species included terrestrial descendants of mesosuchians, such as the pholidosaurs (family Pholidosauridae; *Dyrosaurus* and *Pholidosaurus*) and the goniopholids (Goniopholidae; *Goniopholis*). Still others belonged to new crocodile groups, such as the notosuchians (family Notosuchidae, with the genera *Libycosuchus* and *Notosuchus*). All these mesosuchians, with the exception of the pholidosaurs, died out at the end of the Crecateous.

Other crocodiles to survive the Cretaceous-Tertiary boundary were the sebecosuchians (suborder Sebecosuchia with the genus *Baurusuchus*). Members of the modern crocodile suborder Eusuchia first appeared in the form of the now extinct hylaeochampsids (family Hylaeochampsidae; genus *Hylaeochampsa* and perhaps *Bernissartia* as well) and the stomatosuchids (family Stomatosuchidae; genus *Stomatosuchus*). True crocodiles (*Leidyosuchis* and *Crocodylus*) also existed during the Cretaceous, as did alligators (*Brachychampsa* and *Prodiplocynodon*). Others included species related to the false gavial of today (*Thoracosaurus*) and some giants with a L of approximately 15 m (e.g., *Deinosuchus* = *Phobosuchus*). Indeed, crocodilian fauna changed considerably during the Cretaceous period, both in terms of the development of previous crocodile types and in the evolution of entirely new groups. The most highly specialized crocodilians did not appear until the Tertiary, however, the true gavials (*Gavialis*) would be an example of this advanced type.

◁
Left, from top to bottom:
Ganoid fish *Dapedius punctatus* from Holzmaden Lower Jurassic slate. The dorsal and anal fins formed a long fin seam. L: 33.5 cm. From the Hauff Museum, Germany;
Pachypleurosaurus edwardsi, as mall Triassic lizardlike nothosaur from Tessin;
Protospinax annectens, a skatelike "angel fish" from Upper Jurassic Eichstätt (Germany). From the Bavarian State Collection for Palentology and Historical Geology, Munich;
Right, from top to bottom:
A small telost fish (*Leptolepis*) in the mouth of the ganoid fish *Caturus* (Upper Jurassic Eichstätt slab limestone). This specimen is from the Museum Bergér, Eichstätt-Harthof;
The short-tailed flying saurian *Pterodactylus kochi* (L≃24 cm) from Upper Jurassic Eichstätt/Solnhofen. The specimen is from the Bavarian State Collection for Paleontology and Historical Geology, Munich;
Pachythrissops macrolepidotus, a teleost fish from Upper Jurassic Eichstätt limestone. It, too, is from the Bavarian State Collection.

Among the dinosaur groups, the ORNITHISCHIANS (order Ornithischia) reached their evolutionary zenith. They proliferated into a wealth of large terrestrial species, reaching a size no reptilian group had ever attained before. Some ornithischians first appeared in the Cretaceous and disappeared toward the end of this period. Examples of such groups included the ankylosaurs (suborder Ankylosauria), ceratopsians (suborder Ceratopsia), hadrosaurs, psittacosaurs, and pachycephalosaurs (families Hadrosauridae, Psittacosauridae and Pachycephalosauridae), the last three families belonging to the ornithopod suborder (Ornithopoda). The fossil record of these ornithischians is so extensive that it delivers a detailed picture of the phylogenetic relationships between these animals and the course of their evolution. With the disappearance of these and other animal groups toward the end of the Cretaceous, paleontologists are left with the problem of explaining why so many animal groups would become extinct at this time in geohistory. We shall return to this problem at the end of this chapter.

STEGOSAURS (suborder Stegosauria), which had been prevalent Jurassic dinosaurs, disappeared in the Lower Cretaceous with the genus *Craterosaurus*. The great development within the ornithischian group was the development of bipedal locomotion and the rise of numerous new ornithischian groups. The IGUANODONS (*Iguanodon*; see Color plate, p. 353) include not only the earliest members of this new bipedal group but also the most famous. Footprints of iguanodons have been found, and these yield important information about their walking pattern. Distributed throughout the world, iguanodons were up to 5 m long and were bipedal herbivores. A strong bony thumb spine often appears on the small forelimbs, and it is thought that this spine was used for defense against carnivorous dinosaurs and was not a secondary sexual characteristic (it is not always found in both sexes). The dentition of these iguanodons (genera *Iguanodon*, *Camptosaurus* and *Rhabdodon*) consisted of several rows of spatulate teeth with jagged margins. Teeth like these could even cope with resistant plant materials. The front section of the jaws lacked teeth and probably bore a horny beak. Ossification in the caudal vertebrae region reinforced and stiffened the tail.

We met the HYPSILOPHODONTIDS (family Hypsilophodontidae) in the Jurassic. The group disappeared in the Cretaceous with one small member, the genus *Hypsilophodon*. The resemblance between this genus and modern tree kangaroos has led to the suggestion that the extinct group may have been arboreal. The long, powerful fingers and toes are certainly suggestive of an arboreal life mode. The Upper Cretaceous PACHYCEPHALOSAURS (family Pachycephalosauridae; *Pachycephalosaurus* and *Stegosaurus*) were bizarre ornithopods. They had a highly arched forehead with buttonlike outgrowths. The massive skull roof could mean that these reptiles were highly territorial and engaged in head-to-head fights as they defended their territories.

The HADROSAURS or DUCK-BILLED DINOSAURS (family Hadrosauridae;

Mosasaurs have only been found in the Cretaceous and were distributed throughout the world:

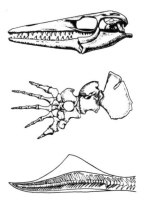

Fig. 16-15. Skull, hand and caudal fin of the North American genus *Clidastes*.

Fig. 16-16. *Plotosaurus* from California. Like *Tylosaurus*, this highly developed form has paddlelike swimming fins.

Fig. 16-17. The mosasaur *Tylosaurus* (L 8–9 m), found in North America and South Africa, has a powerful rudder tail and a cartilaginous protective layer in the auditory region. These indicate that this animal was a diver.

Hadrosaurs or duck-billed dinosaurs

see Color plate, p. 409) were another new ornithopod dinosaur group. They looked somewhat like iguanodons. These medium-sized to large reptiles may have been bipedal. Three or four digits are on each foot and hand. Webbing—possibly used for swimming—was stretched between the toes has been found in unusually well preserved duckbills. The name "duck-bill" (which refers specifically to the genus *Anatosaurus*; see Color plate, p. 348) is a reference to the broad, flattened jaws, which may have been covered by a horny beak, as in the duck-billed platypus. The dentition consisted of up to 600 individual teeth in each jaw half. They formed a very effective chewing apparatus, which would have enabled the duck-billed dinosaurs to feed on hard plant parts. The stomach contents of hadrosaur "mummies" (the unusually well preserved specimens) show that these dinosaurs fed on the twigs, fruits, and seeds of many different terrestrial plants and that they particularly preferred conifer needles. The duck-billed dinosaurs were thus the "ungulates" of the dinosaurs, as it were. They browsed chiefly on land, but as good swimmers they sometimes fled briefly into water to escape carnivorous dinosaurs, their enemies.

Fig. 16-18. Upper jaw fragment of the mosasaur *Globidens*, found in Alabama. The hemispherical tooth crowns suggest that mosasaurs could consume hard-shelled animals.

The picture presented above of duckbill life is supported by a number of specializations occurring in these dinosaurs. Hood- to tube-shaped skull outgrowths have long been known in many members of the duckbill family (e.g., *Saurolophus, Lambeosaurus, Corythosaurus* and *Parasaurolophus*; see Color plate, p. 408). Precise investigations have revealed that these structures were hollow and contained the nasal passages. They were originally interpreted as biological snorkels (hence adaptations to life in water) or as resonators. Actually, it appears that they were related to olfaction (i.e., smelling). As in mammals, the olfactory sensory cells were on the inner surface of the nasal passages. In duckbills, the surface area available to these cells was increased by widening the nasal passages and creating a convoluting nasal pathway. Surface area increase was accomplished in mammals by the developing of bony nasal muscles. The brain structure of duckbills supports the contention that these hollow areas in the skull were used for olfaction: the olfactory lobe of the duckbill brain was huge. The selective advantage of having unusually well developed smell was probably that these herbivores could thereby detect enemy carnivorous dinosaurs from considerable distances.

Fig. 16-19. Jaw fragment of a North American mosasaur, *Globidens alabamensis*. The teeth in mosasaurs were fused with the jaws. Replacement teeth grew on the inside of those teeth and replaced them when they were either worn or lost.

The Cretaceous counterparts to the disappearing stegosaurs were the ANKYLOSAURS (suborder Ankylosauria; family Nodosauridae). In *Polacanthus*, a contemporary of the last stegosaurs, the back and tail bore a double row of large thorns, giving these animals an appearance resembling stegosaurs. In the ankylosaur genera *Ankylosaurus, Nodosaurus* and *Palaeoscincus* a massive bony armor covered the entire upper side of the body, and it was bordered on the sides by a row of bony spines (see Fig. 16-23). The tail was surrounded by bony rings and terminated in some species in a clublike structure. Similar structures have been found in mammals, such as the giant glyptodonts from the Upper Tertiary and the Ice Age.

Ornithischians reached a second phylogenetic high point in the ceratopsians (suborder Ceratopsia; see Color plate, p. 405), which first appeared in the Upper Cretaceous. They probably arose from an ornithopod group in which bipedal locomotion was not fully developed and the hands therefore retained five digits. The earliest known ceratopsian (*Protoceratops*) was a medium-sized four-legged animal with a relatively large head, a neck shield, a parrotlike beak and dentition composed of a battery of teeth. Protoceratopsids lacked bony skull outgrowths characteristic of the ceratopsids (family Ceratopsidae). The number and state of development of these "horns" and the configuration and size of the neck shield are features utilized in analyzing the evolution of the ceratopsids. Using these criteria, several ceratopsid lines can be distinguished in the Upper Cretaceous. We shall cite members of the genera *Monoclonius-Triceratops, Pentaceratops-Torosaurus* and *Anchiceratops*, illustrated in the Color plate, p. 405. The most recent species in these groups were huge animals. One of them, the triceratops (*Triceratops*), is one of the most famous dinosaurs, possibly because it attained a L of at least 6 m and a weight of more than 8 t. The mighty *Triceratops* (see Color plate, p. 354/355) bore a nasal horn and two arched forehead horns. The continuous neck shield was nearly half the 2.5 m skull length. The shield was not only for protection but also was the point of attachment for muscles extending from the neck and trunk to the head. The molars acted like giant scissors. Ceratopsids could not chew like the duck-billed dinosaurs. It is thought that ceratopsids, which probably lived in herds, fed on the fronds of palm ferns and palms.

Of the second major "dinosaur" group of large reptiles, the saurischians (order Saurischia) dating from the Jurassic, some groups were declining (e.g., suborder Sauropodomorpha) while others flourished (the theropods (suborder Therapoda), with the ornithomimids (family Ornithomimidae) and tyrannosaurs (Tyrannosauridae). Some of the saurischians were distributed throughout the world, including some of the brachiosaurs (family Brachiosauridae; *Brachiosaurus* (see Color plate, p. 338), *Camarosaurus* and *Euhelopus*) and titanosaurs (family Titanosauridae; *Titanosaurus, Aegyptosaurus, Mongolosaurus* and *Macrosaurus*). Titanosaurs occurred as late as the Upper Cretaceous. Their distribution, body size, and skeletal structure all indicate that these giant dinosaurs spent most of their time in water and sought their food there. They were most likely herbivores, although we still do not know what they actually ate. The largest animals living today (excepting toothed whales and giant squid) are almost all herbivores with powerful dentition (e.g., elephants) or filter feeders on planktonic organisms like the great baleen whales. Features like these do not occur in the Sauropodomorpha, however. Their weak teeth, which sometimes look like a filter apparatus, are neither suitable for seizing prey nor for dealing with hard or large quantities of food (i.e., plants). The location of the nares on top of the head certainly are suggestive of a life spent in the water, perhaps something like that of the

▷
Archaeopteryx lithographica, the so-called protobird, one of the most famous finds in the entire history of paleontology. This specimen was found in 1877 in a deposit near Eichstätt, Germany. It now reposes in the Paleontological Institute of Humboldt University, East Berlin.

▷▷
Left, from top to bottom: The dragonfly *Aeschnogomphus intermedius* from Upper Jurassic Eichstätt. It is now on display at the Eichstätt Theological School;
The decapod *Aeger tipularius* from Upper Jurassic Eichstätt limestone. This specimen is now at the Museum Bergér in Eichstätt-Harthof;
A perisphinct ammonite shell; the mosslike lines correspond to the septal sutures on the inner wall of the shell. This specimen is at the Paleontological Institute of the University of Tübingen, Germany;
Right, from top to bottom: Serpent star (or brittle star) *Ophiurella speciosa* from Upper Jurassic Solnhofen. It is kept in the Maxberg Museum in Solnhofen;
A ground-dwelling crab, *Palaeopentacheles redenbacheri,* from Upper Jurassic Eichstätt; Coral colony (genus *Ovalastrea*); the original calcareous skeleton and silicified. This fossil is now at the Paleontological Institute of the University of Tübingen, Germany.

Herbivores or marine animals?

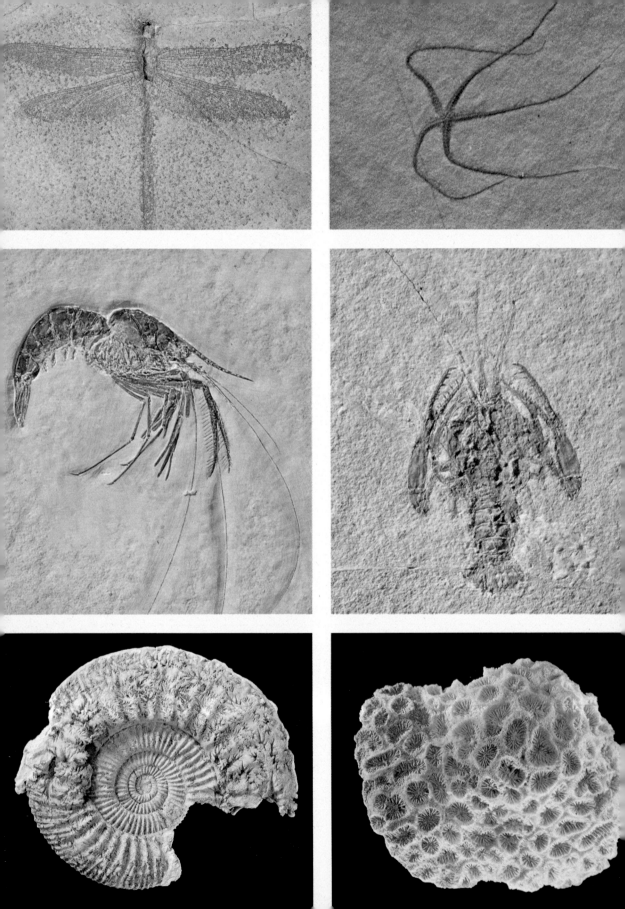

hippo. The huge body size and long neck in brachiosaurs (*Brachiosaurus*) would permit them to live in water 10 m deep. All these giant dinosaurs went on land to lay their eggs.

Water may also have offered these giant herbivores protection from the carnivorous dinosaurs roaming the land at that time. These were the therapods (suborder Therapoda), which were prevalent in the Cretaceous period. Some of them (e.g., *Allosaurus* and *Megalosaurus*) occurred in the Jurassic. Therapods also included the largest terrestrial carnivores that ever lived, the members of the family Deinodontidae. This family included such giant predators as *Gorgosaurus*, *Tarbosaurus* and, undoubtedly the most familiar dinosaur of them all, the tyrannosaurus (*Tyrannosaurus*; see Color plate, p. 347). Their powerful hind legs supported the upright trunk. The vestigial claws on the tiny fore limbs could at best be used for grooming purposes. The head was relatively large, being over 1 m long, and it bore jaws with dagger-shaped jagged teeth. When *Tyrannosaurus rex* stood up, its head was more than 5 m above the ground. This famous dinosaur is from Upper Cretaceous North America. The long tail of *Tyrannosaurus rex* undoubtedly helped balance the mighty body as this leviathan moved about bipedally. Although the skull is huge, it is rather light due to its "crossbeam" structure (see the many openings in the Color plate, p. 347). Tyrannosaurus footprints look like those a tremendous bird might have left.

Tyrannosaurus (in margin)

SPINOSAURS (family Spinosauridae; *Spinosaurus*) were contemporaries from northern Africa. The spinal processes on their dorsal vertebrae were greatly elongated and probably supported skin folds. Similar structures have been found in the Permian pelycosaurs (order Pelycosauria; *Dimetrodon* and *Edaphosaurus*) and in large Triassic suarians (e.g., *Ctenosauriscus*). The function of these skin folds is still under discussion. They may have contributed to temperature regulation.

Spinosaurs (in margin)

Not all Cretaceous carnivorous dinosaurs were huge animals. Others included the COELUROSAURS (infra-order Coelurosauria), which existed as early as the Jurassic with genera such as *Compsognathus* and *Ornitholestes*. These were small to medium-sized, nimble bipedal carnivores. They had well-developed fore limbs, which were of course helpful in capturing prey. Their name (from the Greek κοιλωματ = hollow) stems from their hollow long bones and vertebrae. The Upper Cretaceous period saw the rise of numerous coelurosaurs (*Coelurosaurus*, *Compsosuchus* and *Sarcosaurus*) and a number of coelurosaur descendants (the ORNITHOMIMIDS, family Ornithomimidae; *Struthiomimus*= *Ornithomimus*, and *Oviraptor*). These slender-legged, long-necked reptiles had an ostrichlike appearance (see Fig. 16-28), and they may have had a horny beak as well. The clawed fore limbs bore just three fingers. Of course, these "bird mimics" did not have feathers ! Many questions have been raised about their diet. Some researchers have assumed that they preyed on the eggs of other animals, but this is questionable. They certainly are fine examples of the

Ornithomimids (in margin)

diverse forms that occurred among reptiles in this, their greatest period on earth. The diversity in modern reptiles is not nearly as rich.

Egg clutches and piles of egg shells have been repeatedly found in Upper Cretaceous deposits. The shell structure indicates without doubt that these eggs belonged to reptiles. A number of different egg types have been identified, and Cretaceous strata in Provence (southern France) alone yield nine different egg types. Their volume varies from 0.4 to 3.5 1, which is far less than the volume of the eggs of the giant ostriches from glacial Madagascar. Assigning the various egg types to specific reptilian groups is an extremely difficult task, but recent investigations have nevertheless yielded considerable information relevant to the extinction of these animals at the end of the Cretaceous period.

Fossil dinosaur eggs

We discussed the decline and disappearance of the FLYING SAURIANS (order Pterosauria) during the Upper Cretaceous in our chapter on the conquest of air. With that brief discussion we touched upon one of the most interesting and important aspects of evolution: the extinction of an animal group. Every evolutionary biologist must deal in one way or another with the question of extinction, but paleontologists are particularly concerned with the factors accounting for the disappearance of animal groups. The boundary between the Cretaceous and Tertiary periods, which also coincides with the boundary between the Mesozoic and Cenozoic eras, is a rather clear one because the fauna of the world changed so much between these two periods. The end of the Cretaceous was accompanied by the disappearance not only of many reptilian groups, but also the ammonites, belemnites and rudistids, which even in the Upper Cretaceous were prevalent organisms. Others to die out included the globotruncanes (protozoa), nerineans and actaeonellens (gastropods) and toothed birds (see Color plate, p. 406/407). Extinction struck terrestrial and marine animals; planktonic, crawling, swimming and flying animals. They were replaced in the early Tertiary by mammals, other mollusks, squids, gastropods, and other "modern" groups. The rapid disappearance of many animals and subsequent rapid occupation of the emptied ecological niches show that the rate of evolution was stepped up at the Cretaceous/Tertiary boundary, at least temporarily. However that does not explain why so many animal groups died out at the end of the Cretaceous. A number of recent studies have cast light on this problem. We can distinguish the following hypotheses:

Causes of extinction

1. A climatic change (with or without polar displacement) occurred, and its changes on the plant world destroyed the food resources of herbivorous animals;

2. Mountain formations (e.g., the Laramic Phase at the end of the Cretaceous) led to drying up of the marshes, river mouths and lagoons that supported so much animal life;

3. Degeneration;

4. Appearance of infectious diseases;

Fig. 16-20. The iguanodon *Iguanodon bernissartensis.* Twentynine skeletons of these were found in 1878 in Bernissart, Belgium, and twentythree of them are displayed in the Brussels museum.

Fig. 16-21. The bones of the left hind foot of a Bernissart iguanodon are superimposed on an iguanodon footprint found in Hanover, Germany.

Fig. 16-22. *Hypsilophodon foxi*, a small arboreal ornithopod dinosaur 1.2 m long. The hands had five fingers, while the feet had five toes.

5. Parasitic activity;

6. Extinction of herbivorous dinosaurs by the carnivorous dinosaurs;

7. Appearance of mammals that fed on dinosaur eggs and those of other reptilian groups;

8. Development of pathological egg shells in dinosaurs, inhibiting dinosaur reproduction;

9. A change in atmospheric pressure caused by volcanic gases and other phenomena;

10. A sudden increase in cosmic radiation caused by a supernova (an especially brilliant young star, which explodes and temporarily creates a tremendous amount of radiation), which led to the production of lethal mutations among the organisms on earth;

11. A lack in protection against cosmic radiation caused by a temporary loss of the earth's magnetic field during polar inversion.

As intriguing as some of these hypotheses may seem to be, all of them except the last two deal with local phenomena or are confined to one or a few groups of organisms, and they cannot explain the extinction of all these animal groups satisfactorily. The problem of extinction cannot be resolved without knowing more of the facts involved. Chief among these is the fact that not *all* reptiles died out at the end of the Cretaceous. Turtles, crocodiles and lepidosaurs survived the Cretaceous/Tertiary border and were in fact represented by quite a number of groups. Another important fact is that not all these animals died out simultaneously. Modern dating techniques are only accurate to several hundred thousand years or perhaps even a million years during the Cretaceous period, so at least this much time span must be allowed for the disappearance of Cretacean organisms. The recent studies by the North American paleontologist Edwin H. Colbert have shown that in the most recent Cretaceous deposits from the Lance Formation (U.S.A.), there are fewer genera and species within the same orders as in the next older deposit, the Belly River Formation. In the latter there were sixteen ceratopsian genera, eighteen ankylosaur genera and twentyeight duckbill genera, while the Lance Formation contained seven, six, and seven genera, respectively. However, theropods (suborder Theropoda) suffered a loss of a much smaller magnitude, for they decreased from sixteen genera to fourteen genera. Nonetheless there is a definite trend evident in these two American formations, namely a clear decrease in the number of genera.

Ammonites also experienced a long-term decline in species numbers that terminated in their disappearance. This occurred in the Maestrichtian stage, the last one in the entire Cretaceous period. The next higher (= younger) marine deposits are devoid of ammonites, mosasaurs, plesiosaurs and ichthyosaurs. The evidence indicates that the Lance Formation in the U.S.A. corresponds to the Maestrichtian stage in Europe and that the dinosaurs became extinct at about the same time as the ammonites.

A broader survey, however, shows that not all animal groups died out at the Cretaceous/Tertiary border. "Modernization" of fishes (for when we speak of the disappearance of a group we usually refer to the displacement of more primitive species by more advanced ones) occurred during the Jurassic period, for example, and was completed during the Cretaceous and before the onset of the Tertiary. The whole problem of extinction in the Cretaceous is a complex issue still being researched by paleontologists, and more satisfactory answers may be forthcoming in the future as we gain increased understanding of the geophysical and biological dynamics of the Upper Cretaceous period.

At the conclusion of this chapter we wish to make some comments on avian and mammalian species found in the Cretaceous period. The fossil record of birds is modest but gives nonetheless an interesting picture of these animals in the Cretaceous. The characteristic Upper Cretaceous birds were toothed birds (Odontognathae), which included the TOOTHED DIVING BIRDS (*Hesperornis regalis*; see Color plate, p. 406/407), a flightless swimming and diving species. They looked something like the loons and grebes of today, but the toothed diving birds had a height of approximately 1 m. The fore limbs were largely vestigial, and locomotion in the water came from the swimming (hind) legs. These prehistoric diving birds lived along the coast of Cretaceous seas, such as the Niobrara sea in North America (see Color plate, p. 406/407). Other toothed birds (*Ichthyornis*; see also Color plate, p. 406/407) looked like modern gulls (Lariformes). The few available fossils show that the first radiation of MODERN BIRDS (Neornithes) must have occurred in the Cretaceous, since members of this group (e.g., *Enaliornis* loons, *Baptornis* grebes, pelicans, herons, *Cimolopteryx* plovers, and possibly flamingos) are found from that period. Even these few fossils show that most of what we have of avian Cretaceous fossils are from birds that inhabited the water. The absence of terrestrial birds is due to the fact that conditions on land have always been far less amenable to fossilization. The rise of highly specialized birds in the Lower Tertiary (e.g., penguins) and the paleogeographic situation at the beginning of the Cretaceous suggest that birds diverged into the major groups in the Cretaceous. We can therefore assume that this period was characterized by a rich selection of bird species whose wealth is scarcely preserved in the fossils available today.

The Cretaceous fossil record of mammals is somewhat more extensive. As we described in previous chapters, the first mammalian radiation in the Lower Mezosoic led to the development of various groups, which differed chiefly on the basis of their dentition. They have been divided into two major subdivisions: the "NOT-YET MAMMALS" (Non-Theria) and TRUE MAMMALS (Theria). The "not-yet mammals" were represented in the lowermost Cretaceous by the last triconodonts (order Triconodonta; *Astroconodon* and *Alticonodon*), while true mammal species included the last symmetrodonts (order Symmetrodonta; *Spalacotheroides* and *Manchurodon*) and eupantotherians (order Eupantotheria; *Aegialodon* and *Mela-*

Fig. 16-23. *Ankylosaurus.* During the Cretaceous, ankylosaurs grew to lengths of more than 4 m.

Fig. 16-24. The ceratopsian *Monoclonius* from North America with its nasal horn.

Fig. 16-25. The duck-billed dinosaur (or hadrosaur) *Corythosaurus* with its hood-shaped skull process, in which the nasal passages were located.

Fig. 16-26. Head of the duck-billed dinosaur *Parasaurolophus* with the tubular nasal passage. The dashed line shows the route of air through this passage.

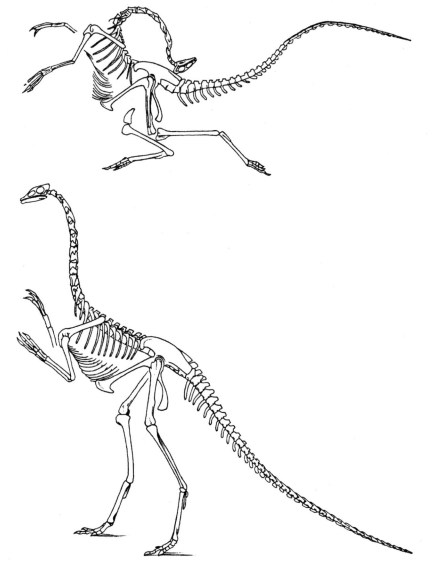

Fig. 16-27. Skeleton of the ornithomimid *Struthiomimus altus* from the Belly River formation in Alberta, Canada, L ≃ 4 m. The skull is toothless and very avian in nature, while the arms are relatively long.

Fig. 16-28. Reconstructed skeleton of *S. altus*.

nodon). Cretaceous triconodonts probably fed on fishes and spent half their lives in water. They and the symmetrodonts became extinct during the Cretaceous. Eupantotherians developed, at the very latest in the Lower Cretaceous, into marsupials (Metatheria) and placental mammals (Eutheria), which first occur with the genera *Holoclemensia* (a marsupial) and *Pappotherium* (a placental) in the late Upper Cretaceous. The first phylogenetic radiation of marsupials took place in Upper Cretaceous North America with the opossumlike animals (Didelphoidea; genera *Alphadon*, *Eodidelphis* and *Didelphodon*). During this time the placental mammals also underwent major radiation.

Thus began the development which in the Cenozoic led to what we commonly call the *Age of Mammals*. The end of the Upper Cretaceous saw the rise of not only the first INSECTIVORES (*Gysonictops*) and HYAENO-DONTS (*Cimolestes*), but also the ancestral UNGULATE group (Condylarthra; with the genus *Protungulatum*) and PROSIMIANS (*Purgatorius*). These animals were small, being mouse- to opossum-sized, and they lived in the shadow of the dinosaurs. Once these large saurians died out, however, the mammals evolved into a vast variety of species, which we describe in the following chapter on the Tertiary. Advanced mammals did not occur in the Cretaceous. The most prevalent mammals were the MULTITUBERCULATES (order Multituberculata; see Chapter 13), with the genera *Locaulax*, *Cimolodon*, *Mesodma*, *Kimetohia* and *Djadochtatherium*. These mammals were between the size of mice and beavers. Their teeth were rodentlike, and it appears that these early mammals occupied niches not unlike those occupied today by rodents. A number of mammalian groups existed in the Jurassic and survived the Cretaceous/Tertiary boundary. These groups did not become extinct until the Eocene. True rodents, on the other hand, and lagomorphs (hares, rabbits and related animals) do not appear until the earliest part of the Tertiary.

Toward the end of the Cretaceous, then, mammals existed and had diverged into a number of groups. All of them were small animals that led secretive lives, and they were not particularly prevalent. We could hardly call the Upper Cretaceous the Age of Mammals.

17 The Tertiary Period—The Age of Mammals

By E. Thenius

The Tertiary Period is the older part of the Cenozoic Era, an era which has lasted 65 million years. Only about 2½ million years of this era belong to the Quaternary Period, however. The Quaternary is divided into the Pleistocene (or glacial) epoch and the present or Holocene. The Tertiary Period comprises a time span of more than 60 million years, and the characteristic fauna and flora of this period were the basis for calling the Tertiary the Age of Mammals. Mammals virtually "exploded" during this period in terms of proliferation, a development related to the widespread disappearance of dinosaurs and other reptilian groups toward the end of the Cretaceous. The extinction of these reptiles made many ecological niches available for occupation by mammals, and it was within these newly accessible niches that the mammals blossomed forth so vigorously.

Origin of the name

The terms "Tertiary" and "Quaternary" stem from a time in which geologists divided the earth's history into four segments. At that time, one spoke of a *terrain primaire, terrain secondaire, terrain tertiare* and *terrain quaternaire*. In modern terminology, the *terrain primaire* is now known as the Paleozoic Era and the *terrain secondaire* as the Mesozoic Era, but the other two old terms have been retained as Tertiary and Quaternary.

Deposits

Many Tertiary deposits are not very firm and are often restricted to small, isolated sites. This is partly due to the relative recency of Tertiary remains and partly because the paleogeographical conditions in the Tertiary did not differ substantially from those occurring today. The location of the continents was much the same as today, at least in the Upper Tertiary.

Flooding did occur during the Tertiary, and these floods often produced new inlets from the sea and even new straits. Fossil-rich shallow sea Tertiary deposits attest to such changes. During the Tertiary, marine fauna came to resemble modern species more and more, and in contrast to the Mesozoic Era the most prevalent marine creatures were mollusks and snails. The progressive predominance of "modern" forms was initially used to subdivide the Tertiary into the Eocene, Miocene and Pliocene epochs. Geologists later designated as Paleocene the oldest Tertiary epoch and added the Oligocene as third oldest epoch in the era. Use is presently made of microfossils in subdividing the Tertiary. These include organisms such as plankton, particularly planktonic foraminifers, radiolarians and coccolithineans. Volcanic rocks are also being evaluated on the absolute time scale measurement technique with the potassium-argon method (see Chapter 3).

Tertiary sedimentation, beginning in the Oligocene, consisted of typical shallow sea deposits with a sequence of gravel, sandstone and marl from the forelands of the Alpidic mountain system. This mountain formation developed during the Lower Tertiary as well and grew more vigorously during the Upper Tertiary. Eroded fragmentary rock from the partially shifted layers deposited in shallow seas, brackish water and fresh water. These sediments are characterized by thin layers, wave ripples, transverse layering and peat deposits formed from Tertiary deciduous, coniferous and palm flora. The macro- and microfossils vary correspondingly with the individual depositions.

Only a few straits still acted as restraints on the propagation of terrestrial animals. The Panama strait in Central America and the Bolivar geosyncline in Colombia prevented an exchange of fauna between South and Central America until the Lower Pliocene. The straits acted to homogenize marine species in the eastern Pacific Ocean and the Caribbean Sea until a land obstacle formed in the Miocene. Even today, the marine organisms from the Atlantic and Pacific coasts (the so-called amphi-american species) show signs of the Tertiary sea bridge between these two oceans.

The Bering Strait existed since the the Upper Tertiary, although interruptions occurred during the glacial period. The Bering land bridge during the Tertiary permitted Asian and North American animals to cross between the two continents. Sea inlets and secondary seas in other parts of the world acted to prevent or inhibit the propagation of terrestial fauna. The Turan Strait east of the Urals in the Lower Tertiary temporarily prevented intermingling of European and Asian animals. The Paratethys was another obstacle during the Upper Tertiary. It extended from the Tertiary Mediterranean across the Vienna Basin, the Pannonic and Ponto-euxinic region and into the Caspian Sea.

During the Paleocene and Lower Eocene there was an important land bridge from North America via Greenland to northern Europe, and modern North American and European animals still reflect the fact that at one time a continuous land mass joined these two continents. The Mediterranean Sea and the Red Sea were never long-lasting borders for terrestrial animals. The Red Sea did not even exist until the Upper Tertiary, while the Mediterranean was to some extent a highly saline inland sea during the Upper Miocene and had no direct connection with the Atlantic Ocean. Deep-sea boring has provided evidence of the partial isolation of the Mediterranean. This means that land animals could cross a Gibraltar bridge and a Sicilian-Tunesian bridge, which would permit African and European fauna to mix.

The Australian continental block and the islands of Tasmania and New Guinea drifted into their present positions during the Tertiary. It was not until the Upper Pliocene and the Ice Age that marsupials (e.g., cuscuses, genus *Phalanger*) could migrate westward to Celebes and Timor and mice could migrate from south-east Asia to New Guinea and Australia.

Climate

In central Europe, the climate grew continuously more inhospitable from the Eocene (when this part of the world enjoyed tropical conditions) to the end of the Tertiary. Subtropical conditions prevailed in the Oligo-Miocene, and even the Pliocene in Europe was warm and temperate. An arctic like climate prevailed during the Ice Age. North America experienced similar climatic changes. But in Australia precisely the opposite occurred, for the climate grew milder and milder during the course of the Tertiary. We also have evidence of alternating moist and dry periods. Before the Red Sea formed in the Miocene, jungle-dwelling animals moved between Africa and Asia (including dwarf deer, mastodons, wild pigs, gibbons and anthropoid apes). Today there are still distinct similarities between Ethiopian and Asian animals.

While repeated exchange of animals occurred between Africa and Eurasia and between Eurasia and North America, South America and Australia remained isolated until the Uppermost Tertiary. Therefore, their fauna evolved within their own gene pools and resulted in the many unusual creatures we see today in South America and Australia. Animals like the koala and marsupial mice come to mind when we think about some of these distinctive animals.

The climatic changes during the Tertiary has been ascribed to a change in the poles of the earth, a theory still being discussed and debated. Extensive glaciation did not occur until the Pliocene, however. Deep-sea boring studies in the Antarctic Ocean have shown that the south polar continent was at least partially iced over as long ago as 4–5 million years. The presence of continental antarctic boulders in marine deposits surrounding Antarctica has not yet been explained in any way other than presuming that they must have drifted into the sea on top of great ice masses. In northern Eurasia there are no signs indicative of a glacial period resembling that found repeatedly for the Pleistocene.

Tertiary flora were primarily characterized by the predominance of angiosperms. However, due to climatic conditions the Tertiary flora were quite unlike the plants of today. We know the Tertiary as the age of mammals, but with reference to plants it could be called the age of lignite (peat coal) flora. Coal from earlier periods is almost exclusively bitu-

Lignite flora

minous coal, while Tertiary flora have been retained as the softer lignite coal. Movements in the earth during the Tertiary caused many fossil flora to turn into anthracite coal. Many fossil specimens even reveal their woody structure, and these are known as xylites (from the Greek ξύλον = wood). Lignitic deposits, which are sometimes quite extensive, have given paleobotanists insights into the plant communities existing when lignite marshes and forests developed as well as into the succession of these communities. We shall survey only the highlights of these findings.

Delta lakes, whose plant communities would most closely resemble the Everglades in Florida, were followed by marsh woods with alders, willows, bog myrtles, palmetto palms and other characteristic marsh plants (see Color plate, p. 438/439). Plant communities inhabiting periodically flooded marshes could be succeeded by swamp cypresses (*Taxodium distichum*), tupelos (*Nyssa*), and aquatic spruces (*Glyptostrobus*), and they were in turn succeeded by dry forests, where the predominant trees were the mighty sequoias (*Sequoia langsdorffi*). The dry forests were the climax flora of the successional pattern; that is, they represented a relatively stable end stage. Their fossil remains regularly contain the trunks of fossilized sequoia trees. Similar plant communities occur presently only in the southeastern U.S.A. (Florida and Louisiana) and on the Pacific coast of North America, where the famous redwoods still impress thousands of visitors year after year.

Fig. 17-1. *Glyptostrobus europaeus.*

Tertiary lignitic forests contained not only "North American" but also "eastern Asian" plants as well, including aquatic spruces (*Glyptostrobus*), Old World sequoias (*Metasequoia*), the fir tree genera *Sciadopitys* and *Keteleeria*, yews (*Torreya*), larches (*Pseudolarix*) and ginkgos (*Ginkgo*). These plant genera were distributed throughout the northern hemisphere during the Tertiary. They were forced southward during the glacial period, but in North America and eastern Asia they returned to their original habitat after the ice retreated. In Europe the Alpine and Carpathian formations and the Mediterranean prevented a similar return.

Lower Tertiary lignitic forests were characterized by a vast diversity of plant species, as we find in our tropical forests of today. Most of them were deciduous plants with large, firm, smooth leaves with partially attenuated leaf tips (e.g., figs, laurels, cinnamon, magnolias). Others included palms (*Sabal*, *Phoenicites* and *Nipa*) and various climbers (*Smilax*). Predominant plants in the Upper Tertiary were deciduous trees with thin, toothed leaves (which are representatives of somewhat moister climatic conditions), such as maples, sycamores, hickories, birches, poplars and many others. Arid regions had small-leaved woody shelled fruit plants such as occur today around the Mediterranean and on savannas. Plains grasses (gramineans) covered the open countryside, and the wide areas of grasses paved the ecological way for the grazing herbivores. Three-toed equids, antelopes and gazelles appeared in the Old World, while North American fauna included gazelle-camels, pronghorn antelopes and horses. South America had various notoungulates (order Noto-ungulata) and litopterns (order Litopterna).

The flourishing flowering plants provided food for the many plant-eating mammals but also were the food base for many insect species and therefore predisposed insect proliferation. Most fossil insects are found in Lower and Upper Tertiary amber. Flowering plants and grasses on the one hand and herbivores and plant-feeding insects on the other are out-standing examples of simultaneous parallel evolution in which the plants and the animals need each other to proliferate.

Some of the most famous amber is from the Baltic region (from the amber pine *Pinus succinifera*) and from Miocene Chiapas, Mexico. The remains of organisms held within amber are not only useful for showing which animals lived at a particular time but are in such good condition that they permit fine analysis often impossible to make on other fossils.

The predominant marine plants were rock algae, and lithothamnians (which have been found as early as the Cretaceous) were prevalent as well. "Grass"-forming dasycladacean green algae were declining. Unicellular organisms, especially the FORAMINIFERS (order Foraminifera; see Chapter 16), were sometimes so prevalent they formed rock deposits and therefore are important key fossils.

The most important foraminifers were the NUMMULITES (*Nummulites* (see Color plate, p. 421) and *Assilina* (see Fig. 17-2)). They attained a considerable size. Unlike most unicellular organisms, which are micro-scopic, nummulites could reach the size of a dime, and some species had a diameter of over 10 cm. Other important foraminifers included alveolins (*Alveolina*, *Borelis*), miogypsins, lepidocyclins and heterostegins. The evolution of their shells permits precise dating of Tertiary deposits. Planktonic foraminifers were major contributors to the formation of mud on the ocean floor, for their shells sank deep into the sand and gradually turned into mud. These particular foraminifers, the GLOBIGERINS, are even better for dating purposes because they underwent very rapid evolution during the Tertiary period. Some of the shell features under-going these rapid changes included the number, size and arrangement of shell chambers and shell convolutions; and the fine structure of the shell, such as the pore system and other features. Planktonic foraminifers reached their phylogenetic zenith during the Tertiary.

SPONGES (phylum Spongia) occur in various forms in Tertiary marine deposits, but they did not form reefs. Fossil spongiid sponges (family Spongidae) have been found; this family includes the once commonly used bath sponge, *Spongia officinalis*.

As in the Cretaceous, COELENTERATES were flourishing in the form of corals (some of which were reef-formers). The northern boundary of coral reefs was pushed southward because of the climatic changes occurring during this period. Members of modern deep-sea coral species (micra-baciids) were largely shallow sea inhabitants during the Lower Tertiary.

Fig. 17-2. *Assilina exponens*, a nummulite foraminifer (Foraminifera).

Fig. 17-3. Shell cross-section (seen cut in half from the outer edge to the middle) of the nummulite *Assilina spira*.

Among hydrozoans (class Hydrozoa), the hydractinids (Hydractinidae) and milleporids (Milleporidae) were flourishing and were fully developed. *Kerunia* (see Fig. 17-4) was an exceptional reef-building hydractinid. After landing on snail shells inhabited by hermit crabs, this coelenterate formed striking antlerlike processes.

BRACHIOPODS (Brachiopoda), which had been prevalent in the Mesozoic Era, were represented in the Tertiary only by the families of terebratulids (Terebratulidae), terebratellids (Terebratellidae) and rhynchonellids (Rhynchonellidae) and by the primitive valveless brachiopod genera *Lingula* and *Crania*. Most of them were deep-sea species, as was the case with members of several other phyla.

Among BRYOZOANS (class Bryozoa), the cheilostomates (order Cheilostomata) flourished and the ctenostomates (order Ctenostomata or Cyclostomata), while still prevalent, were sharply declining compared to their numbers in the Cretaceous period. Bryozoans were particularly diverse in Tertiary shallow seas, and in some places they formed veritable reefs (particularly in the south Russian Paratethys region). Cheilostomates included a great many species with the so-called water sack (suborder Ascophora). These animals developed increasingly complex calcareous processes.

MOLLUSCAN (phylum Mollusca) fauna underwent a major transformation in the Tertiary. Ammonites and belemnites were extinct and had been replaced by the now prevalent squid species and octopus from the suborders Sepioidei, Teuthoidei and Octobrachia. The only shelled cephalopods were the NAUTILOIDS (subclass Tetrabranchiata), which were still cosmopolitan in warmer seas but had become rare. Members of the nautiloid genus *Aturia* are particularly well known; they occurred in Miocene Europe. Little is left of the skeletonless octopus species for the calcareous, unchambered brood shell of *Argonauta*. These shells were once thought to belong to ammonites, and these were considered to be the very last of the ammonites on earth; it is now known that the shells in question belong to an octopus and not to an ammonite. The fact that such a highly specialized octopus like *Argonauta* lived in Tertiary oceans shows that the group was a diverse one during this period. True squids (Sepioidei) are known from the Lowermost Tertiary and include species with a cupshaped hollow calcified skeleton in place of the flat pen or cuttlebone. These squid skeletons bear some resemblance to the internal skeleton or phragmoconus of Mesozoic belemnites.

Soft-bodied Tertiary fauna are characterized by the great diversity of BIVALVES and GASTROPODS (SNAILS). MARINE MUSSELS include not only bottom dwellers but also those living within the sea floor soil. The shell shape or the course of the mantle lines on the inside of the shell reveals that the mollusks had long breathing tubes (siphuncles) used to pull in and expel water. This adaptation enabled them to inhabit not only sandy and muddy floors (genera *Solen*, *Ensis*, *Scrobicularia* and *Mya*) but also to bore into rocks or drifting wood (genera *Pholas*, *Teredo*, *Lithophaga* and *Gastrochaena*).

Fig. 17-4. *Kerunia cornuta*, an unusual hydractinid. It settled and grew on snail shells inhabited by hermit crabs, where it developed antlerlike branches.

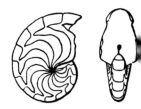

Fig. 17-5. *Aturia angustata*, one of the few shelled Tertiary cephalopods.

Important Tertiary key fossils:

Fig. 17-6. The bivalve *Pecten scabriusculus*.

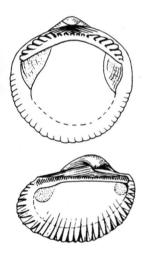

Fig. 17-7. The taxodont bivalves *Glycymeris peucidentata* (above) and *Arca diluvii* (below).

Fig. 17-8. *Limnocardium mayeri*, a characteristic inhabitant of coastal ocean floors.

Teredo has unusually long siphuncles and largely vestigial valves, giving this mollusk a wormlike shape. Various molluscan groups in the Tertiary are used as key fossils, especially the purely marine species (e.g., *Pecten* and *Chlamys*) and certain coastal and brackish water forms (e.g., *Congeria* and *Limnocardium*). The lowered salinity of coastal and brackish water resulted in fewer molluscan species found in those habitats, but this loss was offset by a higher number of individuals occurring there.

No major molluscan taxa appeared during the Tertiary. HETERODONTS (suborder Heterodonta) reached their most diverse stage of development, and they were the most characteristic mollusks of the period along with members of the suborders Leptodonta, (superfamilies Pectinacea and Mytiloidea) and Taxodonta. Two representative taxodonts, *Glycymeris* and *Arca*, are illustrated in Fig. 17-7. The most prevalent mollusks were members of the advanced order Eulamellibranchia and Filibranchia, while the more primitive orders Protobranchia and Septibranchia were declining.

The rudistids and inocerames (see Chapter 16), which had been so diverse and prevalent, respectively, during the Cretaceous, disappeared in the Tertiary. TRIGONIANS (see Chapter 16), which were not uncommon in Cretaceous oceans, were represented in the Tertiary only sparsely by the genera *Eotrigonia* and *Neotrigonia*, and these were restricted to the Australian region. *Neotrigonia* has survived into the present and still occurs only around Australia. It can correctly be called a living fossil. Interestingly in juveniles the shell sculpture (unlike that in adults) retains the patterning found in Mesozoic trigonians.

Just as Tertiary bivalves dwelling in sand and mud proliferated, MARINE GASTROPODS during this period likewise experienced a burst of growth in sand and mud dwellers. These STENOGLOSSANS (suborder Stenoglossa), as "modern" marine gastropods, evolved long siphuncles. And as we saw in the bivalves, these long breathing tubes permitted them to dwell within the soil. Of the many genera of gastropods, which even today populate our seas, we can name only a few of them that arose in the Tertiary period: *Murex*, *Aporrhais*, *Fasciolaria* and *Conus* are all marine snail genera first found in the Tertiary. All of them are highly specialized animals, with poison glands and arrow shaped rasping plates used for procuring prey. The distribution of these and other floor-dwelling snails and bivalves was influenced by their larvae, since the larvae could swim for some two to four weeks. This of course permitted them to increase their range.

The planktonic THECOSOMATES (suborder Thecosomata, members of the subclass Euthyneura), which first occur in the Cretaceous, flourished during the Tertiary and developed into a richly diverse group (e.g., *Clio*, *Cavolina* and *Spiratella*). Their shells often accumulated in huge masses, and they were not too different from the shells of modern species.

Various characteristic Cretaceous snails disappeared entirely. No traces have yet been found of nerineans or actaeonellens (see Chapter 16) in Tertiary sediments, although both were highly prevalent in the Upper Cretaceous in the alpine Gosau deposits. Primitive forms, such as the pleurotomarians (see Chapter 16), became rare (e.g., *Perotrochus* and *Entemnotrochus*), and they had moved into the deep sea.

Fossil TERRESTRIAL GASTROPODS (or SNAILS) became more frequent during the Tertiary period. Most of these are members of modern genera (e.g., *Helix, Cepaea* and *Clausilia*), and they have been used to deduce the prevailing climatic conditions at that time. Other snail genera (e.g., *Arion*) have also been found, although they are represented only by their rudimentary shells. There are no fossils of the purely shell-less snails, the nudibranchs (order Nudibranchia); today they occur in huge numbers in our oceans.

Fig. 17-9. *Aporrhais pespelecani*, a marine snail still occurring today.

Of the other mollusks, the SCAPHOPODS (class Scaphopoda) and PLACOPHORES (class Placophora) declined. Representative Tertiary scaphopods included *Dentalium* and *Cadulus* (see Fig. 17-11), while typical Tertiary placophores included such genera as *Chiton* and *Lepidoplerus*.

The only ANNELIDS (phylum Annelida) found from the Tertiary are tube-building forms related to the serpulimorphs (suborder Serpulimorpha) of today. Annelid tracks have been found, but they are often difficult to assign to specific species.

In the large phylum of ARTHROPODS (Arthropoda), the CHELICERATES or SPIDERS (subphylum Chelicerata) were represented by every major group. SWORDTAILS (order Xiphosura) had become rare, but they were represented in Europe by *Tachypleus* and were much more widely distributed during the Tertiary than they are today. Amber has preserved a great number of ARACHNIDS (class Arachnida), and these fossils reveal that during the Tertiary there were more primitive spiders, which do not build webs (e.g., tarantulas and others) along with those that spin webs and are considered to be highly advanced (e.g., *Epeira*). Amber has even preserved spiders' webs with their gelatinous droplets. Even certain behavioral traits have been revealed in some of these amber deposits, including the transport of scorpions by insects. Some scorpions (e.g., *Chthonius* and *Oligochelifer*; see Fig. 17-13) first occur in Tertiary amber, as have mites (excepting the mite genus *Protacarus* from the Devonian). Amber-preserved fossils are evidence for the great diversity that once characterized mites; during the Tertiary there were not only mites dwelling in the ground (e.g., the oribatids such as *Ceratoppia*) but also parasitic forms (e.g., the ticks *Ixodes* and *Dermacentor*).

Fig. 17-10. The highly specialized gastropod *Conus ponderosus*, equipped with poison glands and a rasping tongue for seizing prey.

Among the CRUSTACEANS (class Crustacea), true barnacles (family Balanidae) existed. They are specialized cirripeds (subclass Cirripedia). Barnacles first appeared in the Upper Cretaceous (with the family Chthalamidae). They proliferated in the Tertiary with the true barnacles. The suborder Cladocera and the AMPHIPODS (order Amphipoda) developed during this period, although both arose prior to the Tertiary.

OSTRACODS (subclass Ostracoda) were prevalent in the oceans and in fresh water. One of the interesting finds in this group is the discovery of the primitive ostracod genus *Puncia* in Tertiary marine strata from New Zealand. On the basis of their characteristics, they and the two New Zealand punciid genera of today must have arisen from the beyrichiids (order Beyrichiida), which were diverse and distributed throughout the Paleozoic world. The most diverse crustacean group were the CRABS (suborders Anomura and Brachyura). Many of them moved from the ocean into fresh water during the Tertiary (e.g., *Potamon*), and they gave rise to many modern terrestrial species such as *Birgus latro*. ISOPODS (order Isopoda) underwent a similar development, and several oniscoidean (suborder Oniscoidea) genera developed. The many crab species are indicative of the success this group experienced and evidence that they are a biologically highly successful animal group.

MILLIPEDES occur almost exclusively as amber-embedded members of modern genera. The Tertiary insect world also resembled that found today. Colonial INSECTS, which includes ants, wasps, bees and specialized termites, reached their high point. Since warm-blooded vertebrates proliferated with a virtual burst during the Tertiary, parasitic insects like lice and fleas developed into a prolific group as well. Other insects found in the Tertiary included the order Phasimida and the families Mastotermitidae and Paussidae.

Tertiary ECHINODERMS differ little from modern species. Compared to the Cretaceous echinoderms, there was an increase in the number of species and diversity of the so-called "irregular" SEA URCHINS, particularly members of the order Spatangoidea. They were adapted to living within loose soil, an adaptation we have already seen in the bivalves and gastropods. Some of the prominent "irregular" sea urchin genera from the Tertiary include *Echinolampas*, *Schizaster*, *Clypeaster* (see Fig. 17-14), *Scutellum* and *Spatangus*. They reached their evolutionary zenith during the Tertiary.

One of the striking developments during this time is the disappearance of sessile CRINOIDS, which by the Tertiary had migrated into the deep sea (where modern crinoids still occur). Most modern crinoids belong to the comatulid order (Comatulida), but there are few fossil comatulid crinoids from the Tertiary. This is probably due to their poor fossilization potential and not to scarcity during the Tertiary. With the comatulid crinoids came a predominant planktonic crinoid group, after numerous planktonic crinoids evolved during the Mesozoic era. The other echinoderm classes of HOLOTHURIDS, STARFISHES and FEATHER or BRITTLE STARS were all prevalent in Tertiary seas, and if the fossil record as now known is representative of these animals they did not differ substantially from modern counterparts.

Only a few fossils are known from the other invertebrate groups, such as the PTEROBRANCHS (class Pterobranchia) *Cephalodiscus* and possibly

Fig. 17-11. The scaphopods *Dentalium* (right) and *Cadulus gracillina.*

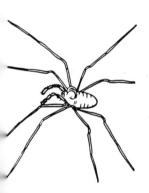

Fig. 17-12. This spider, *Caddo dentipalpus*, is one of the most frequently embedded arachnids (class Arachnida) in amber.

Distichoplax, as well as the TUNICATE (subphylum Tunicata) genus *Micrasci-dites*. These few finds do not permit us to make any conclusions about the evolution of these groups in the Tertiary. Fossil vertebrates are much more substantially represented, and it is to these that we now turn our attention.

The only vertebrates for which we have no Tertiary fossils are the JAWLESS FISHES (superclass Agnatha) with cartilaginous skeletons (i.e., cyclostomes). Jawless armored fishes had long been extinct by this time. RAYS (order Rajiformes), which belong to the class of CARTILAGINOUS FISHES (Chondrichthyes), proliferated during this period and reached their present state of development. All the ray groups (thornback rays, eagle rays, sting rays and electric rays) were present, and such highly specialized (=advanced) forms like devilfishes (family Mobulidae) and electric rays from the family Torpedinidae have been found in the Upper Tertiary. The rays are yet another example of a specific animal type that proved to be highly successful and was able to proliferate into a wealth of different species.

Of the SHARKS there are almost exclusively the "modern" genera such as mackerel sharks (*Lamna* and *Isurus*), fierce sharks (*Carcharias*), great white sharks (*Carcharodon*), tiger sharks (*Galeocerdo*) and others. Primitive forms like comb-toothed sharks (*Notidanus*), horn sharks (*Heterodontus*) and frilled sharks (*Chlamydoselache*) were displaced. The giant basking sharks (*Cetorhinus*) also appeared in the Tertiary, and their filtration teeth show that they were already planktonic feeders, just as they are today. Giant predatory sharks, such as *Procarcharodon megalodon* from the Upper Tertiary, whose teeth were up to 15 cm long (see Color plate, p. 421), were evolutionary end forms. We might note here that tropical sharks like we find today in the Red Sea and in the Indian Ocean occurred during the Lower Tertiary in central European seas.

The modernization trend we saw beginning in the Cretaceous in TELEOST FISHES (class Osteichthyes) continued in the Tertiary period. The sturgeon/paddlefish (Chondrostei) and bowfin/gar (Holostei) super-orders retreated more and more as they were displaced by the true teleost fishes. Ganoid fishes disappeared from the seas with the pycnodontids (family Pycnodontidae) in the Lower Tertiary. Chondrosteins held on only as fresh-water or migrating fishes, as we find in the sturgeons (order Acipenseriformes) of today. Holostein fishes occurring in the Tertiary included the gars (order Lepisosteiformes) and the bowfins (order Amii-formes) in fresh-water regions in Europe and Asia. The gar pikes (*Lepiso-steus*) and the bowfin (*Amia calva*), which still occur but only in remote North American waters, are true living fossils and are remnants of what was once a more significant fish group.

Among TRUE TELEOST FISHES (superorder Teleostei), numerous genera arose during the Tertiary. These included fishes like herring, gadoid fishes, gudgeons, croakers, flounders, giltheads, dentex, mackerals and

Sharks and rays

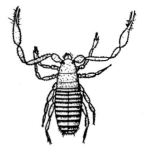

Fig. 17-13. The scorpion *Oligochelifer berendtii* was first found in Tertiary amber.

Teleost fishes

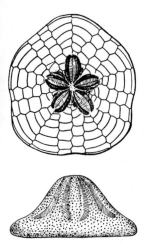

Fig. 17-14. The Tertiary was rich in "irregular" sea urchins: *Clypeaster tesselatus* (above) and *C. altecostatus.*

"Modernization" of fishes

A Tertiary giant salamander was once interpreted as a diluvial man

perch, which are all European and Mediterranean fishes with relatives occurring today in the Indian Ocean and the Red Sea. Other Tertiary teleost fishes have subsequently become extinct (e.g., *Acanthonemus, Odonteus, Phyllodus, Callipteryx* and *Mioplosus*). The only traces of many of these teleost fishes are their otoliths (their auditory bones). These are calcium processes found in invaginations in the inner ear labyrinth, and they often have a specific configuration for each individual fish species. This has made them useful for taxonomic purposes.

Otoliths have been used to gain indirect evidence of Upper Tertiary holothurids (class Holothuroidea). Today there is a fish species, the pearl fish (*Carapus acus*), which is found in the Mediterranean and other warm seas, and it resides within the intestinal tract of holothurids. Since pearl fish otoliths have been found repeatedly in Tertiary marine deposits, it seemed likely that holothuroids also lived there at that time. Direct proof of holothurids has more recently been found, namely in the form of some very small skeletal elements known as sklerites, tiny bones embedded in the body wall.

Otoliths are also used for evidence of evolutionary changes within numerous teleost groups, such as eels, salmon, carp, catfishes, perch, gadoid fishes and flounders. Also, highly specialized spiny-rayed fishes like perch and carp reached their phylogenetic zenith during the Tertiary period, and they have since maintained this status. These include both marine (i.e., salt-water) and aquatic (i.e., fresh-water) fishes, of which there is an enormous variety in the teleost fish group. In fresh water, the predominant species are the carps (order Cypriniformes; see Color plate, p. 416). These species are equipped with the Weberian apparatus, which is a series of bony pieces joining the wall of the swim bladder with the inner ear. Although a few intact or nearly intact skeletons of such fishes have been found, most of the fossil record of them consists solely of their teeth which luckily are highly characteristic features of the various species.

Since no fossils have yet been found of LOBEFIN FISHES (Crossopterygii; also known as CROSSOPTERYGIANS or COELACANTHS), it is believed that these fishes, which inhabited shallow seas during the Cretaceous, migrated into greater depths during the Tertiary. The deeper waters do not have the conditions needed for fossilization. Tertiary fossils of LUNGFISHES (Dipnoi) have only been found in southern continents, and they all belong to modern genera.

Every major AMPHIBIAN group was present in the Tertiary, and the fossil record has permitted paleontologists to trace amphibian evolution in numerous lines during this period. Primitive proteids or olms (family Proteidae) have been found in Middle Eocene Germany (in the Geisel valley). These amphibians had both primitive features and others indicating that they were not cave dwellers (*Palaeoproteus*), as is the case today. One of the most famous amphibian finds was the discovery of the giant salamander *Andrias scheuchzeri*, from Upper Miocene Europe. The salamander is named after the Swiss naturalist Johann Jacob Scheuchzer, who in the early 18th Century erroneously interpreted this salamander to be a "diluvial man" (see Fig. 3-8 and adjacent text). This extinct salamander belongs to the same genus as the modern giant salamander of Japan and China, and other specimens have been found since Scheuchzer's time. These finds show that the giant salamander inhabited all Eurasia during the Upper Tertiary. Its present occurrence in eastern Asia represents the small area remaining after the salamander was displaced elsewhere. Related species ("*Plicagnathus*") have been found in Tertiary North America. Other extant species once found in Tertiary Europe include the axolotls (family Ambystomatidae; genus *Ambystoma*) from the New World and the eastern Asian crocodile salamanders (*Tylotriton*).

Numerous fossil FROG and TOAD genera have been found in Tertiary deposits, but they show little more than the fact that all major frog and toad lines were present in the Tertiary. The most prevalent fossils belong to the families of pipids (Pipidae), discoglossids (Discoglossidae), spadefoot toads (Pelobatidae) and true toads (Bufonidae). Phylogenetic lines of descent can be traced in the spadefoot toads from primitive, Lower Tertiary *Eopelobates* species from Central America leading on the one hand to the genus *Pelobates* and on the other hand to the genus *Scaphiopus*. Of zoo-geographical interest is the finding of pipids from Europe (*Palaeobatrachus*), since these particular fossil frogs are related to the extant South American pipid toads (*Pipa*) and the African clawed toads (*Xenopus* and other genera). These paleobatrachids died out in the Upper Tertiary. It is not yet possible to determine when individual frog and toad lines arose due to the present lack of the necessary fossils from Cretaceous deposits.

At the turn of the Cretaceous to the Tertiary, mammals and REPTILES both underwent significant changes. Groups like the so-called dinosaurs (the term is not a taxonomic one), flying saurians (pterodactyls, etc.) and ichthyosaurs, animals which had been so representative of the Mesozoic era, disappeared along with the plesiosaurs and mosasaurs. The only reptiles to survive the Cretaceous/Tertiary boundary were turtles, crocodiles and lepidosaurs (Lepidosauria), and evolutionary changes occurred even within these surviving lines.

Among the TURTLES, the giant marine species became almost entirely extinct, leaving just a few descendants (*Toxochelys*, *Protosphargis*). They were replaced by various leatherback turtles (*Dermochelys*, *Psephophorus*). In *Psephophorus*, a mosaic of bony plates formed a secondary shell, which was distantly related to the kinds of shells occurring in armadillos. In fact, the first finds of these turtle shells were interpreted as belonging to fossil armadillos from Upper Tertiary Europe. This erroneous interpretation was made because of the fragmentary nature of these finds, and when more complete specimens were located their true nature was revealed. Tertiary oceans were filled with numerous turtle genera that have since died out

Frogs and toads

Coral fishes from Eocene oceans:

Fig. 17-15. *Semiophorus velifer.*

Fig. 17-16. *Mene rhombus.*

Fig. 17-17. *Palaeobatrachus,* a tongueless frog from Middle Eocene Germany.

(e.g., *Eochelone, Procolpochelys* and *Carolinochelys*). Soft-shelled turtles (Trionychidae; see Color plate, p. 438/439) occurred throughout Europe, as did giant terrestrial turtles (*Testudo vitodurana, T. schafferi*). The *Testudo* species have also been found in Asia along with *Colossochelys atlas*. Freshwater European turtles from the Tertiary include forms still occurring there today (e.g., *Emys* and *Clemmys*) and others now found only in southeastern Asia (e.g., *Geoemyda*) or North America (e.g., *Chelydra*).

One primitive turtle suborder—Amphichelydia—was widely distributed in the Mesozoic but is now extinct. One Tertiary descendant of this group, a South American genus named *Crossochelys*, is of phylogenetic and zoogeographical interest because its closest relatives are the meiolaniids (family Meiolaniidae) from Quaternary Australia.

Tertiary CROCODILES included true crocodiles (*Crocodylus*), alligators (*Diplocynodon*; see Color plate, p. 438/439) and relatives of the false gavial (*Tomistoma*), all of which were found in Europe at that time. The Cretaceous primitive crocodiles had been displaced by more modern species belonging to the order Eusuchia, and the only known member of the more primitive order Sebecosuchia is a Lower Tertiary South American genus, *Sebecus*, which is also known from the Upper Cretaceous. Individual teeth from this crocodile genus were once thought to be the remains of Tertiary dinosaurs; this is yet another example of how the fragmentary nature of the fossil record has sometimes led to provisional conclusions that were later confirmed or rejected by later analysis or more thorough fossil documentation.

The very last EOSUCHIAN (order Eosuchia) was the CHAMPSOSAUR (*Champsosaurus*) from Lower Tertiary Europe and North America. RHYNCHOCEPHALIANS (order Rhynchocephalia) disappeared forever from Europe, and today their only representative is a New Zealand species. In contrast, the TRUE REPTILES (order Squamata) proliferated tremendously and blossomed into the diversity similar to that which we know today.

European lizards

During the Tertiary, every major lizard group was on the face of the planet, from the bony-plated, footless anguid lizards (Anguidae, genera *Ophisaurus, Placosaurus*) to the family Cordylidae (genus *Gerrhosaurus*). The helodermatid lizards also occurred in Lower Tertiary Europe, and both the helodermatids and cordylids are still found today but only in North America and Africa, respectively. Many other Tertiary fossil lizard genera have been found, including members of the monitor groups (*Varanus*), agamids (*Agama, Tinosaurus*), *Leiosaurus*, true lizards (*Dracaenosaurus*), geckos (*Cadurcogekko*) and skinks (*Eumeces*). The agamids, which today are found solely in the Old World, were also distributed in the New World during the Tertiary, while some New World species (e.g., *Arretosaurus*) were then living in the Old World as well.

Snakes

In contrast to the situation with lizards, SNAKES (excepting boids) did not become prevalent until the Upper Tertiary. The boids were prevalent in the Lower Tertiary and were represented by many different genera such as *Paleryx* (see Color plate, p. 438/439), *Pterosphenus* and *Gigantophis*. Their Lower Tertiary habitat also included Europe. Cobras, vipers and pit vipers do not occur, until the Upper Tertiary, and anatomically they are the most highly developed snakes in the world. Although these fossil finds cannot be construed as representing the very first appearance of each of these snake groups, all the evidence indicates that the boids are the more primitive snakes. Their skull and dental anatomy suggests this, as does the fact that boids have vestigial hind limbs and the other snakes do not. The pit vipers (family Crotalidae) are among the most recent snakes, and they have only been found as fossils in the New World.

Excepting the large boids and the crocodiles, reptiles evolved into no giant species during the Tertiary period, and even the crocodiles are hardly comparable to the true giant reptiles of dinosaurian proportions occurring in the previous geological period, the Cretaceous. Counterparts to the pteranodons and mosasaurs did not exist in the Tertiary, either.

Our understanding of the Tertiary avian world is much more modest than what is known about mammals from that same period. This is undoubtedly due to the poor ability of birds' bones to fossilize. The toothed birds (Odontognathae) had died out, and all the Tertiary fossils are either from the Palaeognathae or the Neognathae.

Flightless paleognathans related to the Madagascar ostrich (Aepyornithiformes) lived in Africa during the Lower Tertiary (*Stromeria*, *Eremopezus*), and others (related to the true ostriches) inhabited Europe. They must have arisen no later than the Cretaceous; this is suggested by the discontinuous distribution of paleognathans today. Ostriches from the genus *Struthio* occurred not only in Africa during the Upper Tertiary but also in southern Europe and southern and eastern Asia. Their widespread distribution, which lasted until the Lower Pliocene, was probably due to the extensive savanna habitat characteristic of the world at that time.

Every major group of "NEW" (NEOGNATHAN) BIRDS existed in the Tertiary. Penguins occur as early as the Lower Tertiary, and fossil penguins have been found in Australia, New Zealand, South America, Seymour Island and the Antarctic. The fossil record suggests that they evolved in temperate climates in the southern hemisphere. Related species capable of flight and ancestral to penguins may have developed prior to the Tertiary; Alfred S. Romer places these ancestors close to the tube-nosed swimmers (Procellariiformes). Man-sized giant penguins have been recovered from Tertiary strata in Patagonia.

Flightless giant cranes belonging to the modern family of seriemas (Cariamidae; genera *Phororhacos* and *Brontornis*) lived in South America during the Upper Tertiary. Similar, unrelated giant birds (*Diatryma* and *Gastornis*) inhabited North America and Europe during the Lower Tertiary. They confirm once again the close ties between the animal world of both continents in the Lower Eocene.

Interestingly, Tertiary central and western Europe was the home of

Birds

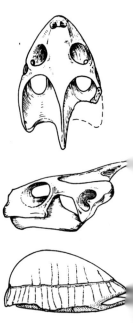

Fig. 17-18. The giant Tertiary terrestrial turtle *Testudo peragrans* from Montana: the skull is seen from the top and side, and the shell is shown from the side.

the rhinoceros bird (see Color plate, p. 438/439), parrots, turacos, New World vultures, mousebirds, flamingos and pelicans. The fossil record shows that flamingos reached their height in the Tertiary. However the record is too sparse to determine the origin of numerous major avian groups, such as parrots, raptors and several others. Likewise we have little information on the radiation of these groups following their origin. For example, of the fiftyseven families of sparrows identified today, only nine of them have also been found in the Tertiary.

Mammals

The fossil record of MAMMALS, thanks to the many finds available to paleontologists, is much more thorough and has given us considerably more detailed knowledge of their phylogeny than we have of birds. As we stated previously, mammals proliferated explosively during the beginning of the Tertiary. This is especially true of placental mammals, which are the higher members of the group. In Chapter 16 we mentioned that various insectivores, ungulates (*Protungulatum*) and primates (*Purgatorius*) first appeared in the Upper Cretaceous. The wealth of Paleocene species indicates that these groups must have radiated during the earliest part of the Tertiary period. This radiation led to the rise of all modern mammalian orders. What we are discussing, then, is a phylogenetic development of substantial proportions, and this "explosive" proliferation of mammalian species has often been linked with the extinction of numerous reptilian groups toward the end of the Cretaceous. The disappearance of these reptiles meant not only the loss of a competitor for food but also the availability of ecological niches heretofore occupied by reptilian species.

Fig. 17-19. Reconstruction of the giant crane *Diatryma steini* from Eocene North America.

Actually, the Tertiary saw a continuous change in the occupation of specific ecological niches, some of which were occupied by mammals that were not closely related to each other. For example, subterranean burrowing forms appear in such diverse mammalian groups as insectivores (moles, golden moles, rice tenrecs; another burrowing insectivore group, the proscalopids, is now extinct), Australian marsupials (the marsupia mole, *Notoryctes*) and in various rodents. The presence of golden moles in Africa prevented moles from the northern hemisphere from penetrating further southward. On the other hand, North American moles acted as a restraint on proscalopids (Proscalopidae), which occupied the same ecological niche. Similar biological controls existed among the South American marsupial borhyenids (Borhyaenidae), which were carnivores, and the carnivore*like* hyenodonts (order Hyaenodonta; see Color plate, p. 438/439) toward the end of the Tertiary. Both these groups were displaced by true carnivores (order Carnivora). Thus, the niches vacated by reptiles were not simply occupied by specific mammals and maintained by those mammals from that time onward. Considerable "maneuvering" occurred among various mammals, one group displacing another from the same niche or preventing the propagation of some other species into a similar, but occupied, habitat.

Fig. 17-20. *Gastornis edwardsi*, a European giant crane.

▷▷
A coastal landscape of an Upper Cretaceous sea with toothed diving birds (1. *Hesperornis*) and other toothed birds (2. *Ichthyornis*), a giant turtle (3. *Archelon*), a mosasaur (4. *Tylosaurus*), a plesiosaur (5. *Elasmosaurus*), and a pteranodon (6. *Pteranodon*).

Many mammalian groups (e.g., marsupials, insectivores, rodents, carnivores, the extinct desmostylians and, of course, whales) developed swimming forms. Gliding flyers evolved independently among marsupials, rodents and others. These convergent developments (i.e., evolutionary changes occurring in parallel but not interrelated) show that these particular life modes had great adaptive value, for they were preserved among several different mammalian lines.

Tertiary mammalian evolution was significantly influenced by the isolation of South America and Australia and by the existence of the Bering land bridge, which joined northeastern Asia and Alaska until the Pliocene. South America and Australia provided the conditions for the evolution of mammals quite unlike those found anywhere else. The Bering land bridge, on the other hand, had quite the opposite effect. It permitted the repeated exchange of terrestrial animals between Asia and North America. Marine mammals (seals, whales and sirenians) had access to a comparable bridge in the Panama strait, which joined the Atlantic and Caribbean Sea with the eastern Pacific Ocean until the Pliocene. This bridge not only influenced the distribution of monk seals and other fur seals but also Tertiary sirenians, which made their way from the Mediterranean via the Caribbean Sea and the Pacific Ocean to the Californian coast. Walrus, which arose in the northern Pacific Ocean, also passed through the Panama strait into the Atlantic; during the Miocene there were no cold-water walrus. Similar possibilities existed for whales (although their early evolution is still very poorly understood).

South America's isolated position, which lasted almost throughout the entire Tertiary, provided the impetus for the evolution of a distinctive

▷
Evolution of the Ceratopsia, an ornithischian (Ornithischia) suborder from the Upper Cretaceous.

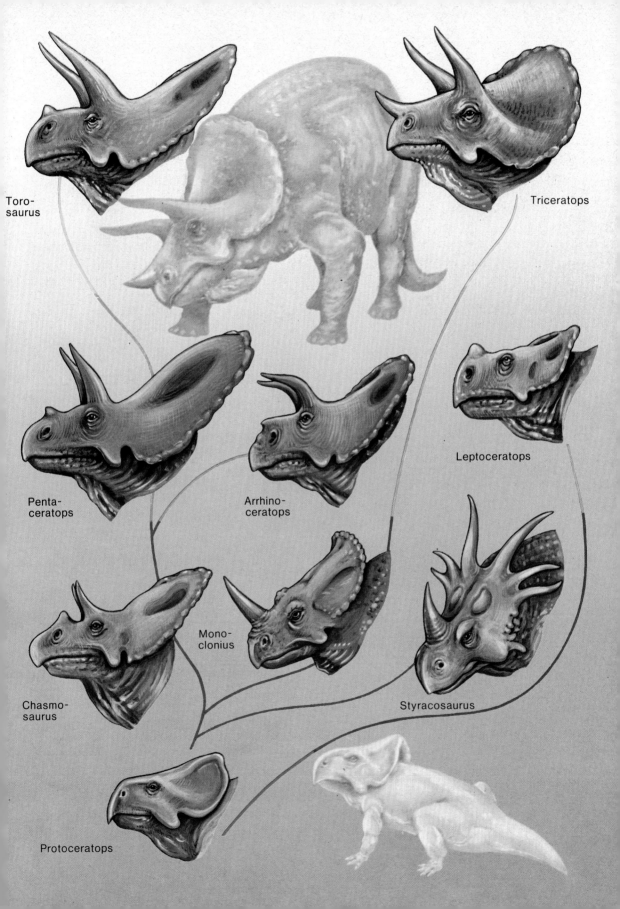

Toro-
saurus

Triceratops

Penta-
ceratops

Arrhino-
ceratops

Leptoceratops

Chasmo-
saurus

Mono-
clonius

Styracosaurus

Protoceratops

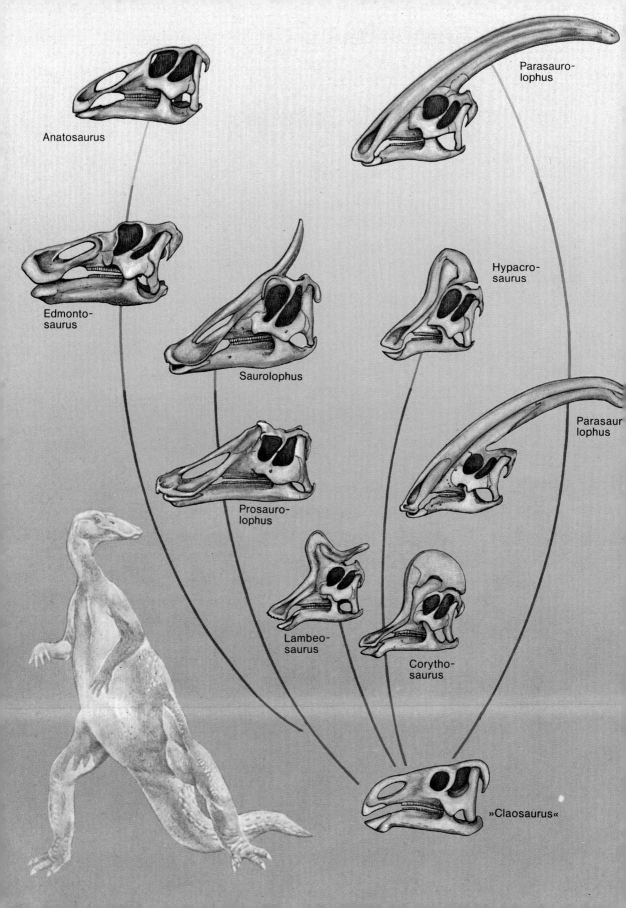

Anatosaurus

Parasauro-
lophus

Edmonto-
saurus

Saurolophus

Hypacro-
saurus

Prosauro-
lophus

Parasaur
lophus

Lambeo-
saurus

Corytho-
saurus

»Claosaurus«

mammalian fauna. These mammals could not cross into Central America until the Panama bridge formed in the Upper Tertiary, at which time they moved into Central and North America. The Tertiary mammalian fauna of South America included not only the extant opossums, edentates, New World monkeys and guinea pigs, but also a series of ungulates, which have since become extinct but which produced a vast variety of animals on that continent. We shall report on these ungulates, which are splendid examples of convergent evolution, in the following chapter.

Tertiary South American MARSUPIALS also played an interesting role. Due to the absence of true carnivores in South America, they evolved into a number of species resembling Australian carnivorous marsupials (Dasyuridae) and others closely resembling true carnivores; yet these South American species were, of course, related to neither group! *Cladosictis* and *Borhyaena* were comparable to martens and hyenas, respectively. The similarity between the saber-toothed marsupial *Thylacosmilus* (see Fig. 17-21) from Pliocene South America and the true saber-toothed cats, such as *Eusmilus* from the Oligocene (see Fig. 17-22) and *Smilodon* from the Pleistocene (see Fig. 17-23), is absolutely amazing. The three species of opossum mice living today (Caenolestidae), which are "living fossils" that have been displaced into parts of South America, are but a remnant from what was a diverse family in the Tertiary (with genera such as *Polydolops* and *Garzonia*), a group that even gave rise to rodentlike animals (e.g., *Groeberia*). Other marsupials, such as *Argyrolagus*, were comparable (but unrelated to) Australian jerboa marsupials and others.

OPOSSUMS (Didelphidae) were widely distributed in the Tertiary. They lived in North America in the Upper Cretaceous and in the Lower Tertiary, and they have been found in Europe since the Eocene. During the Miocene, however, they disappeared in Europe and North America, and they did not return to North America again until the Ice Age, when the opossum *Didelphis marsupialis* reached North America and at present even extends into Canada.

The EDENTATES (order Edentata) were particularly characteristic South American mammals. Today the only edentates are sloths, anteaters and armadillos, but in the Tertiary the edentates were a large group. The armadillos were the first to appear, with the genus *Utaetus* occurring in the Paleocene. The glyptodonts (Glyptodontidae) and ground-dwelling sloths (Gravigrada) developed into giant animals the size of rhinos or even elephants. The largest species, however, did not appear until the Pleistocene. Anteaters (Myrmecophagidae, with the genus *Neotamandua*) occur beginning in the Upper Tertiary. Fossil tree sloths have never been recovered. But modern tree sloths share many features with the Tertiary ground-dwelling sloths (*Hapalops, Planops, Nematherium*), and it appears that they evolved from ground dwellers. During the Upper Tertiary, ground-dwelling sloths reached Cuba and Haiti, and

◁
Hadrosaur (Hadrosauridae) evolution. The family belongs to the ornithopod suborder (Ornithopoda) of the ornithischian order (Ornithischia).

there they evolved into climbing forms such as *Megalocnus*. They did not reach North America until the Upper Pliocene and the Pleistocene.

New World monkeys and GUINEA PIGS (suborder Caviomorpha) first occur in Oligocene deposits from South America, and this had led paleontologists to postulate that they migrated into South America in the Upper Eocene. During the Tertiary they developed into numerous forms, including bear-size giants like *Eumegamys*, a relative of the modern pacarana (*Dinomys branickii*). The GUINEA PIGS (in the narrower sense: superfamily Cavioidea) evolved within this group, and in the Upper Tertiary they were long-legged animals that were widely distributed. All modern guinea pigs in fact evolved from these long-legged ancestors, and proof of this can be found in the structure of their limbs. Their digitigrade manner of locomotion was a subsequent development.

Unfortunately, little is known about the Australian mammals. There are no finds at all dating from the Lower Tertiary, so a vital gap remains in the fossil record from this region. All modern marsupial groups were in existence in the beginning of the Upper Tertiary, and the extinct nototherians also lived at that time. This tells us that MARSUPIALS underwent their initial radiation at some point prior to the Upper Tertiary (probably much earlier). The few fossils we have of MONOTREMES (order Monotremata), the egg-laying mammals, tell us little about this incredible group. We can only presume that monotremes were much more diverse before the rise of the marsupials. No remains of placental mammals have been found in Tertiary Australia.

While South America and Australia played special evolutionary roles due to their geographical isolation, the Old World and North America formed a closed block. We now describe those mammals that were particularly characteristic of the Tertiary period.

Among INSECTIVORES, the hedgehoglike animals (superfamily Erinaceoidea) were distributed throughout the world. Hairy hedgehogs (subfamily Echinosoricinae), which today are found only in southern and southeastern Asia, also inhabited Europe, Africa and North America during the Tertiary. The geologically younger spiny hedgehogs (subfamily Erinaceinae) occurred in North America during the Upper Tertiary. Shrews and moles were also a diverse group in this period. We mentioned moles earlier in this chapter; they included the then prevalent desmans (subfamily Desmaninae). ELEPHANT SHREWS (family Macroscelididea) first appear in Africa. During the Miocene they developed into the myohyracids (Myohyracidae), which were characterized by a molar structure otherwise found only in ungulates. We discussed the fossil evidence of the flying lemur order (Dermoptera) and bats (order Chiroptera) from the Paleocene and Lower Eocene in Chapter 15. A number of bat groups now found only in the tropics lived in Europe during the Tertiary.

Discussion of primate evolution is found in Chapter 21, which deals

An example of convergent evolution: saber-toothed marsupials and saber-toothed cats.

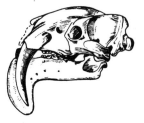

Fig. 17-21. The saber-toothed marsupial *Thylacosmilus atrox*, from Pliocene South America, was the size of a panther.

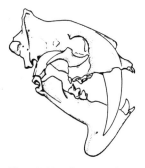

Fig. 17-22. The true saber-toothed cat, *Eusmilus sicarius*, stems from the White River formation of North America. The skull reached a L of 20 cm.

Fig. 17-23. Another true saber-toothed cat, *Smilodon californicus*, is a Pleistocene species from Rancho La Brea, California. The L of the skull is approximately 30 cm.

Fig. 17-24. Paleocene North America was the site of the oldest rodent, *Paramys delicatus*.

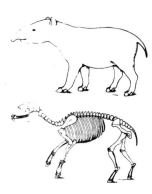

Fig. 17-25. Georges Cuvier described the first fossils of the tapirlike equid species *Palaeotherium magnum*, which was discovered in Montmartre, Paris. Shown here are Cuvier's own depiction of the animal and a skeletal reconstruction.

Carnivores

solely with that animal group. Fossil RODENTS (order Rodentia) are particularly prevalent in fissure fillings, and recent diggings in fresh-water deposits have also yielded many rodent specimens. These recent finds have enabled paleontologists to reconstruct many lines of descent within various rodent groups, and this information is useful not only for gaining insight into rodent evolution but also for dating the layers from which the fossils were recovered. The oldest rodents, which were members of the genus *Paramys* (see Fig. 17-24), stem from the latter part of the Upper Paleocene in North America. They had true rodent dentition, consisting of a pair of rootless incisors in the upper and lower jaws. Thus they were quite unlike the lagomorphs (rabbits and hares; order Lagomorpha), which arose from another animal group. The structure of the skull and the chewing musculature in the oldest rodents (which are all classified into the suborder Protrogomorpha) are very primitive. The sole extant survivor from this group is the sewellel (*Aplodontia rufa*) from western North America. It is another of those few animals that can correctly be called a "living fossil".

Of the many rodent groups, we come now to the MICE (suborder Myomorpha). They appear in the Lower Tertiary with hamsterlike species, and they developed during the Upper Tertiary and Quaternary into such a wealth of different species that they became the most prevalent rodent group and have remained so into the present. The youngest groups to appear were from the subfamily Microtinae and the family Muridae, both of which rose in the Pliocene; we shall discuss them in our chapter (19) on the Ice Age. They confirm a phylogenetic rule that geologically younger groups typically appear with more diversity (and hence more different species) then geologically older groups. These two groups formed lines that even today are flourishing.

The rodents formerly classified as "porcupinelike" rodents are a particularly problematic group. The major question dealing with them is whether the Old World HYSTRICOMORPHS or PORCUPINELIKE SPECIES (suborder Hystricomorpha) and the New World CAVIOMORPHS or GUINEA PIGLIKE SPECIES (suborder Caviomorpha) form a natural entity that arose from a common ancestral group, or whether they arose independently and developed in parallel. Recently there has been an accumulation of evidence supporting the former possibility, specifically that the hystricomorphs and caviomorphs share a common African ancestry. It now appears that the ancestors of caviomorphs "drifted" in the Eocene from western Africa to South America. R. Lavocat claims that the two groups evolved from the Lower Tertiary African phiomyids (Phiomyidae). Phiomyids have several characteristics also found in caviomorphs, and these are not interpreted as primitive features but as special homologies.

Along with ungulates, proboscideans and primates, the CARNIVORES (order Carnivora) are another major mammalian group. They first appear in the form of the miacids (superfamily Miacoidea) in the Middle Paleocene. Members of the major modern carnivore groups appear in the Eocene: mustelids, viverrids and felidlike species. The Oligocene included racoons, bears and canids, while pandas and hyenas arose in the Miocene. CANIDS evolved entirely on North American soil until the Pliocene, when the canid genera *Canis* and *Nyctereutes* appeared in the Old World, where they radiated in the Pliocene and Pleistocene. All other carnivores, with the possible exception of felids, arose in the Old World, and they spread to North America (pandas being the only major carnivore group to remain in Asia and not to migrate to North America). Even HYENAS reached the New World, coming in the lowermost Pleistocene (or Villafranchian) in the form of the "cheetah-hyena" genus *Chasmaporthetes*. However, their stay in the New World was only a temporary one, a fact that is perhaps related to primitive canids (or borophagins, members of the subfamily Borophaginae) in North America assuming the role of the major scavenger species in that continent. Their competition with the likewise scavenging hyenas drove the latter out.

Fig. 17-26. Lower jaw of the Old World pig *Sus erymanthius*.

FELIDS (Felidae) have produced saber-toothed species more than once. There were two subfamilies of true saber-toothed cats: the hoplophoneins (Hoplophoneinae) produced *Eusmilus* in the Oligocene and *Sansanosmilus* in the Miocene/Pliocene, while the machairodontin subfamily (Machairodontinae) gave rise to *Machairodus* in the Pliocene and *Homotherium* and *Smilodon* in the Pleistocene (see Fig. 17-22 and 17-23). Small cats (generic group Felini) appeared with *Felis* in the Lower Pliocene, while big cats (generic group Pantherinii) did not appear until the Pleistocene.

Saber-tooth cats

The MUSTELID group was represented in the Oligocene by the highly aquatic fish otter (*Potamotherium*). One Lower Pliocene otter, *Semantor macrurus*, was even more highly adapted to living in water than the sea otter of today. Badgerlike species have developed ever since the Miocene, and skunks, which arose in the Old World, reached North America in the Pliocene.

Fig. 17-27. *Sus strozzii*—the skeleton of a pig from the lowermost Pleistocene of France.

BEARS (subfamilies Ursinae and Tremarctinae) of today arose from the Lower Tertiary amphicyonids (Amphicyonidae), which first appeared with *Ursavus* in the Lower Miocene. SEALS also stem from amphicyonids, and the oldest fossil seals are found in the Middle Miocene in the northern Pacific Ocean (*Allodesmus*, *Desmatophoca*) and Europe (*Phoca*, *Miophoca*). Blood serum similarities confirm the phylogenetic relationship between bears and seals, and this is a superb example of how modern analytical techniques have confirmed the fossil record.

Seals

WHALES are first represented by primitive cetaceans known as archeocetes (suborder Archaeoceti) in the Lower Eocene. They were marine mammals but had several features characteristic of terrestrial animals, such as some aspects of the vertebral structure and a pelvic girdle with vestigial hind limbs. Furthermore, these archeocetes had differential dentition (unlike the homogeneous dentition of modern cetaceans), including multirooted molars and a skull with several primitive characteristics. All these

Primitive (archeocete) whales

features suggest that whales arose from creodonts, a prehistoric animal group originally classified as primitive carnivores but now understood as not belonging to a unified systematic entity. Some creodonts were hyenodonts (order Hyaeonodonta), while others belong to the primitive ungulate order Condylarthra. It is from the primitive ungulate order that whales evolved. Toothed (odontocete) whales first appear in the Upper Eocene, while the baleen whales arose in the Oligocene. The latter arose from archeocetes, but whether toothed whales did so as well is not yet certain. Early cetacean evolution is one of the cloudier parts of the fossil record, for transition fossils between fully developed whales and their terrestrial ancestors have not been found.

Primitive ungulates

Fig. 17-28. The American giraffe-camel *Alticamelus* from the Upper Tertiary reached a height of 4 m.

The earliest ungulates are from the Upper Cretaceous (see the end of Chapter 16). These belonged to the order Condylarthra we mentioned above. This ungulate order gave rise not only to even-toed ungulates and odd-toed ungulates but also to the tubulidentates (order Tubulidentata), which first appeared in the Miocene in Africa (*Myorycteropus*). The last survivors of this prehistoric ungulate order are the aardvark, sirenians, proboscideans, hyraxes, whales and a number of extinct ungulate orders.

ODD-TOED UNGULATES flourished in the Lower Tertiary, where they proliferated into a rich variety of different forms. There were numerous subgroups within the tapir and rhinoceros lines, and other curious ungulates included the brontotheres (Brontotheriidae) and chalicotheres (Chalicotheriidae), on whose origin, flourishing and disappearance we report in Chapter 18. Finally, there were equids such as the paleotheres (see Color plate, p. 438/439) and true equids (= horses). Since this time the odd-toed ungulates have experienced a steady decline. In the Pliocene/Pleistocene, only the single-toed ungulates, inhabitants of the open plains, were distributed nearly throughout the world; this was after three-toed equids (anchitheres and hipparions) had migrated from North America and inhabited all Eurasia and even northern Africa. The isolated distribution and low number of species of modern odd-toed ungulates are both signs that the group is a dying one.

In contrast, the EVEN-TOED UNGULATES, which includes the ruminants or cud-chewers, are still at their height of development today, as a survey of all the antelopes and antler-bearing animals reveals. Fossil finds also confirm the predominant position of the even-toed ungulates and the gradual rise of the ruminants: during the Lower Tertiary, virtually all even-toed ungulates were non-ruminating species (pigs, peccaries, anthracotheres and related species, the latter having since disappeared). Their role was assumed in the Upper Tertiary by ruminants such as camellike species, chevrotains and antlered ungulates. This is a sign that the ruminants had a biological advantage over the non-ruminants. Interestingly, ruminating as such also arose independent of ungulates among herbivorous marsupials and hyraxes.

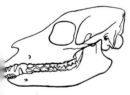

Fig. 17-29. The antlerless deer *Archaeomeryx optatus*, one of the earliest ruminants, is from Mongolia. This specimen has intact dentition.

The Old World PIGS (Suidae) and the New World peccaries (Tayassuidae) probably originated in the Old World, but by the Oligocene they had become separate lines. Tertiary tayassuids appear repeatedly in Eurasia, and true pigs never reached the New World. Apparently the Bering bridge acted as a sort of filter and did not permit them to cross. HIPPOPOTAMUS (Hippopotamidae) reached southern Europe in the Lower Pliocene; they probably arose in Africa, although Europe and southern Asia were part of their habitat during the glacial period.

CAMELLIKE SPECIES (suborder Tylopoda) were restricted to North America until the Pliocene, and in the Upper Tertiary on that continent they developed into some striking species such as the giraffe-camel (*Alticamelus*) and the gazelle-camel (*Stenomylus*), which could probably survive because of the absence (hence competition) of true giraffes, antelopes and gazelles in that part of the world. Not until the Pliocene did camels (specifically, *Paracamelus*) cross the Bering and reach Eurasia. During the ice age, llama ancestors and large camels (*Camelus*) moved into South American and Africa, respectively.

The history of the "true" RUMINANTS began with very primitive forms in the Upper Eocene, namely the genus *Archaeomeryx*. The group proliferated in the Miocene, when such familiar animals as deer, giraffes, and horned animals arose, as well as members of the merycodontid family (*Merycodus* et al., relatives of the pronghorn antelope of today). DEER initially lacked antlers; antlered species arose from the muntjac deer group, animals with a forklike antler spread (*Euprox, Dicroceros*). They proliferated in the Lower Pliocene, and these early deer are classified with the subfamily Cervocerinae.

A similar development occurred among the HORNED UNGULATES. They first appeared in the form of small species (*Eotragus*), tiny creatures resembling duikers of today. Later, in the Pliocene, horned ungulates proliferated and evolved rapidly, developing into the various antelopes and gazelles we see today. Only a few goatlike species (subfamily Caprinae) and bovines (Bovinae) appeared in the Upper Tertiary; they did not proliferate until the Ice Age.

GIRAFFES are of African descent. They were distributed across much of Eurasia during the Miocene/Pliocene, just as ostriches were during the Pliocene. The Pliocene was characterized by short-necked giraffes (*Palaeotragus, Giraffokeryx*), long-necked giraffes (*Giraffa, Decennatherium*) and a few powerful, bovine-giraffes with prominent forehead processes (*Sivatherium, Bramatherium* and other genera). The okapi evolved from plains-dwelling short-necked giraffes that subsequently wandered into the central African jungles.

A close phylogenetic relationship, surprisingly as it may seem, exists between the elephants (order Proboscidea), sirenians (Sirenia) and hyraxes (Hyracoidea), and because of this all three orders are placed together as a superorder of NEAR UNGULATES (Paenungulata), a group stemming from Africa. The American paleontologist George Gaylord Simpson also

Camels

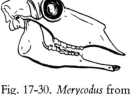

Fig. 17-30. *Merycodus* from America has some extant relatives.

▷
Mesozoic fossils.
Left: the flying saurian *Rhamphorhynchus gemmingi* from Upper Jurassic limestone (Eichstätt). L 53 cm. This specimen is from the Willibaldsburg, Eichstätt; Upper right: Upper Cretaceous mollusks (*Inoceramus lamarcki*, left, and *Inoceramus balticus*, right); Middle right: Upper Cretaceous sea urchins (*Micraster glyphus*, left, and *Echinocorys vulgaris*, right); Below: *Enchodus gracilis*, a carnivorous teleost fish from the Upper Cretaceous. All the above Cretaceous fossils are from the Paleontological Museum of Münster University, Germany.

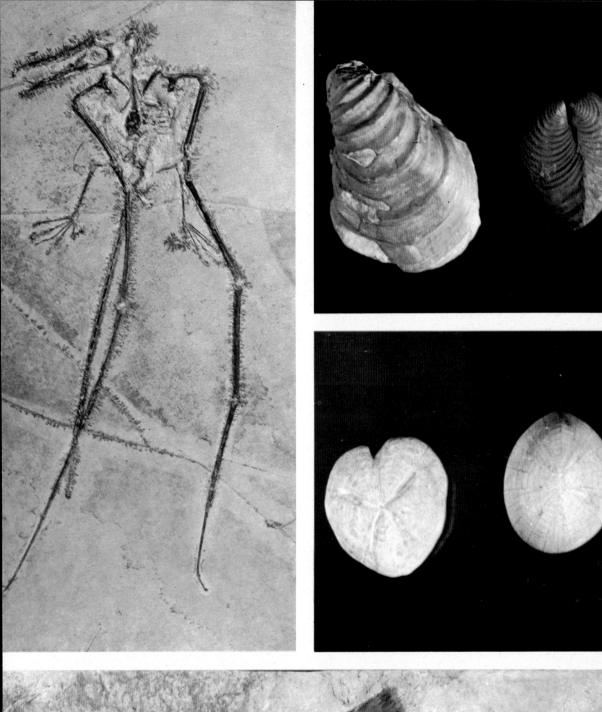

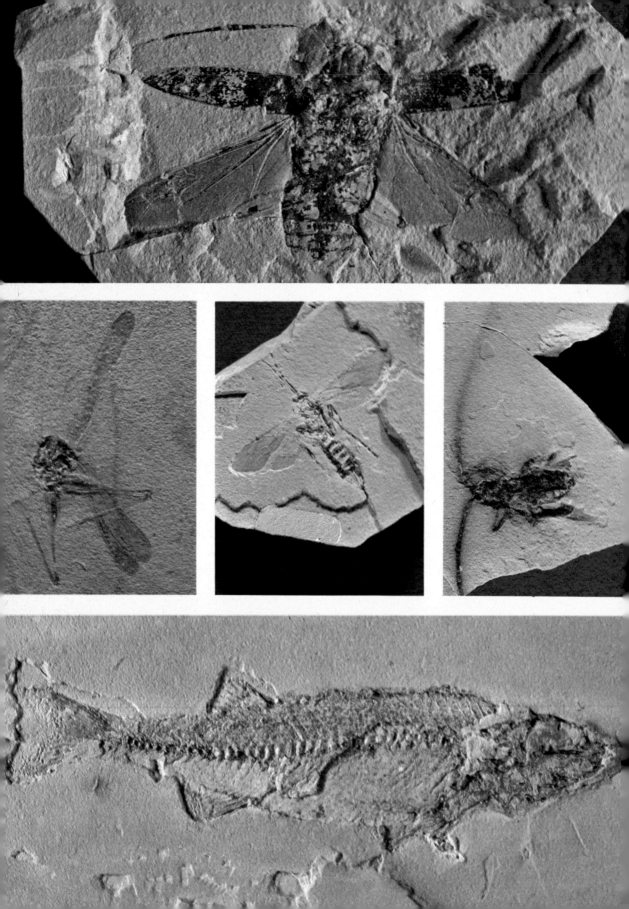

Fig. 17-31. The short-necked giraffe *Giraffokeryx punjabiensis* had two pairs of horns.

Fig. 17-32. A Pleistocene giraffe from southern Asia, *Sivatherium giganteum*, had striking forehead processes.

◁

Upper Pliocene fossils. Top: The long-horned beetle *Saperdopsis robusta*; Middle: a member of the locust genus *Platycleis* (left); *Pimplites preparatus*, a wasp (middle); the long-horned beetle *Monochamoides willershausensis* (right); Below: *Tinca furcata*, a tench.

includes the AMBLYPODS (superorder Amblypoda) with the near ungulates. The amblypods are subdivided into three orders of hippotamus- to rhinoceroslike animals dwelling in Eocene North America and Europe.

The oldest ELEPHANTS (or PROBOSCIDEANS) and SIRENIANS are from Middle Eocene Fayum, Egypt. Moeritheres (Moeritheriidae), which were the size of pygmy hippos and probably lived much like hippopotamus, lacked a trunk, but their dentition and brain structure reveal that they were undoubtedly primitive proboscideans. Fayum sirenians (*Eotheroides*) also possessed numerous primitive features; they were the ancestors of the dugongs and (presumably) manatees of today. However, the distinguishing skull and dental characteristics found in these modern sirenian families are missing in the prehistoric species from the Lower Tertiary. Sirenians had spread to the Atlantic by the Lower Tertiary, where they lived in the Caribbean Sea. By the Upper Tertiary they had crossed the Panama strait and reached the northern Pacific Ocean, where they later gave rise to Steller's sea cow. Manatees have been found in Miocene northwestern South America. The DESMOSTYLIANS (Desmostylia) from the North Atlantic were once thought to be sirenians, but they have since been classified within an independent order. These four-legged mammals lived a seallike amphibian life off the coast, and the group became extinct in the Upper Tertiary.

Species related to moeritheres gave rise (with *Palaeomastodon*) in the Oligocene to the first MASTODONS (Mastodontidae), and mastodons in turn produced two phyletic lines, one of which retained the crowned (bunodont) structure of the molars and the other the transverse ridges (the structure being called zygodont). *Mastodon* (see Color plate, p. 437) existed with several species and was distributed (with other genera that developed from it) across Eurasia. By the Miocene, mastodons had reached North America. The STEGODONS (Stegodontidae; *Stegolophodon* and *Stegodon*) were a more highly developed proboscidean family from the Upper Tertiary. Stegodons were not, however, the ancestors of elephants, which appeared at the end of the Tertiary.

HYRAXES (order Hyracoidea), called CONIES in the Bible, are small, rodent-looking animals that are actually close relatives of elephants. Today there are only a few hyrax species, somewhat marmotlike in appearance, found in Africa and western Asia. Some Tertiary hyraxes were quite large (e.g., the three-toed *Megalohyrax*), and the ecological role of these giant hyraxes was not unlike that of the equids. Giant hyraxes as large as tapirs (*Pliohyrax*) spread into many parts of Eurasia.

Many plant and animal groups disappeared during the Tertiary or toward the end of the period. The reasons for this disappearance will be discussed in our chapter on the ice age.

18   Extinct New World Ungulates

An entire chapter has been reserved for a discussion of four ungulate orders occurring for the most part in Tertiary South America and of which only a few species lived into the Pleistocene. These four orders deserve attention because they are particularly impressive examples of the phenomenon of convergent evolution (the independent but parallel acquisition of similar traits in different phyletic lines). The orders are: LITOPTERNS (Litopterna), NOTOUNGULATES (Notoungulata), ASTRAPOTHERES (Astrapotheria) and the PYROTHERES (Pyrotheria). As we mentioned in the previous chapter on the Tertiary, South America was cut off from the rest of the world almost throughout this period. A terrestrial link with North America did exist from near the end of the Cretaceous to the lowermost Tertiary, and it was during this time that the ancestors of what became distinctive South American fossil fauna migrated southward from North America. These animals gave rise to creatures like marsupials, edentates and the ancestors of notoungulates.

This land bridge sank into the sea during the Lower Eocene, and no land bridge existed between South America and the rest of the world until the narrow Panama bridge formed during the last part· of the Tertiary. During the entire time that South America was isolated, there was of course no intermingling with terrestrial animals from other parts of the world; the South American terrestrial fauna were geographically and genetically isolated and were left to develop alone for some 40,000,000 years. During this period there were no carnivores, lagomorphs, hyraxes, proboscideans, odd-toed ungulates or even-toed ungulates in South America. In other words, those large terrestrial mammals characterizing the Old World fauna were absent on the great "island" of South America for millions of years, and with this fact in mind it is really no surprise that such peculiar animals evolved in South America. The few groups of South American terrestrial mammals that did exist took the ecological place of those mammals that did not occur there, and they produced a fauna fascinatingly unlike the animal life of North America and Eurasia.

By H. Wendt

Faunal isolation in Tertiary South America

Some of these South American animals that survived into the present almost look like caricatures.

Only a few of the notoungulates evolved into giant forms (e.g., edentates like the giant sloths and glyptodonts), but there was nonetheless tremendous diversity among Tertiary South American animals. The wealth of variety among these herbivorous mammals (the sloths and the four ungulate orders we cited above) probably developed because these animals lacked any significant predators, something Old World herbivores had to contend with; and one German naturalist has described their situation as "paradise". The selective pressure exerted by predators was largely absent among the Tertiary South American herbivores, and the absence thereof permitted the rich proliferation that occurred on that continent over the 40,000,000 years of isolation. Some predation did occur, however. Until the Middle Pliocene (and in some cases even into the Pleistocene), some of the carnivores absent in South America were "replaced" by carnivorous marsupials, such as the saber-toothed marsupial *Thylacosmilus* (see Fig. 17-21), the "marsupial hyena" *Borhyaena*, and the "marsupial wolf" *Prothylacinus*. Some of these were bear-sized, but their influence on the Tertiary herbivores was not terribly great because they fed chiefly on carrion and probably could not even deal with some of the powerful herbivores found in South America at that time. Incidentally, these marsupials were not related to the carnivorous marsupials of Australia; they simply arose in parallel with their Australian counterparts.

One of the most intriguing aspects of the four South American ungulate orders is that, with a few exceptions from the Upper Tertiary, none of their members developed horns, antlers or conelike skull processes similar to the ones commonly seen in Old World and North American ungulates. Whether the lack of these structures was due to the absence of carnivores in South America is open to question, since antlers and horns are used far more for intraspecific (i.e., "within the species" or "occurring only within the individual species") display rather than for fighting enemies like predators. Nevertheless, it is interesting and illuminating to find that only in three toxodont genera (*Prototrigodon* and *Haplodonterium* from the Miocene and *Trigodon* from the Pliocene of Monte Hermoso in Argentina) do we find skull processes, with actual horns occurring only in *Trigodon*.

The peculiar South American notoungulates were discussed by Charles Darwin. As he collected a number of fossils when he was in the South American pampas, he came across a number of late notoungulates and litopterns, including *Toxodon*, which was larger than a rhinoceros, and the somewhat camellike (but with a long, movable trunk) genus *Macrauchenia* (see Color plate, p. 454). Darwin was so surprised by the apparent similarity between these animals and other mammals that he wrote the following sentence in his diary: "This wondrous relationship between extinct and living animals on the same continent, a relationship of which

Herbivorous mammals in "paradise"

I have no doubt, will subsequently cast more light on the appearance of organic creatures on our earth and their disappearance than any other class of facts." These South American animals were second only to the Galapagos fauna in impressing Darwin as visible evidence of organic evolution.

Later it was shown that these ungulate groups, which had appeared to be evidence of an evolution, were not "wondrous relatives" of extant groups but had died out with descendants during the Tertiary or the Pleistocene. However, the many similarities between these and other mammalian groups were still astonishing. The litopterns, which included three-toed and one-toed species, had assumed the biological role played elsewhere by horses. The notoungulates included rabbit*like*, hyrax*like*, tapir*like* and rhino*like* creatures. The astrapotheres resembled hippos, and pyrotheres were comparable to proboscideans (= trunked animals). However, the superficial similarities existing between the extinct South American ungulates and their Old World counterparts did not reflect a phylogenetic tie between these groups. Instead they were indicative of convergent evolution, and Tertiary South American ungulates are still among the most noteworthy examples of this form of evolution.

No relationship to living groups

Before we describe these extinct ungulate groups individually, we shall briefly describe their habitat, which was comparable to the glacial loess plains of Europe and Asia. Many parts of South America were grassy plains similar to the more recent North American prairies and the pampas. These plains had arisen in South America during the Oligocene, and the only part of the northern hemisphere having anything like them at the time were the central Asiatic highlands. Thus, South American mammals became adapted to plains life much earlier than their North American and Old World counterparts. The vast grasslands were penetrated by mighty rivers. Numerous Andes volcanos periodically spewed out their contents, spreading ashes over the countryside and creating the habitat conditions for some of the most unusual animal species and communities in the world.

Habitat

Nothing is known about the origin of the LITOPTERNS (Litopterna). Their condylarthran (order Condylarthra; see Color plate, p. 454) ancestors probably migrated into South America from the north across the Paleocene land bridge in existence then. Litopterns have "medial-axles" (mesaxonal) limbs and were once thought to be odd-toed ungulates because of that. Actually they are closely related to condylarthrans and notoungulates and developed in astonishing convergence with the odd-toed ungulates, to which they have no close relationship. Litopterns tended to develop into three- to one-toed animals and are thus interesting counterparts to equids. It is very likely that they developed the single-toed condition much earlier than equids (horses) did; we speak here of the Middle Miocene litoptern family of PROTEROTHERIIDS (Proterotheriidae; genera *Thoatherium* (see Color plate, p. 454), *Proterotherium* et al.). Horses

Litopterns as equid counterparts

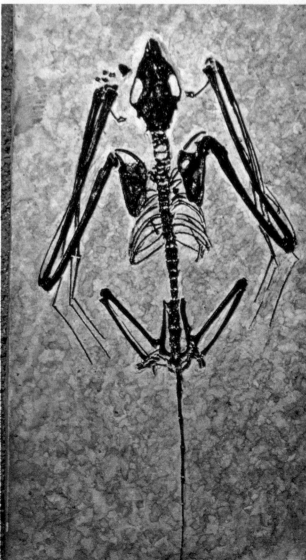
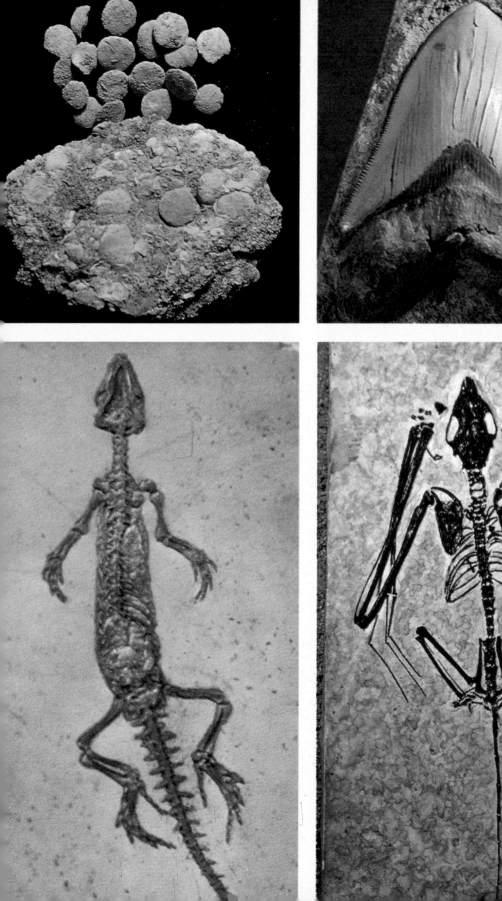

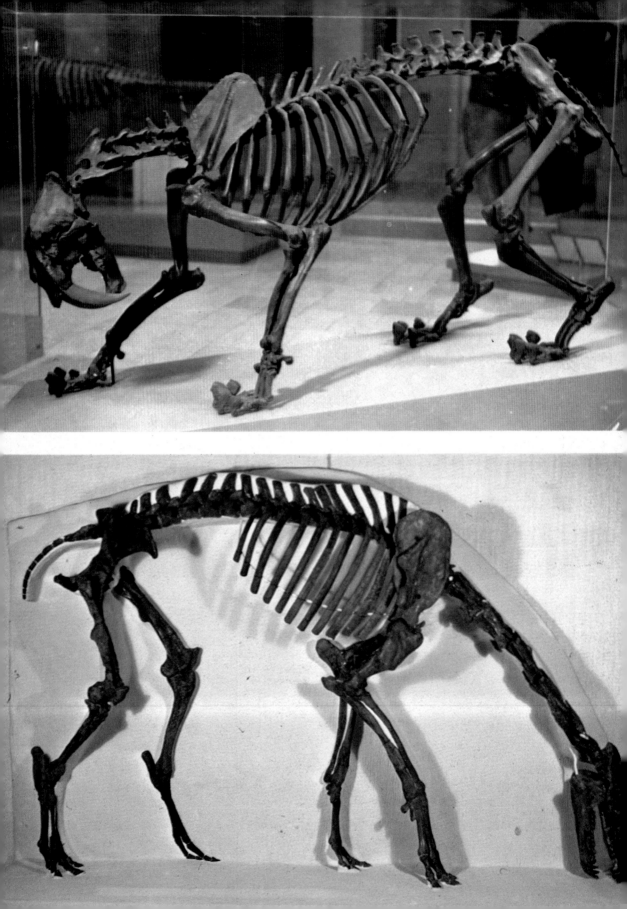

◁◁
Mesozoic and Cenozoic
fossils:
Upper left:
Conglomeration of the
rock-forming foraminifer
genus *Nummulites* ("num-
mulitic limestone") from
Eocene Austria. The lens-
shaped shells reached
diameters of up to 10 cm;
Lower left:
The Upper Jurassic lizard
Homoeosaurus pulchellus.
From the collection of the
Bavarian State Collection
for Paleontology and
Historical Geology in
Munich;
Upper right:
A large single tooth from
the giant shark *Carcharodon
megalodon* from Miocene
Hungary;
Lower right:
The single fully intact
skeleton of the oldest
known bat, *Icaronycteris,*
from the Lower Eocene
of the famous Green River
Formation in Wyoming.
From the collection of the
Princeton University
Museum.

◁
Above:
Saber-tooth cat *Smilodon
californicus* from the late
glacial period of Rancho
La Brea, California. The
specimen is in the Bavarian
State Collection for
Paleontology and Historical
Geology in Munich;
Below:
Theosodon garettorum, a
litoptern (order Litopterna)
ungulate from Miocene
Patagonia.

did not become single-hoofed until the Lower Pliocene. The reason for
this reduction in toe number is that the litopterns were becoming adapted
to plains life, and they evolved toward the three- to one-toed stage earlier
than horses because the plains in South America were older than elsewhere, as we mentioned earlier.

The second litoptern family, the MACRAUCHENIIDS (Macraucheniidae),
contains many more species. All these llama- to camel-sized animals
retained three toes. Deep pits have been found above the nasal openings
of *Macrauchenia patagonica,* which suggests that this species had a long,
muscular trunk, which may have been an adaptation to the arid, sand-
storm-ridden habitat of these macraucheniids. Other members of the
family also possessed a trunk. The somewhat older genus *Theosodon* had
at very least a highly mobile upper lip. *Marauchenia* and *Macraucheniopsis*
were the only macraucheniids to live beyond the Tertiary, for they did
not become extinct until the Pleistocene, the time when the newly formed
Panama bridge permitted superior competitors and powerful predators
to enter South America and eliminate them.

The NOTOUNGULATES (Notoungulata; see Color plate, p. 454) were
still more diverse. They assumed the ecological role of odd-toed ungulates,
lagomorphs, hyraxes and even-toed ungulates. George Gaylord Simpson
distinguishes no less than 15 families among them, and he categorizes
them into four suborders: NOTIOPROGONIANS (Notioprogonians), TOXO-
DONTS (Toxodonta), TYPOTHERES (Typotheria) and HEGETOTHERES (Hegeto-
theria). The groups are all illustrated on the Color plate, p. 454. The
notioprogonians are the oldest of the notoungulates, and their teeth are
strongly reminiscent of condylarthrans. Specimens from Mongolia and
the U.S.A. show that in the Lower Tertiary, notioprogonians wandered
from eastern Asia and North America into South America.

The toxodonts are a mixed group. They include equidlike forms, as in
litopterns, such as the NOTOHIPPIDS (Notohippidae; genera *Rhynchippus,
Nesohippus* and *Notohippus,* as well as others). Even the cement deposits
in their teeth are like those in one-toed ungulates, but all the toxodonts
were three-toed.

Thomashuxleyia, a member of the ISOTEMNIID family (Isotemnniidae),
is one of the smaller, more primitive toxodonts. The HOMALODOTHERIIDS
(Homalodotheriidae) were atypical in one regard, namely that their forelimbs had claw-shaped tips, something found elsewhere among ungulates
only in the chalicotheres, an extinct odd-toed ungulate suborder.

The geologically youngest and also the most specialized members of
this suborder are the TOXODONTIS (Toxodontidae), whose dentition and
shape are similar to tapirs and rhinoceros. They include the only toxodonts
with bony skull processes. The last member of the family, the more than
rhino-sized *Toxodon,* was the genus discovered by Darwin and whose
disappearance in the ice age brought an end to these powerfully built,
short-legged animals.

Typotheres include a number of small notoungulates, some of which look like lagomorphs and some like hyraxes. Hegetotheres were also small, and their large front incisors, elongated feet and short tail gave them an even greater similarity with lagomorphs. They occupied niches in Tertiary South America that would otherwise be filled by pikas and true lagomorphs. The notoungulates are undoubtedly a stunning example of the evolution of an ungulate group in which all sorts of convergent (parallel) features appear.

Parallels to hyraxes and lagomorphs

ASTRAPOTHERES (order Astrapotheria) also arose from condylarthrans, and probably from a group closely related to the ancestors of litopterns. The structure of their auditory region and their teeth justify their separation into a distinct order. Astrapotheres were distributed from Patagonia to Venezuela and Columbia. We cite only *Astrapotherium magnum* from the Santa Cruz strata of Patagonia. It was the size of a rhinoceros and had five toes on each limb. The molars were like those in rhinos. However, quite unlike a rhinoceros, this animal had a powerful trunk, and it may have led an amphibious life something like modern hippopotamus do.

Astrapotheres

The last of these peculiar ungulate orders, the PYROTHERES (Pyrotheria), were formerly connected with proboscideans due to their trunk, the development of the molars and their tusklike incisors. However, all these similarities, disarming as it may seem, are presently interpreted as simply being the result of convergent evolution—something we have seen repeatedly among the South American Tertiary ungulates. One of the pyrotheres, *Pyrotherium*, did resemble an elephant, but it had a much shorter trunk, and its two pairs of upper jaw incisors and one pair from the lower jaw were rootless, weakly arched tusklike teeth. The molars of this animal have been compared to the fossil Australian marsupial *Diprotodon* and with the prehistoric proboscidean *Dinotherium*.

Pyrotheres

When carnivores and "modern" ungulates moved into South America across the newly formed Panama bridge toward the end of the Tertiary, most of the ungulates we described above had already died out. This was primarily because they were not nearly as well equipped to deal with changing conditions as were their Old World and North American counterparts. The intense volcanic activity in the South American Tertiary may have killed many of these animals; at least, this is suggested by many of the skeletons that have been found. Simpson discovered many skeletons of the toxodont *Scarittia canquelensis* in a crater lake in central Chubut. It is possible that the animals sought out the lake to quench their thirst and died during an eruption. Also, volcanic activity could have put a layer of carbonic acid on top of the water, and when the drinking animals put their heads under water they perished.

Victims of volcanic eruptions?

Tertiary ungulates from South America can only be compared with the Pleistocene marsupials from Australia, based on their various mammalian anatomical configurations, life modes and ecological niches occupied. Australia experienced conditions similar in many ways to those of

South America. Most importantly, the great southern continent was isolated, and no placental mammals existed there. In fact, the only mammals occurring in Australia were monotremes like the platypus, spiny echidnas and various marsupials. Ten marsupial families existed in Australia as early as the Tertiary. During the Pleistocene, these marsupials proliferated into a wealth of different forms, some of which bore a remarkable resemblance to the higher (i.e., placental) mammals. Some of these similarities have crept into the vernacular names of many of these species. There are still a great number of Australian marsupials, some of which look like insectivores, some like rodents, and others like carnivores and ungulates.

The nearly lion-sized "marsupial lion", *Thylacoleo*, lived in the Pleistocene. Despite its carnassial dentition, this relative of the phalanger was an herbivore. Other Pleistocene Australian species included various canidlike carnivorous marsupials, most of them larger than the recently exterminated marsupial wolf; the short-headed kangaroos (subfamily Sthenurinae), which when upright were over 3 m tall and which included browsers and grazers (as in other ungulates throughout the world); and finally sturdy-legged giant marsupials (family Diprotodontidae) larger than rhinoceros, which probably led a life much like the tapirs and elephants of today do among their likewise luxuriant plant supplies. The colossal species *Diprotodon optimum* lived as recently as 6000–7000 years ago and was therefore a contemporary of man.

We wish to add here another ungulate group, one characteristic of the North American Lower Tertiary (and also Mongolia). These BRONTOTHERES or TITANOTHERES (superfamily Brontotherioidea, family Brontotheriidae; see Color plate, p. 440) are an example of the rapid development of a Tertiary mammalian group and, in geological perspective, its sudden disappearance, with the youngest members of the line being giant forms. The oldest brontotheres, which lacked horn structures on the head, are from the last part of the Lower Eocene in North America. During the Lower Tertiary a number of lines arose with various skull processes, some of which were very elaborate (see the Color plate, p. 440). However, the North American species became extinct in the Lower Oligocene, while the Asian brontotheres died out in the Middle Oligocene, perhaps just 10,000,000 years after the brontotheres had arisen.

Brontotheres are not as isolated as the extinct South American ungulates, however, for they belong to the order of odd-toed ungulates (Perissodactyla). Although the last, large brontotheres had prominent "rhino horns" (which in reality were not horns at all but bony processes possibly covered with fur) and looked very much like rhinoceros, they belong to the equid suborder (Hippomorpha, which contains horses). The huge bones of these gigantic animals were often swept out during flash floods in the badlands of the western U.S.A., and legends of the Ogallalla Sioux ascribed these bones to "thunder horses". The Ogallalla believed that "thunder horses" sprang from the clouds during rainstorms to drive the bison toward the Indians. After the storm was over, these legendary horses disappeared into the prairie ground.

As he dug for fossils in the bad lands in 1873, the American paleontologist Othniel Charles Marsh (1831–1899) heard about these myths, which played a large part of the cult life of the Ogallalla. He and other paleontologist uncovered the bones of many different brontothere species. These "titanotheres", as they were called, appeared in the North American Eocene as small, hardly sheep-sized forms, and they developed in the Oligocene into sometimes gigantic creatures between the size of a rhinoceros and an elephant.

The white, shimmering soil in the arids of the Dakotas, Nebraska and Wyoming preserved brontothere remains so well that it has sometimes been possible to reconstruct the musculature of these animals. With that began a whole new branch of paleontology: the attempt to gain insight into the soft parts of fossils and their function. The head structures of brontotheres were just as important to paleontologists, and the evolution of these processes has demonstrated the interrelationship between selection and adaptation.

The development of bony processes on the skull

In the oldest brontotheres, the head structures merely existed as thickenings of the skull bones that were not even very noticeable. Even assigning a function to these structures proved to be difficult. "But it is possible that they rammed heads with rivals of the same species and with enemies, even at this early stage", wrote G. G. Simpson in his book *Tempo and Mode in Evolution*. "As they grew into powerful, sturdy creatures with thick skulls, they had practically no other means of fighting. Increasing thickening of the bones in the skull region, the part of the body they used in aggressive and defensive behavior, was an advantage for them."

Thus, what appear to be useless skull processes may actually have been of selective advantage. The trend in brontotheres was to make the skull processes larger, yet more functional structures. Simpson continues, "Thus, surprisingly rapidly, large, effective weapons formed as soon as the size of the animal permitted it to carry such weaponry. The increase in horn size and body size both accelerated, and both trends ran together because both were advantageous and served to adapt the animal to specific habitat conditions." Finally, paired structures were formed. They grew into double horns and looked different in each different species. For example, the genus *Brontotherium* bore a giant bony forked structure on its nose (see Color plate, p. 440).

As we mentioned earlier, brontotheres were one of those mammalian groups that died out without descendants after a relatively short flourishing period. They lived in huge herds, particularly in the flood plains of large rivers, and it is thought they they fed on leaves and buds much as elk do. With their powerful head weaponry, they had few enemies. Furthermore, they were not threatened with volcanos or flooding catastrophes. This has led the Austrian paleontologist Othenio Abel to postulate that the

Victims of a tsetse fly?

sudden disappearance of brontotheres may have been due to a parasite. The very same deposits containing the last brontotheres also contain a relative of the modern tsetse fly, *Glossina oligocenica*. The dry, shrubby plains, which replaced the moist forests in North America at that time, facilitated the proliferation and spread of these flies. In fact, tsetse flies still occur chiefly in arid shrubland. Thus it seems plausible that Oligocene tsetse flies bore trypanosomes, the flagellate protozoans that bear such deadly diseases as sleeping sickness. North American brontotheres first lived in the floodplain forests, where no tsetse flies occur, but during the Oligocene the shrub-covered plains grew, and with them came the tsetse flies. According to Abel's hypothesis, the defenseless brontotheres finally came within range of the flies and succumbed to them.

Extensive climatic changes

Abel's postulated account differs with the opinion of most paleontologists, however, for the majority feel that extensive climatic changes were the cause of the death of these huge animals. These changes caused the moist habitats to turn into arid plains. Brontothere teeth were adapted to chewing foliage and soft plants, not the hard grasses of the prairie. This is also true of most Lower Tertiary South American ungulates (excepting the grassland forms): they all lived on foliage. As the climate and therefore the flora changed, some ungulates "solved" their problem of feeding by changing their dentition in various ways. This evolutionary change was effective in proboscideans, rhinoceros, horses and even-toed ungulates. Why did it not work for brontotheres?

The American paleontologist Henry Fairfield Osborn feels that brontotheres had maintained their developmental trend so powerfully that their particular evolution reached its greatest expression. For some time, their teeth were useful in the moist forests and meadows, and in any other habitat they were inappropriate. Since it was initially useful, brontotheres evolved into more and more specialized foliage feeders, with dentition designed solely for that purpose. They could feed only on foliage, and it was over-specialization that was their doom, according to Osborn. When brontotheres were forced to begin feeding on grasses, they experienced a biological catastrophe. The animals were in their death throes phylogenetically speaking, and they could not recover.

It is also possible that brontotheres died out because of the increasing competition with other ungulates, animals better adapted to the environment. The even-toed ungulates would be particularly superior in this regard. Brontotheres are one of many examples of the odd-toed ungulates, which had been so numerous in the Tertiary, and are now a dying order. The even-toed ungulates have proven to be better adapted to conditions in this geological period, and they have continued to thrive and are still at their height.

19 The Ice Age

By E. Thenius

The Quaternary is the youngest geological period. The name, as we mentioned at the very beginning of Chapter 17, has been handed down from the time when the history of the earth was divided into four great time spans. The period itself is divided into two epochs, the Pleistocene (commonly called the Ice Age and sometimes known as the Diluvium) and the Holocene or Recent, the epoch that began 10,000 years ago and continues today. The Pleistocene, the subject of this chapter, is characterized by extensive glaciation in the polar regions and by glacier movements in the mountains. The Ice Age has not even disappeared from Greenland and Antarctica, so in a sense the geological present is only an intraglacial interim.

Our notions of the time scale of the Ice Age have changed drastically during recent years. This is because the Tertiary cannot be subdivided in any simple, regular way. Opinions on the temporal boundaries of the so-called Villafranchian (named after the Italian village Villafranca) differ among various authorities. It is either considered to be the most recent part of the Tertiary or the earliest part of the Pleistocene, and actually it has both Pliocene and Pleistocene components. Originally it was thought that the Pleistocene lasted 600,000 years, but geologists and paleontologists later discovered that the Pliocene actually ended much earlier, and this led them to date the beginning of the Pleistocene at 1,000,000 years ago. However, still more recent research employing dating of the oldest glacial rocks has caused the Tertiary-Quaternary border to be pushed back even more, and today the beginning of the Pleistocene is dated at approximately 2,500,000 years ago. Since the Holocene began about 10,000 years ago, the Pleistocene would be the shortest completed epoch in the earth's history. It is hardly more than 1/5th the length of the Miocene, the second-shortest epoch. Yet despite the relative brevity of the Pleistocene, the changes wrought on the flora and fauna of the world were so profound that this epoch deserves an entire chapter within this volume.

The major Pleistocene changes were climatological and paleogeo-

Specifying the time period

graphical in nature, and they occurred throughout the world. Since this epoch is not at all in the deep past, it has left a much clearer, more complete record than any other geological epoch. The Pleistocene was a time of continuous alternation between warm and cold periods, and climatic changes occurred even within the glacial parts of the Pleistocene, so it is impossible to strictly separate the colder and warmer periods as individual geological entities.

It was not until the 19th Century that geologists learned that glaciation occurred over many of the continents during the course of the Pleistocene. Glacier traces (called erratic blocks) in northern Germany were ascribed to floating icebergs until their true nature was understood. Glaciation occurred throughout northern Europe as far south as central Germany and in North America as far south as the Great Lakes. Parts of the Alps were glaciated as well.

The Alpine deposits were the first ones used to date the Pleistocene, and since that time deep-sea borings in the Atlantic have been utilized in evaluating data on an absolute time scale. Microfossils have also been useful in learning about the climatic changes occurring during this epoch. Recent deep boring samples from the inland ice of Greenland have proven that this region, covered with ice throughout the year, is a superb fossil "calendar". However, the entire Pleistocene is not represented in Greenland, so the fossils found there are not indicative of the entire epoch.

There are many different opinions about the causes of the Ice Age. During the Tertiary, central Europe began experiencing a cold trend, which gradually led from what was initially a tropical climate to a temperate one. Although there are only sporadic signs of glaciation in the northern hemisphere during the Upper Tertiary, the Antarctic at that time was covered, at least for a temporary period, by an ice cap. This was before there was any indication of cold climate among the terrestrial fauna and flora of what are now temperate or subtropical zones. Boreal (= northern) species (e.g., the Iceland mollusk *Arctica islandica*, the snail *Buccinum undatum*, and the unicellular foraminifer *Hyalinea baltica*) did not appear in the Mediterranean until the beginning of the Pleistocene. This period in central Europe is characterized by the presence of lemmings and ptarmigans. Boreal animals also occurred in the Atlantic and Pacific oceans.

Some scientists feel that extraterrestrial factors were decisive in producing the Ice Age. These would include changes in solar radiation caused by alterations in the earth's orbit or a weakening of solar radiation. Other researchers emphasize terrestrial factors, while still others combine both in their theories. It is important to realize that geological periods with glaciation (e.g., the Permo-Carboniferous and Pleistocene) alternated with those not characterized by glaciation (e.g., the Mesozoic era). During the Pleistocene there was repeated alternation between warmer and colder periods. Three conditions are necessary prerequisites for the formation

Erratic blocks

Causes of the glacial periods

of polar ice caps: 1. A continental block must exist around the pole; 2. An ocean must be present to give off humidity; 3. Specific marine currents are necessary. An ice cap cannot form in an open sea at the pole of a planet, since any ice would simply drift away. Even a largely closed polar sea (like the Arctic Ocean of today) has only a thin ice cover; the Arctic Ocean does not have an ice cap. On the other hand, a polar continent not surrounded by an ocean would not develop an ice cap, either, since it would not have the humidity necessary to form one. Thus, in a period in which large parts of northern Europe and North America were covered by ice shields, many parts of Siberia remained free of ice. The lack of glaciation in Siberia was due to the absence of the necessary humidity.

Therefore, an ice age can only occur when land is in the vicinity of the pole and when that land is surrounded by a polar sea. Recent finds have provided evidence of this phenomenon. They show not only that continental drift did occur but also that during the Permo-Carboniferous periods the south pole was in the area of South Africa and from the Ordovician to the Devonian extended from the Sahara and Congo to South America (eastern Brazil). This pathway can be followed using the glacial signs left behind. These finds also confirm that South America and Africa were once joined together in the great southern continent known as Gondwana (see Color plate, p. 76).

Glaciation of the southern continent

During the Mesozoic and the lowermost Tertiary the polar regions were found outside the continents and in the oceans, so of course no ice ages could occur during that time. During the Tertiary, the antarctic continent drifted into the south polar region and led to the ice age of the Quaternary. The full effect of this ice age made itself known in the Pleistocene, partly because of the fact that terrestrial animals and plants adapted to cold living conditions arose very slowly.

The climatic conditions in the arctic and the central location of Antarctica were probably responsible for the fact that cold and warm periods alternated several times during the Pleistocene. As the ice shield was formed, the northern polar sea was sometimes entirely cut off from the Pacific Ocean by the Bering bridge, and the Iceland swell largely separated this sea from the North Atlantic. In the antarctic, the ice cap grew from the actual shelf region out toward what is called the antarctic convergence, where warmer ocean currents prevented further marginal growth of the ice cap and where less precipitation prevented glacier formation. As the marginal glaciers melted, they may have resulted (as A. T. Wilson believes) in raising the sea level. This eventually would have led to joining the polar sea with the Atlantic and the Pacific oceans. The Gulf Stream was also important, and we must recall that the Panama bridge formed during the uppermost Tertiary. All the warm sea currents of the world are in fact largely responsible for retaining the heat on our planet.

Glacial depositions

Most ice age deposits are sedimentary in nature. They are distributed throughout the world but are most prevalent as continental erosion sediments. Sea formation declined substantially during the Pleistocene, and the continental coasts were largely like they are today. Only around continental shelf regions were sea levels changing that were related to variations in warm and cold periods.

Loess is one of the most characteristic glacial deposits. It is a fine, dusty sand, which was blown into the treeless glacier regions. Loess landscapes formed as a result of this wind activity, and many of them still exist.

The tracks of the Ice Age remain behind as glacier scratches of various kinds, the moraines or glacial walls, and characteristic U-shaped valleys formed by glaciers. Areas outside the glacial regions experienced ground compression caused by the alternating freezing and thawing of the upper levels. Gravel-laden loess and river terraces are other glacial signs, and they sometimes have beds scratched by glaciers as the huge ice bodies moved past. The loess formations are indications of warm periods; depending on the duration and effectiveness of those periods, the loess soil can be brown to a deep red.

Scarcely any mountains formed during the Pleistocene. There are signs of local upswellings near the Alps, underwater movements in the Mediterranean and Pacific, or horizontal soil shifts (which still occur). Volcanic activity during the Pleistocene was no more intense than it is today.

The sea level changes we mentioned earlier were significant paleogeographically, even though they altered the face of the earth very little. These changes were due to the fluctuations in temperature during the warmer and colder periods. The formation of vast ice caps involved considerable amounts of water and of course led to a lowering of the sea level. Entire shallow seas like the North Sea and the Sunda Sea laid dry during the cold periods, and numerous islands became joined to the mainland. This was important for the radiation of terrestrial animals and for the development of insular species. The rise of the oceanic sea level in the warmer parts of the Pleistocene led to flooding of those same places that had been dry before. England was once part of the European continent, as was the Indonesian island region with southeastern Asia, New Guinea with Australia, Japan with eastern Asia and Alaska (via the Bering bridge) with northeastern Asia.

Zoogeography of the
Ice Age

The fossil record regarding the propagation and migrations of Pleistocene terrestrial animals supports the above schema. During the Upper Pleistocene, Alaska was joined with Siberia and was the home of the mammoth, early equids, the yak and the saiga antelope. The interchange of animals between New Guinea and Australia can also be understood in this light. The marsupials, which were "island hoppers", spread from New Guinea to the west as far as Timor and Celebes, while the mice

(Muridae) from southeastern Asia crossed from New Guinea to Australia during the Ice Age and then spread across the entire Australian countryside.

Paleogeographic changes in the form of land bridges do not fully explain the spread of all plant and animal groups, however. Climatic changes also played an important role in species distribution. These included not only temperature fluctuations but also changes in annual precipitation. In tropical regions, moist rainy seasons were followed by drier seasons, although it is not certain whether these seasons corresponded to the warmer and colder periods affecting the temperate zones. During the rainy seasons, the parched, arid deserts of today were covered by savannas or even jungle. The most famous example of this is the Sahara, which consisted of a tree-covered savanna during the Pleistocene and even into the post-glacial period. The diversity of animal and plant life frequenting that hospitable habitat is depicted on many Stone Age rock drawings in the Sahara but only occurs today south of that region. Similar events occurred in Australia; the large inland lakes grew (Lake Callabonna, Lake Eyre), and forests once covered a large, continuous stretch of land. Today the forests of Australia are isolated. The insular distribution of many water-dependent animals (such as amphibians) is further evidence of the Pleistocene climatic changes and their influence on the living organisms of the modern world.

The Pleistocene plant world did not differ greatly from that of today, with the exception that at that time there were many plants in areas now covered by desert and that plants occurring today in the mountains and arctic regions were found in the lowlands during the Pleistocene. What has changed, then, is the distribution of the plants and not their individual species. During cold periods the temperature declined approximately 8–12° C. New measuring techniques (such as the carbonate method) have resulted in determining absolute values of earlier marine temperatures. Of the unicellular organisms of the seas in the Pleistocene, we mentioned only the planktonic foraminifer *Globigerina pachyderma*. At low temperatures its coiled shell rotates to the left, while at warmer temperatures the convolutions wind to the right !

As temperatures declined, the forest border moved downward (southward) 1000–1200 m. Many parts of Europe lacked forestation during the cold periods, and there were extensive cold steppe regions with dwarf birches (*Betula nana*), dwarf willows (*Salix herbacea* and *S. reticulata*) and numerous small plants (*Dryas octopetula* and *Thalictrum alpinum*) covering huge spaces. As we mentioned earlier, the loess steppe was the most characteristic glacial formation. Fossil loess gastropods show that this drifting sand was deposited under different climatic conditions. Loess extends from warm plains with the so-called banatica fauna (named after the gastropod *Helicigona banatica*) to cold plains with a pupilla and columella fauna (named after the gastropods *Pupilla loessica*, *P. muscorum* and *Columella columella*). The loess steppes were the habitat of many mammals, which will be discussed subsequently.

Fig. 19-1. Characteristic flora of glacial steppes of central Europe: dwarf birch (*Betula nana*; 1), dwarf willow (*Salix herbacea*; 2) and small plants such as *Dryas octopetala* (3).

The flora of the Ice Age

The earliest Pleistocene plant communities included a number of remnant species from the Tertiary, such as the fern *Azolla*, the hemlock *Tsuga*, the hickory nut *Carya* and the winged nut *Pterocarya*. They and the other flora of this period show that the beginning of the Pleistocene was characterized (in European latitudes) by a warm, temperate climate. The cold trend occurred slowly and very gradually, and it was then succeeded by a warm phase. The warm-cold fluctuation led not only to the shifting vegetation borders, but also (at least in Europe) to the disappearance of certain plant groups that had survived during the Tertiary. In North America and eastern Asia, plants "moved" southward during cold periods and "returned" to the north when the warm trend returned. However, the Mediterranean and the Alpina-Carpathians served as barriers in Europe, and the European fauna declined in number of species during the fluctuating climate. Since reforestation following cold periods did not always occur with the same successional sequence, the individual pollen communities found in Pleistocene deposits in Europe can be used to determine specific warm periods. Pollen was retained in what once were lakes and seas.

During the warm parts of the Pleistocene, European flora were diverse and consisted of such plants as hazel nut, linden, oak, walnut, conifers (*Thuja*) and ivy. Warmth-loving plants, like the rhododendron (*Rhododendron ponticum*), which today occur in Asia Minor, were also found in Europe. The forests housed elephants, Merck's rhinoceros, the aurochs, fallow deer and wild pigs. They occurred in Europe as late as the last intraglacial period.

Marine flora, as far as is known, did not differ substantially from those found in Tertiary seas. Only a few species disappeared or arose during the Pleistocene. However, the distribution of those plants wavered considerably and changed with climatic changes. In the Atlantic, a few planktonic calcareous flagellates became extinct (e.g., *Discoaster challengeri*, *D. pentaradiatus* and *D. brouweri*). *Coccolithus pelagicus* died out at the beginning of the Pleistocene.

Only a few animal groups can be followed phylogenetically during the Pleistocene, and we shall describe only those groups differing structurally from Tertiary and modern groups. In some cases the differences were solely geographical.

Protozoans

About all of interest concerning UNICELLULAR ORGANISMS is their geographical distribution. During the cold parts of the Pleistocene, northern marine forms moved far southward, reaching in Europe the Mediterranean and in the New World the Caribbean. As warm climatic conditions returned, these organisms retreated poleward again and back to those regions where they are found today. Thus, the beginning of the Ice Age is associated with the extinction of numerous foraminifers (e.g., *Globigerinoides fistulosus*) and radiolarians (e.g., *Pterocanium prismaticum*) in the Atlantic and Pacific, respectively, and with the first appearance or prevalent distribution of *Hyalinea baltica* in the Caribbean Sea and *Globorotalia trun-catulinoides* in the southern North Atlantic.

A similar situation existed for MOLLUSKS. At the beginning of the Pleistocene we suddenly find many northern species in the Mediterranean region. Particular examples include the ocean guahog (*Arctica islandica*) and the snails *Buccinum*, *Natica montaculi* and *Trophonopsis muricatus*. The sudden appearance of North Atlantic snails and bivalves at the very beginning of the Pleistocene in the Mediterranean region was caused by a substantial cooling of the weather in that part of the world. The same climatic change led to the disappearance of coral along the North Atlantic coast, where they had been thriving as late as the Upper Tertiary.

The composition of echinoderm and marine fish fauna also are indicative of the influence of cooler marine currents. However, this influence did not lead to genetic changes (i.e., evolution). Only a few animal groups show modifications that are phylogenetic in nature; this is not surprising in view of the relative brevity of the Pleistocene. We shall turn first to these animals and then to a discussion on a topic we brought up in the chapter on the Cretaceous: the disappearance of numerous animal groups. This happened at the end of the Pleistocene, just as we witnessed the disappearance of so many species toward the end of the Cretaceous.

The most striking evolutionary advances taken during the Pleistocene were undoubtedly within the ELEPHANTS (Elephantidae). Elephants are the most advanced proboscideans (order Proboscidea), a group widely distributed during the Tertiary in the form of mastodons, stegodons and dinotherians. The ancestors of elephants were Upper Tertiary mastodons, and the fossil record suggests that elephants may have existed as early as the Upper Pliocene, although these would be very primitive elephants. The earliest confirmed elephant fossils are from the oldest Pleistocene deposits. Whether they arose in Africa or southern Asia is still undecided.

The earliest elephants (*Archidiskodon et al.*) have several primitive characteristics, the most prominent ones being the largely vestigial low jaw tusks and the low-crowned molars with their few ridges. During the glacial period in the earliest part of the Pleistocene, these warm-plains elephants (e.g., *Archidiskodon meridionalis*) gave rise to both the early elephants *Palaeoloxodon* (with the species *P. antiquus*), which in some respects look like modern African elephants and which were forest dwellers occurring in Europe during warm parts of the Pleistocene; and to the MAMMOTHS (*Mammonteus*). Mammoths, which were plains dwellers, had high-crowned molars and more tooth ridges (now lamellae). In the last (Upper Pleistocene) mammoth, *Mammonteus primigenius*, there are thirty lamellae on the third molar (see Color plate, p. 480). Unlike its precursor from the Lower and Middle Pleistocene (the plains mammoth *M. trogontherii*), the Upper Pleistocene mammoth lived exclusively on cold plains and was covered with a thick fur of wool and bristles. It had small ears; a high, dome-shaped skull and powerful, highly coiled tusks.

Mollusks

Development of plains and forest elephants

These mammoths were distributed throughout the northern parts of the Old and New World, and they died out toward the end of the Pleistocene or even (in some regions) thereafter. The role of the mammoth, the most well-known Pleistocene animal, is discussed in the following chapter, where we also discuss the finds of intact mammoths frozen in the Siberian permafrost zones.

During the Pleistocene, elephants were in Africa, Eurasia and North America, and they may have reached northern South America as well. They were practically as widely distributed as the mastodons in the Upper Tertiary. Their rapid, far-reaching propagation was related to their diet. The transformation (a phylogenetic one!) from low-crowned ridged teeth to high-crowned lamellar teeth enabled these animals to feed on harder plant materials such as the grasses found in plains and savannas. A parallel evolutionary development is found in both equids and in rodents (cricetid mice). The dinotherians (superfamily Dinotherioidea), which were restricted to Africa, became extinct toward the end of the Lower Pleistocene, as did European mastodons. Mastodons did not disappear in North America until the end of the Pleistocene. Asian stegodons also survived until the Upper Pleistocene.

Cricetid rodents   The most recent rodent group are the CRICETIDS. They first appeared during the Upper Tertiary and developed into a diverse group whose variety continues to express itself. The most striking evolutionary modifications occurring in cricetid rodents are the alterations in the structure of the molars. Originally, mice had low-crowned, rooted teeth (e.g., *Mimomys, Cosomys*). During the course of their evolution they first developed high-crowned teeth and finally rootless molars (as in the lemmings and voles). These rootless teeth consist of individual prisms whose numbers can multiply. They are superb for chewing hard plant materials, and these rodents continue to display their chewing talents for thousands of angry human beings. The fossil record shows that within numerous lines there was a rapid surge in evolution of cricetids during the Pleistocene. Pure terrestrial forms were supplemented by species adapted to living in water (e.g., the muskrat, *Ondatra zibethica*).

Climatic changes, and the habitat modifications caused by these changes, were also significant for the cricetid group. During cold periods, boreal species such as lemmings (*Lemmus* and *Dicrostonyx*) and voles (*Microtus gregalis* and *M. nivalis*) inhabited central Europe, and plains species such as the steppe lemming *Lagurus* spread into western Europe. Other plains rodents also occurred in central and western Europe, including such animals as dwarf hamsters, dwarf jerboas, ground squirrels, and plains marmots. Lagomorphs were represented by animals such as pikas.

True mice   MURID RODENTS (Muridae) assumed the ecological role of cricetids in southern parts of Eurasia, Africa and Australia, where they proliferated into a richly diverse group. Cricetids differed little from one another in

▷▷
A European Middle Eocene lignite moor. Flora include reeds and pondweeds (f, *Potamogedon*), myrtles (d, *Myrica*), willows (e, *Salix*), fan palms (a), aquatic spruces (b, *Glyptrostrobus*) and swamp cypresses (c, *Taxodium distichum*). Fauna: A giant boid (1, *Paleryx*), soft-shelled turtle (2, *Trionyx*), alligator (3, *Diplocynodon*), rhinoceros bird (7, *Geiseloceras*), hyena (5, *Prodissopsalis*), a primitive horse (4, *Propalaeotherium*) and a prosimian (6, *Adapis*).

▷
Phylogeny of the proboscideans (Proboscidea): Middle Eocene African moeritheres, animals resembling hippos, gave rise to the most primitive proboscideans (based on dentition and brain structure) and then continued through the primitive mastodons (*Palaeomastodon*, the ancestor of the Miocene genus *Mastodon*), which diverged into two major lines, one leading to elephants (we show the African elephant, *Loxodonta*) and the other to African-Asian stegodons (not shown), a group that later became extinct. One mastodon branch includes *Stegomastodon*, from Pliocene/Pleistocene North America.

comparison to the variety that developed among the mice. The evolution of Australian mice is particularly interesting. The modern wealth of Australian mice species, which comprises mouse- and ratlike animals (*Pseudomys* and *Uromys*), squirrel- and rabbitlike creatures (*Mesembriomys* and *Leporillus*) and vole- and jerboalike species (*Mastacomys* and *Notomys*) all originated during the Pleistocene. The process occurring in Australia was simply adaptive radiation of mice (i.e., phylogenetic diversity coming about as a result of interacting with the environment and natural selection). Large insular species also evolved, including *Canariomys* from the Canary Islands, *Spelaeomys* and *Papagomys* from Flores and *Coryphomys* from Timor.

The counterparts to the Old World mice and also to the voles are the NEW WORLD MICE (generic group Hesperomyini), which developed into a tremendously varied group, especially in South America. New World mice migrated from North America to South America during the Upper Tertiary, and they developed into numerous new species during the Pleistocene. These were adapted to specific habitats, and some of them bore an astonishing resemblance to voles. The similarity existing between those specific New World mice and the voles of Europe was not solely one of appearance and burrowing life habits but also a similarity in such fine details as the anatomy of the molar teeth. In the New World mice, the molars were high-crowned and either rootless or built of prisms, as in the voles. New World mice included species adapted to open land (such as the cotton rat and related species) and marsh- and water-dwelling forms (the *Scapteromys* and *Ichthyomys* groups), including even some fish-eating species (*Rheomys*). Other rodents from the Pleisto-

New World mice

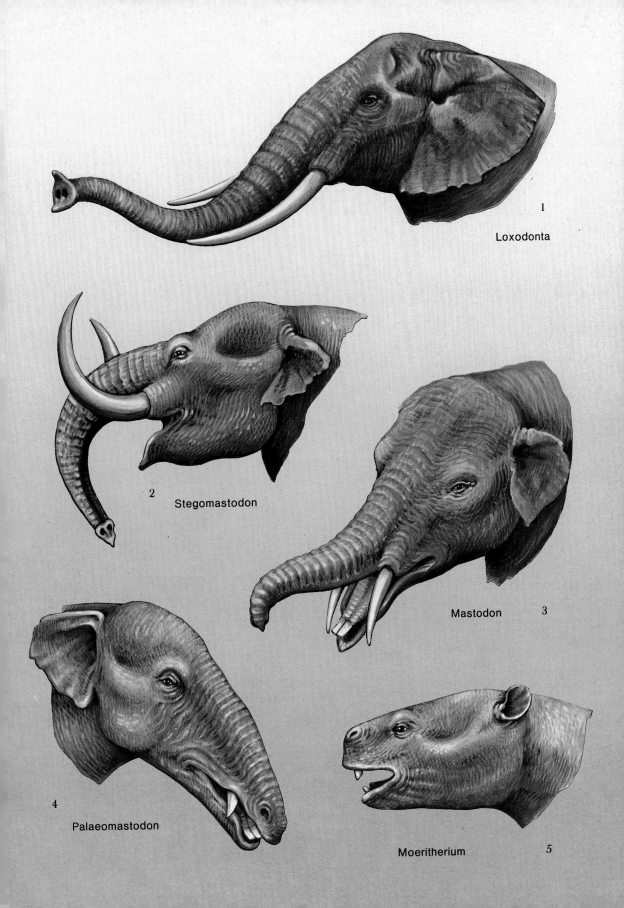

1 Loxodonta

2 Stegomastodon

3 Mastodon

4 Palaeomastodon

5 Moeritherium

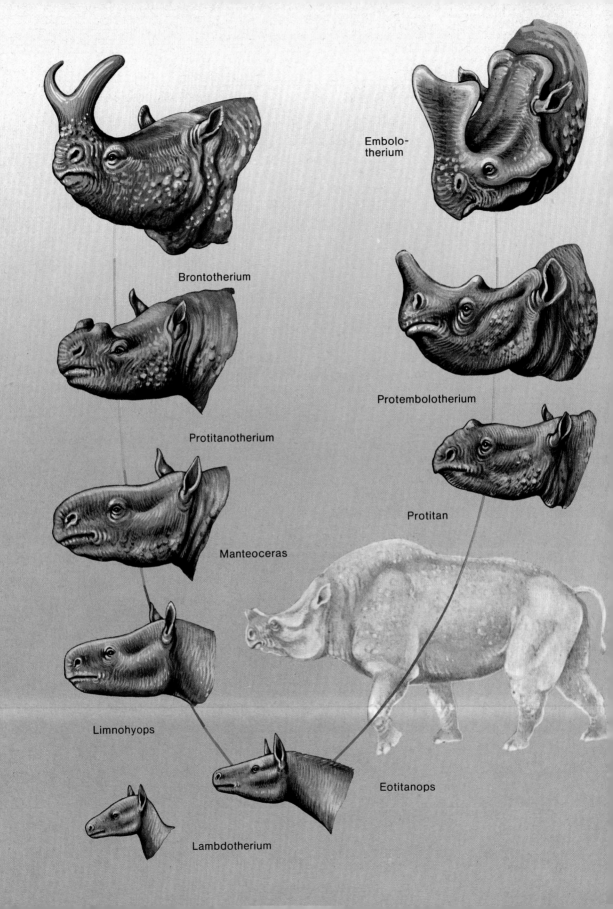

Embolo-
therium

Brontotherium

Protembolotherium

Protitanotherium

Protitan

Manteoceras

Limnohyops

Eotitanops

Lambdotherium

Giant beavers and giant doormice

cene include the GIANT BEAVERS (*Trogontherium* in Eurasia, *Castoroides* in North America) and the GIANT DOORMICE (*Leithia*) from Mediterranean islands; these are all extinct.

Other mammalian groups that did not fully develop until the Pleistocene include the big cats, wild canids, wild cattle and goats (including sheep and goats, generic group Caprini). They first arose in the last part of the Tertiary, but all these animals truly proliferated in the Ice Age.

The BIG CATS or LARGE FELIDS (generic group Pantherini), which presently occur in every continent save Australia and Antarctica, first appeared in the Lower Pleistocene with a panther-sized animal (*Panthera gombaszögensis*), which gave rise to the leopard and the lion of today. Lions first occur in the Lower Quaternary, and they were found in Europe during the Ice Age. The cave lion (*Panthera leo spelaea*; see Fig. 19-2) was found in Europe as late as the Upper Pleistocene, and lions did not disappear from Europe until historical times. The tiger (*Panthera tigris*), on the other hand, has apparently always been restricted to Asia. Alleged fossil tigers from Europe and Africa have consistently been discounted (see also Othenis's erroneous interpretation of a cave lion drawing in Fig. 19-2.) *Panthera cristata*, a big cat from lowermost Pleistocene southern Asia, may have been related to the tiger. Snow leopards (*Uncia uncia*) presently occur only in Asia, but during the Ice Age they were also found south of their present distribution. Jaguars make their earliest appearance in the lowermost Pleistocene of North America (*Panthera palaeonca*). One extinct jaguar, *P. atrox*, may have been a kind of giant jaguar, but it did not inhabit the jungle like the modern jaguar; it was found on the open plains.

◁
Phylogeny of brontotheres (or titanotheres) from Lower Tertiary North America and eastern Asia: The genus *Lambdotherium*, which resembled the oldest equidlike animals (paleotheres), gave rise to two major groups, the North American and the eastern Asian genera of brontotheres. The similarity between these two groups is most likely due to American brontotheres migrating to Asia across the Bering land bridge. In both brontothere lines we see prominent nasal outgrowths, particularly in the most recent members of each line (*Brontotherium* and *Embolotherium*). The American genera died out in the Lower Oligocene, while the Asian brontotheres survived into the Middle Oligocene.

The puma (*Puma concolor*), once erroneously classified as a big cat, is actually a member of the SMALL FELID group. Its immediate ancestry dates to a North American puma from the earliest Pleistocene, *Puma studeri*. *Felis lunensis*, from lowermost Quaternary Europe, is related to modern wild cats and is thought to be the immediate ancestor of the European wild cat, *Felis silvestris*, a species that appeared in the Middle Pleistocene. Two felids lived in Europe during the Ice Age and are still extant but no longer found there; the manul or Pallas' cat (*Otocolobus manul*) occurred in central Europe during the Upper Pleistocene but is now found in central Asian mountains; likewise, the jungle cat (*Felis chaus*) was in Upper Pleistocene Europe but now chiefly occurs in many parts of Asia and north Africa. Its westernmost occurrence outside of Africa is west of the Caspian Sea. Cheetahs (*Acinonyx*) were also found in Europe during the Lower Pleistocene.

Saber-tooth cats

The Pleistocene was a time not only of big cats as we know them today but also of the SABER-TOOTHED CATS, whose precursors we met in the chapter on the Tertiary. The most striking characteristic of these short-tailed big cats are the dagger-shaped elongated, laterally compressed canine teeth of the upper jaws. The edges of the fangs were finely jagged.

The saber-toothed cats were members of the machairodontins (subfamily Machairodontinae), which arose in two separate groups. They include the stocky-legged genus *Smilodon* and the slender-legged genus *Homotherium*. Saber-toothed cats disappeared in Europe during the Pleistocene, but they survived until the end of the epoch in North America. *Smilodon*, the last of the plump-legged forms, was a lion-sized cat that arose from *Megantereon* from the lowermost Pleistocene. *Homotherium* was ancestral to the most advanced slender-legged saber-toothed cat, *Dinobastis*. Various opinions have been expressed about the diet of saber-toothed cats. Since a surprisingly high number of these animals were victims of certain bone diseases (chronic joint ailments and pathological changes in the vertebrae), it is sometimes thought that saber-toothed cats fed largely on carrion and were not true predators. However, justified objections to this theory have been advanced. The group died out during the uppermost Pleistocene (see Color plate, p. 422 and Fig. 17-23).

Wild canids

The modern wealth of WILD CANIDS (subfamily Caninae) stems from one or more adaptive radiations during the Ice Age. The first canids appear as raccoon dogs (*Nyctereutes*) and with members of the genus *Canis* in the Upper Tertiary. Primitive foxes and wolves arose in the earliest part of the Pleistocene. They distributed themselves almost all over the world during the Ice Age, and in the post-glacial period the dingo (a primitive domestic dog, which later became feral) was brought to Australia by humans. The evolution and disappearance of fox and wolf species is well represented in the fossil record. A number of the unusual South American canids (e.g., *Dusicyon*, *Cerdocyon* and *Atelocyon*) arose during the Ice Age; their ancestors did not reach South America until the Panama bridge was formed late in the Tertiary. The bush dog (*Speothos venaticus*) is a specialized member of the generic group Speothonini, while the maned wolf (*Chrysocyon brachyurus*) occupies a distinct position. These are both "ecological niche forms", which arose by adapting to certain ecological conditions or niches where they lived. Of geographical interest is the fact that red dogs or dholes (*Cuon*), arctic foxes (*Alopex lagopus*) and corsac foxes (*A. corsac*) occurred in Pleistocene Europe; none of them do today. Raccoon dogs were distributed as far west as western Europe during the Ice Age, and only today are they beginning to reappear in Europe.

Mustelids

Among the MUSTELIDS, all the major groups appeared during the Tertiary, so only developments within these groups and the distribution of individual species is of interest here. The Tertiary ancestors of the wolverine were apparently plains animals (*Plesiogulo*), and the wolverine did not spread across northern Eurasia and North America until the Pleistocene. The wolverine and other mammals from the far north moved into central Europe during the cold parts of the Pleistocene and even into the post-glacial period. Martens, ferrets, polecats, weasels, badgers and river otters have been found in Pleistocene central Europe along with

Fig. 19-2. The wall of the cave in Combarelles, France (in the Dordogne) contains a drawing by Magdalenian man of what is called the "lion of Combarelles". Cave lions were a subspecies of the modern lion; the old opinion that the stripes on the animal's head and lower jaw here signified that this animal was a tiger or a "liger" has been discredited.

Fig. 19-3. Ice Age cave drawing in the grotto Les Trois Frères, near Montesquieu, France. It depicts a brown bear killed by arrows and rocks that "cave" men threw at it. The drawing is 60 cm long.

Fig. 19-4. Ice Age relief drawing of a cave bear near Poncin, France.

the marbled polecat (*Vormela*) and the extinct genus *Baranogale*. There were also mustelids during the earliest Pleistocene related to the grisons and tayras of modern South America (*Pannonictis*). Closely related contemporaries (*Trigonictis*) have been described from North America, and they were probably the ancestors of the modern South American mustelids. In addition to otters of the *Lutra* group, which includes the modern river otter of Europe, members of another otter line lived in Ice Age Europe. B. Kurtén classifies them as relatives of the clawless otters of Africa (*Aonyx*-group). Pleistocene skunks are only known from the New World, and they inhabited South America during the Pleistocene epoch.

With the extinction of *Agriotherium* in the earliest Pleistocene, the bear branch line known as agriotheriins (subfamily Agriotheriinae) died out. The short-snouted bears (Tremarctinae), whose last survivor is the spectacled bear, lived in North and South America with some large species (*Arctodus*); they disappeared toward the end of the Ice Age. In the Old World, bears (subfamily Ursidae, including the subgenera of black bears, *Euarctos*, and brown bears, *Ursus*) were distributed throughout Eurasia. They reached North America during the early Ice Age. One of the largest of them was the cave bear from the Upper Pleistocene (*Ursus spelaeus*; see Fig. 19-4), which died out toward the end of the epoch. Cave bears and brown bears evolved from a common ancestor, *Ursus etruscus* from the earliest part of the Quaternary. Polar bears (genus *Thalarctos*), which first appeared late in the Pleistocene, are specialized descendants of brown bears (see Color plate, p. 477).

The New World RACCOONS (family Procyonidae) and the Old World PANDAS (Ailuridae) are of paleogeographical interest. As late as the end of the Tertiary, the genus *Parailurus* inhabited Europe and then retreated to Asia. The giant panda (*Ailuropoda*) also had a much more extensive distribution during the Pleistocene; it was found throughout southern China. The Pleistocene climatic changes powerfully affected the habitat of these animals, and they were forced to retreat into the small area where they are found today.

VIVERRIDS (Viverridae) have not been found in Pleistocene Europe, so it is assumed that the genet (*Genetta genetta*) came from Africa after the Ice Age, crossing Iberia and spreading into southern France. HYENAS (Hyaenidae), on the other hand, lived in Europe throughout the Pleistocene, and with the CHEETAH-HYENAS (*Euryboas* = *Chasmoporthetes*) in the Villafranchian they temporarily became the only hyenas in North America. Striped and spotted hyenas lived in Europe until the end of the Pleistocene. *Hyaena* (*Pliohyaena*) *perrieri* from the early Pleistocene is a relative of the modern brown hyena. The cave hyena (*Crocuta crocuta spelaea*), in contrast, was only a subspecies of the spotted hyena. It disappeared from Europe toward the end of the Ice Age.

Among the FUR SEALS of the northern seas, the fluctuations in cold and warm periods and the sea level changes caused by these led to repeated separation of the distributions of fur seal populations and therefore to speciation (and the formation of subspecies as well). WALRUS included cold-water forms, which belong to the modern genus *Odobenus*, and more southerly warm-water species (*Trichecodon*).

Fig. 19-5. The European bison, *Bison priscus*, with its large, powerful horns.

As we mentioned earlier, the CATTLE or BOVINES (Bovinae) and SHEEP and GOATS (generic group *Caprini*) are phylogenetically very recent groups that developed during the Pleistocene. The first bovines were apparently buffalo- and antelopelike forms from the Pliocene. The Asiatic water buffalo arose during the Pleistocene (genus *Bubalus*), and for a time it inhabited Europe and northern and eastern Africa. Other new Pleistocene forms included the African buffalo (*Syncerus*), bisons (*Bison*), and finally true cattle with the subgenera koupreys (*Bibos*) and oxen (*Bos*). Koupreys, gaurs and bantengs remained in southeastern Asia, while the true cattle spread from Asia to Europe and north Africa. One subgenus, the yak (*Poëphagus*), temporarily inhabited Alaska. Yaks later became extinct in Alaska, but bison moved into Eurasia and North and Central America during the Pleistocene, using the Bering land bridge to cross into the New World. Plains bison, with their huge powerful horns, lived in Europe (*Bison priscus*) and North America (*B. crassicornis* and *B. latifrons*). The last surviving species of this group of wild cattle, which were so numerous during the Pleistocene, are the North America bison (*Bison bison*) and the European bison or wisent (*B. bonasus*). Water buffalos, being dependent on the presence of water, were limited in the extent they could spread throughout the world and consequently did not move into nearly as many habitats as the true wild cattle. Numerous wild cattle species lived in woods and on plains, sometimes covering huge areas. One of them, the aurochs (*Bos primigenius*), was the ancestor of our domesticated cattle.

Fig. 19-6. The Aurochs (*Bos primigenius*), the ancestor of our domesticated cattle.

GOATLIKE SPECIES (subfamily Caprinae) probably arose during the Tertiary, i.e., somewhat earlier than wild cattle. We distinguish the sheep and goats (belonging to the tribe *Caprini*) from the chamois and other members of the subfamily. Sheep and goats are known from the Upper Tertiary in primitive forms (*Tossunnoria*, *Pachytragus*). Modern genera (e.g., tahrs, ibex, Barbary sheep, blue sheep and true sheep) arose during the Pleistocene. Their original habitat was in Asia, and from there they spread into Europe, many parts of Africa, and North America. The only species reaching North America, however, were members of the sheep genus *Ovis*. As in bison, sheep probably migrated more than once to the New World. The present disjunct occurrence of various goat species is an indication that all these species once had a continuous distribution. Ibex (*Capra ibex*), for example, occurred throughout central and western Europe during the Upper Pleistocene. Temporary geographical isolation of goat populations later occurred and led to the development of subspecies.

MUSK OXEN, which are now confined to the far north, occurred during the Pleistocene with the modern genus *Ovibos* and the genera *Praeovibos*

Fig. 19-7. Musk oxen were widely distributed across Eurasia and North America during the Ice Age, and they existed in numerous varieties. Today there is just one species left of this cattlelike animal, which is actually a relative of the sheep. That species, *Ovibus moschatus*, is found in northernmost North America and in Greenland, and it has been artificially introduced into Spitsbergen and elsewhere in Norway.

and *Symbos* throughout much of Eurasia and North America. The PRONGHORNS, which now occur with just a single species in North America (the pronghorn antelope, *Antilocapra americana*), gradually declined. There were several pronghorn genera during the Pleistocene (*Capromeryx, Tetrameryx, Antilocapra*), but the merycodonts and oreodonts, which were prevalent in the Upper Tertiary and North American Tertiary, respectively, had both become extinct. Of the ANTELOPE group, close relatives of the modern Indian blackbuck lived throughout Asia (*Spirocerus*) and southern Europe (*Gazellospira*).

Giant elk

The DEER family in the Pleistocene was represented by the giant elk (*Praemegaceros* and *Megaloceros*; see Color plate, p. 464 of the giant extinct Irish elk, *Megaloceros giganteus*). Their antlers spanned more than 3.5 m. These giant elk died out after the end of the Pleistocene. Others occurring during the Pleistocene included modern deer species (e.g., roe deer, American deer, muntjac, axis deer, fallow deer, red deer, reindeer and elk). Their distribution at that time was considerably larger than it is today, and most deer genera in the Pleistocene were represented by species other than those we know today). The fossil record permits us to follow the course of moose antler evolution quite clearly. A phylogenetic line of descent runs from "*Libralces*" (*Praealces*) *gallicus* through the moose *Alces latifrons* to the modern moose, *Alces alces*; the trend occurring in the antlers is a shortening from very extensive, broad antlers to shorter ones. In contrast, the red deer group developed more extensive antlers, which in *Cervus elaphus* reached their highest state of development in the form of a true set of cup-shaped antlers. Reindeer (*Rangifer*) played an important role in the life of late ice age man; these reindeer occurred in central and western Europe, even during cold parts of the Pleistocene.

Fig. 19-8. Black wall drawing from Lascaux, France: a group of swimming red deer with upheld heads.

During the Pleistocene, the TYLOPOD group (order Tylopoda) spread from their original North American habitat to southern Europe and northern and eastern Africa in the form of camels (*Camelus*). Llamas (*Lama*), also descended from tylopods, spread into South America and have remained there into the present. Camels remained in North America during the Pleistocene (*Camelops, Tanupolama*) and did not disappear from there until the end of the Pleistocene or the early post-glacial period.

The evolution of PECCARIES (Tayassuidae) is more complex, for the group apparently fluctuated repeatedly between Central and South America. During the Pleistocene, there were several North American peccary genera (*Platygonus, Mylohyus, Tayassu*), whereas today they are all found in the tropics of the New World. Among the OLD WORLD PIGS (Suidae), wart hogs were at their phylogenetic height. Several genera (*Metridiochoerus, Phacochoerus* and *Notochoerus*) with numerous species lived in Pleistocene Africa, and as late as the Upper Pleistocene there were wart hogs in northern Africa and Asia Minor.

Fig. 19-9. Rock drawing from Le Gabillou in France.

A similar situation existed for HIPPOPOTAMUS (Hippopotamidae). During the Pleistocene they were found in central Europe and in Southeastern Asia, and pygmy forms occurred on Mediterranean islands (*Hippopotamus pentlandi*) and on Madagascar (*H. lemerlei*). Neither of these was closely related to the modern pygmy hippopotamus; they descended from the large hippopotamus genus (*Hippopotamus*).

With that we turn now to the exclusively Pleistocene mammalian groups. In Australia, the MARSUPIALS radiated into many different ecological niches, something they could do because of the absence of placental mammals in that part of the world. Some of these marsupials evolved into huge giant forms (now extinct). Their disappearance was in part related to climatic changes, which altered the fauna on which they fed. The nototherians (*Diprotodon*, *Nototherium*, *Euowenia*, *Euryzygoma*) of Australia assumed the role taken elsewhere by ungulates, and some of these reached the size of a rhinoceros. The short-snouted kangaroos (*Sthenurus*, *Procoptodon*) were also very large. Wombats also produced a giant form, *Phascolonus gigas*. One of the most interesting of all Australian Pleistocene marsupials was the marsupial lion (*Thylacoleo carnifex*), which was nearly as big as a modern lion (see Fig. 19-11). Its scientific and common names reflect the fact that biologists initially thought this animal was a carnivore, but recent evidence has disproven this hypothesis. True, the enlarged, sharp molars look very much like the kind of dentition we find in carnivores. However, the "marsupial lion" is now classified as a member of the phalanger family (Phalangeridae); its modified molars were merely cutting tools used on the kinds of plants on which this animal fed.

The EDENTATES (order Edentata), which are restricted to the New World, also developed into giant forms in Pleistocene South America (see Fig. 19-12). GLYPTODONTS (Glyptodontidae; see Color plate, p. 493) included species with a L of 3.5–4 m. It would be more appropriate to call these animals "giant armored animals" and not "giant armadillos" to avoid confusing them with the true armadillo group (Dasypodidae), which also contains a giant member (*Priodontes giganteus*). The armadillos and other glyptodonts form an infraorder within the edentate group known as Cingulata. Glyptodont armor consisted of stiff, heavy, polygonal bony plates. The bony tail terminated in some species (*Doedicurus*) in spines and may have been used to defend against enemies. The giant armored animals were savanna and plains inhabitants, and their high-crowned molars suggest that they fed on plains grasses.

The GIANT SLOTHS (Gravigrada), known now since the mid-19th Century, are even more well known than the glyptodonts. During the Upper Pleistocene, giant sloths entered North America. They were as large as cattle or even elephants (*Eremotherium*, *Mylodon*, *Megalonyx*, *Nothrotherium* and *Megatherium*; see Color plate, p. 453), and some had a L of 7 m. Like glyptodonts, giant sloths were herbivores. Their fossilized feces (found in caves along with skeletal remains and sometimes even pieces of their fur) contain twigs, leaves and plants. Some of these purely ground-

Fig. 19-10. Skeleton of a giant marsupial (*Diprotodon*) from Australia.

Fig. 19-11. The marsupial lion (*Thylacoleo carnifex*) was formerly classified as a carnivore due to its sharp molars, but today it is recognized to be a member of the phalanger family.

Fig. 19-12. Reconstruction of *Glyptodon clavipes* (right) and *Doedicurus clavicaudatus* (L over 4 m) from the Ice Age pampas loess of Argentina.

Fig. 19-13. Reconstruction of the Ice Age giant sloth *Mylodon robustus* from Argentina's pampas. It is somewhat smaller than *Megatherium*. Preserved fur from *Mylodon domesticum* found in the cave Ultima Speranza on the Argentinian coast permitted paleontologists to reconstruct the entire animals.

Fig. 19-14. A painting of woolly rhinoceros from Rouffignac, France.

Fig. 19-15. Woolly rhinoceros from the Aurignacian period, etched on a limestone disk and found in the rocky cave of Colombière (France). The neck, nape and flanks are shown to be heavily fur-covered, and perhaps the fur was particularly long on these parts of the body.

dwelling giant sloths even reached Alaska. The bones, fur and droppings of what has been called the "domesticated giant sloth" (*Mylodon domesticum*) were found in a cave subsequently sealed off by contemporary men in Ultima Esperanza in South America. Biologists initially thought that since the cave was sealed off, early men must have kept the giant sloths as domestic animals. However, it is more probable that post-glacial prehistoric Indians (who must have encountered these animals) closed up the cave containing the giant sloth in question and then subdued the creature by smoking it out.

Although ground-dwelling sloths spread successfully across North America, they never reached the Old World. Some sloths were restricted to the Antilles island group (*Megalocnus, Acratocnus, Microcnus*). They were much smaller than their continental relatives and probably could climb. In some of them the front teeth were modified into rodentlike teeth. Another group of scaled animals unrelated to edentates, the members of the order Pholidota, evolved into giant forms that once lived on Borneo.

Just as the giant armored animals and giant sloths were among the typical New World fauna of the Pleistocene, Madagascar contained a number of giant PROSIMIANS. We discuss them in detail in our chapter on primate evolution (see Chapter 21), where we also discuss the giant African baboons and the giant anthropoid apes of eastern Asia. Some of the giant lemurs on Madagascar lived in historical times and were driven into extinction by humans.

The once flourishing ungulate fauna of the Americas, with so many distinctive orders found throughout the Tertiary, were either driven into extinction or were displaced by newly arriving species (see Chapter 17). The last members of the NOTOUNGULATES (order Notoungulata) and the LITOPTERNS (order Litopterna) survived in the Pleistocene (*Toxodon* and *Macrauchenia*, respectively). The number of species of odd-toed ungulates (order Perissodactyla) also declined, and the only ones to flourish in the Pleistocene were the EQUIDS.

Inhabitants of the open plains, equids spread during the Pleistocene into every continent on earth save Australia and Antarctica. They developed into a great wealth of species like zebras, asses, horses, and some extinct groups (e.g., *Hippidion* and *Onohippidium* from South America).

RHINOCEROS (Rhinocerotidae) lived in Europe during the Pleistocene, where there were half-armored rhinos (*Dicerorhinus etruscus, D. kirchbergensis*), woolly rhinoceros (*Coelodonta antiquitatis*; see Color plate p. 478/479) and members of an extinct side branch, the mighty elasmotheres (*Elasmotherium caucasicum*). *Elasmotherium*, whose skull was nearly one meter long, was a plains dweller. The woolly rhinoceros lived in a cold habitat and was a contemporary of the Upper Pleistocene mammoth. The well-preserved bodies of woolly rhinoceros from Siberian permafrost zones and the detailed cave drawings of them executed by Upper Pleistocene man (see Fig. 19-15) are evidence that these animals had a thick woolly fur covering, small ears and broad lips. The lips resemble those of the modern wide-mouthed rhinoceros, but this is not indicative of a close relationship between the two species, merely the occurrence of convergent evolution. In this giant animal, *Elasmotherium*, it is noteworthy that the molars lacked roots and that the enamel was ridged. Both these features suggest that the elasmothere fed on hard food (namely plains grasses).

We shall only briefly mention the development and spread of the other mammals of the Pleistocene. TAPIRS (Tapiridae) crossed the Panama bridge during the Pleistocene into South America. Another group of odd-toed ungulates, the CHALICOTHERES (suborder Ancylopoda, family Chalicotheriidae), had disappeared from Europe and North America in the Upper Tertiary, and their last members (*Circotherium* and *Ancylotherium*) are from glacial deposits in Asia and southern and eastern Africa. HYRAXES (order Hyracoidea) lived in eastern Asia as late as the Lower Pleistocene, where they developed into large forms (*Postschizotherium*). The only African hyraxes known from this period are members of the family Procaviidae (the modern hyrax family), although some of these hyraxes (*Gigantohyrax*) were considerably larger than their modern relatives.

The ungulate order of DESMOSTYLIANS (Desmostylia) from marine deposits of the North Pacific Ocean had become extinct during the Pleistocene along with so many other ungulate groups, and there are no confirmed finds of Pleistocene desmostylians. Among the SIRENIAN order (Sirenia), relatives of Steller's sea cow spread as far as California.

The avian fauna of the Pleistocene are much better represented in the fossil record than they were in the Tertiary, but we shall only mention a few noteworthy extinct groups of birds. Among them were the GIANT OSTRICHES of Madagascar (superorder Aepyornithes) and the New Zealand moas (Dinornithidae). Ancestors of the Madagascar giant ostriches occur in Lower Tertiary deposits; the last species from Madagascar (*Aepyornis maximus*; see Fig. 19-16) attained a standing height of well over 3 m and laid eggs with a diameter of 35 cm. Like the giant lemurs of Madagascar, the giant ostriches were encountered by man.

MOAS, which with kiwis are classified into the flightless order Apteryges, were highly specialized running birds lacking wing bones. Moa species came in varying sizes and included slender-legged and plump-legged species (*Dinornis*, *Anomalopteryx* and *Pachyornis*; see Color plate, p. 494). The last moas were extirpated in historical times by the Maoris (the New Zealand aborigines) prior to arrival by Europeans. Even today we still find moa skeletons, feathers, pieces of skin and soft body parts. The giant ostriches of Madagascar and the moas only survived into historical times because of the lack of competing species for them on their isolated islands.

The long period of isolation of New Zealand was also the reason for the development of still other flightless birds. The most well-known of them are the GIANT RAIL (*Aptornis*) and the GIANT GOOSE (*Cnemiornis*; see Color plate, p. 494), the latter of which reached a standing height of nearly one meter.

RAPTORS (i.e., predatory birds) also evolved into a number of giant species. The La Brea pits of central Los Angeles contain, among many other fascinating prehistoric animals, a condor from the end of the Pleistocene (*Teratornis merriami*), a bird with the magnificent wingspan of almost 5 m. It is of zoogeographical interest that the lappet-faced vulture (*Torgos*), which today is found only in Africa, lived in central Europe during the Pleistocene. The great auk (*Pinguinus impennis*), which disappeared forever in the 19th Century, occurred on the coast of Europe during the Pleistocene.

Pleistocene REPTILES differed little from modern species, so we shall only mention a few striking forms. Pleistocene Australia was the homeland for the giant monitor (*Megalania priscus*), which with its L of nearly 7 m was twice the size of the Komodo dragon of today. Giant turtles (*Miolania= Meiolania*) also existed in Australia; they had bony skull processes. Their closest relatives are from Tertiary South America. The meiolaniids (Meiolaniidae), now extinct, were the last descendants of the Mesozoic ancestral turtle group, the suborder Amphichelyida. All these large animals have disappeared, and many other species died out along with them at the end of the Pleistocene. Thus we are faced with the great question of what caused so many organisms to die out; we were faced with the same question at the end of the Cretaceous.

We can determine the living conditions and the time of disappearance of animal groups in the Quaternary much more precisely than for any earlier geological period, since we are talking about animals dying out just a few tens of thousands of years ago. In fact the use of carbon (C^{14}) dating techniques (see Chapter 3) has enabled paleontologists to determine the last appearance of important species with great accuracy. J. J. Hester found that all the large North American mammals to die out during the Quaternary disappeared 8,000–12,000 years ago. In Australia, the giant marsupial *Diprotodon australe* was alive just 6500 years ago. On Madagascar, the giant ostriches (*Aepyornis*) and the large prosimians (*Megaladapis, Hadropithecus* and others) became extinct just a few thousand years ago, and some even lived into historical times. The last moas on New Zealand (*Megalapteryx didinus*) disappeared a few *centuries* ago!

The animals that died off were the larger mammals and birds, while many small mammals survived into the present. This has led to the hypothesis, particularly in recent times, that man's appearance on earth was related to the disappearance of these animals. That is, that man has been responsible for the mass disappearance of large animal species, both today and in the Pleistocene. Extinction was not only caused by hunting pressure, but also by displacement and by disruption of the environment. This hypothesis is undoubtedly true for extinct insular species, which would include the prosimians and giant ostriches of Madagascar and the New Zealand moas. In these parts of the world the ceaseless hunting activities and ecological disturbance caused by humans undoubtedly increased the death rate of such animals to well above the normal one, and this resulted in the disappearance of what would have been viable animal species.

A different situation existed on the continents, where animals could still avoid the presence of humans. It is unlikely that post-glacial human beings, who lacked firearms, could have extirpated a great many animal groups. One need only consider the huge buffalo herds the Indians lived off before white men came to America or the rich fauna of Africa that existed prior to the advent of Europeans on that continent.

The most important piece of evidence that climatic changes were not responsible for the extinction of so many animal groups is that a comparable reduction in species did not take place in the transitions between the various cold and warm parts of the Pleistocene. However, mammals are affected not so much by climatic changes *per se* as by the modifications occurring in plants as a result of new weather conditions.

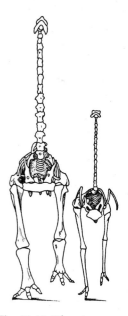

Fig. 19-16. The giant ostrich (*Aepyornis maximus*) from Madagascar reached a height of over 3 m. A modern ostrich is shown to the right for comparison.

A survey of the Pleistocene mammals of Europe shows that quite a few large mammals died out *during* the Pleistocene, *NOT* subsequent to the appearance of well-armed hunting peoples who would have been in a position to cause great changes in the fauna. In discussing the extinction of Pleistocene species we do not consider those species that were simply replaced ecologically by others. For example, the plains mammoth (*Mammonteus trogontherii*) did not disappear in the strict sense, for it was the phylogenetic precursor to another mammoth species, *Mammonteus primigenius*.

Some of the most significant large mammals that disappeared during the Pleistocene without evidence of dying off because of human interference are the mastodons and the primitive elephants belonging to the genus *Palaeoloxodon*, the equids known as hipparions, the elasmotheres from the rhinoceros group, and finally the saber-toothed cats, cheetah-hyenas, and early bears (*Agriotherium*). Numerous small mammal genera disappeared during the Pleistocene as well, including the shrews *Beremendia fissidens* and *Petenyia hungarica*, the beaver *Trogontherium cuvieri*, the rodents *Allophaiomys pliocaenicus* and *Pliomys coronensis*, and finally the lagomorphs *Hypolagus brachygnathus* and *Oryctolagus lacosti*. These are just a few of the wealth of species and entire genera that left the face of the earth during the course of the Pleistocene.

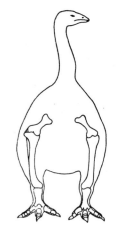

Fig. 19-17. The stocky moa, *Moa pachyornis*, showing only the leg bones. All moa fossils lack wing bones.

Other mammals died out after the Pleistocene (or at least disappeared from Europe during that time). Examples include the Irish elk, the lion, leopard, and animals driven to the far north like musk oxen, reindeer and wolverines, the equid *Equus* (*Asinus*) *hydruntinus* (related to the ass), the chamoislike horned *Myotragus balearicus*, and finally the pika *Prolagus sardus*, with which an entire phylogenetic line of these lagomorphs disappeared.

Nonetheless it is striking that with the end of the so-called Würm cold period in Europe more species died than with the end of previous cold Pleistocene periods. This may have resulted from the sudden improvement in climatic conditions that came about in the northern hemisphere 12,000 years ago (whose existence has been proven by techniques measuring absolute temperature). It is noteworthy that the extinct species were consistently highly specialized ones within their own phylogenetic lines. Most of them were inhabitants of the open plains, including the mammoth, the plains bison, Irish elk, the woolly rhinoceros, cave bear and the cave hyena. Species that survived were forest dwellers (e.g., aurochs, red deer, wild pig, brown bear and lynx) or mountain animals like the chamois, ibex and the marmot.

When all these facts are considered, it appears even more questionable that the extinction of numerous mammals at the end of the Pleistocene was really due to the efforts of humans. Changes in the plant kingdom, caused by climatic alterations, seem to be the real underlying cause behind the disappearance of so many animal species. Extensive plains and tundras became forested, and this restricted the amount of habitat for plains-dwelling animals. As a consequence of that, plains animals declined, and as these populations were restricted to smaller and smaller areas they finally disappeared.

Comparable changes occurred in North America toward the end of the Tertiary (long before the appearance of *Homo sapiens*), and these led to the widespread disappearance of savanna and plains animals. Even-toed ungulates that died out included the oreodonts, merycodonts and numerous tylopods, while odd-toed ungulates that became extinct included the three-toed horses (hipparions) and rhinoceros; the rodent group of mylagaulids disappeared, as did their relatives, the aplodontoids (superfamily Aplodontoidea) with the exception of the mountain beaver or sewellel, a "living fossil". In South America, Africa, and Australia there were also floral changes caused by climatic changes, and these led on the one hand to the forestation of open areas and to the drying up of previously humid areas, which were the direct causes behind the disappearance of many animal species from these parts of the world. In historical times there was a rich mammalian fauna in northern Africa including elephants, giraffes, buffalos, rhinoceros, zebras, hippopotamus, antelopes, gazelles, carnivores and numerous rodents. Nearly all these animals have disappeared because of the gradual drying up in northern African habitat and the formation of what we now know as the Sahara (which continues to grow). At the end of stratum IV in the east African Olduvai gorge, the site famous for its hominid finds, there is a similar prominent change in the animal species; this is again due to a time period in which a dry, arid climate prevailed in that area.

The highly specialized species became extinct

Climatic changes, the growth of forests, and increasing aridity

Thus, when we survey the entire Pleistocene and the climatic and floral changes occurring during that time, it is evident that the widespread extinction of animal species toward the end of the Pleistocene was a consequence of a major climatic change (which led to changes in the vegetation, as we pointed out). Man merely "helped" this whole extirpation process along, especially on islands like Madagascar and New Zealand. He was not, at least at that time, the main cause of animal species extinction.

▷
Skeleton of the late Ice Age giant sloth *Megatherium americanum* from Argentina (Field Museum, Chicago).
▷▷
Extinct South American ungulate groups:
- I. Condylarthrans (order Condylarthra):
    1. A condylarthran
- II. Pyrotheres (order Pyrotheria):
    2. *Pyrotherium* (elephantlike)
- III. Litopterns (order Litopterna):
    3. *Thoatherium* (one-toed ungulatelike)
    4. *Theosodon*
    5. *Macrauchenia*
- IV. Astrapotheres (order Astrapotheria):
    6. *Astrapotherium*
- V. Notoungulates (order Notoungulata):
    - A. Suborder Hegetotheria (lagomorphlike):
        7. *Propachyrukhos*
        8. *Paedotherium*
    - B. Suborder Typotheria (often hyraxlike):
        9. *Protypotherium*
        10. *Miocochilius*
        11. *Typotheriopsis*
    - C. Suborder Toxodonta (often tapir- or rhinolike):
        12. *Thomashuxleyia* (primitive form)
        13. *Homalodotherium* (with parallels to the odd-toed ungulate suborder of cahlicotheres)
        14. *Scartittia*
        15. *Toxodon* (rhino-sized)
        16. *Nesodon*
        17. *Rhynchippus*)

20   In the Deep-freeze of Nature

By H. Wendt

Bones from large Pleistocene animals, especially from proboscideans, have been found ever since antiquity. They have played a very special role in the legends and traditions (and hence the cultural history) of man, even into recent times.

In ancient Greece, these bones were thought to belong to the bodies of dragons or giants. An elephant skull with its prominent nasal opening (see Fig. 20-1) could easily lead someone in those days to believing that the skull was from some sort of cyclops. It is important to remember that the Greeks had no knowledge of African elephants, so they could not associate fossil proboscidean skulls with these animals. Fossil elephant skulls were dug from the Sicilian soil as late as the second half of the 17th Century. The specimens located there were probably from the pygmy elephant *Palaeoloxodon falconeri*, a close relative of the Pleistocene forest elephant and one inhabiting the Mediterranean islands.

The impression created by these gigantic bones was so powerful that many fossils were set up in churches, castles and public buildings or even as omens above portals. The alleged dragon and giant bones were thought to bring good luck and drive off evil spirits, and these bones often brought a high price.

When the site for St. Stephan's cathedral in Vienna was being excavated in 1443, the upper leg bone of a mammoth was dug out, and it was hung at the city gate. A stone mason chiseled the date 1443 into the bone along with the motto of Emperor Friedrich III, A.E.I.O.U. (*Alles Erdreich ist Österreich untertan* = all the earth is subordinate to Austria). Finally the bone was sent from the Hofburg in Vienna to the geological institute of Vienna University.

Some of the giant bones specifically mentioned in old tales have been retained in various collections, including one called the "Lucerne bone" found among the roots of an oak tree near Lucerne in 1577. The people who found this bone thinking it had belonged to a human being, wanted to put it to rest in the local church cemetery. A physician from Basel came

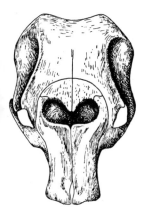

Fig. 20-1. Skull of an Indian elephant, resembling that of a dwarf elephant with a nasal opening that looks surprisingly like fused optic openings. Homerian sailors finding fossilized dwarf elephant skulls in caves on the Sicilian coast did not know the elephant, and the fossil skulls may have given rise to the legend of the cyclops.

to investigate this bone, and he declared that it was the remains of a giant 6 m tall. Since no giant could be buried with a Christian ceremony, the bone was displayed in the Lucerne city hall. Two centuries later a zoologist looking at it recognized it as a mammoth bone.

Almost every country has its own "national giants", legendary creatures whose alleged existence was verified by giant fossilized bones. The remains of fossil elephants in Belgium were thought to belong to giants, and similar bones found in Spain were thought to be those of St. Christopher. The mammoth itself gave added impetus to stories about giants and fabled animals, and mammoth diggings created a great deal of excitement. One such digging took place in 1645 near Krems, Austria by Swedish troops, an episode depicted two years later by a famous artist, Matthäus Merian, in his *Theatrum Europaeum*:

"National giants"

The soldiers encountered a number of large, fragile and partly fragmented bones, which "scholars examined and recognized to be the bones of humans." According to Merian, the bones were "revered in Antiqua and sent to Sweden and Poland." The artist reports, "The actual size of the person's body is incredible, for the head alone is as large as a table, the arms as thick as the body of a normal human being, and a single tooth weighs 5.5 pounds. . ."

Merian drew two of the teeth and called them "giant's teeth". The picture of his clearly shows that the teeth belonged to a mammoth. Some of the Krems finds were taken by the Swedes, and others were displayed in numerous Austrian collections and churches. Not until a century later did people realize that the Swedes had discovered a Pleistocene elephant and not the remains of a giant. The last of the intact Krems teeth was discovered by a person knowledgeable in paleontology in 1911 in the Benedictine cloister in Kremsmünster. The astronomer at the cloister had been using the tooth as a paper weight, unaware of what the object actually was!

A vast collection of mammoth bones found by a grenadier in 1700 along the Neckar River in Cannstatt was a particularly singular sensation. The Duke of Württemberg ordered that the bones be systematically dug out, and six months later the digging team had recovered seventy pieces of fossilized ivory. Scholars met in Stuttgart, on the command of the duke, to write a paper on whether "these bones and horns are simply a quirk of nature or whether they are from animals or man".

Canstatt bones

The scholars discussing these questions soon got into a great debate. One claimed that the bones were left from Hannibal's elephants that died after the great crossing of the Alps was made. Another believed that these bones were from an old Roman sacrificial site. A third interpreted them as animal parts remaining from the Great Flood. The differences among these scholars were so great that these individuals remained bitter enemies until their deaths (as we often find even today among scholars of differing opinions).

However, these great intellectuals were able to agree on one major point: the Cannstatt bones were not mineral deposits but were animal parts. In 1703, a physician from the town of Öhringen burned some of the bones, and as they burned he noted "a fleeting urinelike substance and a stinking oil smell", and the animallike odor was conclusive evidence that these objects were indeed from animals.

The much-discussed Cannstatt bones were sent to the Stuttgart *Naturalienkabinett*, partly as a noble gift to the curiosity collections of the nobility in this part of the country. Most of the teeth were ground down in the Stuttgart court apothecary and were used as medical powder. This was done because mammoths were incorporated into the legends that had grown up around the unicorns (of Biblical origin), and unicorn

"Supercure" unicorn   horn was supposed to be the most effective of all medicines. The findings of mammoths gave renewed impetus to unicorn legends, and unicorn apothecaries flourished throughout Germany. Even today, the unicorn remains as the symbol of the apothecary in Germany.

"Unicorns" from the Middle Ages and post-medieval period were usually narwhal teeth, and this wonder medicament brought a high price. When mammoth bones and mammoth tusks were dug out of the ground, the situation changed dramatically. Suddenly, mammoth teeth were interpreted as true unicorn horns, and narwhal teeth were fakes and depreciated rapidly in value. Since pulverized unicorn horns were thought to heal all diseases, combat every poison and work unfailingly against impotence and infertility, mammoth teeth were suddenly being ground down and sold. Fossil bones and teeth, called "fossil ivory", were purchased by

"Fossil ivory"   healers and apothecaries everywhere.

Westerners believed in unicorns as late as the 17th Century, and many attempts were made to reconstruct the animals using mammoth parts. Numerous mammoth bones and teeth were discovered in 1663 near Quedlinburg, Germany. The Madgeburg mayor, a renowned physician known for his invention of the air pump, attempted to rebuild a unicorn skeleton from these bones. The drawing he executed of his reconstruction, mentioned by Leibniz in his work "Protogaea", is shown in Fig. 3-9 and is uproariously far from reality. Nevertheless, Mayor von Guericke was the first person to attempt to reconstruct a prehistoric beast.

The drawing executed by the Swedish horseman Tabbert von Strahlenberg around 1720 (see Fig. 3-10) when he was a Russian prisoner of war was similarly fantastic. It is also based on mammoth finds from the Siberian taiga and depicted a unicorn. For centuries people have known that mammoths were in the Siberian ice. Tungus and other tundra hunters repeatedly encountered mammoth cadavers in this natural freezer. They usually waited for the ice block containing the cadavers to thaw enough for them to extract the tusks. They fed the flesh to their dogs (Pleistocene meals for dogs!), unless of course hungry bears, wolves and foxes did not beat them to it.

Eventually there was a flourishing mammoth trade. Thousands of kilograms of mammoth ivory were collected every year, especially in northeastern Siberia. Much of the ivory stemming (until very recently) from northern and eastern Asia originated from mammoths. It was only a matter of time before ivory merchants realized that the mammoth had nothing to do with the unicorn and other legendary animals. One merely had to examine the tusks closely, and it became evident that the animal in question was in fact an elephant. But how was one to explain elephants (or, more precisely, proboscideans) living in Siberia?

A great deal of discussion on this point ensued in the Petersburg academy. Two centuries before Alfred Wegener had propounded his theory of continental drift, Russian zoologists toyed with the idea that the poles could have migrated. They felt that the Siberian elephants were identical with modern elephants of southeastern Asia. Where now the arctic tundra and its permafrost extended was once the site of a warm, tropical climate. When the poles of the earth changed, Siberia was iced over and the elephants there died off.

The question of continental drift

The theories circulating among Russian zoologists scarcely traveled beyond the borders of Russia. Until the beginning of the 19th Century, reports of the Siberian mammoths were extremely fragmentary, and it was the European mammoth finds that were far more prominent among scientists. In 1799, German zoologist Johann Friedrich Blumenbach recognized that they belonged to a species different from that of the two living elephants. He called these mammoths *Elephas primigenius* (=first born elephant). Today, the Siberian mammoth is known under the scientific name of *Mammonteus primigenius* (see Color plates, pp. 478/479 and 493).

"First-born elephant"

Blumenbach could not have known that the mammoth was by no means the "first" elephant but a rather late descendant of what was formerly a diverse, cosmopolitan proboscidean group. Initially he also did not suspect that his "first elephant" was identical with the mysterious Siberian mammoth. However, in 1799, the year Blumenbach found a scientific name for the European mammoth, Ossip Shumakhov gave important impetus to the solution of the Siberian mammoth puzzle.

Shumakhov found a fully intact mammoth cadaver in the ice at the mouth of the Lena. At first he could not get at the animal, for it was frozen in a tremendous ice block. The local Tungus tribes believed that the mammoth was a subterranean monster, a kind of giant mole that died as soon as it caught the sight of light. Anyone finding a mammoth cadaver, so they believed, was condemned to death along with his entire family.

The Siberian mammoth puzzle

Shumakhov was in conflict. He wanted to examine the animal's tusks, but at the same time he feared death, for he himself was a Tungus. In the following two years he visited the ice block several times. But each time he found that the block had just began to thaw, and he could not get at the frozen corpse. His superstitious fear grew to such an extent that he actually became seriously ill.

Fig. 20-2. Thus did the ivory trader Boltunov represent the mammoth, the remains of which Ossip Shumakhov found in 1799 in the ice at the mouth of the Lena. Boltunov sent his drawing to the Petersburg academy.

Fig. 20-3. Boltunov's drawing resulted in an expedition to the mammoth find at the Lena. The expedition was organized by the English zoologist Henry Adams, who was at the Petersburg academy at the time. The nearly intact skeleton was transported to Petersburg and displayed in the natural history museum there. The specimen was identified as a fossil elephant ("*Elephas*" *primigenius*) and not a mammoth. The above drawing was executed by Georges Cuvier.

Adam's mammoth

The disease he contracted did not last long, however. As it disappeared Shumakhov's tribal fears also dissipated, and he grew more and more interested in attempting to reach the frozen mammoth cadaver. He began doubting the legends of his own landsmen, and in 1803 (four years after the mammoth was discovered) he led the merchant Boltunov to the ice block. The ivory of the mammoth had appeared by this time, and in fact the head of the animal was exposed and began decomposing.

Shumakhov sold the Russian the tusks for fifty rubles. Boltunov examined the animal closely and made a drawing of it. It was not a beautiful drawing and not even a correct one, for Boltunov could only draw what he could see of the animal. The skull in Boltunov's drawing looks like a giant pig's head, and the eyes are depicted where in truth the openings of the auditory passages are located. However, Boltunov's rendition of the whole body was fairly accurate.

The ivory merchant did something else, however. He sent his drawing and a verbal description to the academy in Petersburg, where it was passed into the hands of the British zoologist Henry Adams. Adams sent the picture on to Göttingen, Germany, and asked his German colleague Blumenbach for an opinion on what was portrayed. Blumenbach knew at once what he saw: it must have been *Elephas primigenius*! Famed French zoologist Georges Cuvier also looked at the picture and shared Blumenbach's opinion.

Adams in the meantime was not sitting idly. In 1806 he equipped an expedition, went out to the mammoth corpse, and examined those parts of the body that had not decayed or had not been consumed by wolves, wolverines or foxes. He traveled to the Lena mouth by reindeer sled and met with Shumakhov.

The corpse still contained three-fourths of the skin, one ear, and all the bones excepting a fore foot. A thick, woolly hair covering surrounded the body. Adams had a great deal of difficulty packing all these remains of this giant animal onto his sled, for ten men were hardly able to move the loosened skin from this site. On the long way to St. Petersburg, "Adam's mammoth" suffered so greatly that the skin soon lacked any hair. However, the animal was finally transported to the natural history museum in St. Petersburg and put on display, and within a short time it created a sensation throughout the world. Adams sent the skeleton, skin, and a number of mammoth teeth to St. Petersburg. He correctly saw that "the tusks of this prehistoric elephant are much more greatly arched and also much longer than in either of the two living elephant species."

Adams's mammoth, as it was called, made history. Its hair covering, as Adams explained, was "clear proof that the animal was adapted for living in cold regions". The same was true for a second thickly hair-covered animal, which occurred together with the mammoth in Europe and in Siberia: the woolly rhinoceros (*Coelodonta antiquitatis*). In 1773, the great naturalist Peter Simon Pallas discovered the body of a woolly rhino-

ceros in the frozen tundra ground at a tributary of the Lena. Blumenbach recognized that this was the same species as the rhinoceros that had
been found in Pleistocene European deposits. Thus, large animals caught in
the Siberian permafrost and which had been remarkably well preserved
were identical with Pleistocene European fauna that had heretofore only
been known as fossils.

For a long time there was a great deal of controversy over whether
Blumenbach's opinions regarding the mammoth and woolly rhinoceros
in Europe and Siberia were actually correct. His interpretation was at odds
with the then prevailing creationist and catastrophic theories. Some zoologists felt that mammoths, woolly rhinoceros and other animals of Pleistocene Europe still existed in remote parts of northern Asia. Once ideas about
the Pleistocene became more refined, the problem of these species was
resolved. Mammoths and woolly rhinos, it was finally recognized, are
characteristic animals of the cold parts of the Pleistocene. Since they had
adapted fully to the harsh climate and the living conditions in the Pleistocene tundra, they could make a living in both Europe (during its cold
periods) and in the cold parts of northern Asia. Mammoths even wandered
into Alaska, where frozen cadavers have also been found in the ice.

Once paleobiologists realized that there were alternating cold and
warm periods within the Pleistocene, it was appreciated that the Pleistocene fauna and flora alternated rhythmically and in unison as the climate
changed. Mammoths, woolly rhinos, musk oxen, reindeer and other
"cold-weather specialists" inhabited Europe during the cold parts of the
Pleistocene, and whenever a warm period entered Europe they migrated
to the north and the northeast. Species characteristic of warm climates then
migrated northward during this time, and these included animals such as
the forest elephant (*Paleaoloxodon antiquus*), the plains rhinoceros (*Dicerorhinus hemitoechus*) and the forest rhinoceros (*Dicerorhinus kirchbergensis*).
When a cold period returned, the forest elephants, plains and forest rhinos
and other warm-loving animals retreated to the south, while mammoths,
woolly rhinoceros and other similar species came back from the north and
northeast. Even the plains mammoth (*Mammonteus primigenius*) and the
woolly rhinoceros in the Upper Pleistocene became true cold-weather
forms.

With the discovery of the Ice Age began also the discovery of human
prehistory. And there arose vigorous debates about the relationship between primitive men and the mammoth. The first scientists researching
early historic man repeatedly found the remains of mammoths and other
Pleistocene animals along with the signs of early Stone Age man. Early
Stone Age cultures even left behind carvings and small drawings on mammoth ivory. This led scientists to conclude that early Stone Age man must
have lived during the Pleistocene, together with mammoths and other
Pleistocene animals, and these animals undoubtedly served as prey for
early man.

Fig. 20-4. Portion of the
drawing in the Galerie des
Fresques of Font-de-
Gaume. The drawing
clearly depicts a mammoth,
a bison and a reindeer.

Cave drawings and carvings were enormously useful in permitting paleontologists to gain an accurate picture of mammoths. These drawings also demonstrated that early Stone Age humans must have lived during the Ice Age.

Fig. 20-5. Mammoth on the cave wall of the Galerie des Fresques in the grotto of Font-de-Gaume (Dordogne, France). The relief was covered later by a painting.

Fig. 20-6. Attacking mammoth, etched in a fragment of a mammoth tusk, from the La Madeleine cave (Dordogne).

Fig. 20-7. Mammoth drawing from the early Magdalenian on the cave wall in the grotto of Combarelles (Dordogne). Note the protruding portion of the high skull roof and the nape indentation.

This interpretation was repeatedly attacked for more than half a century by a number of scientists, including the German naturalist Rudolf Virchow (1821-1902) and the Danish zoologist Johann Japetus Steenstrup, an expert on arctic fauna and arctic prehistory. Steenstrup claimed that there were no Pleistocene men and that therefore humans never saw Pleistocene animals like the mammoth.

Much of the debate was fired by a find near the village of Předmost, Czechoslovakia. Farmers there found a loess hill 34 m high containing numerous fossils. They were using the find, crushing the ancient bones and fertilizing their fields with them. A paleontologist accidentally learned of the great fossil mound, and he convinced the government to declare the site restricted and to undertake excavation thereof. Great quantities of mammoth bones were uncovered there, but the site also contained human bones from the Old Stone Age (about 20,000 years ago).

Systematic digging continued, and scientists found all sorts of human implements as well as crushed and split mammoth bones with clear signs of being burnt. Similar finds were uncovered elsewhere in Austria and Czechoslovakia. Some of the most interesting were small sculptures of humans and animals, because they were made almost exclusively from mammoth ivory.

The Czech and Austrian researchers working these finds had no doubt they had come upon the remains of the camping sites of nomadic mammoth hunters. Even with this evidence, Steenstrup remained adamant in his conviction. He studied the Předmost finds and in an exhaustive report noted that even today the Tungus and other hunting peoples from northern Asia slaughter and feed on mammoth cadavers they recover from the ice. Why could not a similar event have taken place in prehistoric Czechoslovakia?

However, Steenstrup was soon contradicted. Just after his studies in Předmost in 1895 were concluded, another naturalist found a mammoth figure an early Stone Age man had carved from a mammoth tusk. It was a well-executed, accurate representation. Even the large protuberance was there, and this anatomical feature was not known to zoologists until cave drawings were found after 1900.

When the many mammoth drawings were found throughout Europe, it was evident beyond doubt that man and mammoth had coexisted during the Pleistocene. Early Stone Age man depicted the mammoth in herds, showing the bulls, cows and calves and sometimes showing how these huge animals were hunted or caught in carefully laid traps (see Figs. 20-4, 20-5, 20-6, and 20-7). These prehistoric cave artists were good observers, for their drawings depict the coarse hair covering, the mighty dorsal hump and the position of the arched tusks with great accuracy. Zoologists did not discover with direct evidence that the cave drawings were accurate until later.

In one respect there is nevertheless some possibility that Steenstrup was correct in his contention. That Ice Age man hunted mammoths is known to be fact. But did he kill the mammoth herds near Předmost? There are a number of questions about this site and its meaning. The remains of 500–600 mammoths have been recovered at Předmost, and it is difficult to conceive that early man was sufficiently well equipped to bring down so many giant, powerful animals. Of course the early Stone Age hunters could have been killing mammoths sporadically throughout the surrounding area, but it would hardly have been possible for them to drag all these hundreds of mammoths to one great collecting site, and it is extremely unlikely that they did so. For this reason, some zoologists feel that the Předmost mammoths died due to some natural event, like an unusually severe snowstorm. They were driven together and perished at the place where they collected.

Large collections of mammoth bones and teeth have also been recovered in loess regions in Austria and Germany, in northern Asia, on northern Siberian islands, in the Arctic Ocean, and even on the floor of the North Sea. More than 2,000 molars from mammoths were found between 1820 and 1833 by oyster fishermen around the Dogger bank off the European coast. The bones or teeth of over 20,000 mammoths have so far been found in Europe; Siberia has yielded the remains of nearly 50,000 of them. In addition, several thousand mammoths have been found in North America.

The mass mammoth bones sites in certain areas led the Pleistocene researcher Wolfgang Soergel in 1912 to conclude that these sites were gigantic cemetries sought out by diseased or "elderly" mammoths. Soergel compared these death sites with the "elephant cemeteries" reported by African game hunters. The so-called elephant cemeteries are mere fable. The reason for the accumulation of mammoth bones in certain areas is still unknown, and the mass mammoth collections remain a mystery.

It is in the harshest, coldest, most inhospitable parts of Siberia where we find the greatest numbers of mammoth teeth and frozen mammoth cadavers. Mammoths actually lived somewhat longer in Siberia and Alaska than in other parts of the world. They died out during the Pleistocene elsewhere in the world but lived into the post-glacial period in Siberia and Alaska.

In the Siberian and Alaskan tundra, mammoths found the same ecological conditions they had encountered in Pleistocene Europe. Many parts of northern Asia were free of glaciation during the Pleistocene, and glacial conditions did not occur until the end of the Pleistocene and in the Holocene. Mammoths fed chiefly on grasses and other tundra plants, on moss and on young coniferous shoots. Since these plants still occur today, zoologists and paleontologists are inquiring into the reasons that mammoths disappeared from the face of the earth, as did the woolly rhinoceros (which fed on the same plants). Reindeer and musk oxen, two other

▷
Fossil Cenozoic plants. Above: imprint of an oak leaf from the interglacial Travertine formation near Bad Cannstatt, Germany: Middle left: Pollen granules of beech, hazelnut, birch, oak, linden (clockwise). Senckenberg Museum, Frankfurt; Middle right: Fruits and cones from Miocene peat deposits of the lower Rhine. A pinecone is at the far left, and beside it are six fruit cones from sequoias. On the bottom and to the right are the seeds from four genera of subtropical deciduous trees (*Rehderodendron, Tectocarya, Ganitrocera, Mastixia*). Senckenberg Museum; Lower left: Branch of the evergreen *Metasequoia* from Tertiary Spitsbergen (Senckenberg Museum); Lower right: Leaf fragment from the palm *Sabal* of Miocene central Europe (Senckenberg Museum).

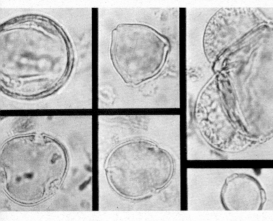

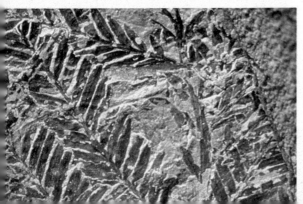

Skeleton of the Upper
Quarternary Irish elk,
Megaloceros giganteus, found
in the moors of Bally-
betagh, Ireland (Bavarian
State Collection for Pa-
leontology and Historical
Geology, Munich).

Pleistocene animals, also migrated northward toward the end of the
Pleistocene, and these two animals are still found today.

It is possible that large fauna like the mammoth and the woolly rhinoceros were particularly sensitive to the unusually intense climatic changes
occurring in the late Pleistocene and post-glacial period, and that they
sought out the most inhospitable (but also driest) parts of northeastern
Asia since they could survive nowhere else. Vigorous rainstorms ensued
in Europe after the last glaciars disappeared, bringing conditions completely alien to animals which heretofore lived in dry regions swept by the
practically germ-free polar winds. Avalanches developed in the mountains
of Europe. The Siberian tundra, which had experienced relatively mild
climate, was now subjected to snowstorms of an intensity without pre-
cedent in that part of the world. Snow was piled many meters thick on
top of the ice covering the rivers and lakes, and if the ice sheet was thin, a
large animal could easily fall through if it did not detect the danger. Huge
clefts in the ground also acted like giant traps for unsuspecting animals.

The mammoth and woolly rhinoceros finds in the Siberian tundra
have repeatedly caused sensations. In August, 1900, a well preserved
mammoth cadaver was found in the Siberian ice. The body was investigated one year later by an expedition of the St. Petersburg Academy. The
scientists examining the animal found from the stomach contents and
from food fragments on the teeth that this animal has fed on plants. They
also learned that this mammoth died suddenly in the spring.

Freshly fallen snow had filled a deep cleft in the earth, transforming it
into a deadly trap. The snow concealed this depression from the mammoth, which walked right over the great hole and fell in. In falling, the
mammoth broke its pelvis and right fore leg. This kind of accident was
not unique among the Pleistocene large fauna. Other Pleistocene elephants
and rhinoceros died in a similar manner. Many proboscideans from North
America have been dug out the ground and have been in the same kind
of posture the aforementioned mammoth was in. Many prehistoric elephants fell off cliffs or into chasms.

Swamps, marshes and dirt pits often became transformed into death
traps. A Polish paleontologist found the bodies of two woolly rhinoceros
calves and one mammoth calf, all three fully saturated with paraffin, near
Starunia, Poland. However, all these are individual incidents that could
indeed become more frequent under certain environmental conditions but
would not lead to the extinction of those species. Only a catastrophe that
leaves the species as a whole weakened and unable to survive brings
extinction.

Who survives?

As we mentioned in the chapters on the Tertiary and the Pleistocene,
the problem of the disappearance of whole animal groups is one of the
chief areas of investigation of paleontology. Why, for example, did certain
Pleistocene groups die out while others somehow survived into the present? The extinct groups include the Pleistocene elephants and rhinoceros

a few herbivores like the Irish elk, and several carnivores like the cave bear, cave lion and cave hyena. The extinct groups are in the minority, for almost all the animals that lived during the Pleistocene also live today, and they have scarcely changed (if at all) since the days of the Ice Age.

The far north is the home not only of the reindeer and the musk ox but also of wolves, lynx, wolverines, arctic foxes, arctic hares and lemmings, all species that lived during the Pleistocene. In central Asia we still have the wild horses, wild asses, saiga antelopes and pikas. Ibex and chamois still inhabit the mountains (along with marmots). The modern European and American buffalos are not identical animals, but they are close relatives of the Pleistocene plains and forest bison. Aurochs and wild horse were the ancestors of the modern domesticated cattle and horse, respectively.

Why must some species leave the earth while others remain? We have already discussed the impact on the animal kingdom of changes in the plant species, something which is based ultimately on changing climatic conditions. Modifications in existing plants have been principally responsible for the extinction of large animals. Other possible reasons for the disappearance of species have been discussed: natural events, epidemics, and even extirpation by stone age man. Finally, prehistoric life and death has sometimes been compared to modern life and death, a kind of comparison drawn by the father of modern paleobiology, the Austrian paleontologist Othenio Abel.

Othenio Abel, founder of paleobiology

Paleobiologists attempt to determine how prehistoric animals looked, what sort of environment they lived in, and for what reasons individual species died off. Paleobiological deductions are not drawn from pure fantasy but from logical bases, especially with regard to the animals of the glacial and post-glacial period. There are many fossilized clues descriptive of animals' appearance and their habitat: tracks, traces of feeding, crawling and scratch marks, fossilized bits of food and excretions, dwelling tubes and burrows, other kinds of passages dug in the ground, and finally even injuries, traces of diseases, and skeletons with signs of being in the last stages of the fight with death. To those "in the business", all these signs are fossils. While we commonly may think of a fossil as a shell or skeleton, a paleobiologist thinks of any sign of prehistoric life as a fossil, and all this evidence is used to draw paleobiological conclusions about the life of the past.

Early paleobiological investigations gave rise to a whole new field, the study of prehistoric diseases. Many fossils are indicative of animals suffering from broken bones, caries, arthritis, cancer, rachitis, jaw inflammations, or other maladies. Dinosaurs may have sometimes contracted bone cancer, based on available evidence. When pathological conditions occur in a great many individuals, it is often possible to tell what the reasons were for the extinction of an animal group.

Fossil diseases

Fossils have also yielded information about the degenerative processes in prehistoric animals. The mammoth is particularly informative in this regard. We always associate the very word "mammoth" with something of tremendous dimensions. This is an accurate description of the temperate plains elephants from the earliest Pleistocene (*Archidiskodon meridionalis*) and the plains mammoth of somewhat later (*Mammonteus trogontherii*). Bulls of the Middle Pleistocene had a shoulder height of 4–4.5 m. In later mammoth species—the ones that first led us to coin the term "mammoth" —the situation was quite different.

Abel found that mammoths displayed numerous degenerative processes during the later Pleistocene. Bulls had a height of just 3 m, while cows were scarcely 2.5 m tall. Even modern elephants are larger than this. In some regions, mammoths developed into veritable dwarf forms, and this was probably a result of their ever dwindling habitat space (the cold plains) something caused by increasing forestation during the Upper Pleistocene and post-glacial period. Mammoths were adapted to feed on specific cold plains plants, and as the forests intruded on the plains the resulting loss in plant food subsequently resulted in the diminished size of mammoths. Abel even described some mammoths, paradoxically, as "dwarfs". He wrote, "The mammoth at the end of the Pleistocene was not the mammoth of earlier times."

A decrease in body size is often an indication of decreased viability in the particular animal group so afflicted. It would not be difficult to conceive that the dwarf mammoths could not deal with the climatic and floral changes occurring in the post-glacial period. And thus it is possible that accidents, epidemics or other diseases became more significant as causes of death and thereby contributed to the extinction of these once mighty animals. Formerly, the mammoths may well have been able to fight off these diseases, but in their weakened state they succumbed to them.

21 The Evolution of Primates

Human beings are members of the order of primates (Primates); hence the evolution of this mammalian group is of particular importance for us. The primate order includes insectivorouslike species appearing at the turn of Upper Cretaceous to the lowermost Tertiary (Paleocene) and a host of other animals all the way to the human being. Indeed, the primate order encompasses the broadest evolutionary span of all the vertebrates. Primatology has become an important area of inquiry in recent years. Its methods include not only comparative studies of living species but also systematic investigations of fossils.

The pioneers of evolutionary study from Darwin's and Haeckel's time knew that primates had evolved from insectivores. A comparison of the most highly developed primates (anthropoid apes and humans) and modern insectivores (e.g., shrews, moles or hedgehogs), two animal groups strikingly different from each other, may make it seem surprising that in some fossil species we do not know for sure whether this fossil is "still" an insectivore or "already" a primate, but this problem does occur in the fossil record. Some of the surviving descendants from the transition field between insectivores and primates (descendants that are now on their own evolutionary pathways) include elephant shrews, tree shrews and gliding phalangers, animals we will discuss in more detail. The fact that tree shrews, for example, are classified by some zoologists as insectivores and as primates by others is itself evidence of how indistinct the border between insectivores and primates actually is.

Until recently, fossil and modern insectivores were placed within a single zoological order. Today paleontologists tend to recognize a transition order. In his work *Generelle Morphologie der Organismen* (1866), Ernst Haeckel placed the elephant shrews and tree shrews into a distinct order, Menotyphla, while all other insectivores were classified as Lipotyphla. The two Haeckelian menotyphlan groups (elephant shrews and tree shrews actually are not closely related) differ from the "true" insectivores (order Insectivora) in having a caecum and a highly developed brain, the latter related to an enlargement of the optical sensory organs.

By E. Thenius and H. Wendt

Transition stage: insectivores

The FLYING LEMURS or COLUGOS (order Dermoptera), which in some ways resemble primates, also occupy an intermediate position. These creatures have been classified in various mammalian orders in the past: marsupials, insectivores, carnivores, bats or primates. Obviously there was a great deal of confusion about just what flying lemurs were! Today it is thought that they arose from an insectivore group that also gave rise to the tree shrews and primates. The prehistoric genus *Planetetherium*, which contained ancestors of the modern flying lemurs living in southeast Asia today, appeared in Paleocene North America. It is of evolutionary interest that the external parasites of these prehistoric animals are closely related to the parasites that attack tree shrews!

The ELEPHANT SHREWS (Macroscelidea) also caused problems for systematists, and they have been variously classified as marsupials, insectivores, protoungulates (in the broadest sense) and subprimates. Like the giant phalangers, elephant shrews are a very old animal group, but in some ways they are highly specialized. The fossil genus *Metoldobotes* occurred in Lower Oligocene Egypt. Like tree shrews, elephant shrews have a caecum and a highly developed brain (they, too, have well developed eyes), but they have not developed the keen sense of smell we find in the insectivores. They have a Jacobson's organ on the inner wall of the nostrils (it is used to perceive the odor of foods). The young are precocial. It is possible that elephant shrews are neither insectivores nor subprimates but a unique mammalian order confined to Africa.

The "tree shrew problem"

The most fascinating living members of the transition field between insectivores and primates are undoubtedly the TREE SHREWS (Tupaiidae). Once thought to be insectivores, these almost squirrellike animals were classified in the 1920s as transition forms betwween insectivores and primates. And in 1945 no less than the respected paleontologists W. Le Gros Clark and George Gaylord Simpson "promoted" them into the prosimians. Although we might look at these animals in a zoo and insist that they are rodents, tree shrews do indeed share some characteristics with primates. Like prosimians, tree shrews have a bony ring around the eyes, an elastic, partly cartilaginous lower tongue, a lemurlike ridge in the fur, and (behaviorally) the phenomenon of marking in males. The well developed color vision, binocular vision, the development of the optical centers in the brain and the structure of the female sexual organs are all primatelike features. Because of this it is thought that tree shrews are slightly "modified" proto-primates, which have scarcely changed in the more than 50 million years that have passed since the Lower Tertiary when they first appeared.

Yet, in other respects, the tree shrews differ substantially from prosimians. They reach sexual maturity early in life, have a short generation sequence, and have altricial young (i.e., young that stay in the nest a long time). The jugosternal scent glands of tree shrews are not built like those of prosimians. Their thumbs can be spread but not opposed to the other fingers, so the tree shrew hand is not like the hand of a primate. One of the most decisive anatomical criteria in determining whether these perplexing animals are "still" insectivores or "already" primates is the structure of the auditory region and the protective bulla tympani. On the basis of the bulla tympani structure, tree shrews cannot be classified as "true" primates, although most zoologists today place them in a superfamily of the prosimians.

The only known fossil relatives of tree shrews have been found in Lower Tertiary deposits in Mongolia and China. These ANAGALIDS (Anagalidae) had a complete insectivorelike dentition with the dental formula $\frac{3 \cdot 1 \cdot 4 \cdot 3}{3 \cdot 1 \cdot 4 \cdot 3}$, which is quite unlike the dentition of tree shrews. Anagalids might be considered either members of the tree shrew group or "protoprimates", or perhaps even advanced insectivores. The structure of their bulla tympani resembles that of elephant shrews and gliding phalangers. Anagalids may have developed parallel to tree shrews before they died out in the Lower Tertiary.

In connection with the "tree shrew problem" arises the question of whether there is any distinctive characteristic that would show us which fossils belong to the transition field or advanced insectivores or primitive primates. Any such delineation drawn between the two groups is of necessity an artificial one and ultimately will depend on how one defines a primate. We can, for example, define primates as mammals that developed grasping hands or feet as a result of their arboreal life. Characteristics shared by all primates include the hairless underside of the hands and feet with their characteristic wrinkles. According to this definition tree shrews would not be a prosimian superfamily but instead would be placed within an independent mammalian order like the gliding phalangers and elephant shrews. All three groups can be considered as survivors of the radiation of leptictoid insectivores, a phylogenetic change that began in the Upper Cretaceous and during the course of which numerous primatelike features developed. The tree shrews are without doubt closer to the primate model than either phalangers or elephant shrews, and tree shrews may well have evolved parallel to primates.

These LEPTICTOIDS (superfamily Leptictoidea) from the Upper Cretaceous and Lower Tertiary North America are the geologically oldest and in many ways the most primitive insectivores; the ancestors of primates most likely arose from the leptictoids. One prehistoric genus, *Adapisoriculus* from Paleocene Europe, appears to be closely related to tree shrews but we do not know exactly whether this animal was a leptictoid or an AMPHILEMURID (Amphilemuridae) from Lower Tertiary North America and Eurasia. Until very recently, amphilemurids were thought to be the first primates, but today we class them with the hedgehog superfamily (Erinaceoidea). No doubt the early ancestors of primates were nocturnal ground dwellers that looked something like modern hairy hedgehogs (subfamily Echinosoricinae). They had complete dentition, a hedgehoglike

Fig. 21-1. Above: Whether *Anagale gobiensis* from the Lower Tertiary is a tree shrew is uncertain. Below: Skull of a modern feather-tailed tree shrew (*Ptilocercus lowii*).

auditory region, an unconvoluted brain with a particularly large olfactory lobe, and small, lateral eyes and five-digit limbs.

From an olfactory orienting to a visual orienting animal

The first radiation occcurred toward the end of the Upper Cretaceous, as various macrosmate (olfactory orienting animals, such as reptiles) gradually developed into optically orienting animals with less well developed olfaction (these were the microsmates). The eyes grew larger, while the skull grew shorter and the cranial portion of the skull increased in size. The olfactory lobes became smaller and smaller. Eventually these animals took up an arboreal life. As they did so their claws developed into nails; the thumbs and big toes became opposable to the other digits. The original ground-dwelling insectivores thus gradually evolved into various lines of tree dwelling, climbing animals relying on vision rather than olfaction and which eventually became animals active at day instead of night.

In addition to tree shrews, a number of extinct families from the Lower Tertiary (e.g., the APATEMYIDS—family Apatemyidae—MIXODEC-TIDS—family Mixodectidae—and the MICROSYOPIDS—family Microsyopidae) show us how difficult it is to separate the first primates from the insectivores. As with the amphilemurids, these animals were initially classified as primates, but the absence of a bony auditory vesicle and their skull structure suggest that these families are "still" insectivores. The systematic position of the likewise Lower Tertiary CARPOLESTIDS (Carpolestidae) and PICRODONTIDS (Picrodontidae) is also uncertain. Their teeth are primate-like, and the specialized incisors resemble those of the plesiadapids, the ancestors of lemurs. However, the existing specimens of these mammalian groups are too few in number to permit a satisfactory classification of the carpolestids and picrodontids. It appears that picrodontids were fruit-eating mammals whose teeth are batlike.

All we have of these fossils are the remains of their teeth and jaws, and none of the animals were direct ancestors of primates. In 1965 the teeth of the small, geologically older mammal *Purgatorius ceratops* from Upper Cretaceous North America were described. They are quite primatelike and suggest that the earliest subprimates may have appeared toward the end of the Cretaceous. Today, *Purgatorius* is classified as a member of the PAROMOMYIDS (Paromomyidae), of which several genera existed in the Paleocene and which are ancestral to tarsiers. But on the basis of tooth structure it is impossible to make any final judgement about the precise systematic position of *Purgatorius* and hence impossible *at present* to determine when the first primates appeared.

Formerly two suborders within the primates were distinguished: PROSIMIANS (Prosimiae) and APES (Simiae). Prosimians are those primates that share more primitive characteristics, so the question arises as to whether this is a natural systematic unit or a series of transition stages characterized by a similar degree of development. On the basis of the structure of the male sexual organs we can distinguish three groups within the modern prosimians: LEMURS (Lemuriformes), LORISES and GALAGOS

(Lorisiformes) and TARSIERS (Tarsiformes), the tarsiers apparently closer to apes than the others. Radiation in the Lower Teriary produced a number of other side branches of primates, which became extinct and left no descendants.

The PLESIADAPIDS (Plesiadapidae; see Color plate, p. 503) are highly specialized members from the first Tertiary radiation; they are from Paleocene and Eocene North America and Europe. Some researchers even classify them as the "base stock" of primates, while others put them in a direct line with lemurs. Their molars are structured much like primates, and their auditory region even resembles that of higher primates. However, they have true claws on their fingers and toes, and this feature disqualifies the plesiadapids as true primates. They were probably climbing arboreal animals living much the way tree shrews do. Like the other aforementioned extinct groups, plesiadapids display a tendency to enlarge the first pair of incisors. They formed a dead-end evolutionary pathway, which disappeared in the Lower Tertiary.

Fig. 21-2. Skull of *Plesiadapis*, an early prosimian with rodentlike incisors.

Before we turn to the tarsiers, which are highly significant in primate evolution, we shall go into the origin of LEMURS in somewhat more detail (infra-order Lemuriformes). These are the animals that are thought of by most laymen as the "prosimians", although they are but one group of prosimians. Lemurs occur today solely on Madagascar, where apes are absent and where they exist in numerous genera and species. A great many fossil lemurs have been found in Eocene North American and European deposits, however, showing that the group has not always been restricted to such a small area. The prehistoric Old World lemurs are collectively classified as ADAPIDS (Adapidae), while all the New World species plus the European genus *Cantius* are placed in the NOTHARCTID family (Notharctidae).

Fig. 21-3. Skull of an adapid from Tertiary Europe. Adapids are the best understood fossil prosimians.

The fossil record of these Lower Tertiary lemurs includes not only teeth and fragments of jaws and skulls but also partial and entire skeletons, making the adapids and notharctids the presently most fully understood fossil prosimians. C. L. Gazin claims that lemurs arose in the Old World and that the notharctids migrated during the Eocene into North America. It was once thought that the notharctids were the ancestors of the New World monkeys, but this theory has been discredited with new evidence. We still do not know whether the lemurs on Madagascar evolved directly from adapids, but the adapids were the original prosimians. *Notharctus* (see Color plate, p. 503) was a lemurlike animals (though somewhat more heavily built) with hands and feet that could grasp. Adapids and notharctids both lacked the nearly horizontal placement of the lower incisors and canines as we find in modern lemurs. They undoubtedly lived in trees, where they moved about by grasping branches and climbing about. Since their hind limbs were longer than their fore limbs, we can assume that they were good jumpers. There are no fossil transition species between the adapids and the three lemur families occurring today on

Fig. 21-4. Skull of the extinct Madagascar giant lemur *Megaladapis edwardsi*. This is a slightly simplified drawing of the original specimen from Madagascar.

Madagascar. It is believed that prosimians from the adapid group inhabited Madagascar during the Lower Tertiary, and they survived since then
on this isolated island since they were not subject to competition with apes.

All three lemur families gave rise in the Upper Tertiary and Pleistocene to large and even giant forms, of which a few survived into historic
times. One of the LEMURS (Lemuridae) was a GIANT LEMUR (†*Megaladapis*),
which survived into the period when man occupied Madagascar and
which had a skull over 30 cm long. There were long- and short-headed
giant lemurs, and variability in the skull shape was a phenomenon also
occurring in the Upper Pleistocene cave bears. Some skull features have
led a number of zoologists to surmise that the giant lemurs had soft nasal
processes like the modern proboscis monkeys, but it could be equally likely
that their nose and lips were simply very well developed and had a great
deal of mobility. A number of researchers, including the English primato-
logist Osman W. C. Hill, maintain that these giant prosimians lived in
swamps or even partly in water, but this seems unlikely in view of the
fact that their bones have always been recovered in caves. Furthermore,
the grasping limbs and the shape of the fingers and toes strongly suggest an
arboreal form of life. The first human occupants of Madagascar, who
arrived there from Indonesia, encountered the giant lemurs, for reports
written by the French during the colonial period include descriptions of
giant lemurs made by the native peoples of the island. It was these natives
who also caused the extinction of these animals.

The second Madagascar lemur family, Daubentoniidae, is represented
today by just a single endangered species, the AYE-AYE (⊕ *Daubentonia
madagascariensis*). This creature caused a great deal of trouble for systematists, for with its rodentlike teeth it was initially classified as a rodent and
later placed within its own order. Other zoologists felt that aye-ayes arose
directly from plesiadapids, since some of them (*Eochiromys, Chiromyoides*
and others) shared the rodentlike (diprotodont) incisors and vestigial
dentition with the aye-aye. Today, aye-ayes are recognized as particularly
highly specialized descendants of primitive lemurs; indeed, juvenile aye-
ayes have fully lemurlike milk dentition! Another extinct giant form, the
GIANT AYE-AYE (*Daubentonia robusta*), lived during the Pleistocene. This
animal is of little help in determining the evolutionary origin of aye-ayes,
however, since there are no other known fossil aye-ayes at present and
thus nothing with which to make any comparisons.

With their short skulls and large crania, the INDRISOID LEMURS (Indri-
idae) look much more like apes than the long-snouted lemurs. Indrisoid
lemurs were a large, diverse group during the Pleistocene and on Madagascar into historical times. Those still extant—the INDRIS (*Indri indri*),
the two SIFAKA species (both in genus *Propithecus*) and the WOOLLY INDRIS
(*Avahi laniger*)—are but remnants of what were the prevailing large
prosimians that once lived on Madagascar. While *Neopropithecus* was
closely related to the sifakas of today, *Mesopithecus pithecoides* bore a greater

resemblance to apes than to sifakas and indris, for *Mesopithecus* had a less well-developed snout and a large, arched cranium. This giant indris resembled primitive Old World monkeys in other respects as well. These similarities are due to convergent evolution (parallel evolution) and not to a phylogenetic relationship between the indris and Old World monkeys, for the brain of the giant indris was structured as in lemurs.

In 1901, zoologist and student of Madagascar G. Grandidier described a number of subfossil limb bones (under the name "*Bradytherium madagascariense*"), which he thought were the remains of a Madagascar sloth. Later research showed that the "sloth" was actually an indrid. The genus is now called *Palaeopropithecus* (see Color plate, p. 503), and two species in this genus existed on Madagascar until the present era. The apparent similiarity to sloths is due to the claws on *Palaeopropithecus*, and the placement of the head on the vertebral column shows us that this indrid even moved about as it dangled, upside down, in the trees, just like sloths do. This anatomical peculiarity led to a great deal of confusion among the zoologists who were trying to understand this animal. As late as 1938 the Italian paleontologist G. L. Sera thought that *Palaeopropithecus* and the aforementioned giant lemur *Megaladapis* inhabited swamps, where they moved about by swimming. Other zoologists compared this "half sloth, half ape" with the orang-utan in terms of locomotion. Still others maintained that *Palaeopropithecus* was a climber like the African pottos and the southern Asian lorises. Present knowledge indicates that *Palaeopropithecus* occupied a similar ecological niche on Madagascar as the pottos in Africa, the lorises in southern Asia, and the tree sloths in South America. Once again, the anatomical similarities between the extinct indris and these other animals is due to convergent evolution.

A fourth Madagascar lemur family is now extinct. These ARCHEO-LEMURS (Archaeolemuridae), in the form of the genera *Archaeolemur* (see Color plate, p. 503) and *Hadropithecus*, were also discovered in a subfossil condition. Like some of the other lemurs, they also existed into modern times and were driven into extinction by the human settlers in Madagascar. Although the archeolemurs arose from primitive indris species, they bear a stronger resemblance to apes than the indris do. Their short skulls and various dental features give them a guenon-like appearance. Their skull is larger than that of a macaque and bears a strong resemblance to the skull of New World monkeys, although it does have indris characteristics as well (see Fig. 21-6). Archeolemurs were not ancestral to apes but were a separate line within the prosimian group and one that developed in parallel with apes. Archeolemurs probably moved on two legs, as their short fore arms indicate, and thus they were not primarily climbers like the sifakas.

Little is known about the evolution of the LORISES (infraorder Lorisiformes). The group includes jungle dwelling climbing animals like the African pottos and the southern Asian lorises and bipedal jumping,

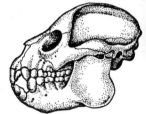

Fig. 21-5. Skull of the apelike early lemur *Archaeolemur edwardsi* from Madagascar (a reconstruction).

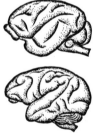

Fig. 21-6. Comparison of the brain of the fossil species *Archaeolemur* (above) and a modern macaque (below) indicates that the prosimian line of Old World lemurs evolved parallel with that of the apes. The brains of prehistoric lemurs reached the degree of development we find in modern apes.

large-eyed, long-snouted shrub and savanna inhabitants like the galagos, which are found only in Africa. The few fossil lorises from Miocene and Pliocene east Africa and India differ little from modern lorises. According to E. L. Simons the adapid genus *Anchomomys* from Eocene Europe may be an ancestor of lorises. If this were true, it would mean that there are phylogenetic links between lorises and Madagascar lemurs, and both superfamilies would probably have evolved from a single root stock within the adapids. Further fossil evidence will be needed to clarify this question.

The "tarsier problem"

In addition to the "tree shrew" problem, evolutionists must also deal with the complex "tarsier problem". The three extant species of TARSIERS (*Tarsius*), which are distributed from Indonesia to the Philippines, were long thought to be "living fossils" closely related to the immediate ancestors of modern apes. Whether the tarsiers are prosimians or an independent suborder between prosimians and apes is still under discussion. Many tarsier fossils from the Lower Tertiary (Paleocene to Eocene) have been recovered. Tarsiers share many important features with apes: the upright posture; the bony septum leading to the temporal fossae; the large, frontal eyes; the greatly enlarged optical center in the brain; the structure of the female sexual organs and the placenta; and, most importantly, the absence of a nasal ridge. Because of these features, zoologists such as Osman W. C. Hill classify living and fossil tarsiers and the apes into one large group, HAPLORHINI. F. Wood Jones presented one of the most exciting theoretical interpretations for some time; he claimed that the tarsiers represented an evolutionary step toward the hominoid superfamily (Hominoidea).

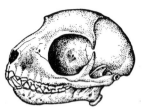

Fig. 21-7. Old World necrolemurid: skull of *Necrolemur antiquus* (from Europe).

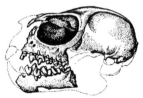

Fig. 21-8. A New World anaptomorphid: skull of *Tetonius homunculus*. The original specimen (from Lower Eocene North America) is shown with what is thought to be the rest of the skull (hashed lines).

But even if we assume that tarsiers are closer to apes than to prosimians, the fact is that they are bipedal, hopping nocturnal animals with a head unique among mammals in the extent it can be turned. This means that tarsiers are so highly specialized they could not be considered as ancestors for apes or hominoids. The similarity they bear to the higher apes could once again be a result of convergent evolution (the parallel but independent development of similar features). Furthermore, Lower Tertiary tarsiers are so different from each other that they do not form a single phylogenetic entity. Paleontologists will have to continue examining new fragments of skulls, jaws and other fossil remains to more fully clarify the tarsier problem.

The Paleocene paromomyids we mentioned before are often thought to be the ancestors of all the tarsioids (infraorder Tarsiiformes). Then, in the Eocene or possibly Paleocene, the Old World NECROLEMURIDS (Necrolemuridae) and the New World ANAPTOMORPHIDS (Anaptomorphidae) appeared. Necrolemurids were widely distributed in the Old World Eocene with genera such as *Necrolemur*, *Nannopithex*, *Microchoerus* and *Pseudoloris*. They developed many apelike features (again convergent evolution!). One of the most notable Upper Paleocene and Eocene

Fig. 21-9. Lower jaw of *Omomys*. The specimen is from North America. Omomids are probably ancestral to all monkeys.

American anaptomorphids is the species *Tetonius homunculus* (see Color plate, p. 503), a species of which there are some very well preserved skull fragments. The striking similarity between *Tetonius* and apes has led many zoologists to classify the animal as an ancestor of apes or even of hominoids. Existing fossil evidence suggests that this anaptormorphid was a bipedal hopper or jumper and moved about much like the tarsiers and galagos of today. However, its teeth are quite different from tarsier dentition, for *Tetonius* (see Fig. 21-8) has enlarged upper incisors (diprodont teeth) and reduced lower incisors and canines.

If necrolemurids and anaptomorphids really belong to the tarsier line (and further evidence will be needed to confirm or reject this hypothesis), they would also be evolutionary side branches. Tarsiers would then be understood as the last living representatives of a Lower Tertiary primate radiation, which produced a wealth of species during the Eocene and of which some species became nearly as highly developed as apes. The other prosimians and the apes would have originated from other radiations occurring in the Eocene and later.

One of the most apelike Lower Oligocene species, one recovered in Texas, is the genus *Rooneyia*. Paleontologists are debating whether this is a necrolemurid or a member of another family related to tarsiers. This other family, the OMOMYIDS (superfamily Omomyoidea, family Omomyidae) was separated as a distinct tarsioid group by the paleontologist C. L. Gazin in 1958. Omomyids were small primates with little or no specialization and slender jaws (see Fig. 21-9) whose dental formula $\left(\frac{2 \cdot 1 \cdot 3 \cdot 3}{2 \cdot 1 \cdot 3 \cdot 3}\right)$ corresponds to New World cebus monkeys. Some omomyids include *Omomys* and *Hemiacodon* from North America, *Lushius* and *Hoanghonius* from eastern Asia, and *Periconodon* from Europe. *Teilhardina belgica* from Lower Eocene Belgium, once thought to be a necrolemurid, is now placed with the omomyids even though it probably has the same dental formula as the primitive mammals. The lower jaw formula is $\left(\frac{}{3 \cdot 1 \cdot 4 \cdot 3}\right)$.

Omomyids were most prevalent in the Eocene, the most recent genus being *Macrotarsius* from Lower Oligocene North America. They figure importantly into ape evolution, but whether omomyids evolved from anaptomorphids (e.g., the still disputed genus *Navajovius* from Upper Paleocene North America) or from the older (Lower Paleocene) paromomyids is uncertain. In 1958, C. L. Gazin designated the older, unspecialized omomyids as the ancestors of New World monkeys, while in 1965 G. E. Quinet claimed that *Teilhardina belgica* was ancestral to Old World monkeys. In view of the primitive dentition and other primitive features in the omomyids, it seems highly probable that omomyids were the ancestors of both Old and New World monkeys.

Designating the omomyids as ancestral to all modern monkeys answers a long discussed question of whether the monkeys of the world comprise a single phylogenetic entity. That is, did New World monkeys

▷▷
Late Ice Age steppe habitat showing mammoths (*Mammonteus primigenius*), a woolly rhinoceros (*Coelodonta antiquitatis*) and a reindeer herd (*Rangifer tarandus*); in the foreground are dwarf shrubs and various plains plants.

▷
Evolution of the true bears (*Ursus*). The brown bear (subgenus *Ursus*) and the cave bear (†*U. spelaeus*, extinct by the end of the Pleistocene) have evolved from a common ancestor (*U. etruscus*) from the lowermost Quaternary. Polar bears (genus *Thalarctos*) are specialized descendants of brown bears. The black bear subgenus *Euarctos*, which comprises the central Asian sun bear (*Ursus thibetanus*) and the North American *Ursus americanus*, have scarcely changed since the Lower Quaternary. The phylogenetic separation between brown and black bears must have occurred at a very early time.

Mono- or polyphyletic primate evolution?

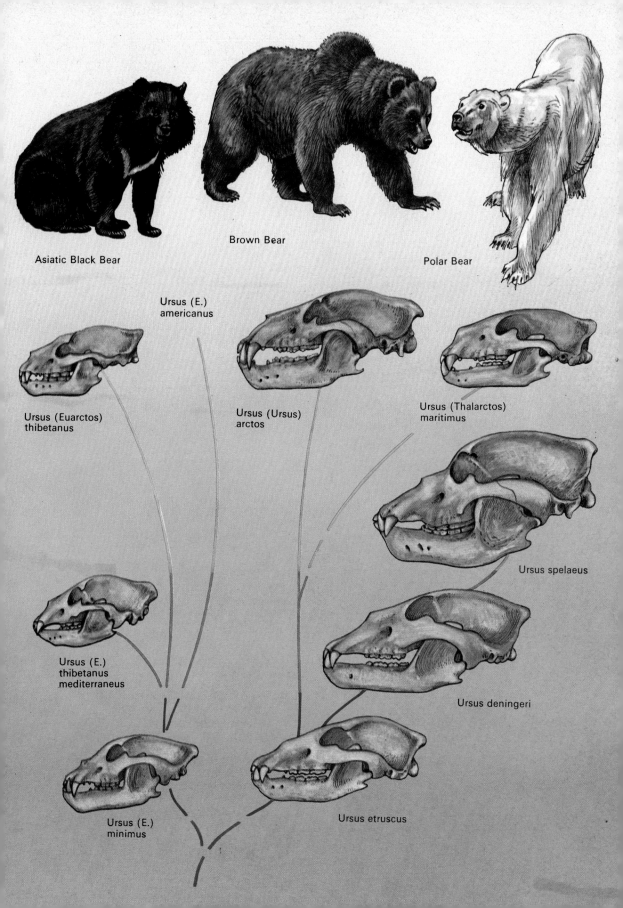

Asiatic Black Bear

Brown Bear

Polar Bear

Ursus (E.)
americanus

Ursus (Euarctos)
thibetanus

Ursus (Ursus)
arctos

Ursus (Thalarctos)
maritimus

Ursus spelaeus

Ursus (E.)
thibetanus
mediterraneus

Ursus deningeri

Ursus (E.)
minimus

Ursus etruscus

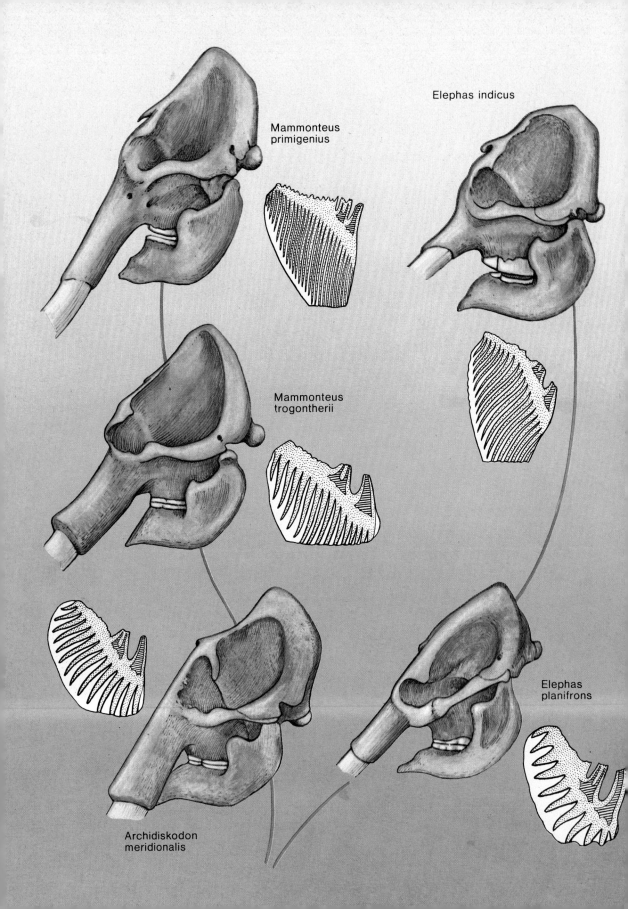

Mammonteus
primigenius

Elephas indicus

Mammonteus
trogontherii

Elephas
planifrons

Archidiskodon
meridionalis

and Old World monkeys evolve from a single ancestor (monophyletic evolution) or as separate lines (polyphyletic)? The polyphyletic interpretation was based on an assumption that the similar features found in New and Old World monkeys are due to the independent evolution of those features, just as some of the Madagascar lemurs (like the archeolemurs from the family Archaeolemuridae) display a "trend" towards the apes. However, the similarities between Old and New World monkeys are so prominent that we cannot describe them as two phylogenetically distinct groups. However, the evolutionary division into the broad-nosed monkeys of the New World and the narrow-nosed Old World monkeys occurred no later than the Eocene. And since that time the two groups have gone their own evolutionary paths. Another indication of a common ancestry is in the fact that the oldest omomyids gave rise to monkeys in the northern hemisphere, occurring at a time when a subtropical to tropical climate prevailed in that part of the world.

Few fossil finds are available of NEW WORLD MONKEYS (infraorder Platyrrhina), and these are from the Oligocene, the Miocene and the Quaternary. There are currently two theories about their origin. One of these was developed by the great American paleontologist, George Gaylord Simpson. Since South America was not joined, with North America during the Eocene and Oligocene, Dr. Simpson maintains that the first primates migrating into South America and giving rise to the various New World monkey lines must have gotten there by "island hopping" from North America. New World monkeys at that time would scarcely have been beyond the omomyid phase.

French paleontologist R. Lavocat makes a different interpretation. He sees the ancestors of New World monkeys in the Lower Tertiary monkeys of Africa, since these species bear a number of characteristics (e.g., the dental formula and the structure of the brain) found elsewhere only in the New World monkeys. Lavocat believes that monkeys "drifted" from western Africa to South America during the Eocene. Support for his theory comes from evolutionary patterns occurring among the cavies or guinea pig-like rodents during that time and from the paleogeographic situation during the Eocene. According to Lavocat's theory, the division into Old and New World monkeys did not occur at the prosimian level (i.e., with the omomyids) but at the very base of primate evolution. In other words, this great cleavage occurred during the Oligocene, and the great similarities between New and Old World monkeys would be largely due to convergent (independent) evolution.

One Lower Oligocene South American species, *Branisella boliviana*, was a monkey with some prosimian features but a dental formula indicating descent from the broad-nosed monkeys. Fossil New World monkeys that can be classified with the CAPUCHIN monkey family (Cebidae) have been found in Upper Oligocene and Miocene Argentina and Colombia. Among them, the genus *Homunculus* has a particularly interesting history

◁
Elephant (including mammoths) evolution, illustrated by the phylogenesis of the molars from low-crowned teeth to the high-crowned lamellar structure in the more recent species.

because the Argentinian paleontologist and geologist Florentino Ame-
ghino (1854–1911) classified it as a direct ancestor of the hominoids. A
number of these "little men", as the name translates, have been found and
studied in detail. These studies have shown that *Homunculus* is a specialized
member of a primitive capuchin line. According to Dr. Philip Hershkovitz, the same is true for *Dolichocebus gaimanensis* from Upper Oligocene
South America; this animal was previously thought to be the oldest New
World monkey. The dental formula of the badly preserved skull's upper
jaw is, unlike the original configuration, ($\frac{2 \cdot 1 \cdot 3 \cdot 3}{}$), and this is the
same as the capuchin dental formula. It is clues like the teeth configuration
that have yielded so much information to paleontologists, and here we
have yet another example of how details like tooth formula can be of
importance to those who study the life of the past. The above bit of
information tells us that *Dolichocebus* is not, as we thought until recently,
a member of the MARMOSET/TARMARIN family (Callithricidae).

This provides us with an opportunity to note that the marmosets,
which are the smallest monkeys of all, are by no means primitive New
World monkeys (as we formerly thought). Instead they are a geologically
very young group whose great number of species tells us that the marmosets are in the process of very active evolutionary development. Their
"claws" arise, as embryological investigations have shown, secondarily
from modified nails. The marmosets evolved from primitive capuchin
species (Cebidae) and they, like howler monkeys and several other broad-
nosed monkeys, can oppose their thumbs to the other fingers. During the
Tertiary, they occupied the same ecological niche in South America as
the aroboreal rodents and small carnivores did in the Old World tropics.

Upper Miocene Colombia was the habitat of *Cebupithecia* and other
species belonging to the modern SAKIS (Pitheciinae), which Hershkovitz
claims belong to an independent line. The lower jaw of one of these
fossils, *Weosaimiri*, combines characteristics of the capuchin SQUIRREL
MONKEY (*Saimiri*) and the clawed monkeys. This suggests that during the
Upper Miocene the most important New World monkeys (excepting
the marmosets) had made their appearance. Other fossil New World
monkeys are from Pleistocene deposits, and all of them except the Jamai-
can species *Xenothrix mcgregori* are closely related to modern species.
Xenothrix, of which all we have is a lower jaw with two molars, does
not conform to the other monkeys and cannot be classified with them.
Hershkovitz feels it belongs with a separate family, Xenothricidae. Interestingly, neither fossil nor living monkeys have been found on any
other Caribbean island save Cuba! This could mean that the Upper
Tertiary jungles in South America formed a continuous land mass subject
to periodic flooding. Some primates migrated into what is now the
Caribbean region during the dry periods, and when floods came they were
isolated on islands such as Cuba and Jamaica.

A great deal more is understood about the evolution of OLD WORLD

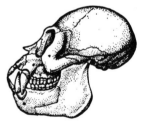

Fig. 21-10. Skull of *Cebupithecia sarmientoi* from
Colombia; this species is
closely related to the sakis
(Pitheciinae) of today.

Important finds of fossil primates in North and South America.

Fig. 21-11.
1. *Notharctus osborni*
2. *Tetonius homunculus.*
3. *Omomys.*
4. *Cebupithecia sarmientoi.*
5. *Homunculus patagonicus.*

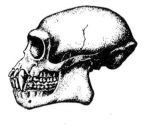

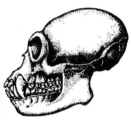

Fig. 21-12. Skull of the fossil gibbon *Pliopithecus* (above) and a modern gibbon (below).

or CATARRHINE MONKEYS (Catarrhina). Two superfamilies are typically distinguished among modern catarrhine monkeys: CERCOPITHECOIDS (Cercopithecoidea) and HOMINOIDS (Hominoidea). These two groups, which probably originated from Old World omomyids, diverged at an early time and probably in the Eocene. However the claim that cercopithecines and hominoids arose from different lines is now highly questionable, for primitive cercopithecines display a number of hominoidlike features. One of these is their molar anatomy. And similarly, the oldest hominoids have several cercopithecinelike features. The Old World apes probably arose in Africa; the vast diggings in the Fayum layers (in northern Egypt) would tend to confirm this hypothesis. Fayum specimens have included the teeth, jaws and skulls of monkeys with cercopithecine features (*Oligopithecus*), others with gibbonlike characteristics (*Aeolopithecus*) and true hominoids (or anthropoid apes: *Aegyptopithecus* and *Propliopithecus*). Primitive members of all main groups of Old World apes occurred during the Upper Oligocene in Africa.

Among the fossil cercopithecines, *Oligopithecus savagei* has both New World features and prosimianlike characteristics, something we would expect at this level of evolution. Today we distinguish two cercopithecine groups: GUENON-LIKE MONKEYS (Cercopithecidae), which includes guenon monkeys, mangabeys, macaques and baboons; and the LEAF MONKEY group (Colobidae), with the Asian leaf monkeys and proboscis monkeys as well as the colobus monkeys of Africa. Although modern leaf monkeys and colobus monkeys are highly specialized (they are adapted to feeding on foliage and have large, subdivided stomachs), they are more primitive than guenon-like monkeys in a number of characteristics. Some of the fossil leaf monkeys include *Prophylobates tandyi* from Lower Miocene Egypt, *Victoriapithecus* from Miocene eastern Africa, and *Mesopithecus pentelici* (see Color plates, p. 503), known for a longer time than any of the others. *Mesopithecus* was discovered in 1838 near Pikermi, Greece, and was the first fossil monkey. Later, Lower Pliocene deposits yielded nearly intact skulls and skeletons of this fossil monkey. Other prehistoric leaf monkeys have been recovered in Pleistocene southern Asia and eastern Africa.

The systematic position of the macaque-sized *Libyithecus markgrafi* from Upper Pliocene Egypt is uncertain. Some zoologists class it with leaf monkeys, while others place it with the guenon-like monkey group. *Dolichopithecus*, from Pliocene/Pleistocene Eurasia, was formerly interpreted as a leaf monkey but is now understood as a ground-dwelling Old World monkey related to macaques. Baboons probably evolved their specializations during the Lower Pleistocene. They share a common ancestor with the macaques and were represented during the Pleistocene by some giant baboon species (*Gorgopithecus* and *Dinopithecus*; see Color plate, p. 503). The latter, from southern Africa, was the size of a medium-large gorilla! *Simopithecus* is another baboon-sized Old World monkey from Pleistocene Africa. It is actually related to the modern gelada baboon (*Theropithecus*), which is not a close relative of the other baboon species.

Gibbons, which today are found only in southern and southeastern Asia, were distributed across Africa and Europe during the Miocene and Pliocene. Skull and skeleton fragments have been recovered of such prehistoric gibbons as *Pliopithecus* and *Limnopithecus*. One of these, namely *Pliopithecus vindobonensis* from Middle Miocene Europe (see Color plate, p. 504), differed from all modern gibbons in that it had a tail. Miocene gibbons did not have the greatly elongated arms we find in modern species, and this means that Miocene gibbons (or at least *Limnopithecus* from eastern Africa) were not the great acrobats our modern species are. Thus, the extreme arm elongation is a relatively recent adaptation.

Fig. 21-13. Skeleton of the tailed gibbon *Pliopithecus vindobonensis* from Miocene Austria (original find plus postulated missing bones).

One particularly well preserved lower jaw (with complete dentition) specimen from Lower Oligocene (Fayum) Egypt was long thought to belong to a transition species intermediate between prosimians and Old World (catarrhine) monkeys. Later studies showed that this prehistoric primate (*Parapithecus fraasi*; see Color plate, p. 504 and Fig. 21-14) is a link between Old World monkeys and anthropoid apes. This was a very small monkey about the size of a squirrel monkey. Its dental formula corresponds with that of New World monkeys, but this is merely a primitive feature because *Parapithecus* naturally is unrelated to the monkeys of the new world. The hypothesis that this small ape stood phylogenetically at the base of catarrhine monkeys and anthropoid apes is no longer deemed accurate, chiefly because the oldest known anthropoid ape (*Propliopithecus haeckeli*; see Fig. 21-16) is from the very same deposits and thus comes from the same time period as *Parapithecus*. *Parapithecus* belongs to an evolutionary side branch pursuing its own phylogenetic pathway. It may be related to *Apidium phiomense* (another disputed species), of which all we have are jaw fragments from Fayum. E. L. Simons places *Apidium* into the anthropoid ape family of OREOPITHECIDS (Oreopithecidae). The sparsity of data (i.e., fossils) on these apes is the reason for our uncertainty on their meaning, and future zoologists will hopefully unearth further working material in the form of more and better fossils and thus shed additional light on the phylogenetic significance of these fossil apes.

Fig. 21-14. A row of teeth from *Parapithecus fraasi*.

Regardless of how *Apidium* is classified, parapithecids and oreopithecids show that catarrhine monkeys developed anthropoid apelike lines during several periods within the Tertiary, all of which save the one leading to hominids died out. In 1872, French paleontologist P. Gervais described the lower jaw of an ape from Lower Pliocene lignite in Monte Bamboli, Italy. The jaw belonged to *Oreopithecus bambolii* (see Color plates, pp. 503 and 525), an ape about the size of a small chimpanzee (Fig. 21-15 shows the skull of *Oreopithecus*). The species was classified as a catarrhine monkey until 1954, when Swiss paleontologist J. Hürzeler found a badly crushed but nearly complete skeleton of the same species at the same site where the first one was found. He was able to determine that *Oreopithecus* was an anthropoid ape-man (Homididas). Hürzeler's finding met with a great deal of debate from other paleontologists, since this fossil ape was about ten million years old and thus much older than any known anthropoid ape-man. Later it was found that these seemingly anthropoid features were only primitive characteristics. *Oreopithecus* was indeed an anthropoid ape, but one that was on its own evolutionary pathway and not in the mainstream of anthropoid evolution. It is closer to the great apes of today (*Pongidae*) than to humans (Homininae).

In 1967, E. L. Simons described from the Upper Oligocene of Fayum what he called "the oldest known precursor of anthropoid apes and man". This *Aegyptopithecus zeuxis*, of which we have a nearly intact skull and many skeletal parts, is an example of what is called Watson's Rule, according to which some characteristic can develop independently and at different times throughout geological history. The sum of all characteristics of a living organism is actually a mosaic of many different kinds of features: some primitive and conservative and others advanced and specialized. Not every feature of an animal is at the same stage of evolutionary development! Those species with marked evolutionary differences among various features best exemplify Watson's Rule and hence mosaic evolution, and *Aegyptopithecus* is one of these kinds of animals. Some of the prosimian characteristics of this fossil include the not quite forward placed eyes and the relatively small cranium, while catarrhine features include the structure of the snout (prognathy); similarly, there are also anthropoid features like the structure of the molars! In *Aegyptopithecus* we have a splendid mosaic of prosimian, catarrhine and anthropoid characteristics all rolled up into one organism.

The lower jaw fragments of a very similar primate from Lower Oligocene Fayum were known since the first decade of the 20th Century, and the German paleontologist Max Schlosser named the animal to whom the jaw belonged *Propliopithecus haeckeli* (see Color plate, p. 504 and Fig. 21-16). The primate appeared to be a gibbon precursor, and since Haeckel considered gibbons to be close to the direct ancestors of man, Haeckel's name was incorporated into the scientific name of this fossil. Although Haeckel's gibbon theory has since been disproven, *Propliopithecus* has retained its place in the evolution of early hominids. Its systematic position is still being debated. Most paleontologists feel that *Propliopithecus* and *Aegyptopithecus* are the ancestors of the anthropoid apes; but one of the most renowned interpreters of the Fayum primates, E. L. Simons, sees in the tooth anatomy of this Lower Oligocene primate a preliminary stage for the human dental structure, and the American D. R. Pilbeam even claims that *Propliopithecus* is the "earliest predecessor of man".

In the following chapter we discuss the question of whether human evolution extends back to the Oligocene or whether early man developed later (Miocene/Pliocene) from "subhuman" anthropoid apes. True

Fig. 21-15. Reconstruction of skull of the anthropoid ape *Oreopithecus*.

Fig. 21-16. Row of teeth from the human ancestor *Propliopithecus haeckeli*.

ANTHROPOID APES (Pongidae) did exist and were highly diverse in the Miocene and Pliocene of Africa and Eurasia. A great many generic names were developed for all these apes, and this no doubt gave the impression of there being more phylogenetic lines than in truth existed. Often a single bone fragment or tooth was enough "evidence" for an investigator to name the animal in question, and eventually (excluding the parapithecids, gibbons, oreopithecids and *Gigantopithecus*) there were more than 20 fossil anthropoid ape "genera". In 1965, Simons and Pilbeam published a provional revision of all these genera, and in it all the Miocene/Pliocene anthropoid apes were classified as members of the genus *Dryopithecus*, which had three subgenera (*Proconsul* from Africa, *Dryopithecus* from Europe and *Sivapithecus* from Asia). These divisions do not actually reveal phylogenetic relationships between these apes; they arc a reflection of geographical distribution.

In all probability, the anthropoid apes (along with catarrhine monkeys and gibbons) arose in Africa. They were forest dwellers and migrated into Eurasia in the Miocene. They lacked the elongated fore arms characteristic of modern species, which indicates that the prehistoric anthropoid apes did not move by brachiating (swinging hand over hand). If it is assumed that the ancestor of man was among these fossil species, we have the answer to a long-asked question: man did not evolve from brachiators or other species of the humid tropical rain forest habitat but from "prebrachiators", animals that could run on the ground as well as in the trees and that could in emergencies even run on two legs.

Of the wealth of Tertiary anthropoid apes, the genus *Proconsul* (see Color plate, p. 504 and Fig. 21-18) has a particularly prominent position. It existed in several species from Middle and Upper Miocene Africa. The most well known of them is *Proconsul africanus*, which was found on Rusinga island in Lake Victoria and later in Uganda as well. This primate was about the size of a bonobo and was characterized by well-developed frontal sinuses, a short face, a forward set atlo-occipital articulation, and the absence of supraorbital swellings and bony ridges on the skull (as shown in Fig. 21-18). The snout was not particularly prominent. A typically apelike bony bridge in the chin is not evident. Unlike modern anthropoid apes, its arms were shorter than its legs. The hand, with an opposable thumb, looked more like our hand than the "hooked hand" of brachiating apes. Thus *Proconsul africanus* is an early anthropoid ape that moved largely on four legs and in which some seizing ability (permitting the body to hang down while the animal climbs about) is evident for the first time. The largest species within the genus, gorilla sized *Proconsul major*, was probably a pure ground dweller.

In many respects, *Proconsul* is an ideal "ancestral model" for all the precursors of anthropoid apes, or at least of chimpanzees, and hence of early man. Some paleontologists felt that this primate could not be part of the evolutionary line leading to humans, because it had prominent

Important finds of fossil primates in Európe, Asia and Africa.

Fig. 21-17. 1. *Anagale gobiensis* (not a primate probably an advanced insectivore).
2. *Plesiadapis tricuspidens* and *Teilhardina belgica*.
3. *Necrolemur antiquus*.
4. *Archaeolemur* and *Megaladapis*.
5. *Mesopithecus*.
6. *Pliopithecus*.
7. *Oreopithecus bambolii*.
8. Fayum: *Parapithecus fraasi, Oligopithecus, Aeolopithecus, Aegyptopithecus* and *Propliopithecus*.
9. East Africa: *Limnopithecus, Victoriapithecus Dryopithecus* and *Proconsul*.
10. *Dinopithecus*.
11. *Sivapithecus*.
12. *Gigantopithecus*.

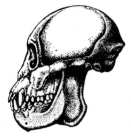

Fig. 21-18. *Proconsul major*.

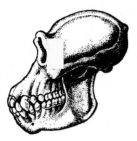

Fig. 21-19. Chimpanzee.

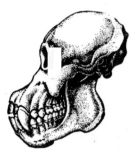

Fig. 21-20. Orang-utan.

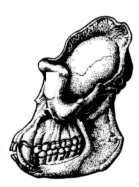

Fig. 21-21. Gorilla (extreme development of the crest and the canine teeth).

daggerlike canines. In any event, this is indeed a primitive anthropoid ape, which may have scurried about on all fours from one group of trees to another on the African savanna and may have occasionally run bipedally.

The first fossil anthropoid ape was discovered in 1856. French paleontologist Edouard Lartet, digging in Saint-Gaudens on the northern slopes of the Pyrenees, found the upper arm bone and other fragments of an animal whose dentition resembled a gorilla but which was only the size of a small chimpanzee; the animal was named *Dryopithecus fontani* and was thought to have lived in the Upper Miocene. 20th Century finds have now shown that *Dryopithecus* also lived in other parts of Europe and in Asia as well. Other *Dryopithecus* species were subsequently described. In contrast with *Proconsul*, they do have a bony bridge in the inner corner of the chin, and the rows of molar teeth are parallel as in modern anthropoid apes.

Only scattered fragments of limbs exist of some of the Tertiary anthropoid apes, such as *Austriacopithecus weinfurteri* from Middle Miocene Austria and *Paidopithex rhenanus* from Lower Pliocene Germany. The latter looked so humanlike that one German philosopher (Schleiermacher) thought the bones belonged to a child! Today it is believed that *Austriacopithecus* and *Dryopithecus* are related and that a phylogenetic relationship may also exist between *Paidopithex* and *Proconsul*.

No less than nine genera of anthropoid apes have been described from Miocene and Pliocene deposits from Eurasia and southern and eastern Asia. This apparent diversity is no doubt due to the fact that anthropoid apes within even single species could vary considerably anatomically from one individual to another and that males in general are larger than females. Furthermore, the record of some of the "genera" consists of just a few bone fragments, hardly enough evidence to use in naming yet another fossil ape genus. Today, all these apes are placed within a single genus, *Sivapithecus*, and Simons and Pilbeam distinguish two species, *S. sivalensis* and *S. indicus*, within this genus. Whether these species, which are from Tertiary Asia, gave rise to the orang-utan (today occurring only on Sumatra and Borneo) is questionable. Giant orang-utans have indeed been found in Pleistocene deposits from the southeastern Asian continent, but no intermediate species linking them with *Sivapithecus* have been unearthed.

The orang-utan is classified within a separate genus (*Pongo*) in the anthropoid ape family, while the gorilla and chimpanzee are more closely related. Unlike the orang-utan, they are not "dangling" arboreal species but probably arose from ground-dwelling dryopithecines. Male gorillas (and there are no fossil precursors of gorillas!) have great difficulty climbing a tree. Possible ancestors of chimpanzees have been found in Olduvai gorge in eastern Africa. Adriaan Kortlandt and Jane van Lawick-Goodall have recently found that among modern chimpanzees there are not only jungle inhabitants but also chimps on the savannas, and this could indicate that chimpanzees secondarily moved into the jungle. Although chimpanzees can move very adroitly in the treetops, they do not swing through them; instead they "run" upright along the tree trunks, using their elongated arms to help. The habit of standing on two legs when threatened does not indicate that gorillas and chimpanzees evolved from jungle dwellers, because in the jungle there would be no advantage to standing upright amidst all the tangled vines and leaves. In the savannas, however, such behavior would be adaptive (i.e., of evolutionary advantage in terms of adapting to the environment), because standing on the hind legs would permit the animal to peer above the tall grasses for signs of danger.

We close this description of the fossil subhuman primates with an extinct side branch of anthropoid apes, one that for a time created a great deal of excitement among paleontologists. The fossil evidence involved includes over 1000 individual teeth and several jaws with nearly intact dentition, all found in deposits from Lower and Middle Pleistocene southern China. The lower jaw and molars of this *Gigantopithecus blacki* (see Color plate, p. 504 and Fig. 21-22) are larger than those of any living or extinct anthropoid ape or human; the incisors and canines, however, are smaller and less sharp than those of modern anthropoid apes. Some thought the animal was a "giant man", while others more correctly believed that *Gigantopithecus* was a huge anthropoid ape with large jaws and teeth and some humanlike features. Recently, Simons found another lower jaw of this species; it was from Pliocene Sivalik deposits at the foot of the Himalaya, and this meant that the "giant" existed in the Tertiary. This Chinese giant anthropoid ape was presumably a ground-dwelling herbivore. It is doubtful that the animal could walk upright, and this is a vital clue to determining if a fossil species might be included in the phylogenetic line leading to man.

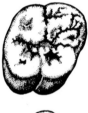

Fig. 21-22. Molar of the giant anthropoid ape *Gigantopithecus blacki* (above) and the molar of a modern human.

22 Man and His Ancestors

By G. Heberer

The evolution of man extends over a much longer period of time than we previously thought. Many new fossil finds in the last few decades have provided decisive evidence that our phylogenetic heritage has its beginning millions of years ago. As we described in the previous chapter, the oldest group of hominoids (superfamily Hominoidea) was discovered in the first decade of the 20th Century. The find consisted of a primitive, apparently gibbonlike lower jaw found by the renowned "fossil hunter" Markgraf in El Fayum, Egypt; it stems from the Oligocene, the last period of the Lower Tertiary.

Max Schlosser, German paleontologist named this fossil *Propliopithecus haeckeli* (see Color plate, p. 504). Zoologist Ernst Haeckel's name was used because Haeckel had long maintained that man—genus *Homo* —evolved from gibbonlike species. In 1961, Swiss paleontologist J. Kälin thoroughly restudied this specimen, and soon thereafter Simons and others unearthed two more Fayum Oligocene lower jaws plus a number of teeth. In 1967 a skull was added to the growing collection. This new find, named *Aegyptopithecus zeuxis*, combined characteristics of gibbons, anthropoid apes and early hominids. The fact that the eyes were not quite frontally located even indicated prosimian features! On the basis of present knowledge it appears that *Propliopithecus* and *Aegyptopithecus* belong to the ancestral group from which man evolved. This means that our human beginnings extend back more than 20,000,000 years!

Other Tertiary predecessors of humans have been found in the succeeding epochs of the Upper Miocene and the Lower Pliocene. These RAMAPITHECINES (subfamily Ramapithecinae) belong to the same family— Hominidae—as we and thus can be considered as early hominids. The only record we have of ramapithecines are a few lower and upper jaw fragments, but even this sparse evidence suffices to demonstrate that ramapithecine jaw anatomy differs from that of contemporary anthropoid apes and bears a greater resemblance to human jaw structure. From

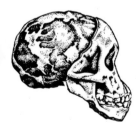

Fig. 22-1. The "child of Taung", the first find of an australopithecine skull (1924).

the Sivalik strata in northern India comes *Ramapithecus punjabicus* (see Color plate, p. 521); from eastern Africa (near Fort Ternan on Lake Victoria) comes *Kenyapithecus wickeri*, discovered by that renowned digger L. S. B. Leakey and which may belong to the same species. Ramapithecines may even have utilized tools. And if they indeed produced and used tools, this would mean that what we call the "animalman-transition period" was reached during the ramapithecine stage. With this "period" begins man.

Recent work by Adriaan Kortlandt, Baroness Jane van Lawick-Goodall and others has shown that tool production and manipulation occurs in anthropoid apes (especially in chimpanzees). Thus, the borderline between animal and man (or better: between anthropoid apes and humans) cannot easily be determined solely on the basis of tool use. This criterion is particularly difficult to apply when we go back so far in prehistory as the ramapithecines. However, there was a group of early hominids that we know utilized tools. These hominids, somewhat unfortunately, are called AUSTRALO-PITHECINES (subfamily Australopithecinae), a name meaning southern ape. The name arose from the first person to describe these hominids, anatomist Raymond A. Dart, who in 1924 found the skull of an australopithecine infant in Taung, Bechuanaland. Dart first thought that the skull belonged to an infant anthropoid ape (hence the name).

About 100 australopithecine specimens have been found since this first discovery. Important work was carried out by Robert Broom (1860–1951), John T. Robinson, and P. V. Tobias (who also described an australopithecine skull from the Olduvai gorge in eastern Africa). All these people worked in southern Africa at sites such as Sterkfontein, Kromdraai, Swartkrans (these three near Johannesburg) and Makapansgat in central Transvaal. Later it was found that australopithecines also lived in eastern Africa, such as at the Olduvai gorge, in northern Kenya, and in the Omo region on Lake Rudolf.

Olduvai, which is on the edge of the Serengeti, became particularly prominent. This was where Dr. Louis S. B. Leakey found not only skulls and other remains of australopithecines but also of other early hominids. A whole series of finds was made, and these were responsible for a basic reevaluation of the time span involved in hominid evolution. First an excellently preserved australopithecine lower jaw was found at Lake Natron, and later other jaw fragments from this same group were discovered at other sites. Many of these were considerably older than any previous fossil hominids. Richard Leakey, Louis's son, was fortunate enough to find a relatively well-preserved skull at Lake Rudolf of the same type as was found at Olduvai. Other similar specimens had been discovered at Swartkrans and Kromdraai. These hominids appeared to be herbivores, and they were called *Australopithecus boisei* (those from eastern Africa; see Color plate, p. 526) and *Australopithecus robustus* (South Africa; see Color plate, p. 521). They were distinguished from a

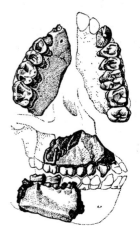

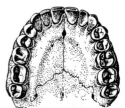

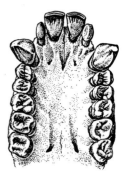

Fig. 22-2. Above: upper and lower jaw fragments of *Ramapithecus*, put together from various finds. These fragments are evidence of the existence of hominoids some 12 million years ago. Middle: the jaw of a modern human, shown for comparison. Below: the jaw of an anthropoid ape, the orangutan.

third type, *Australopithecus africanus* (see Color plate, p. 521), a more delicate form from southern and eastern Africa.

Thus we have two varieties of these African early hominids: the slighter so-called A-type, with an omnivorous dentition, and the more robust P-type, with heavier teeth (indicative of an herbivorous diet) and a ridge on top of the skull to which powerful chewing muscles were attached (see Fig. 22-3). Other skull features of the P-type australopithecine are more developed than in the A-type. A recent hypothesis explains this difference by interpreting the two types as members of the same species but different sexes; in other words, the differences in stature between the A- and P-types would be due to sexual dimorphism. The P-type would be the male and the A-type the female. However, this hypothesis has been criticized by some researchers. Also, the precise geological age of the australopithecines in question has not always been determinable. Modern dating techniques suggest that the oldest finds from the region north and east of Lake Rudolf, especially from the Omo region, are about 5½ million years old. Leakey felt that further diggings between Nairobi and Olduvai could yield specimens up to 7 million years old. On the other hand, the south African australopithecines are about 1 million years old. The oldest traces we have of hominids are in the Pliocene (which ended 2½ million years ago), and presumably (and hopefully!) the cleft between the ramapithecines and the australopithecines will eventually be closed by additional future finds.

In the east African australopithecine country there are also fossil finds in which it is difficult to determine whether these are australopithecines or members of the genus *Homo*. Thus, some finds may be identified with both generic names (i.e., *Australopithecus habilis* and *Homo habilis*; see Color plate, p. 521), depending on the researcher's interpretation. The species name *habilis*, which means able or capable, refers to the fact that these hominids were capable of tool production and use. The ungulate bones found near Makapansgat in Transvaal suggest that australopithecines, too, may have used those bones as tools. A number of authorities doubt that australopithecine tool use was a very likely possibility, but the weight of the evidence is in favor of that. If you were to examine all the available finds and compare them at their original sites, as I did at the side of Raymond Dart himself, I believe you would conclude that the bones there were actual tools and were not the remains of hyena activity or of other predators who fed on ungulates and left their bones lying about. Even stone implements may have existed among the australopithecines.

A debate still revolves around the question of whether the australopithecines were confined to Africa (making it the true original homeland of mankind) or whether the finds from other parts of the world can likewise be classified as australopithecines. J. T. Robinson, who found several jaw fragments on Java and named them *Meganthropus palaeojavanicus*,

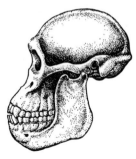

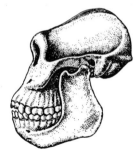

Fig. 22-3. Reconstruction of *Australopithecus*. Above: Side and front view of the A-type (named after *Australopithecus*). Below: Side and front view of the P-type (after the earlier generic name *Paranthropus*).

claims that this species is an australopithecine. His viewpoint is contested by P. V. Tobias and G. H. R. von Koenigswald, who describe *Meganthropus* as a primitive precursor of the *erectus* type (described by us subsequently). We do know that *Meganthropus* of the Djetis fauna of Java was associated with the older early hominid type, *Homo erectus modjokertensis* and thus can be excluded as an ancestor of man.

The aforementioned early hominid group (the "*Pithecanthropus* group") became known to science through the Javan "ape men" ("*Pithecanthropus*") found by the Dutch scientist Eugen Dubois between 1891 and 1893. Dubois initially excavated caves in Sumatra before turning to research in Java. His interest in Java work was sparked when the skull of an ice age modern man was sent to him from Java while he was still working on Sumatra. Dubois chose to research the region around Trinil on Java's Solo River, because this site was long known for its fossils of Pleistocene mammals. It was one of anthropology's greatest strokes of luck for Dubois to find here, at a depth of about 15 meters, a skull roof. Dubois initially thought the skull roof was from a chimpanzee, but a short distance away he found a fossil human femur. Later other femur fragments were found at this site, but Dubois did not at first believe that this anthropoid ape-like skull roof was from the same species as the manlike leg bones. Not until 1894 did he consider both skeletal elements to be the remains of an "upright walking anthropoid ape" he called "*Pithecanthropus erectus.*"

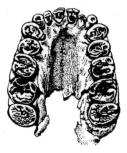

Fig. 22-4. Herbivorous dentition of the P-type *Australopithecus*; note the enlarged molars.

The term "*Pithecanthropus*" had been coined by Ernst Haeckel in 1866, who used it as a reference for the as yet undiscovered missing link between anthropoid apes and humans. For decades, Dubois's Javan hominid was the subject of a vigorous scientific controversy, and Darwinians viewed "*Pithecanthropus*" as an ape-man, the direct precursor of modern humans, until the African australopithecines were bestowed with this position on the basis of later research. Today, this so-called "*Pithecanthropus*" is classified within genus *Homo* as *Homo erectus erectus* (see Color plate, p. 521).

The early hominid character of Java man was not confirmed until other finds were discovered in the 1920s in China, in Pleistocene deposits. Between 1927 and 1939, scientists from numerous nations working in a model cooperative effort near Choukoutien, China (40 km west of Peking), found the remains of some forty early hominids, which like Java man had a relatively flat forehead and pronounced supraorbital processes. These "Peking men", who lived during the middle Pleistocene, were initially dubbed *Sinanthropus pekinensis* and today are known under the scientific name of *Homo erectus pekinensis* (see Color plate, p. 524). They are thus members of the same species as Java man and simply belong to another race or subspecies. Some of the most prominent individuals responsible for digging and researching Peking man include Canadian anatomist Davidson Black, Chinese paleontologist Pei Weng-chung, and the German-American anthropologist Franz Weidenreich.

▷
Above:
Skeleton of the Upper Pleistocene giant armadillo *Glyptodon asper* from Argentina (Institut de Paléontologie, Paris).
Below:
Skeleton of the Upper Pleistocene plains mammoth *Mammonteus primigenius* (L = 5.40 m; body height = 2.80 m) (Paleontological Museum of the University of Münster, Germany).

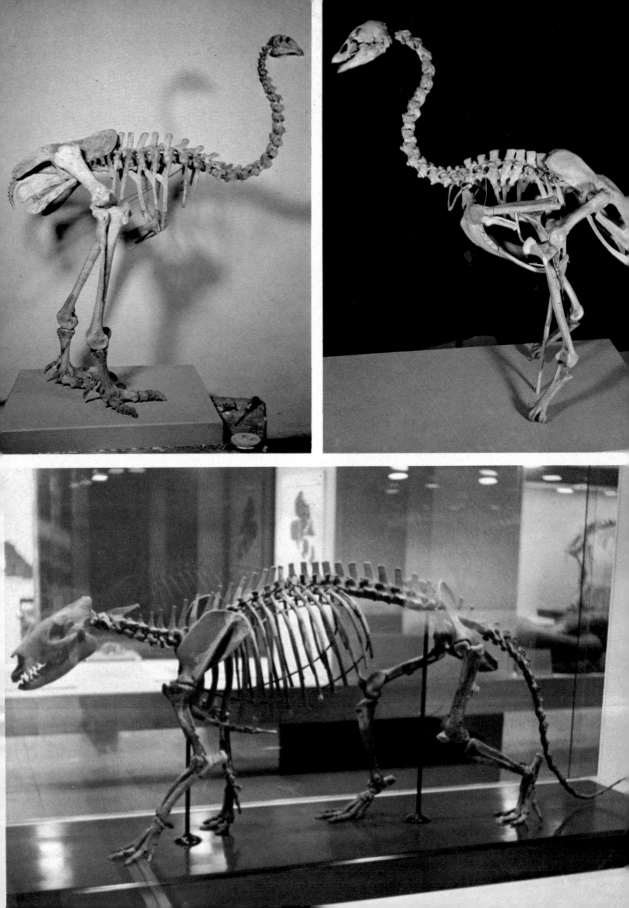

The Choukoutien find was particularly significant because it is here that we find indisputable signs of culture, in the form of knives hewn from stone. This is essentially the same kind of implement used by australopithecines.

Early hominids of the *erectus* type were at this time a diverse group occurring on three continents. On Java, G. H. R. von Koenigswald found not only additional Java men but also the fossil remains of earlier, more robust hominids known today as *Homo erectus modjokertensis* (= *robustus*; see Fig. 22-7). These heavy-set hominids also appeared in China, possibly as early as the Lower Pleistocene. After the Peking man specimens were lost during the course of World War II, Chinese scientists dug further in the Shansi district and discovered portions of the skull and jaws (including the teeth) of an early hominid equipped with a particularly prominent supraorbital process. This hominid is now called *Homo erectus lantianensis*. In 1959, renowned anthropologist Louis Leakey, digging in stratum II in the Olduvai gorge, also came across an unusually robust forehead skull fragment with a very prominent supraorbital process. The implements used by this hominid, named *Homo erectus leakeyi* (see Color plate, p. 521), look like hand wedges (as shown in the color plate), and absolute time dating shows that this hominid lived more than 500,000 years ago.

Germany has also yielded an early hominid, this in the form of a lower jaw discovered in 1907 and belonging to Heidelberg man (*Homo erectus heidelbergensis*; see Color plates, pp. 524 and 526). We do not know exactly when Heidelberg man lived, but recent investigations indicate that this hominid occurred quite early in the Pleistocene and is the oldest hominid known from Europe. Heidelberg man is characterized by the massiveness of its jaws, within which however rests typically human dentition. Hominid-made implements may well have occurred in Europe in still earlier times, for some have recently been described that appear to stem from the Upper Pliocene (over $2\frac{1}{2}$ million years ago!).

Saldanha man (*Homo erectus capensis*), discovered in the Transvaal, may be a relative of Heidelberg man. North Africa has yielded a number of lower jaws and other fossil bone fragments (particularly sites in Ternifine, Algeria, and Casablanca), and these have been designated as parts of *Homo erectus mauretanicus*.

As demonstrated by numerous skull fragments from Java and Africa, the *erectus* type (formerly called the *Pithecanthropus* type) survived in some places as late as the period when *Homo sapiens*, our species, first appeared. These late *Homo erectus* forms were initially called "tropical Neanderthal men", but many of their anatomical features are characteristic of *Homo erectus* and bear a closer correspondence to the *erectus* type than to the Neanderthal European hominids. On the other hand, some features, as for instance the dentition, are of more modern structure. In Ngandong, Java, not far from the site where Dubois found the first "*Pithecanthropus*"

◁
Above left:
Skeleton of the Quaternary giant moa (*Pachyornis elephantopus*), a gigantic running bird from New Zealand.
Above right:
Skeleton of a giant duck (*Cnemiornis calcitrans*) from Pleistocene New Zealand.
Below: A Lower Tertiary precursor of the ungulates: the skeleton of the primitive ungulate or condylarthran *Phenacodus primaevus* (order Condylarthra) from Lower Eocene U.S.A. (Senckenberg Museum, Frankfurt, Germany).

specimens, a total of eight skulls were found, all of which had been smashed and most likely were the leftovers from a head hunt. Most researchers agree that these hominids were Upper Pleistocene descendants of early Java man, and the name given them is *Homo erectus soloensis*. Saldanha man, which we mentioned earlier, probably gave rise to the Rhodesia man of Broken Hill (*Homo erectus rhodesiensis*; see Fig. 22-10 and Color plate, p. 524). The Rhodesia man has a skull with a great many features also occurring in Saldanha man. Another skull, found at Lake Eyassi in eastern Africa, may also belong to this group. All these forms are late descendants of the early hominids. In Europe there are also finds, albeit fragmentary, that nonetheless suggest that early hominids may also have existed in this part of the world. These fossils were located in Vertésszöllös (near Budapest) and have been researched by the Hungarian scientist Lászlo Vértes. Vértes found not only implements and signs of true camp fires but also the occipital region of a human skull and a few human teeth.

It was during the second half of the Ice Age that the first traces of modern man—*Homo sapiens*—appear. Here, too, there are puzzling questions just as we encountered with the Neanderthal hominids. Although the Neanderthals were often described as the original hominids in earlier literature, they were by no means the first ones. Hominids of the *erectus* group lived still earlier, and before them were the first true hominids, the australopithecines. The Neanderthals could more appropriately be called early men, not early hominids. It is important to recognize that the Neanderthals were not a uniform, single type that gave rise to the Pleistocene races of *sapiens* men. The "classic Neanderthal", in contrast, was a terminal form adapted to cold weather, and this last Neanderthal was not a direct predecessor of modern man.

The first Neanderthal find, discovered by Carl Fuhlrott in 1856 in the Feldhofer grotto of the Neander Valley near Düsseldorf, created a major sensation and has made the Neanderthal one of the most familiar "cave men" names to laymen (see Color plate, p. 526). Scientific arguments about the significance of Neanderthal men were voiced continually to practically the turn of the century. Actually, Neanderthal and other fossil hominids had been found before this time, but pre-Darwinian scientists thought that there was no such thing as fossil men, and of course any Neanderthal specimens found were interpreted in a much different light. Also, finds of Neanderthals were not given much attention. One Neanderthal skull was found in a cave on Gibraltar in 1848, and later (also on Gibraltar) a child Neanderthal skull was dug up. In the 1830s, a Belgian physician found another infant Neanderthal skull near Lüttich, and 50 years passed before this skull was properly identified as a member of the Neanderthal race. The geologists Fraipont, Lohest and De Puydt, excavating between 1885 and 1886 near Spy (in the vicinity of Namur), recovered the skeletons of two adult Neanderthals.

The Neanderthal find was an exciting, highly important event his-

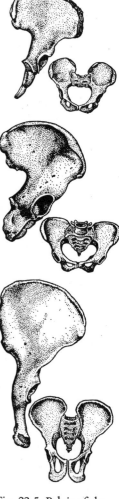

Fig. 22-5. Pelvis of the bipedal *Australopithecus* (top) and modern man (middle). A gorilla pelvis (bottom) is shown for comparison.

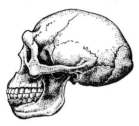

Fig. 22-6. Reconstruction of Peking man, *Homo erectus pekinensis* (formerly *Sinanthropus*).

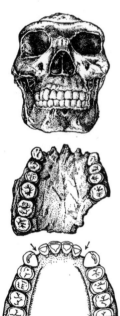

Fig. 22-7. Skull, jaw fragment and reconstructed jaw of *Homo erectus modjokertensis* (Java). The arrows point to the so-called "ape gap" in the dentition. Generally called Java man.

torically because it was the first confirmed prehistoric *sapiens* discovery. More succinctly, this was the first confirmed prehistoric man. As with many significant scientific discoveries, a lot of sheer good luck was involved with finding the Neanderthal man. It was extremely fortunate that the stone workers who located the Neanderthal remains reported their find to Carl Fuhlrott. Bear in mind that during this time there was by no means a consensus among the scientific community that prehistoric men even existed! Many parts (but not all) of the skeleton of Fuhlrott's Neanderthal were recovered, most important of these being the parts of the skull. This skeletal component shows us that Fuhlrott's specimen was a typical Neanderthal and also that Neanderthal was an early man and not simply a hominid. Indeed, the discipline of paleanthropology (the study of fossil men) was born in the year 1856 when Fuhlrott examined this Neanderthal.

More recent excavation activity has yielded a total of some 130 more or less intact Neanderthal skeletons. This gives us a relatively clear picture of this early group of men. The Neanderthals were by no means a highly uniform group, and they are divided into a number of different phylogenetic classes. The so-called "classic Neanderthals" (*Homo sapiens neanderthalensis*; see Color plate, p. 524 and Fig. 22-13) are represented not only by the aforementioned skull from the Neanderthal Valley and the Belgian and Gibraltar specimens but also by a number of western European examples, such as those found in 1905 near La Chapelle aux Saintes, in 1908 in Le Moustier, between 1908 and 1921 in La Quina (France) and 1920–21 in La Ferassie. A beautifully preserved Neanderthal skull was found in a cave in Monte Circeo, Italy (see Fig. 22-12), where it had been closed off from the outside world since the Ice Age. Two other Neanderthals were found in Saccopastore, near Rome. Still others stem from Petralona, Greece and Dschebel Irhoud, Morocco.

Further to the east, in Asia Minor, we find a problematic Neanderthal group, a group interpreted in various ways by different authorities and one with a number of characteristics seen in modern men. We shall discuss this group closer in our subsequent discussion of the pre-Neanderthal men. We refer to the Mount Carmel population of Tabun and Skhul, which also include the finds from the caves of Kafzeh and Amud and from Shanidar in Iraq.

The "classic" development of Neanderthal man is particularly well portrayed by the skeleton of La Chapelle aux Saintes (see Color plate, p. 526). The cranium is particularly large, suggestive of course of a large brain capacity, and in fact of dimensions exceeding those in modern man! One Belgian Neanderthal skull has a cranial capacity of 1753 cc, compared to the dimensions in modern man of (an average of) 1400 cc in males and 1350 cc in females. Other skull measurements, excepting the height of the skull, are also quite large. All adult males had a skull length exceeding 15 cm, in some cases even attaining 20 cm. Skull width varies but averages some 15 cm. The length–width index is approximately 70–75 cm. This means that Neanderthal man had a relatively long skull. Other skull measurements are above the average values of modern human races, the only exception being the short height of the skull.

These large, thick skulls are characterized by pronounced prognathy (jaw projection). The supraorbital ridge extends without interruption across the forehead. Its medial and lateral sections form a unified structure, while in modern man (with few exceptions) there is but a weakly developed orbital arch directly above the inner corner of the eye. In its central portion, the cranial curve of Neanderthal is long and flat, with a slight ridge. In other words, this early man had a short, receding forehead. The occipital (hind) portion of the skull extended posteriorly and was tapered; a relatively steep angle led to the nape. Seen from directly above, the Neanderthal skull displays its great length and the indentation immediately behind the eyes.

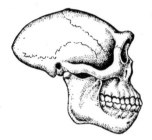

Fig. 22-8. Reconstruction of the Javan hominid *Homo erectus erectus.*

Thus, we have here a highly characteristic skull type, one that can be readily distinguished from others. The orbital cavities are relatively broad and high, and the nasal opening also appears broad. The zygomatic fossa of modern man (i.e., the hollow area of our cheeks) was filled with bone in Neanderthal man. Seen head on, the sides of the face have no indentation from the point of attachment of the jaw arch to the upper jaw segment (a phenomenon termed "extension"). This gave the Neanderthal a tapering face. The impression of a "pointy face" is amplified by the outward extension of the lower jaws (both upward and posteriorly). This and the short, receding forehead give the Neanderthal a sort of "pointy" face that we often associate with "cave" men. They almost appear to have snouts.

The 300–400 Neanderthal teeth we currently have indicate that the classic Neanderthal had massive dentition, which however correspond closely to that in modern man. In contrast to the cranial portion of the skull, the facial skeleton was larger than in modern man.

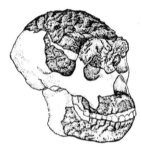

Fig. 22-9. Skull fragments (and supplementary lines) of *Homo erectus lantianensis* (Shansi province, China), bearing a strong resemblance to early Java man (*Homo erectus modjokertensis*).

We also have a reasonably detailed picture of the trunk and limbs of Neanderthal-man. The arms and legs were heavy, powerful structures. The thigh bone was bent outward. The forearm possessed a wide gap between the ulna and the radius bones. Earlier it was believed that Neanderthal man walked with the body arched forward, a posture we often see in caricatures and other illustrations: the head is bent down, the shoulders pulled forward and lowered slightly, with the arms dangling downward as in some anthropoid apes, and the gait is depicted as a clumsy one with heavy footing. Recent studies of the bones (such as the joints of the tibia, the shinbone) have shown that *this was not the case in Neanderthal*! Our understanding now is that Neanderthal man walked just as we do, upright and bipedally. This incorrect impression of the gait of Neanderthal man had even penetrated our museums, and when new research broadened our knowledge of this aspect of Neanderthal man, many

Fig. 22-10. Skull of the Rhodesia man of Broken Hill (Zambia): *Homo erectus rhodesiensis.*

Fig. 22-11. 1–4: Finds of early hominids:
1. Trinil, Sangiran and Ngandong (Java).
2. and 3. Choukoutien and Lantian (China);
4. Olduvai and Eyassi Lake (eastern Africa), Saldanha and Broken Hill (southern Africa).
5. European early hominids (Heidelberg man) and presapiens men (Steinheim and Swanscombe man).
6. North African early hominid finds, from Ternifine and Casablanca.

museums had to immediately disassemble their "cave man" displays and otherwise alter their materials. The beautifully executed Neanderthal displays in Chicago's Museum of Natural History, which had been lauded for decades for the particular excellence of their execution and accurate portrayal, were among those displays to be dismantled because they depicted this mistaken interpretation of Neanderthal posture!

With a height of only 155–165 cm, Neanderthal man was indeed a tiny example of *sapiens* humans; modern human males usually attain a height of more like 180 cm. The famous infant skulls—the most important being those from Teschik Tasch, Uzbekistan; Pesch de l'Azé, France; and La Quina, France—display the characteristic Neanderthal features at the beginning of their development.

Typical Neanderthal men have not been found in eastern Africa or Asia. The origin and the fate of the "classic" Neanderthal group is not well understood. It appears that they arose in northern Europe. One major question is still whether the classic Neanderthal man is to be classified as a distinct species, *Homo neanderthalensis*, or as a subspecies of our own diverse species, *Homo sapiens*. The classic Neanderthal is not apparently a direct descendant of the *erectus* hominid. However, one or more of the known transition forms could link Neanderthal and *erectus*.

Such transition forms, which are older than Neanderthal but which have more modern, more *sapiens*-like features than Neanderthal, are collectively termed pre-Neanderthals, *Praesapiens* or *Primisapiens* man. They lived into the Middle Pleistocene. From Germany stems a well preserved and presumably female skull (minus lower jaw) from one of these transition men. This is Steinheim man (see Fig. 22-14 and Color plate, p. 524), *Homo sapiens steinheimensis*, which has the distinct supraorbital ridge also seen in Neanderthal but in other respects has a more modern appearance. For example, Steinheim man has zygomatic fossae (Neanderthal did not), and the skull is much higher. The occipital portion of the skull bears a greater resemblance to modern man than to Neanderthal; it lacks the conical protrusion of Neanderthal. The snoutlike appearance we mentioned earlier as characteristic of Neanderthal man is likewise absent in Steinheim man. However, there are features more primitive in Steinheim man than in Neanderthal man! The cranial volume was between 1100 and 1200 cc. In cross-section, the occipital portion of the skull is roughly pentagonal. The presence of supraorbital processes need not necessarily indicate a phylogenetic relationship with Neanderthal man, for this could have been an example of parallel evolution.

Steinheim man also was found in a cave near Tautavel, France. Another member of this same *Praesapiens* group is Swanscombe man. found in a sand pit (the Barnfield Pit) some 30 km southeast of London, This pit had long been famous for the bones of Pleistocene mammals and prehistoric implements. But in 1935 a human occipital skull portion was recovered there, and one year later (at the same depth) a parietal bone

Fig. 22-12. The Neanderthal skull found in a cave in Monte Circeo (north of Rome).

was found just seven meters from the occipital bone site. Later the parietal bone from the other side of the skull was also located, and these three pieces all fit together (see Fig. 22-14). Anthropologist E. Breitinger has attempted to reconstruct the entire skull of Swanscombe man, and he believes that Swanscombe man had a supraorbital ridge as in Steinheim man, even though the three skull bones we have of Swanscombe man are from the top and rear of the skull and thus cannot provide physical evidence of a ridge above the eyes (see again Fig. 22-14). The best representation we can make of Swanscombe man is the rear view shown in the Color plate, p. 524 (Item #6), which includes Dr. Breitinger's supraorbital ridge.

Whether Swanscombe man had a ridge above the eyes or not, this reconstruction shows us that there were indeed *Praesapiens* men, which may or may not have been Neanderthals, that correspond more closely to the anatomy of modern man than to the classic Neanderthal. This transition group may also include the skull fragments found by L. S. B. Leakey between 1931 and 1932 in Kanjera, eastern Africa and which are from the Middle Pleistocene. Morphologically, Leakey's specimens seem to belong to the *sapiens* group. There is no supraorbital ridge. And these finds are geologically older than the "classic" Neanderthals.

Along with these *Praesapiens* men there are a number of finds predating the Neanderthals that are collectively known as pre-Neanderthals, because these men do possess Neanderthal characteristics but not in their "classic" development. The fossil record of pre-Neanderthals includes two lower jaws, fragments of the occipital region of a skull, and a number of other bone fragments from Ehringsdorf (near Weimar), all of which stem from the middle of the last inter-glacial period. The reasonably well preserved frontal bone region displays Neanderthal features, albeit in a less pronounced extent than in the classic Neanderthal. Somewhat older are the skeletal remains from Krapina (in Croatia), which were earlier interpreted as the "leftovers" from a cannibalistic meal. They consist of nearly 20 bone fragments, all of which were smashed up; traces of fire have also been found at Krapina. The lower jaw pieces from Ochos and Schipka (Czechoslovakia) indicate that there was a human being from the last inter-glacial period living in eastern and central Europe who was somewhat more "advanced" than the (geologically younger) "classic" Neanderthal. This human type is also represented by the aforementioned pair of skulls from Saccopastore.

Another find dating from 1947 in Fontéchevade, France, is yet another indication that precursor humans resembling *Homo sapiens* of today lived long before the classic Neanderthal appeared. The French find consists of the very fragmentary remains of a skull roof and a frontal bone fragment, which are from the last temperate period. Seen from above and from the side, this skull roof resembles that of modern man and has thick walls. Investigations by French anthropologist H. V. Vallois show that Fonté-

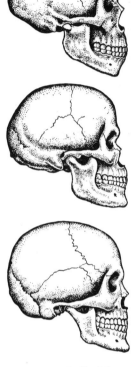

Fig. 22-13. Skull of the "classic" Neanderthal (top), the late Pleistocene Cro-Magnon man (middle), and modern man (bottom).

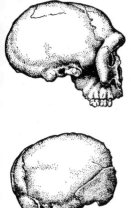

Fig. 22-14. Presapiens men: above is Steinheim man; below is the hind skull fragment of Swanscombe man from England.

Skulls of pre-neanderthal types:

Fig. 22-15. Fragments from Ehringsdorf, Austria.

Fig. 22-16. The "women of Tabun".

chevade man did not have the prominent supraorbital ridge as we find, for example, in Steinheim man. We have here a picture, albeit not a carefully drawn one, of how a precursor of modern man looked during the last temperate period in Europe.

Early men of the pre-Neanderthal type have also been found in the Middle East. Parts of some 20 humans were found in Palestine (in Tagbah, on Lake Genezareth and in the caves of Tabun and Skhul in the Carmel Mountains) in 1925 and 1931–1935. The parts include skeletal remains, lower jaws, and teeth. The Shanidar cave in Iraq yielded four skeletons, which are somewhat older than the Palestinian finds (which are thought to be from the beginning of the last ice age). These western Asiatic men (see Figs. 22-16, 22-17 and 22-18), of which there are still more recent finds from the cave of Amud (near Lake Genezareth) and also in Kafzeh, have skulls with supraorbital processes that are not very well developed and which therefore should not be blindly classified as being of the Neanderthal type. The supraorbital ridges have a medial and a lateral segment, something that is suggested in Neanderthals but which in these humans looks entirely different. One gains the impression that these ridges are actually substantially well-developed supraorbital arches. The skull has a powerful, steep forehead, is considerably higher than in Neanderthal man, and has a rounded occipital portion (as compared to the typical Neanderthal tapered occipital bone shown in Fig. 22-19).

In spite of the supraorbital arches, these men look very *sapiens*-like, and they particularly resemble the very recent Cro-Magnon man. Some authorities, such as Wilhelm Gieseler, do not classify these western Asian men as pre-Neanderthals but as members of the *Praesapiens* group (and hence in the class containing modern man). In comparing all the skulls from the Dschebel-Kafzeh series, one indeed gets the impression that these men belong in the diverse class of modern man. Instead of the Neanderthal supraorbital ridge, which extends all the way across the forehead, these Asiatic men have just a single eyebrow ridge. The reconstruction executed in 1953 of the Skhul V skull (from Carmel) gives us a more complete picture of these men (see Fig. 22-17). The evidence indicates that in Asia Minor there existed a human group belonging to the pre-Neanderthal class but with close ties to modern man. All these Asiatic men are now classified as *Homo sapiens shanidarensis*, named after the Iraq find.

Diggings in Jabrud (Palestine) carried out by A. Rust have shown that Neanderthal and *sapiens* cultures were at this location deposited one on top of the other in alternating layers. This makes it highly probable that the two populations intermingled! On the other hand, there is also a strong likelihood that this part of Asia was the home of a distinct human group we might colloquially call "Not-Yet-Modern-Men".

It was after the flourishing period of these early men, whose significance is still not well understood, that we find *Homo sapiens sapiens* (modern man), in deposits from the last glaciation dating from 40,000 years ago.

One of the major representatives of this group is named after a find at Cro-Magnon, France, and is the well-known Cro-Magnon man (see Color plate, p. 524). It was here in 1868 that the remains of at least five humans were found in layers of the Lower Aurignacian. One of these five humans was nicknamed the "Old Man of Cro-Magnon" (see Color plate, p. 526), and it achieved worldwide fame. Since the time of Cro-Magnon man, the anthropological human type has not varied. Cro-Magnon man has, since 1868, been located in a wide number of different sites. The somewhat younger find from Chancelade, France, only deviates in fine details from the Cro-Magnon skulls.

One Cro-Magnon skull, found in 1909 in France in the cave of Combe Capelle, was sent to a museum in Germany and was almost destroyed by the fires of World War II. When the old skull was dug up it displayed many of the same kinds of deformities that occur in modern skulls subjected to fire. Two other finds of the many Cro-Magnon discoveries also include that at Laugerie-Basse, France, and the double set (a 50–60 year old man and a 20–30 year old woman) from Bonn, Germany.

Skulls and bones found in Czechoslovakia differ somewhat from Cro-Magnon man. It is possible that the eastern members of these early *sapiens* humans simply differed somewhat in appearance. Important Czech finds include the remains of three skulls from Brünn (dating from the Gravettian culture) and the skull fragments of a 20-year old woman and an older man from Mladêk (with implements of the Aurignac culture). The diggings conducted in Předmost from 1884 to 1924 were particularly interesting; an elliptical mass grave here contained the remains of 30 men, some in a stooping posture, and the find included the fully intact skeletons of several young men and women. A Gravettian deposit in Dolni Vestonice yielded the remains of ten men belonging to the same or similar anatomical type.

These Czech finds give no indication that there were other than *sapiens* men alive at this particular time in the earth's history, although these men differ in some ways from Cro-Magnon man. Of Cro-Magnon we can say that this was a large form with a long, moderately broad skull and deeply recessed nasal cavity. Cro-Magnon man had a low, wide face and broad jaws; the rectangular orbital cavities were low set. The Combe Capelle find was perhaps somewhat smaller and possessed a relatively longer, higher skull with a long face. The Combe Capelle man looks like a Mediterranean *sapiens* man. Although there are minor detail differences, *sapiens* men of the Late Pleistocene, as far as we can tell from the available skeletons, did not differ substantially from the appearance of modern man. Most authorities now agree that if, hypothetically, a Cro-Magnon man were transported into the present and walked about on our city streets, we would not recognize the fact that this man dates from the Ice Age. Cro-Magnon man differs no more than our own modern races from each other.

In contrast to Neanderthal man and still earlier forms, Cro-Magnon

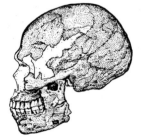

Fig. 22-17. Skull of Skhul V.

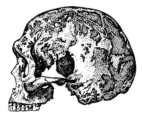

Fig. 22-18. Skull from Kafzeh (near Nazareth).

▷
Primate Evolution I: Prosimians and Old World monkeys.
1. *Plesiadapis tricuspidens* (primitive prosimian; Upper Paleocene France);
2. *Notharctus osborni* (New World lemurs; Eocene North America);
3. *Tetonius homunculus* (an anaptomorphid from Lower Eocene North America);
4. *Megaladapis edwardsi* (giant lemur from Quaternary Madagascar);
5. *Palaeopropithecus* (sloth-like indris; Quaternary Madagascar);
6. *Archaeolemur edwardsi* (monkeylike indris; Quaternary Madagascar);
7. *Mesopithecus pentelici* (leaf monkey; Lower Pliocene Greece);
8. *Dinopithecus* (gorilla-sized giant baboon; Pleistocene southern Africa).

Fig. 22-19. The particularly characteristic Neanderthal from La-Chapelle-aux-Saints.

◁

Primate Evolution II: Parapithecids, gibbons and anthropoid apes.
1. *Parapithecus fraasi* (a parapithecid from Lower Oligocene Egypt);
2. *Propliopithecus haeckeli* (pre-anthropoid ape; Lower Oligocene Egypt);
3. *Pliopithecus vindobonensis* (gibbon species from Middle Miocene Europe);
4. *Oreopithecus bambolii* (an oreopithecid; Upper Miocene Italy);
5. *Gigantopithecus blacki* (giant anthropoid ape; Pliocene northern India and Pleistocene southern China);
6. *Proconsul africanus* (an anthropoid ape from Miocene east Africa).

man carried on what is called a flint culture, and flint cultures are found relatively early in the Middle Old Stone Age. Cro-Magnon man was most likely the creator of those stunning cave drawings known throughout the world, the artistic representations of Pleistocene animals as well as various sculptured works and carvings.

One difficult question to resolve is the reason for the sudden appearance of Cro-Magnon man in the fossil record and hence the rapid displacement of Neanderthal man by Cro-Magnon man. The question also arises of whether the modern human races likewise arose suddenly, or, as anthropologist Carleton Coon postulates, if there was a differentiation of hominid types as early as the *erectus* stage, with the various hominid-human lines developing independently since that time. Coon counts five racial groups: Caucasoid, Mongolian (this includes Polynesians), Australoid (including Melanesians), Congoid (i.e., Negros) and Capoid (Bushmen and Hottentots), lines he claims all existed in the *Homo erectus* stage of evolution. However, Coon's theory is very unlikely from a genetic standpoint. It is more likely, as E. Breitinger and others believe, that modern races of man arose within the species *Homo sapiens* no earlier than the Middle Pleistocene.

The modern races of man cannot actually be differentiated clearly from one another, and this is a suggestion that the birth of these races does not lie in the distant past. Many differences existing in humans are not truly racial differences but adaptations to the prevailing climate and other ecological conditions. Furthermore, none of the previous human races seems to have changed (evolved), although the latest research is now suggesting that evolution has not stopped in our species. Outward appearance and gross anatomy have remained constant over the past 40,000 years, but evolution may be now occurring on the biochemical level. Some evidence indicates that certain blood groups are better adapted to the environment than others, for the incidence of some diseases may be higher in particular blood groups than in others. Specific blood groups may also be associated with increased resistance to specific diseases. This is one of the most active lines of current research, and any specific information given in this text is bound to be outdated soon after publication, so we merely note that more and more evidence is being accumulated that evolution has not stopped in man but that natural selection is still taking place. One last point, a brief one, should be made about language. The development of the human languages is intimately associated with the evolution of human races. It is certain to assume that Late Pleistocene men were communicating verbally with one another. At which *prior* stage of evolution the beginnings of language can be found is a far more difficult problem, for fossils do not always permit us to make any conclusion about language abilities. Efforts are being made at present to examine the muscle attachment points in fossil skulls, and these investigations are shedding some light on the evolution of language in prehistoric man. This is yet another line of research that may prove tremendously exciting in the near future.

23   The Evolution of Domestic Animals

Before Lamarck and Darwin founded the discipline of evolutionary biology, almost all naturalists held that animal species were fixed, mutually exclusive units produced by a Creator. The same interpretation was made for domesticated animals, and each breed of animal was thought to have had its own individual lineage. But a comparison of, for example, various domestic dog breeds quickly shows that the dachshund, boxer and greyhound differ more from each other than wild canids like the wolf, jackal or coyote differ from each other. Thus it appears that changes have occurred within domestic breeds that are more pronounced than natural species differences, and these changes may have been the basis or model for phylogenetic modifications. This hypothesis, then, predicts that the changes we see in domestic animals are the basis or operating principle for the lesser differences existing among wild species. According to yet another hypothesis, the differences existing among domestic breeds came about by hybridization (or crossing) of various wild species.

Today we know beyond question that wild species and domestic animals are not immutable, unchanging entities. Considerable differences among individual animals can occur even in wild species within a single population inhabiting a common range. Animal species populating adjacent ranges can develop such pronounced differences that they are designated as subspecies. The geological record also shows that animal species can change over the course of time. It is extremely difficult to identify the specific events and processes that lead to evolutionary changes of animal species in the wild. Zoologists have developed a number of experimental approaches to get at basic questions like the nature of evolutionary processes, and one of these includes an extremely long-term "experiment", which mankind has long conducted unconsciously: the domestication of wild animals (= the creation of domestic animals).

All domesticated animals have evolved from wild species, and they have developed during the time man has been on earth. During the course of domestication, the dissimilarity between the parent wild species and

By W. Herre and
M. Röhrs

Phylogeny of
domesticated animals

the domesticated animals becomes greater. As early as 1868, Charles Darwin gained basically accurate insights into the biological nature of domestic animals based on his observations of domestication (breeding): "One can say that mankind has conducted an experiment of gigantic proportions, an experiment that even Nature has not interrupted throughout the long course of domestication. Thus, it follows that the biological bases of domestication are significant processes. The chief consequence of the phenomenon of domestication is that organisms that are domesticated can vary considerably and that the variations occurring in them can be inherited." Since Darwin's time, more than 100 years ago, zoology has witnessed impressive advances in our understanding of domestication.

Distinguishing from the wild animal

Humans produced domestic animals from wild ones by isolating small groups of animals, preventing their sexual access to the wild populations, and bringing them out of the state of nature and into the home life of humans. Individual animals that proved superior in some specific traits were bred together over the course of many generations, eventually producing a large number of populations of domesticated animal groups. These animals improved the economic standard and actually resulted in man becoming dependent on them for maintaining his new way of life. The intraspecific relationships among domesticated populations were quite different from those relationships in wild animals: vast herds became the rule, replacing the small packs of wild animals. This was possible because of behavioral changes effected by systematic breeding. In wild animals, hierarchy contests (usually between males) are a part of life and generally take place over a relatively large area used for all the displays occurring as the contestants seek to upgrade their status within the group (or to protect their status). Large herds provide little room for such contests, and furthermore the desire for such fighting decreased in domesticated animals; breeders generally selected for more docile, weaker animals. Thus, behavioral changes also occurred during the course of evolution. Breeders systematically attempted to heighten those characteristics they considered ideal for their purposes, and this could be done by differential breeding of the available stock.

Humans actively modified animals—anatomically and behaviorally—and took over Nature's role as the selector. In doing so, mankind also altered his own social milieu. The relationship between humans and domestic animals differs in many ways from those life communities we define as symbiotic. Symbiotic (from the Greek συμβίωσις = living together) refers to a life relationship between different species that results in a benefit to each species (as opposed to parasitism). However, no animal species has made others as reproductively dependent as man has done with his domesticated animals.

The question of species determination

The transformation of one species into another is the basic process of evolution. A clearly defined concept of species is also particularly important with domestic animals. Darwin proposed that a species is comprised of all those animals with similar characteristics. His criterion was similarity in the anatomical sense, a criterion used by biologists into this century. With this kind of definition of species, it was quite fascinating to observe the manifold variations existing among domestic animals (even within a single species) in comparison to the comparable wild species. In fact, this definition of species became unwieldy precisely when biologists attempted to explain the diversity existing among domesticated animals.

It was long before Darwin's time (in 1829) that Georges Cuvier proposed a definition of species using a biological basis. His biological definition was not given a great deal of attention at that time, but it has met with more and more interest by modern researchers. According to the biological definition, a species is comprised of all the individual animals that form a reproducing community (= gene pool) creating fertile offspring. A species, therefore, is not an artificial entity proposed by biologists; it is a biological reality. Species do not interbreed; their are biological (often behavioral) barriers preventing this. The few exceptions to this rule are quite rare. It is from this biological definition that species characteristics are to be evaluated. Animals differing considerably from each other, even in the wild, are often conspecifics. Our understanding of just what comprises a species is still being expanded, and some of the problems of our biological definition of species are traced with great lucidity by Ernst Mayr in his monumental *Animal Species and Evolution*. The important point here with regard to domesticated animals is that if they are given a choice of potential mates, they will reproduce with just one wild species, namely the parent species from which they descend. In other words domestic animals form a voluntarily reproducing community or gene pool with but a single wild species.

As we mentioned earlier, striking individual differences can occur within the same reproducing community, even among wild species. These intraspecific differences are often related to geographical barriers within the range of that community. In 1861, Henry Walter Bates proposed the term "subspecies" to designate these geographical differences occurring within a species, and this term has been officially recognized since 1898. Since that time, species and subspecies are being named according to a unified system of rules adopted by biologists throughout the world. We can gain some valuable insights by applying the species concept to domesticated animals.

Species and subspecies

Thorough understanding of the parental species and its relatives is vital for an evaluation of the biology of domestic animals. The fact that many domesticated animals cannot live in the wild is often interpreted by laymen as an indication that these animals are degenerates or that the characteristics of domesticated animals come about by degenerative breeding. Those who make this claim are missing a basic evolutionary phenomenon that Charles Darwin recognized a century ago: "Since

Domestication as a process of adaptation

man's will is involved with the production of domestic animals, it is clear that domestic breeds are adapted to man's needs and desires. This explains why domesticated races of animals and plants often are abnormal when compared in some ways with natural (wild) species; it is because they have developed and have been modified for man's purposes and not their own." Domestic animals are much more adapted to living with or around human beings than they are to living under the same natural conditions as their wild relatives. The great experiment of domestication gives us insight into the developmental possibilities of an animal species under a particular set of living conditions; and subspeciation shows what diversity can occur and be sustained under natural conditions. Domestication and subspeciation are two different evolutionary phenomena, and both can be examined to help us gain increased understanding of evolutionary processes.

Domestic guinea pigs and rabbits

The number of domestic animals that are important for human cultures is surprisingly small. The guinea pig is the most significant domesticated rodent; pre-Columbian Andes Indians utilized guinea pigs for food and for sacrifices. The guinea pig belongs to the same species as the wild guinea pig, *Cavia aperea*, and according to Bohlken is designated as *Cavia aperea porcellus*. The only domesticated lagomorph was the wild rabbit (*Oryctolagus cuniculus*), and the scientific name of the domesticated rabbit is *Oryctolagus cuniculus domestica*. Of the carnivores, the most important domesticated species are the ferret, dog and cat. Only recently (1970) did

Ferret

U. Rempe demonstrate that the ferret is descended from just one wild species, *Mustela putorius*, and the ferret is now designated as *Mustela putorius furo*. The phylogeny of the domestic dog is a complex subject

The domestic dog

about which a great deal of discussion is still taking place. Three wild species have been implicated in the development of the dog: the wolf (*Canis lupus*), the golden jackal (*Canis aureus*) and the coyote (*Canis latrans*). Most authorities feel that an analysis of the structure of the skull, teeth, brain, behavior, the heart, blood protein, and other aspects of biology indicate that all domestic dogs have descended from the wolf, and dogs are presently identified as *Canis lupus familiaris* (or, as Dr. Michael Fox has proposed, *Canis lupus overfamiliaris*—only partly in jest !). The dog is one of the most spectacular examples of how a wild species can be modified under the selective pressures of domestication. Less mysterious is the evolution of the domestic cat, for there is virtually universal agreement

The domestic cat

that all our tabbies have descended from the wild cat, *Felis silvestris*, and the house cat is known as *F. silvestris catus*.

Domesticated asses and horses

From the ungulates descend the domestic horse and ass. Sumerian-Babylonian drawings suggest that the Asiatic wild ass may have been domesticated into the Persian onager (*Equus hemionus onager*), but there is no scientific proof of this and it is possible that these ancient peoples depicted occasional tame animals instead of truly domesticated ones. The ancestral species of the donkey is the African wild ass (*Equus africanus*), with possibly several subspecies involved; the donkey is designated as *Equus africanus asinus*. The descent of the domestic horse was a subject of great controversy until the 1970s. However, a series of recent studies, especially that of Günter Nobis (whose findings were published in 1970), has resolved all ambiguities. Nobis believes that a number of Pleistocene horses belong to one and the same species with Przewalski's horse and the domesticated horse, all bearing the scientific name *Equus ferus*. If only extant equids are considered, the nomenclature would be *Equus przewalskii*, with the domesticated horse breeds being *E. przewalskii caballus*.

Horse and donkey evolved from odd-toed ungulates; the even-toed ungulates have given rise to a larger number of domesticated forms, including such familiar animals as swine, camels and the cud-chewers (ruminants). All domestic pigs—from both Europe and Asia—stem from subspecies of the European-Asiatic wild pig (*Sus scrofa*) and are termed *Sus scrofa domestica*. The camel group gave rise to a number of domestic forms. The South American genus *Lama*, for example, gave rise to one domestic animal, the guanaco, which in turn was developed into the llama (used as a work animal) and the woolly, smaller alpaca. The scientific terminology for these animals is *Lama guanicoë glama*, and the alpaca is thought to be a race of the llama. The phylogenetic relationships in camels are less clear; we refer specifically to the double-humped, primarily Asiatic camel and the single-humped dromedary, which is used primarily in Africa. There are probably no wild camels alive today, and the evidence at present strongly suggests that camels and dromedaries belong to the same species, *Camelus ferus bactriana*. The reindeer, which is the only domesticated deer species, is a highly unusual animal, for it occurs only in the far north regions of Eurasia (an unusually inhospitable climate for a domesticated animal). All wild reindeer belong to one species, and the domesticated form is thus termed *Rangifer tarandus domestica*.

Cattle, which arose from several wild species, have undoubtedly been one of the most influential domestic animals in terms of altering man's way of life. The "taurine" cattle are the most widely distributed, and all these descended from the European aurochs; their scientific name is *Bos primigenius taurus*. One Himalayan species gave rise to another domesticated cattle form, the cold-loving yak (*Bos mutus*). The domesticated yak bears the name *Bos mutus grunniens*. In Indonesia, the banteng (*Bos iavanicus*) was domesticated into the Bali cattle. A closely related, southeastern Asian species, the gaur (*Bos gaurus*), led to the domesticated cattle form called the gayal (*Bos gaurus frontalis*). Yet another wild ruminant, the Asiatic water buffalo (*Bubalus arnee*), has played a role in domestication. Its descendants, the domestic buffalo (*Bubalus arnee bubalis*), are more efficient than "taurine" cattle in tropical and subtropical regions. They are useful in rice propagation, where the climatic conditions would not permit the cattle we commonly see to perform efficiently.

The domestic pig

Camels

Reindeer

Cattle

Sheep and goats

Two of the "classical" domesticated mammals are the sheep and the goat. All European wild sheep related to domestication belong to one species, *Ovis ammon*, and thus all domesticated sheep are classified as *Ovis ammon aries*. Domestic goats are more difficult to understand phylogenetically. Wild goats, which form a naturally reproducing community, exist in such diverse strains that zoologists hesitate placing them all within one species. This is one of those instances where the notion of "superspecies" is used. This concept refers to differences in populations of separate geographical regions even though no sexual species boundary exists between these populations. With goats, then, we are dealing with some highly complex evolutionary relationships due to the existence of superspecies; here we cannot go into all the details of the origin of domestic goats. A comparison of all the evidence presently suggests that the Bezoa goat (*Capra aegagrus*) is the ancestral form of all domesticated goats, which are designated as *C. aegagrus hircus*.

Poultry

The species of "classical" domestic poultry are also characterized by a wealth of diversity, but all of them have descended from one wild species. For example, all domestic pigeons have descended from the rock dove, *Columba livia*, and all the European-north African geese arose from the greylag goose (*Anser anser*). East Asian domestic geese were produced from a different species, *Anser cygnoides* (the swan goose). Domesticated ducks all arose from the northern mallard (*Anas platyrhynchos*), while the so-called Muscovy ducks are descendants of the wild Muscovy duck, *Cairina moschata*. Turkeys have been domesticated from the wild turkey, *Meleagris gallopavo*, and our chickens all arose from the red jungle fowl (*Gallus gallus*). Domestic guineafowl have been developed from the helmeted guineafowl (*Numida meleagris*).

Motives for domestication

These "classical" domestic animals were not the only ones that came into man's household. Others have been taken into the home and the laboratory or the farm, and domestication is still occurring in a lot of these animals. They include fur animals, laboratory animals, parakeets, canaries, and many other animals that are being kept as pets. Attempts are being made to produce new breeds of domestic antelope and deer for use in marginally productive areas, and these breeding attempts amount to the beginnings of domestication. Keeping wild animals within specific preservations (which is often done for hunting purposes) often results in the development of new strains, particularly in terms of coloration patterns.

Domestic animals play a role in man's economic life and in satisfying his esthetic desires. Most wild species lack those qualities that are valuable in domesticated animals (such as, for example, docility). Early domesticating man could not have had any suspicion of this; the earliest domestications were not the result of systematic breeding programs. Domestic animals only attained their enormous importance over a very long course of development. Some of the major features of various domestic animals include wool and fur production, early maturation, high meat yield, efficient use of food, and high and rapid reproduction rate, production of marketable quantities of milk, etc., and less practical (more esthetic) qualities such as various coloration and anatomical patterns.

The earliest human beings subsisted on a hunting and gathering culture. At a later stage of development, agriculture and animal breeding became important, beginning in areas where the natural food resources were low. Peoples in game- and food-rich regions in Africa, the Americas, and Australia did not domesticate animals. Domestication brings many problems with it, and it has only occurred where man was forced to domesticate. The "force" was usually increasing numbers of people to maintain.

Economic changes

Domestication of animals took place over a period of centuries we now identify as the neolithic (= "New Stone Age"). revolution. During this time, the skills of pottery-making, agriculture and animal husbandry were developed, and communities arose. These three major skills did not all develop simultaneously nor in the same way throughout the world. For example, the same wild animals were not domesticated in all parts of the world. It does appear that domestication of animals began in various parts of the world, utilizing those appropriate wild animals living in those regions. The specific regions where animal domestication began include Asia Minor, northern Africa, central and southern Europe, eastern Asia, the highlands of Mexico, and the Andes of South America. The first domesticated animals were put to many different kinds of uses, and this suggests that the reasons behind domesticating animals were different in different parts of the world.

When did domestication actually begin? To answer this question we have to turn to paleontologists. 19th Century diggings indicate that sheep and goats are the oldest domesticated animals. Each species, as wild animals, is relatively easy to handle. Also, these two animals can readily utilize plant materials and convert these metabolically into meat and fat. Sheep and goats had been domesticated in Asia Minor by 8000 B.C. From there, they quickly spread into new areas where wild sheep and goats did not exist. The domestic pig also has a long history. English researchers have found true domesticated swine dating from about 8000 B.C. in the Crimea. Wild pigs were also domesticated in other parts of the world, of course, as for instance in eastern Asia. This has been verified by studies on the anatomy of skull bones dating from that period. Unfortunately, there is no precise information as to when domestication efforts began and how long it took for early man to domesticate these species.

Early forms of domestication

The earliest domestic cattle date from Greece (ca. 6500 B.C.) and Anatolia (ca. 5800 B.C.), and these cattle were developed from the European aurochs. The horse is an even later development than cattle. The earliest domesticated horses appear around 3000 B.C. in southern Russia.

The history of the domestic dog (i.e., the domestication of the wolf) is still poorly understood. For a long time, the dog was thought to be the oldest domesticated animal, for dog bones have been found in Middle Stone Age deposits in northern Europe. But recent advances in dating techniques have shown that many of these sites actually date from the Upper Stone Age. Also, our knowledge has increased about the oldest domestic dogs. The single oldest confirmed find, made by Star-Carr in England, dates from 7500 B.C. Recently, a North American find of a dog has been estimated to date from 8500 B.C., but there is doubt on both the dating accuracy and even whether the find involved is actually a true domesticated dog.

Early history of the domesticated dog

We are still much in the dark about the domestication of this familiar animal. One important point that is indicated by all our current understanding is that the dog did *not* play a significant role in the economic life of early man. Dogs were used as food in Middle Stone Age northern Europe (and still are by some peoples in Asia and Africa).

A few words are in order here about so-called early "race" or "breed" development in dogs. One can only speak of of a "breed" or "race" in the true biological sense when man has kept a group of very similar individual animals in sexual isolation from other groups of animals belonging to the same species. All domestic breeds (whether dogs, cats, cattle, or whatever) are the result of breeding programs. The resultant diversity of individual groups is simply the product of how strictly these groups have been sexually isolated from other animals. Controlled breeding leads eventually to certain highly bred, distinct lines, and the members of these lines can be utilized to develop a distinct breed. Different breeds of dogs did not develop during the neolithic revolution because the dogs kept interbreeding. This important point was missed by many early students of dog breeds, who claimed that there were a number of prehistoric "species" or "races" of dogs from which our modern breeds have been developed. We read about early dog breeds such as *"Canis palustris"*, *"Canis intermedius"*, *"Canis matris-optimae"*, *"Canis leineri"*, and others, reported in the early literature on dog breed development. These prehistoric "races" were allegedly the basis for all our modern breeds, thus suggesting that present-day breeds have a long history and an important cultural role as early as thousands of years ago.

Modern zoology has provided evidence that this interpretation is inaccurate. While early dogs do indeed display diversity, we do not find particular types of dogs concentrated in any one area. Thus there is no indication that particular lines were systematically developed; various kinds of dogs existed, but they interbred to a large extent and were not subject to sexual isolation. The breeds we know today are the result of relatively recent systematic breeding programs originating with dogs displaying highly diverse anatomy and behavior. Domestic animals have undergone changes in both anatomy and behavior, something that is important to consider when we compare domestic breeds with naturally occurring subspecies of wild animals.

Under the conditions of domestication, animals undergo changes in their shape, their vigor, and their behavior. No organ or organ system remains uninfluenced during the course of domestication. However, some sets of changes go hand in hand. For example, one result of domestication is that the overall size of the domestic animal can vary drastically in contrast to the size of the wild ancestral form, and we can produce animals that are either larger or smaller than the ancestor. However, we are limited in what changes we can make because of inherent biological limitations in animals. As body size increases, the bones must adjust to support the body. No elephant can have the slender legs of a gazelle. Thus we speak of sets of changes coupled to each other that limit the degree to which we can vary the anatomy of a domesticated animal. The proportional differences occurring in different sized animals are known as allo- metric differences (from the Greek αλλος = other), and in general we find that proportions change fairly regularly with changes in body size. Inter- specific proportional differences (differences among a number of species) are much less pronounced than differences within a single species (i.e., in- traspecific differences), and this factor has permitted the development of such diversity within domestic animals. The many diverse dog breeds impress us as dissimilar, but the differences occurring among them are simply the result of allometric factors associated with their body size.

Intraspecific and interspecific allometry

Domestic animals can come in very narrow, elongate forms and in short, compact shapes. These differences are far less pronounced among wild animals. Extreme variation can occur in the skull of domestic dogs. In compact animals, the skull has a short, upward turned snout, while slender domestic dogs have a long, narrow skull. The teeth are relatively unchanged, so that dogs with long skulls tend to have spaces between their teeth.

Various growth forms

This leads us to another kind of phylogenetic change we call "mosaic inheritance". This refers to changes in individual characteristics, not sets of characteristics we mentioned earlier, and it is mosaic development that most domestic animals have undergone. Limb length is relatively de- pendent on body size (allometry), but the length and texture of the fur is a relatively independent entity and can be varied regardless of body proportions (mosaic development). Mosaic development can also occur in the metabolic organs and among the internal secretory glands and in reproduction physiology. Many such modifications associated with sys- tematic breeding can lead to disturbances in the animals subjected to intensive inbreeding, and this is a problem we are now experiencing with a number of highly inbred dogs, for example. The frantic rush to produce registered dog breeds, which for many people serve as ego-inflating status symbols, is producing a flood of genetic degenerates that make much poorer pets than the mixed breeds.

Mosaic inheritance

Changes in brain size

One of the greatest modifications occurring in domesticated animals is in brain structure. Darwin found in 1868 that the size of the cranium in domestic rabbits and pigeons is smaller than in their wild counterparts. Other studies have shown that practically no domestic animals have crania as large as their wild ancestral forms. Furthermore, brain size decreases during the course of domestication; in other words, more highly domesticated lines have yet smaller brains. However, since brain size is dependent on body weight, comparisons between wild and domesticated animals must also take into account the factor of body weight. That is, intraspecific relationships between brain and body weight must also be considered if any meaningful comparison between wild and domesticated animals is to be made. The intraspecific dependence of the brain weight on body weight holds for both wild and domestic species, and on the basis of such intraspecific relationships, the following brain weight decreases have been calculated: laboratory mouse 0.0%; laboratory rat 8.7%; domestic rabbit 8.9%; domestic cat 23.4%; ferret 32.4%; domestic dog 31.1%; domestic pig 34.0%; domestic sheep $\simeq$30.0%. These figures show that brain weight decrease differs among various domesticated species. The lab mouse, for example, has no smaller a brain than the wild mouse, while the domestic dog has a brain 31.1% smaller in weight than the wolf. Less well developed brains, like those in rodents or lagomorphs, show no brain weight decrease or a decrease of less than 10%. In contrast, the relatively highly developed brains of carnivores and even-toed ungulates decrease in weight by 19–34%. Thus, there is apparently a relationship between the evolutionary stage of development of the brain and the extent of its weight decrease in domestication.

Degeneration of forebrain and neocortex

A quantitative comparative study of all the brain segments (forebrain, midbrain, mesencephalon, cerebellum, and the upper portion of the spinal column) in mammals has shown that in all instances it is the most highly developed portion of the mammalian brain (the forebrain) that undergoes the greatest decrease in size during the course of domestication. The smallest decreases occur in the midbrain and the spinal column. An even greater decrease than in the forebrain as a whole takes place in the part of it known as the neocortex. Since in domesticated animals the greatest size decrease occurs in the neocortex, it is not surprising that an animal like the mouse, which has little neocortex, experiences the least brain size decrease. Apparently, then, domestication does not affect the brain of lower mammals as much as in the higher mammals. These changes in the central nervous system lead us to ask what the consequences are of such modifications.

Major changes occur in the large sensory organs of the head: the nose, ear and eye. Finer differences can also occur within parts of these systems, but in any case what results is a decrease, to a greater or lesser degree, of the sensitivity of those systems. Similar brain modifications and consequences have been found in the avian brain during the course of domestication.

Domestic animals are also "different" from wild animals behaviorally. These behavioral modifications are no doubt closely related to the neural changes occurring in the brain during domestication. Domestic animals are largely characterized by their docility in contrast to the excitable nature of their wild relatives. The individual behavior patterns do not all change uniformly during domestication; many of them may be carried out less frequently or less intensively, while others may disappear entirely in the domestic animal. The typical social organization of wild animals tends to dissolve during domestication; most domestic animals are less active and show greatly decreased escape behavior. Domestic animals are rarely as aggressive as their wild relatives. Other components of food seeking and prey behavior may also disappear. Reproductive behavior in domestic animals often lacks the rivalry contests occurring among wild animals; nest-building and defense of the young are much less pronounced as well. On the other hand, some components of food seeking behavior and sexual behavior are intensified in domestic animals. Behavioral sequences often lose their integrity in domestication. This disintegration even occurs in behavioral patterns having a genetic basis in the wild species.

In many instances it is difficult to determine to what extent behavior is under genetic influence. Certainly, some behavioral changes in domestic animals are genetic in nature, produced by the artificial selection conducted by man. However, domestic animals do not display any new behavior patterns; behavior modifications are all quantitative in nature and simply represent a different intensity of the same behavior we observe in wild species. The modifications occurring in the central nervous system and in behavior often give us the impression that domestic animals are "degenerate wild animals". This conclusion is unjustified, however. All the changes we have described above are adaptations to the "environment"—the living conditions—created by man as he selectively breeds for specific characteristics. Domestic animals have adapted to the *Umwelt* in which they live just as wild animals are adapted to their habitats. To what extent domestic animals can actually be modified is still an open question.

The changes taking place in the many domestic breeds as they are transformed from their wild ancestors have often been interpreted as models of the evolutionary process. Darwin wrote, "the cultivated races of one and the same species differ among each other in exactly the same way as related wild species belonging to the same genus". However, this is not quite correct, for the critical factor in evolution is species formation; only with species formation do we have evolution, and domesticated animals are not new species, as we pointed out earlier. The sexual boundaries existing between wild and domestic animals are purely artificial ones, and they disappear if man permits a domestic animal to freely associate with wild members of the same species. Domestication does NOT show us how new species are produced; it shows us to what extent a wild

Behavior of domestic animals

In wild rabbits, the same structures are usually inherited, so that offspring have the same fur covering as their parents and other similar characteristics. However, major variations among individual genes sometimes occur, and the resulting mutants can arise in both wild and domesticated animals. A mutation can cause a drastic change in hair growth patterns. It was in this way that the short-haired wild rabbit gave rise to something like the long-haired domestic angora rabbit.

Fig. 23-1. Three examples of the development of the Angora rabbit, which took place within two centuries: the angora rabbit as seen in 1775, 1900 and today. The modern angora rabbit has well developed ear and forehead tufts, furry cheeks, and foot tufts.

species can be modified under selective breeding conditions. Furthermore, domestication shows us that there is no strong connection between anatomical modification and speciation. This is a fine example of how 19th Century naturalists were erring when they defined species on the basis of appearance ("all animals that look alike belong to the same species" kind of reasoning). Domestication has never produced a new species of animal, and therefore it follows that the process of domestication is not any sort of model of speciation.

Of course, anatomical modification is no doubt a preliminary step toward speciation, and the causes responsible for substantial intraspecific anatomical modification are worthy of research and shed light on the process of evolution in its entirety. Characteristics of animals are the result of various developmental processes that are generated and guided by the genetic apparatus of the individual animal. Therefore, some genetic change must have taken place in domestic animals, even if no species formation (speciation) took place. In general, the genetic constitution of a species remains intact and is passed on, intact, from one generation to the next, subject to modification caused by gene recombination during reproduction. The changes taking place in domestic animals can only have come about from a new distribution of the genetic material in wild species or by a change in the genetic material itself. We must not forget that most characteristics are not the subject of a single gene but of a whole complex of genes, interacting to produce the final characteristic as we see it (the phenotype). Many genes remain "hidden" because they are recessive, and they can only be expressed when the particular kind of gene jumbling occurring during reproduction permits them to be expressed. A redistribution of the genetic material can eliminate the inhibitory factors that prevent recessive genes from being expressed in the phenotype, and this is often the kind of development that is occurring in domestic animals. The development of anatomical and behavioral characteristics, of course, is not only a genetic problem but also a physiological one.

Early naturalists interpreted the characteristics of domestic animals from a purely genetic standpoint, and the diversity taking place among domestic breeds was thought to be the result of mutations in the genetic apparatus of these animals. Since mutations can be elicited by various environmental factors, and since many domestic animals live under conditions deviating substantially from the normal life of wild species, naturalists thought that domestication greatly increased the mutation rate.

This problem—of whether there was a relationship between mutations and the diversity in domesticated animals—had to be solved by putting domestic animals into relatively "natural" conditions: instead of being put into a stall, the domestic animal was allowed to roam about freely and seek its own food, for example. This has been done on a large scale with such domestic animals as the reindeer and the llama. Reindeer freely

Transformation from wild form to domestic form and alterations within the domestic form, as seen in the Berkshire swine, which at the end of the 18th Century was one of the most prevalent English domestic swine breeds.

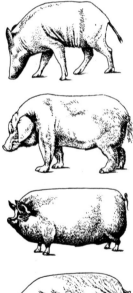

Fig. 23-2.
From top to bottom:
Wild swine *Sus scrofa scrofa*;
Berkshire swine in 1780,
1880 and 1940.
Crosses (hybrids) create
new gene combinations
and increase the degree of
variability within a breed.

seek their food, and the people "keeping" them follow the reindeer about, living a nomadic existence. Earlier naturalists thought that there was no difference at all between the domesticated reindeer and the wild species, but our current understanding gives us a different picture. Domestic reindeer live in vast herds, not in the small packs we see in wild reindeer, and young male reindeer are castrated so they will not engage in the rivalry fights that take place in wild reindeer. Selective breeding is thus occurring in domestic reindeer, and this leads merely to a redistribution of the genetic material found in wild reindeer, not to any kind of selective breeding of mutants. No mutation occurs in wild reindeer to make them live in herds. Free-roaming domestic reindeer are no different from domestic reindeer kept in a stall. Thus, all the changes occurring in domestication of the reindeer involve selective breeding of the wild species for specific characteristics, not the exploitation of a particular mutant strain.

A study of domesticated guanacos leads to the same findings. Guanacos are a good example of how a "thoroughbred" line can be developed in an extraordinarily short time by selective breeding. All these findings show that the genetic material in wild and domesticated animals is essentially the same, but the environmental conditions under which domestic animals live have led to new combinations of those same genes and hence to the appearance and sustenance of characteristics not seen in the wild. Man selects specific individuals and has them reproduce in such a way as to maximize the characteristics he seeks in those animals. Domestication shows us convincingly how influential selective breeding can be, and in this we can recognize one of the major factors responsible for evolution, even though the guiding forces in nature and under domestication are not the same ones.

Darwin also recognized the importance of domestication, and he wrote: "The principle of natural selection can be regarded as mere hypothesis, but its existence seems likely on the basis of what we know of the variability of organic creatures in nature, of the struggle for existence and the retention of advantageous variations, as well as the analagous relationships occurring in domestic races. However, for a long time it remained a question as to how the modification could be achieved, and this would have been an open question much longer if I had not studied the results of domestication." Indeed, the numerous examples of selective breeding are impressive evidence of the influence of differential reproduction on the characteristics of an animal species (and plant species, too!). In this perspective, domestication can be understood as a model of evolutionary processes.

Further zoological studies of domesticated animals have raised important evolutionary questions. A contrast drawn between interspecific differences of wild related species of ancestral forms with the intraspecific modifications of domestic animals shows that differences exist

Darwin's natural selection

not only between wild vs. domestic animals but also that the differences among domestic animals lie in another realm from those in wild animals. The degree of variation is much greater among domestic animals than among related wild species, for one. Furthermore, domestic animals regularly display the same kinds of changes, even though they belong to entirely different taxonomic categories. We mentioned earlier that brain size decreases in domestic animals, and the reader will recall that we cited a number of quite unrelated domestic species displaying this same phenomenon. The phenomenon of parallel evolution is in fact much more pronounced in domestic animals than among wild species, even though we know that parallel evolution does truly occur in the wild.

Parallel (homologous) structures

Parallel changes occur when homologous genes existing in different animals are expressed through interaction with the environment; when the environmental conditions under which unrelated domestic animals are the same, then the same homologous genes are expressed in the different animals. Homologous genes are very old ones in the genetic structure of animals, having been handed down from a common ancestor (perhaps from a very early period in the evolution of the species involved). Thorough investigations have recently shown, however, that statements about gene homologies are actually on rather unsteady ground, even though the parallel structures (homologies) often are part of identical (homodynamic) processes. The fact that similar characteristics occur in different animals does not mean that the characteristics involved are homologous, for it turns out that different genetic bases can yield similar, non-homologous characteristics. These similar but unrelated characteristics are called analogous characteristics. So the whole problem becomes a complex puzzle of sorting out analogies from homologies, and this is the reason that one must be extremely cautious about concluding that some sets of characteristics in different animals are homologous. The whole problem of homologies is still unresolved. A more thorough understanding of the phenomenon of parallel evolution will help future biologists more completely comprehend the complex set of processes we know as evolution.

To understand the special characteristics of domesticated animals it is necessary to appreciate the fact that a species of animal is a vast mixture of genes, a huge complex of genes. The individual members of a species possess their individual genetic constitution, but individuals do not possess all the different genes occurring within this reproducing community we call a species. Generally, the reproducing community we call a species scatters its genes fairly uniformly among the members of the population, and the result of this is that all the conspecifics are fairly similar and possess the typical characteristics of the species. Furthermore, each of the many characteristics is under the influence of a number of different genes, not just one gene. If a selected group of individuals from the species are sexually isolated, which is what we do with domestic animals, and then these groups are selectively reproduced, one would expect that new gene combinations would appear and that the new combinations would result in changes in some of the "normal" characteristics. If in addition this isolated group of animals lives under conditions unlike the conditions experienced by the original population, these new characteristics can eventually be maintained and will be expressed in all the members of the isolated group.

Processes like this take place in nature on remote islands, where breeding is highly selective due to restricted access to the bulk of the population and where special living conditions may occur. And similar processes take place in domestication. The phenomenon of domestication is by no means fully understood, for there is still a great deal to be learned about developmental physiology. Domestic animals are particularly suited for many studies in this field, and research in this area will help us expand our understanding of what this entire volume is all about: evolution.

▷
Hominoids I:
1. *Ramapithecus punjabicus* (= *Kenyapithecus wickeri*?); about 12 million years ago; northern India and eastern Africa;
2. *Australopithecus africanus*;
3. *Australopithecus robustus* (= *Paranthropus*); #2 and #3, according to latest information, are 5.5 million years old;
4. *Homo* (*Australopithecus*?) *habilis*; 900,000 to 1.5 million years ago; eastern Africa;
5. *Homo erectus leakeyi* (Olduvai early hominid); 500,000 years ago; eastern Africa;
6. *Homo erectus erectus* ("*Pithecanthropus*" or early Java man); 400,000–700,000 years ago; Java.
▷▷
Australopithecine group producing implements from bones.

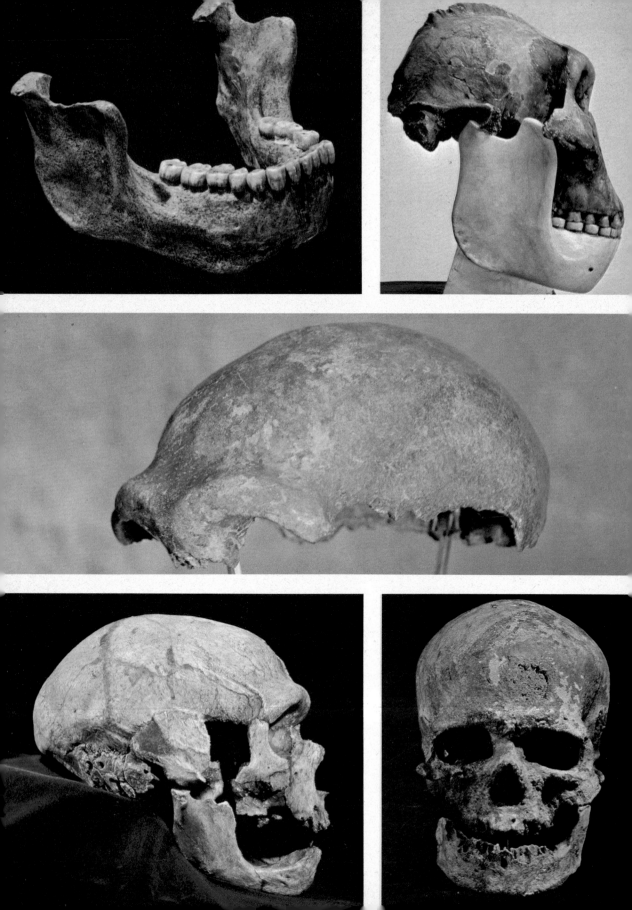

P. 524

Hominoids II:

1. Peking man (*Homo erectus pekinensis*), a fire user;
2. Heidelberg man (*Homo erectus heidelbergensis*), a whittler;
3. Rhodesian man (*Homo erectus rhodesiensis*), a late African form of early hominid;
4. The classic Neanderthal man (*Homo sapiens neanderthalensis*), who carried on a cave bear cult;
5. Steinheim man (*Homo sapiens steinheimensis*), like #6 a precursor of *sapiens*-man;
6. Swanscombe man, whose hind skull has practically the same conformation found in modern man;
7. Cro-Magnon man (*Homo sapiens sapiens* of the late Ice Age); used adornment, clothing, and practiced art.

P. 525

Oreopithecus bambolii; some 10 million years old, embedded in Lower Pliocene lignite from Monte Bamboli, Italy. Its teeth resemble those of hominids: there is no "ape gap" between them; the incisors are vertical; the canines are somewhat longer than the other teeth. The face is short (as in hominoids), but the fore limbs are long. This skeleton, discovered in 1954 by Hürzeler in a mine, was initially thought to be a hominid, but now it is placed with the anthropoid ape family of oreopithecids (Oreopithecidae), which differ somewhat from the true anthropoid or pongid apes (Pongidae). From the Natural History Museum of Basel.

P. 527

Examples of variation of body shape and coloration in domesticated animals. From left to right and top to bottom:

In domesticated cattle: Bavarian piebald bull; brown cow; Watusi cattle (eastern Africa); black and white zebu cattle of India; African zebu with its giant horns;

In domesticated poultry: Dutch chicken; chabos rooster; crested fowl; Japanese yokohama (feathers 8 m long!); Malayan chicken; silver-laced wyandotte.

P. 528

Examples of homologies (parallel structures) in domesticated animals:

1. Long-hairedness: English sheepdog, Persian cat, Scottish highland cattle;
2. Long ears: Rabbit, mamber goat, basset hound;
3. Dwarf forms: Pygmy horse, chihuahua, dwarf goats;
4. Half-body coloration: In swine, goats and sheep.

P. 526

Top left: Heidelbergman (*Homo erectus heidelbergensis*); this is the lower jaw of the oldest human remains found from Pleistocene Europe; it was discovered in 1907 near Heidelberg. The jaw contains typically human dentition; from the Kurpfälzisches Museum, Heidelberg.—Top right: skull of a P-type australopithecine (*Australopithecus boisei*), with the characteristic skull ridge on top; the lower jaw has been reconstructed; from the Museum of Tanzania, Dar-es-Salaam. —Middle: the classic skull roof of the Neanderthal (*Homo sapiens neanderthalensis*); the first of these famous fossil hominids was found in 1856 by a schoolteacher, Johann Carl Fuhlrott in a cave in the Neander Valley near Düsseldorf, Germany. Neanderthal man was adapted to a cold climate from the last third of the Pleistocene in Europe. This skull roof is in the Rheinisches Landesmuseum in Bonn.—Bottom left: another classic Neanderthal skull, found in 1908 in La Chapelle-aux-Saints, Corrèze (France); it is from the Middle Old Stone Age (Musée de L'Homme, Paris).—Bottom right: the well-preserved skull of the "old man of Cro-Magnon"; the name Cro-Magnon stems from fossil finds in 1868 in Cro-Magnon in the Dordogne region of France. Cro-Magnon man was a modern man (*Homo sapiens sapiens*, the same species and subspecies as we) from the last glaciation. This magnificent fossil skull can be seen in the Musée de L'Homme in Paris.

SYSTEMATIC CLASSIFICATION

This classification contains only the largest systematic units and only fossil taxa. Smaller taxa can be found within the individual chapters of this volume.

SUBKINGDOM UNICELLULATES (PROTOZOA)

Class Flagellates (Flagellata)

Class Rhizopods (Rhizopoda)
  Order Testacea   ? Carboniferous, Tertiary to present

Order Foraminifera   Cambrian to present
Order Radiolaria   Precambrian to present

Class Ciliata
  Order Spirotricha   Jurassic to Tertiary, present

SUBKINGDOM METAZOANS (METAZOA)

SUBKINGDOM PARAZOA

Phylum Archaeocyatha

Classes Monocyathea, Archaeocyathea, Anthocyathea   Cambrian

Phylum Sponges (Spongia)

Class Glass sponges (Hexactinellida)   Cambrian to present

Class Common sponges (Demospongia) Cambrian to present

Class Calcareous sponges (Calcarea)   Cambrian to present

SUBKINGDOM EUMETAZOA

Subsection Coelenterates (Coelenterata)

Phylum Cnidaria

Class Hydrozoa   Cambrian to present

Class True jellyfish (Scyphozoa)

  Subclass Scyphomedusae   Precambrian to present

  Subclass Conularia   Cambrian to Triassic

Class Anthozoa

  Subclass Rugose coral (Rugosa)   Ordovician to Carboniferous

  Subclass Tabulate coral (Tabulata)   Ordovician to Permian, ? Tertiary, ? Present

  Subclass Hexacorallia   Triassic to present

  Subclass Octocorallia   Triassic to present, ? Permian

Subsection Bilateralia

Main Branch Protostomia

Phylum Aschelminthes = Nemathelminthes

Class Nematodes (Nematoda)   Tertiary, present

Superphylum Articulata
Phylum Annelida

Class Polychaetes (Polychaeta)   Cambrian to present

Class Clitellata   ? Carboniferous, present

Class Myzostomida   ? Ordovician to present

Phylum Onychophora   ? Precambrian, Cambrian to present
Phylum Arthropods (Arthropoda)
Subphylum Trilobites (Trilobita)

Class Trilobitoidea   Cambrian to Devonian

Class Trilobita   Cambrian to Permian

Subphylum Chelicerata

Class Merostomata
  Order Aglaspids (Aglaspida)   Cambrian to Ordovician
  Order Giant sea scorpions (Eurypterida) Ordovician to Permian

  Order Horseshoe crabs (Xiphosura)   Silurian to present

Class Arachnids (Arachnida)   Silurian to present

Class Pantopoda   Devonian, Jurassic, present

Subphylum Diantennata

Class Crustaceans (Crustacea)

  Subclass Phyllopoda   ? Silurian, Devonian to present

  Subclass Anostraca   Tertiary to present

  Subclass Lipostraca   Devonian

  Subclass Ostracods (Ostracoda)   Cambrian to present

  Subclass Barnacles (Cirripedia)   Silurian to present

  Subclass Copepods (Copepoda)   ? Triassic, present

  Subclass Ascothoracida   Cretaceous, present

  Subclass Higher crustaceans (Malacostraca) Cambrian to present

Subphylum Tracheates (Tracheata)

Class Myriapoda   Silurian to present

Class Insects (Insecta)   Devonian to present

Phylum Mollusks (Mollusca)

Class Polyplacophora   Cambrian to present

Class Monoplacophora = Tryblidiacea   Cambrian to present

Class Univalves (Gastropoda)

  Subclass Streptoneura = Prosobranchia Cambrian to present

  Subclass Euthyneura = Opisthobranchia and Pulmonata   Carboniferous to present

Class Tusk shells (Scaphopoda)   Devonian to present

Class Bivalves (Bivalvia)   Cambrian to present

Class Cephalopods (Cephalopoda)

  Subclass Nautiloids (Nautiloidea)   Cambrian to present

  Subclass Endoceratoidea   Ordovician

  Subclass Actinoceratoidea   Ordovician to Carboniferous

  Subclass Bactritoidea   ? Ordovician, Silurian to Permian

  Subclass Ammonites (Ammonoidea) Devonian to Cretaceous

Phylum Bryozoa   Ordovician to present

Phylum Brachiopods (Brachiopoda)

Class Inarticulata   Cambrian to present

Class Articulata = Testicardines   Cambrian to present

Phylum Echinoderms (Echinodermata)

Subphylum Homalozoa   Cambrian to Devonian

Subphylum Crinozoa

Class Eocrinoidea   Cambrian to Silurian

Class Cystoidea   Ordovician to Silurian

Class Blastoidea   Silurian to Permian

Class Crinoidea   Ordovician to present

Subphylum Asterozoa

Class Somasteroidea   Ordovician, Devonian to ? present

Class Asteroidea   Ordovician to present

Class Ophiuroidea   Ordovician to present

Subphylum Echinozoa

Class Helicoplacoidea   Cambrian

Class Ophiocystioidea   Ordovician to Devonian

Class Cyclocystoidea   Ordovician to Devonian

Class Edrioasteroidea   Cambrian to Carboniferous

Class Sea urchins (Echinoidea)   Ordovician to present

Class Holothuroidea   Devonian to present

Phylum Pentocoela

Subphylum Brachiotremata

Class Graptolithines (Graptolithina)   Cambrian to Carboniferous

Phylum Chordates (Chordata)

Subphylum Tunicates (Tunicata)

One questionable fossil species: *Ainiktozoon*   Upper Silurian

Subphylum Vertebrates (Vertebrata)

Superclass Jawless fishes (Agnatha)

Class Cephalaspidomorphi

  Subclass Osteostraci   Upper Silurian to Upper Devonian

Subclass Anaspida   Upper Silurian to Upper Devonian

Class Cyclostomes (Cyclostomata)   Upper Carboniferous, present

Class Pteraspidomorphi

> **Subclass Heterostraci**   Middle Ordovician to Upper Devonian

> **Subclass Thelodonti**   Silurian to Middle Devonian

Superclass Fishes (Pisces, Gnathostomata)

Class Spiny sharks (Acanthodii)   Upper Silurian to Lower Permian

Class Placoderms (Placodermi)

> **Subclass Arthrodira**
> Order Euarthrodira   Devonian
> Order Ptyctodontida   Middle and Upper Devonian
> Order Phyllolepida   Middle and Upper Devonian
> Order Petalichthyida   Lower and Middle Devonian
> Order Rhenanida   Devonian

> **Subclass Antiarchs (Antiarchi)**   Devonian

Class Cartilaginous fishes (Chondrichthyes)

> **Subclass Elasmobranchs (Elasmobranchii)**
> Order Primitive sharks (Cladoselachida and Cladodontia)   Middle Devonian to Lower Permian
> Order Sharks (Selachii)   Upper Devonian to present
> Order Rays (Rajiformes)   Upper Jurassic to present
> Order Freshwater sharks (Xenacanthida)   Middle Devonian to Upper Triassic

> **Subclass Chimaeras (Holocephali)**
> Order Chimaerida   Lower Carboniferous to present
> Order Edestida   Lower Carboniferous to Lower Triassic
> Order Iniopterygii   Upper Carboniferous

Class Bony fishes (Osteichthyes)

> **Subclass Spiny-rayed fishes (Actinopterygii)**
> Superorder Chondrostei   Upper Silurian to present
> Superorder Lower bony fishes (Holostei)   Upper Permian to present
> Superorder True bony fishes (Teleostei)   Triassic to present

> **Subclass Sarcopterygii**
> Order Coelacanths (Crossopterygii)
> Suborder Actinistia, Coelacanthini   Middle Devonian to Cretaceous, present
> Suborder Rhipidistia
> > Family Holoptychidae   Devonian
> > Family Osteolepididae   Middle Devonian to Lower Permian

> Suborder Onychodontida   Lower Devonian to Upper Carboniferous
> Order Lungfishes (Dipnoi)   Lower Devonian to present

Class Amphibians (Amphibia)

> **Subclass Lepospondyli**
> Order Aistopods (Aistopoda)   Carboniferous
> Order Nectrids (Nectridea)   Upper Carboniferous
> Order Microsaurs (Microsauria)   Upper Carboniferous to Permian
> Order Primitive salamanders (Lysorophia)   Upper Carboniferous to Permian

> **Subclass Labyrinthodonts (Labyrinthodontia)**
> Order Ichthyostegalia   Uppermost Devonian to Lower Carboniferous
> Order Temnospondyli
> > Suborder Rhachitomi   Lower Carboniferous to Triassic
> > Suborder Stereospondyli   Middle Permian to Upper Triassic
> > Suborder Plagiosauria   Upper Permian to Upper Triassic
> Order Batrachosaurs (Batrachosauria)
> > Suborder Anthracosaurs (Anthracosauria)   Lower Carboniferous to Middle Permian
> > Suborder Seymouriamorpha   Lower Carboniferous to Upper Permian
> > Suborder Gephyrostegoidea   Upper Carboniferous

> **Subclass Lissamphibia**
> Order Anurans (Salientia, Anura)   Triassic to present
> Order Urodeles (Caudata, Urodela)   Cretaceous to present
> Order Caecilians (Gymnophiona, Apoda)   Only present

Class Reptiles (Reptilia)

> **Subclass Anapsida**
> Order Cotylosauria
> > Suborder Captorhinomorpha   Upper Carboniferous to Triassic
> > Suborder Procolophonomorpha and Parciasauria   Middle Permian to Triassic
> > Suborder Eunotosauria   Middle Permian
> Order Turtles (Chelonia, Testudines)   Upper Triassic to present

> **Subclass Lepidosaurs (Lepidosauria)**
> Order Eosuchia   Upper Permian to Lower Triassic
> Order Rhynchocephalians (Rhynchocephalia)   Lower Triassic to Lower Cretaceous, present

Order Squamata
Suborder Prolacertilia and Eolacertilia   Middle
to Upper Triassic
Suborder Lizards (Sauria, Lacertilia)   Upper
Jurassic to present
Suborder Snakes (Serpentes, Ophidia)   Lower
Cretaceous to present

Subclass Archosaurs (Archosauria)
Order Thecodontia
Suborder Pseudosuchia   Triassic
Suborder Parasuchia   Triassic
Order Crocodiles (Crocodilia)   Upper Triassic
to present
Suborder Proto- and Archaeosuchia   Triassic
Suborder Mesosuchia   Triassic
Suborder Eusuchia   Cretaceous to present
Order Ornithischia
Suborder Ornithopoda   Triassic to Cretaceous
Suborder Stegosaurs (Stegosauria)   Jurassic to
Cretaceous
Suborder Ankylosauria   Cretaceous
Suborder Ceratopsia   Upper Cretaceous
Order Saurischia
Suborder Theropoda   Triassic to Cretaceous
Suborder Sauropodomorpha   Jurassic to
Cretaceous
Order Flying saurians (Pterosauria)
Suborder Rhamphorhynchida   Jurassic
Suborder Pterodactyls (Pterodactylida)   Upper
Jurassic to Cretaceous

Subclass Euryapsida
Order Araeoscelidia   Permian to Triassic
Order Sauropterygia
Suborder Nothosaurs (Nothosauria)   Triassic
Suborder Plesiosaurs (Plesiosauria)   Middle
Triassic to Upper Cretaceous
Order Placodonts (Placodontia)   Middle to
Upper Triassic

Subclass Ichthyopterygians (Ichthyopterygia)
Order Ichthyosaurs (Ichthyosauria)   Middle
Triassic to Cretaceous

Subclass Synapsida
?Order Mesosaurs (Mesosauria)   Upper
Carboniferous to Lower Permian
Order Pelycosaurs (Pelycosauria)
Suborder Caseoidea   Middle Permian
Suborder Edaphosauria   Upper Carboniferous
to Middle Permian
Suborder Ophiacodontia   Upper
Carboniferous to Middle Permian
Suborder Sphenacodontia   Upper
Carboniferous to Middle Permian
Order Therapsids (Therapsida)
Suborder Deinocephalia   Middle to Upper
Permian

Suborder Anomodontia   Middle Permian to
Upper Triassic
Suborder Theriodontia   Middle Permian to
Upper Jurassic

Class Birds (Aves)

Subclass Archaeornithes
Order Archaeopterygiformes   Upper Jurassic

Subclass Toothed birds (Odontognathae)
Order Hesperornithiformes   Upper Cretaceous

Subclass Recent birds (Neornithes)
Superorder Palaeognathae   Lower Tertiary to
present
Superorder Neognathae   Lower Cretaceous to
present

Class Mammals (Mammalia)

Subclass Prototheria or Non-Theria ("Not-Yet Mammals")
Order Multituberculata   ?Upper Triassic/Lower
Jurassic, Upper Jurassic to Lower Tertiary
Order Triconodonta   ?Upper Triassic, Middle
to Upper Jurassic
Order Docodonta   ?Upper Triassic/Lower
Jurassic, Upper Jurassic
Order Symmetrodonta   ?Upper Triassic, Upper
Jurassic to Middle Cretaceous
Order Pantotheria (= Eupantotheria)   Middle
Jurassic to Lower Cretaceous
Order Monotremes (Monotremata)   Pleistocene
to present

Subclass True Mammals (Theria)

Infraclass Marsupials (Marsupialia)   Upper
Cretaceous to present

Infraclass Higher mammals (Eutheria or Placentalia)
Order Insectivores (Insectivora)   Upper
Cretaceous to present
Order Hyaenodonta   Upper Cretaceous to
Miocene
Order Dermoptera   Lower Tertiary, present
Order Chiroptera   Eocene to present
Order Primates (Primates)
Suborder Prosimians (Prosimiae)
Infraorder Tupaiiformes   Oligocene, present
Infraorder Lemurs (Lemuriformes or Lemuroidea)
?Upper Cretaceous, Paleocene to present
Infraorder Lorises (Lorisiformes or Lorisoidea)
Miocene to present
Infraorder Tarsiers (Tarsiiformes or Tarsioidea)
Lower Tertiary, present
Infraorder Omomyids (Omomyoidea)   Eocene
to Oligocene
Suborder Monkeys, Apes and Man (Simiae)

Infraorder New World monkeys (Platyrrhina or Ceboidea)   Oligocene to present
Infraorder Old World monkeys (Catarrhina)
  Superfamily Cercopithecoids (Cercopithecoidea)   Oligocene to present
  Superfamily Hominoids (Hominoidea)
    Family Parapithecidae   Lower Oligocene
    Family Oreopithecidae   ?Lower Oligocene, Pliocene
    Family Gibbons (Hylobatidae)   Miocene to present
    Family Anthropoid apes (Pongidae)   Oligocene to present
    Family Humans (Hominidae)   Upper Miocene to present
Order Edentates (Edentata)
  Suborder Palaeanodonta   Lower Tertiary
  Suborder Xenarthra
  Infraorder Cingulata   Paleocene to present
  Infraorder Pilosa or Tardigrada   Oligocene to present
Order Pholidota   Oligocene to present
Order Taeniodonta   Lower Tertiary
Order Rodents (Rodentia)
  Suborder Protrogomorpha (Ischyromyoidea and Aplodontoidea)   Upper Paleocene to present
  Suborder Sciuromorpha   Oligocene to present
  Suborder Theridomorpha   Eocene and Oligocene
  Suborder Myomorpha   Oligocene to present
  Suborder Hystricomorpha   Oligocene, Miocene to present
  Suborder Caviomorpha   Oligocene to present
Order Lagomorphs (Lagomorpha)   Paleocene to present
Order Carnivores (Carnivora)
  Suborder Fissipedia   Paleocene to present
    Superfamily Arctoidea   Lower Tertiary to present

Superfamily Canids and Felids (Cynofeloidea)   Lower Tertiary to present
Superfamily Viverrids (Herpestoidea)   Lower Tertiary to present
Suborder Pinnipeds (Pinnipedia)   Miocene to present
Order Cetaceans (Cetacea)
  Suborder Archaeoceti   Eocene, Oligocene
  Suborder Odontoceti   Eocene to present
  Suborder Mystacoceti   Oligocene to present
Order Artiodactyla
  Suborder Nonruminantia   Eocene to present
  Suborder Tylopoda   Eocene to present
  Suborder Ruminantia   Upper Eocene to present
Order Condylarthra   Upper Cretaceous, Lower Tertiary
Order Tillodontia   Paleocene, Eocene
Order Litopterna   Paleocene to Pliocene, Pleistocene
Order Notoungulata   Paleocene to Pliocene, Pleistocene
Order Astrapotheria   Paleocene to Miocene
Order Tubulidentata   Miocene to present
Order Pantodonta   Upper Paleocene to Oligocene
Order Dinocerata   Upper Paleocene, Eocene
Order Pyrotheria   Upper Eocene, Oligocene
Order Desmostylia   Miocene, Pliocene
Order Sirenians (Sirenia)   Middle Eocene to present
Order Proboscidea   Middle Eocene to present
Order Hyraxes (Hyracoidea)   Lower Oligocene to present
Order Embrithopoda   Lower Oligocene
Order Perissodactyla
  Suborder Ancylopoda   Lower Eocene to Pleistocene
  Suborder Ceratomorpha   Eocene to present
  Suborder Hippomorpha   Eocene to present

Conversion Tables of Metric to U.S. and British Systems

U.S. Customary to Metric		*Metric to U.S. Customary*	

—— Length ——

To convert	Multiply by	To convert	Multiply by
in. to mm.	25.4	mm. to in.	0.039
in. to cm.	2.54	cm. to in.	0.394
ft. to m.	0.305	m. to ft.	3.281
yd. to m.	0.914	m. to yd.	1.094
mi. to km.	1.609	km. to mi.	0.621

—— Area ——

sq. in. to sq. cm.	6.452	sq. cm. to sq. in.	0.155
sq. ft. to sq. mi.	0.093	sq. m. to sq. ft.	10.764
sq. yd. to sq. m.	0.836	sq. m. to sq. yd.	1.196
sq. mi. to ha.	258.999	ha. to sq. mi.	0.004

—— Volume ——

cu. in. to cc.	16.387	cc. to cu. in.	0.061
cu. ft. to cu. m.	0.028	cu. m. to cu. ft.	35.315
cu. yd. to cu. m.	0.765	cu. m. to cu. yd.	1.308

—— Capacity (liquid) ——

fl. oz. to liter	0.03	liter to fl. oz.	33.815
qt. to liter	0.946	liter to qt.	1.057
gal. to liter	3.785	liter to gal.	0.264

—— Mass (weight) ——

oz. avdp. to g.	28.35	g. to oz. avdp.	0.035
lb. avdp. to kg.	0.454	kg. to lb. avdp.	2.205
ton to t.	0.907	t. to ton	1.102
l. t. to t.	1.016	t. to l. t.	0.984

Abbreviations

U.S. Customary	*Metric*
avdp.—avoirdupois	cc.—cubic centimeter(s)
ft.—foot, feet	cm.—centimeter(s)
gal.—gallon(s)	cu.—cubic
in.—inch(es)	g.—gram(s)
lb.—pound(s)	ha.—hectare(s)
l. t.—long ton(s)	kg.—kilogram(s)
mi.—mile(s)	m.—meter(s)
oz.—ounce(s)	mm.—millimeter(s)
qt.—quart(s)	t.—metric ton(s)
sq.—square	
yd.—yard(s)	

By kind permission of Walker: Mammals of the World
©1968 Johns Hopkins Press, Baltimore, Md., U.S.A.

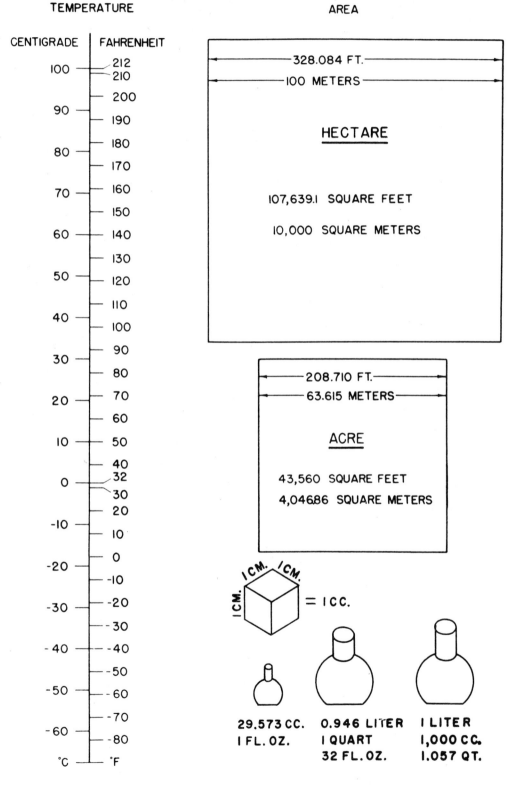

TEMPERATURE

CENTIGRADE	FAHRENHEIT
100	212 / 210
90	200 / 190
80	180 / 170
70	160 / 150
60	140 / 130
50	120 / 110
40	100 / 90
30	80 / 70
20	60 / 50
10	40 / 32 / 30
0	20 / 10
-10	0 / -10
-20	-20 / -30
-30	-40 / -50
-40	-60 / -70
-50	-80
-60	
°C	°F

AREA

328.084 FT.
100 METERS

HECTARE

107,639.1 SQUARE FEET

10,000 SQUARE METERS

208.710 FT.
63.615 METERS

ACRE

43,560 SQUARE FEET
4,046.86 SQUARE METERS

1 CM. 1 CM. 1 CM. = 1 CC.

29.573 CC.
1 FL. OZ.

0.946 LITER
1 QUART
32 FL. OZ.

1 LITER
1,000 CC.
1.057 QT.

WEIGHT

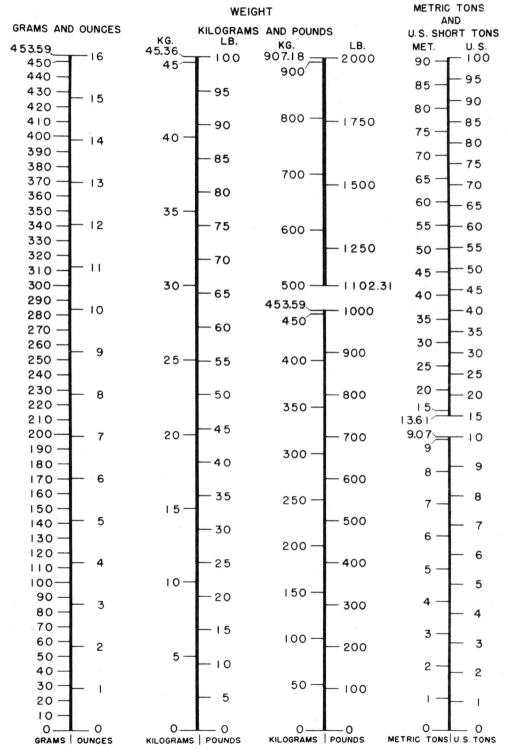

LENGTH: MILLIMETERS AND INCHES

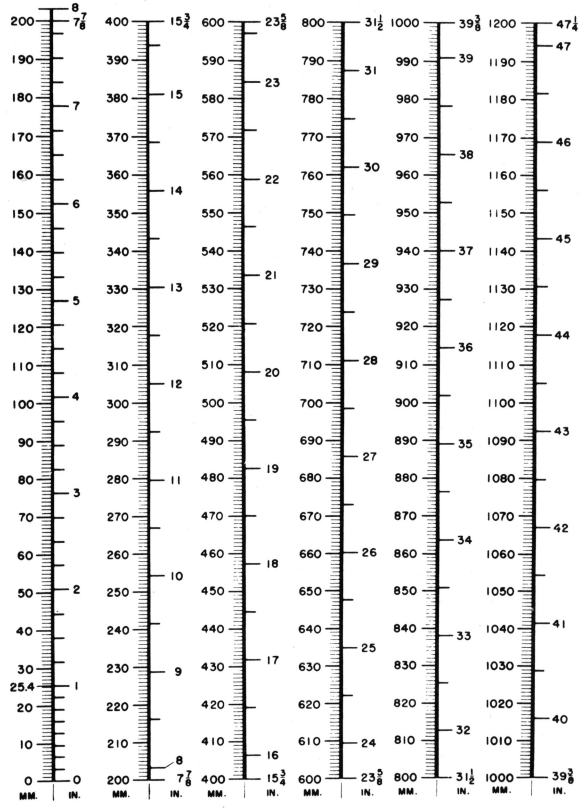

LENGTH

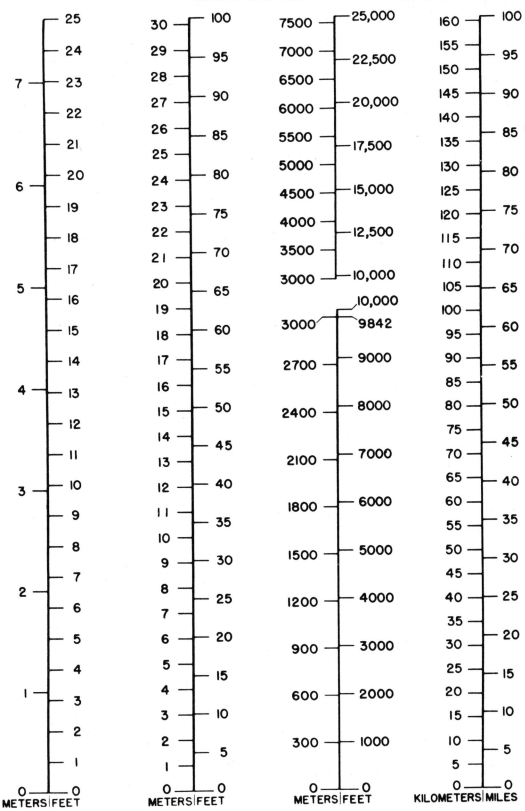

| METERS AND FEET | | KILOMETERS AND MILES |

Supplementary Readings

General References

American Geological Institute, 1973, *Investigating the Earth*. Houghton-Mifflin, Boston.

Bernal, J. D. 1967 *The Origin of Life*. World, Cleveland. 345 Pp.
Blum, H. F. 1968 *Time's Arrow and Evolution*. Princeton Univ. Press, Princeton, N.J. 232 Pp.

Calvin, M. 1969 *Chemical Evolution*. Oxford Univ. Press, N.Y. 278 Pp.
Carlquist, S. 1965 *Island Life: A Natural History of Islands of the World*. Natural History Press, Garden City, N.J.
Colbert, E. H. 1955 *Evolution of the Vertebrates*. Wiley, N.Y.
Cott, H. B. 1940 *Adaptive Coloration in Animals*. Methuen, London. 508 Pp.
Crow, J. F. and M. Kimura. 1970 *An Introduction to Population Genetics Theory*. Harper and Row, N.Y. 591 Pp.

Darlington, P. J. 1957 *Zoogeography: The Geographical Distribution of Animals*. John Wiley, N.Y. 675 Pp.
De Beer, G. R. 1951 *Embryos and Ancestors*. Oxford Univ. Press, Oxford.
Dice, L. R. 1952 *Natural Communities*. Univ. Michigan Press, Ann Arbor. 547 Pp.
Dobzhansky, Th. 1972 *Genetics of the Evolutionary Process*. Columbia Univ. Press, New York. 505 Pp.
Drake, E. T. (ed.). 1968 *Evolution and Environment*. Yale Univ. Press, New Haven, Conn. 478 Pp.

Ehrlich, P. R., and R. W. Holm. 1963 *The Process of Evolution*. McGraw-Hill, N.Y. 347 Pp.

Fisher, R. A. 1930 *The Genetical Theory of Natural Selection*. Clarendon Press, Oxford.
Ford, E. B. 1971 *Ecological Genetics*. Methuen, London. 410 Pp.
Fox, S. W., and K. Dose. 1972 *Molecular Evolution and the Origin of Life*. Freeman, San Francisco. 359 Pp.

Grant, V. 1963 *The Origin of Adaptations*. Columbia Univ. Press, New York. 606 Pp.

Haldane, J. B. S. 1932 *The Causes of Evolution*. Harper and Row, New York.
Hutchinson, G. E. 1965 *The Ecological Theater and the Evolutionary Play*. Yale Univ. Press, New Haven, Conn. 139 Pp.
Huxley, J. S. 1942 *Evolution, The Modern Synthesis*. Harper and Row, New York.

Jepsen, G. L., E. Mayr, and G. G. Simpson (eds.). 1949 *Genetics, Paleontology, and Evolution*. Princeton Univ. Press, Princeton, N.J.
Jukes, T. H. 1966 *Molecules and Evolution*. Columbia Univ. Press, New York. 285 Pp.

Lack, D. 1947 *Darwin's Finches*. Cambridge Univ. Press, Cambridge.
Lerner, I. M. 1958 *The Genetic Basis of Selection*. Wiley, N.Y.
Levins, R. 1968 *Evolution in Changing Environments*. Princeton Univ. Press, Princeton, N.J. 120 Pp.
Lewontin, R. C. 1974 *The Genetic Basis of Evolutionary Change*. Columbia Univ. Press. N.Y. 346 Pp.

MacArthur, R. H. 1972 *Geographical Ecology: Patterns in the Distribution of Species*. Harper and Row, N.Y. 269 Pp.
Maynard Smith, J. 1958 *The Theory of Evolution*. Penguin, Baltimore. 320 Pp.
Mayr, E. 1963 *Animal Species and Evolution*. Harvard Univ. Press, Cambridge, Mass. 797 Pp.
Mettler, L. E., and T. G. Gregg. 1969 *Population Genetics and Evolution*. Prentice-Hall, Englewood Cliffs, N.J. 212 Pp.
Murray, J. 1972 *Genetic Diversity and Natural Selection*. Hafner, New York, 128 Pp.

Pianka, E. R. 1974 *Evolutionary Ecology*. Harper and Row, N.Y. 356 Pp.
Ponnamperuma, C. 1972 *The Origin of Life*. Dutton, N.Y. 215 Pp.

Raymond, P. E. 1958 *Prehistoric Life*. Cambridge, Mass.
Romer, A. S. 1966 *Vertebrate Paleontology*. 3rd ed. Univ. Chicago Press, Chicago.

Salthe, S. N. 1972 *Evolutionary Biology*. Holt, Rinehart and Winston, N.Y. 437 Pp.
Simpson, G. G. 1944 *Tempo and Mode in Evolution*. Columbia Univ. Press, New York.
Simpson, G. G. 1951 *Horses*. Oxford Univ. Press, New York.
Simpson, G. G. 1961 *Principles of Animal Taxonomy*. Columbia Univ. Press, New York.
Simpson, G. G. 1965 *The Major Features of Evolution*. Columbia Univ Press, New York. 434 Pp.
Stebbins, G. L. 1966 *Process of Organic Evolution*. Prentice-Hall, Englewood Cliffs, N.J.
Stokes, W. L. 1960 *Essentials of Earth History*. Prentice-Hall, Englewood Cliffs, N.J.

Washburn, S. L. and R. Moore. 1974 *Ape into Man*. Little, Brown, Boston. 196 Pp.
White, M. J. D. 1973 *Animal Cytology and Evolution*. Cambridge Univ. Press, Cambridge, England. 961 Pp. 3rd ed.
White, M. J. D. (ed.). 1974 *Genetic Mechanisms of Speciation in Insects*. Australia and New Zealand Book Company, Sydney, Australia. 170 Pp.
Wickler, W. 1968 *Mimicry in Plants and Animals*. World Univ. Library, N.Y. 249 Pp.
Williams, G. C. 1966 *Adaptation and Natural Selection*. Princeton Univ Press, Princeton, N.J. 307 Pp.
Woodfort, A. O. 1965 *Historical Geology*. Freeman, San Francisco.
Wright, S. 1968 *Evolution and the Genetics of Populations. Vol. I. Genetic and Biometric Foundations*. Univ. of Chicago Press, Chicago.
Wright, S. 1969 *Evolution and the Genetics of Populations. Vol. II. The Theory of Gene Frequencies*. Univ. of Chicago Press, Chicago.

Evolutionary Theory — Phylogeny — Genetics

Abel, O.: *Paläobiologie und Stammesgeschichte*, Jena (G. Fischer) 1929.

Beer, G. de: *Bildatlas der Evolution*. München (BLV) 1966.

Colbert, E. H.: *Die Evolution der Wirbeltiere*. Stuttgart (G. Fischer) 1965.

Eigen, M.: *Selbstorganisation der Materie und die Evolution biologischer Makromoleküle*. In: Umschau in Naturwissenschaft und Technik, 1970, S. 777–779.

Gregory, W. K.: *Evolution Emerging. A Survey of Changing Patterns from Primeval Life to Man*. 2 Bde., New York (Macmillan) 1951; Neudruck 1957.
Günther, K.: *Systematik und Stammesgeschichte der Tiere*. In: Fortschritte der Zoologie 14, 1962, S. 269 bis 546.
—: *Zur Geschichte der Abstammungslehre. Mit einer Erörterung von Vor- und Nebenfragen*. In: Heberer, G. (Hg.): *Die Evolution der Organismen*, Bd. 1, Stuttgart (G. Fischer) 3. A. 1967, S. 3–60.
Gutmann, W. F.: *Der biomechanische Gehalt der Wurmtheorie*. In: Zeitschrift für wissenschaftliche Zoologie 182, 1971, S. 229–262.
—: *Die Hydroskelett-Theorie*. Frankfurt/Main (in Druck). Aufsätze und Reden der Senckenbergischen naturforschenden Gesellschaft.

Herberer, G. (Hg.): *Die Evolution der Organismen. Ergebnisse und Probleme der Abstammungslehre*. Stuttgart (G. Fischer) 3. A. 1967 ff.
— und Schwanitz, F.: *Hundert Jahre Evolutionsforschung*. Stuttgart (G. Fischer) 1960.

Hennig, W.: *Grundzüge einer Theorie der phylogenetischen Systematik.* Berlin (Deutscher Zentralverlag) 1950.
—: *Systematik und Phylogenese.* In: Bericht der Hundertjahrfeier der deutschen entomologischen Gesellschaft. Berlin 1957, S. 50–71.

Jacob, F.: *Die Logik des Lebendigen. Zwischen Urzeugung und genetischem Code.* Frankfurt/Main (S. Fischer) 1972.

Karlson, P.: *Biochemie.* Stuttgart (Thieme) 7. A. 1970.
Kenyon, D. H. und Steinmann, G.: *Biochemical Predestination.* New York (McGraw Hill) 1969.
Keosian, J.: *The Origin of Life.* London/New York, (Van Nostrand Reinhold) 1964.

Mayr, E.: *Animal Species and Evolution.* Cambridge/Mass. 1963; dt.: *Artbegriff und Evolution.* Hamburg/Berlin (Parey) 1967.
Mohr, H. und Sitte, P.: *Molekulare Grundlagen der Entwicklung.* München/Bern (BLV) 1971.
Monod, J.: *Zufall und Notwendigkeit.* München (Piper) 1971.
Moore, R.: *Evolution.* Time-Life International 1964.

Osche, D.: *Grundzüge der allgemeinen Phylogenetik.* In: Bertalanffy, L. von (Hg): *Handbuch der Biologie,* Bd. 3, Frankfurt/Main 1966, S. 817–906.

Peters, D. S. und Gutmann, W. F.: *Über die Lesrichtung von Merkmals- und Konstruktions-Reihen.* In: Zeitschrift für zoologische Systematik und Evolutionsforschung 9, 1971, S. 237–263.
Portmann, A.: *Einführung in die vergleichende Morphologie der Wirbeltiere.* Basel/Stuttgart (Schwabe) 3. A. 1965.

Remane, A.: *Die Grundlagen des natürlichen Systems der vergleichenden Anatomie und der Phylogenetik.* Leipzig (Geest & Portig) 1956.
—: *Die Geschichte der Tiere.* In: Herberer, G. (Hg.): *Die Evolution der Organismen,* S. 589–678.

Siewing, R.: *Lehrbuch der vergleichenden Entwicklungsgeschichte der Tiere.* Berlin/Hamburg (Parey) 1969.
Simpson, G. G.: *The Meaning of Evolution. A Study of the History of Life and Its Significance for Man.* London (Oxford Univ. Press) 1950; dt.: *Auf den Spurens des Lebens.* Berlin (Colloquium) 1957.
—u. Roe, A. (Hg.): *Behaviour qnd Evolution.* New Haven (Yale Univ. Press) 1958; dt.: *Evolution und Verhalten.* Frankfurt/Main (Suhrkamp) 1969.
—: *Zeitmaße und Ablaufformen der Evolution.* Göttingen (Musterschmidt) 1951.
Stebbins, C. L.: *Evolutionsprozesse. Grundbegriffe der modernen Biologie.* Stuttgart (G. Fischer) 1968.

Träger, L.: *Einführung in die Molekularbiologie.* Stuttgart (G. Fischer) 1969.

Wieland, T. und Pfleiderer, G.: *Molekularbiologie.* Frankfurt/Main (Umschau) 2. A. 1967.
Wickler, W.: *Vergleichende Verhaltensforschung und Phylogenetik.* In: Heberer, G. (Hg.): *Die Evolution der Organismen,* S. 420–510.

Zimmermann, W.: *Evolution. Die Geschichte ihrer Probleme und Erkenntnisse.* Frieburg i. B. (Alber) 1953 (Orbis academicus).
—: *Methoden der Evolutionswissenschaft.* In: Heberer, G. (Hg.): *Die Evolution der Organismen,* S. 61–160.

Paleontology

Brinkmann, R.: *Abriß der Geologie II. Historische Geologie.* Stuttgart (Enke) 8. A. 1959.

Ehrenberg, K.: *Paläozoologie.* Wein (Springer) 1960.

Fraas, E.: *Der Petrefaktensammler.* Stuttgart (Lutz) 1910. Neuauflage Thun (Ott) 1972.

Hölder, H.: *Geologie und Paläontologie in Texten und ihrer Geschichte.* Freiburg i. B. (Alber) 1960.
Huene, F. R. Frhr. von: *Paläontologie und Phylogenie der Niederen Tetrapoden.* Jena (Fischer) 1956.

Keast, A. u. a.: *Evolution of Mammals on Southern Continents.* I–V. London (The Quarterly Revue of Biology 43, S. 225–451) 1968.
Kuhn-Schnyder, E.: *Geschichte der Wirbeltiere.* Basel (Schwabe) 1953.
—: *Paläontologie als stammesgeschichtliche Urkundenforschung.* In: Heberer, G. (Hg.): *Die Evolution der Organismen,* Bd. 1, Stuttgart (Fischer) 3. A. 1967.
Kurtén, B.: *The Age of Mammals.* The World Naturalist 1–250 S. London (Weidenfeld & Nicolson) 1971.

Lehmann, U.: *Paläontologishes Wörterbuch.* Stuttgart (Enke) 1964.
Lippolt, H. J. und Simon, W.: *Geochronologie als Zeitgerüst der Phylogenie.* In: Heberer, G. (Hg.): *Die Evolution der Organismen,* Bd. 1.

Moore, C. R. (Hg.): *Treatise on Invertebrate Paleontology.* 1953 The Geographical Society of America.
Moy-Thomas, J. A. (2. Auflage von R. S. Miles): *Palaeozoic Fishes.* London (Chapman & Hall) 1971.
Müller, A. H.: *Lehrbuch der Paläozoologie.* 3 Bde., Jena (Fischer) 1957 ff.

Romer, A. S.: *Vertebrate Paleontology.* Chicago (University of Chicago Press) 1966.
—: *Notes and Comments on Vertebrate Paleontology.* Chicago (University of Chicago Press) 1968.
Rothe, H. W.: *Kleine Versteinerungskunde.* Bern/Stuttgart (Hallwag) 1969, 2. A. 1971 (Hallwag Taschenbuch 87).

Schindewolf. O. H.: *Wesen und Geschichte der Paläontologie.* Berlin (Wissenschaftliche Editionsgesellschaft) 1948.
—: *Grundfragen der Paläontologie.* Stuttgart (Schweizerbart) 1950.
Simpson, G. G.: *Leben der Vorzeit, Einführung in die Paläontologie.* München (dtv) 1971.

Termier, H. und G.: *Les temps fossiliferes.* Paris (Masson & Cie) 1964 ff.
Thenius, E.: *Paläontologie. Die Geschichte unserer Tier- und Pflanzenwelt.* Kosmos Studienbücher. Stuttgart (Franckh) 1970.
—: *Versteinerte Urkunden. Die Paläontologie als Wissenschaft vom Leben in der Vorzeit.* Berlin (Springer) 1963.

Zeuner, F. E.: *Dating the Past. An Introduction to Geochronology.* London (Methuen) 1952.
Zittel, K. A.: *Geschichte der Geologie und Paläontologie bis Ende des 19, Jahrhunderts.* München/Leipzig (Oldenburg) 1899.
—: *Grundzüge der Paläontologie.* München/Berlin (Oldenburg); Invertebrata, 6. A. 1924, Vertebrata, 4. A. 1923.

Paleobotany

Andrews, H. N.: *Studies in Paleobotany.* New York/London (John Wiley & Sons) 1961.

Gothan, W. und Remy, W.: *Steinkohlenpflanzen.* Essen (Glückauf) 1957.

Kirchheimer, F.: *Grundzüge einer Pflanzenkunde der deutschen Braunkohle.* 1–153. Halle/Saale (Knapp) 1937.
Kräusel, R.: *Versunkene Floren. Eine Einführung in die Paläobotanik.* Frankfurt (Kramer) 1950.

Mägdefrau, K.: *Die Geschichte der Pflanzen*. In: Heberer, G. (Hg.): *Die Evolution der Organismen.* Bd. 1, Stuttgart (Fischer) 3. A. 1967.
— : *Paläobiologie der Pflanzen*. Jena (Fischer) 4. A. 1968.

Zimmermann, W.: *Die Phylogenie der Pflanzen*. Stuttgart (Fischer) 1959.

Mesozoic Age

Arkell, W. J.: *Jurassic Geology of the World*. Edinburgh/London 1956.

British Mesozoic Fossils. London (British Museum) 2. A. 1964.

Hauff, B.: *Das Holzmadenbuch*. Öhringen (Rau) 1953. Neue Auflage in Vorbereitung.
Hölder, H.: *Jura. Handbuch der stratigraphischen Geologie*. Stuttgart (Enke) 1964.

Kuhn, O.: *Die Tierwelt des Solnhofener Schiefers*. Wittenberg (Ziemsen) 3. A. 1971.
Kurten, B.: *Die Welt der Dinosaurier*. München (Kindler) 1968.

Leich, H.: *Nach Millionen Jahren ans Licht*. München (Ott) 1968.

Schmidt, M.: *Die Lebewelt unserer Trias*. Öhringen (Hohenlohe'sche Buchhandlung) 1928; Nachtrag 1938.
Scott, W. B.: *A History of Land Mammals in the Western Hemisphere*. New York (Hafner Publ. Co.) 1962 (reprint).
Slijper, E. J.: *Riesen des Meeres*. Berlin (Springer) 1962.
Swinton, W. E.: *Dinosaurs*. London (The British Museum) 1967.
— : *Fossil Birds*. London (The British Museum) 1965.
— : *Fossil Amphibians and Reptiles*. London (The British Museum) 1965.

Thenius, E.: *Grundzüge der Verbreitungsgeschichte der Säugetiere. Eine historische Tiergeographie*. Jena (Fischer) 1972.
— und Hofer, H.: *Stammesgeschichte der Säugetiere. Eine Übersicht über Tatsachen und Probleme der Säugetiere*. Berlin (Springer) 1960.

Wenger, R.: *Germanische Ceratiten*. Diss. Tübingen 1955.

Recent History (Cenozoic)

Davies, A. M. *Tertiary Faunas. I. The Composition of Tertiary Faunas*. London (Allen & Unwin Ltd.) 1971.

Ebers, E.: *Vom großen Eiszeitalter*. Berlin (Springer) 1957.

Kurtén, B.: *Pleistocene Mammals of Europe. The World Naturalist*. London (Weidenfeld & Nicolson) 1968.

Papp, A. und Thenius, E.: *Tertiär*. Handbuch der Stratigraphischen Geologie III/1 und 2. Stuttgart (Enke) 1959.
Pearson, R.: *Animals and Plants of the Cenozoic Era. Some Aspects of the Faunal and Floral History of the Last Sixty Million Years*. London (Butterworths) 1964.

Termier, H. und G.: *Paléontologie stratigraphique*. Fasc. III & IV. Paris (Masson) 1960.

Evolution of Man

Anthropologie. Hg. Heberer, G. u. a. Frankfurt/Hamburg (Fischer) 2. A. 1970 (Das Fischer Lexikon, 15).

Clark, Sir W. E. Le Gros: *Man-Apes or Ape-Man. The Story of Discoveries in Africa*. New York (Holt, Rinehart & Winston) 1967.

Day, M. H.: *Guide to Fossil Man. A Handbook of Human Paleontology*. London (Cassell) 1965.
Dobshansky, Th.: *Die Entwicklung zum Menschen. Evolution, Abstammung und Vererbung. Ein Abriß*. Hamburg (Parey) 1958.

Glowatzki, G.: *Tausend Jahre wie ein Hauch. Woher kommt der Mensch?* Stuttgart (Franckh) 1968.
Grahmann, R. und Müller-Beck, H.-J.: *Urgeschichte der Menschheit*. Stuttgart (Kohlhammer) 3. A. 1967.

Heberer, G. (Hg.): *Menschliche Abstammungslehre. Fortschritte der Anthropogenie, 1863–1964*. Stuttgart (Fischer) 1965.
— : *Homo—unsere Ab- und Zukunft. Herkunft und Entwicklung des Menschen aus der Sicht der aktuellen Anthropologie*. Stuttgart (DVA) 1968.
— : *Der Ursprung des Menschen. Unser gegenwärtiger Wissensstand*. Stuttgart (Fischer) 2. A. 1969.
Howell, F. C. und die Redaktion der TIME-LIFE-Bücher: *Der Mensch der Vorzeit*. TIME-LIFE International (Nederland) 1970.
Howells, W. W.: *Mankind in the Making*. New York 1959.

Koenigswald, G. H. R. von: *Die Geschichte des Menschen*. Berlin u. a. (Springer) 2. A. 1968.

Wendt, H.: *Ich suchte Adam. Die Entdeckung des Menschen*. Erweiterte Neuauflage: Reinbek (Rowohlt) 1965.
— : *Der Affe steht auf. Eine Bilddokumentation zur Vorgeschichte des Menschen*. Reinbek (Rowohlt) 1971.

Picture Credits

Artists of biotopes, phylogenetic trees and phylogenetic diagrams: Z. Burian, Prague (Pp. 41, 126/127, 150/151, 170/171, 202/203, 222/223, 285, 286/287, 329, 330/331, 406/407, 438/439, 478/479). K. Grossmann, Frankfurt (Pp. 144, 221, 503, 504, 521, 524). E. Hudecek-Neubauer, Vienna (Pp. 405, 408, 437, 440, 477, 480). J. Kühn, Heidelberg (Pp. 133, 172, 454). F. Laube (P. 522/523). H. Losert, Munich (Pp. 73, 74/75, 125, 128, 143, 149, 169, 201, 204). B. Schelhorn, Munich (Pp. 134, 152). G. Wankmüller, Munich (Pp. 76, 224).

Models for the portraits in Chapters 1 and 3: Staatsbibliothek Bildarchiv Berlin: Cuvier, Owen, Buffon, Lamarck, Steno, Maupertuis, Oken, Mendel, Gegenbauer, Linné, Schelling, Wallace. Ullstein Bilderdienst, Berlin: St. Hilaire, Agricola, Scheuchzer, Agassiz.

Photographs: The American Museum of Natural History, New York (Pp. 347, 348, 356 middle). Angermayer, Munich (Pp. 528, 1st row from bottom, left and right). Baird/Museum of Natural History, Princeton (Pp. 241 left top and middle, right top, 421 right bottom, 422 bottom). Dr. Barthel/Paläont. Mus. Humboldt Univ., Berlin (Pp. 338, 381). Campbell/Photo Researchers, England (P. 526 top right). Field Museum of Natural History, Chicago (P. 453). Geol. Museum Muenster (P. 415 top right and middle, bottom, 493 bottom). Goebel/dpa (P. 527, 1st row from top right). Green/ARDEA, London (Pp. 241 bottom left, 2d photo right, 243, 494 top left and right). Grøndal/Inst. fuer Film und Bild, Munich (P. 51 right top and bottom, 52 bottom right, 241 right 3d and 4th photos, 242, 376 middle left, 421 top left and right). Haupt, Offenbach (p. 353). Dr. Klemmer/Natur-Museum Senckenberg (Pp. 356 and 494 bottom). Loeppert, Munich (Pp. 367 top, 376 bottom left, right side, 382 middle left and right). Lossen-Foto, Heidelberg (P. 526 top left). Museum Hauff, Holzmaden (Pp. 366 bottom, 367 middle and bottom, 376 top left). Museum Maxberg, Solnhofen (P. 382 top right). Muséum National d'Histoire Naturelle, Paris (Pp. 493 top, 526 bottom left and right). Larousse/ Muséum National d'Histoire Naturelle, Paris (P. 525).

Myers/Photo Researchers, England (P. 52 bottom left). Okapia, Frankfurt (Pp. 421 bottom left, 527, 3d row from top middle and right, 1st row from bottom middle and right, 528, 3d row from top left, 1st row from bottom middle). Paysan, Stuttgart (Pp. 527, 2d row from top right, 528, 2d row from top middle). Piecko, Chicago (P. 244, all except top left). Ramaekers, Holland (Pp. 527, 3d row from top left, 528, 2d row from top left, 3d row from top right). Reisel/Anthony, Starnberg (P. 527 top right and middle, 2d row middle). Fot. Schafgans, Bonn (p. 526 middle). Schmidt/Bavaria, Munich (P. 528, 3d row from top middle). Schultz/Bayer. Staatsslg. f. Palaeont. und hist. Geologie, Munich (Pp. 337, 422 top, 464). Schuenemann/Bavaria (P. 527, 2d row from top left). Silvester/Bavaria (P. 527, 1st row from bottom left). Smithsonian Institution, Washington, D.C., USA (P. 354/ 355). Soyamoto/Jacana, Paris (P. 528, 1st row from top left and middle, 2d row from top right). Steinhorst/ Landesbildstelle Wuerttemberg (Pp. 52 top left, 365, 366 top left and right, 368, 375, 382 top left and bottom, bottom right, 463 top). Stephan, Sobernheim (Pp. 189, 192). Sternberg Memorial Museum, Kansas, USA (P. 356 top). Prof. Stürmer, Erlangen (Pp. 190, 191). V-Dia-Verlag, Heidelberg (Pp. 51 left, right middle, 52 top right, middle left and right, the lower 4 photos on p. 463). Prof. Walliser, Göttingen (P. 416 all). Weigert, Prögedrucke, Neuberg/Donau (P. 415 left). Dr. Zangerl/Field Museum of Natural History, Chicago, Illinois, USA (P. 244 top left). ZEFA, Germany (P. 528, 1st row from top right).

Black-and-white drawings: Althuber, Wels (Chapter 17 Pp. 400, 411 top, 412 bottom, 446 top and middle). Grossmann, Frankfurt (all drawings in Chapters 21 and 22). Huber, Munich (Pp. 304, 321, 336, 396 top and middle, 397 bottom, 399, 401, 402 top, 403 top, 410 top, 413 top, 443, 445 top, 446 bottom, 447 top and bottom, 459 top, 461). Klein-Rödder, Frankfurt (Chapter 4 drawings). Steffel, Munich (all others).

Photo-Layout: J. Kühn, Heidelberg.

Index

Abbreviations and Symbols

C, °C .... Celsius, degrees centrigrade

ff .... following (pages)

L .... total length (from tip of nose to end of tail

I.U.C.N. . Intern. Union for Conserv. of Nature and Natural Resources

BH ...... body height

♂ .... male

♂♂ ...... males

♀ .... female

♀♀ .... females

+ .... extinct

$\frac{2 \cdot 1 \cdot 2 \cdot 3}{2 \cdot 1 \cdot 2 \cdot 3}$ dental formula (explanation in Vol. 10 of Grzimek's Animal Life Encyclopedia)

▷ .... following (opposite page) color plate

▷▷ ..... Color plate or double color plate on the page following the next

▷▷▷ ... Third color plate or double color plate (etc.)

∅ ⚦ ..... Endangered species and subspecies